Prime Numbers

Springer
New York
Berlin
Heidelberg
Barcelona
Hong Kong
London
Milan
Paris
Singapore
Tokyo

Richard Crandall Carl Pomerance

Prime Numbers

A Computational Perspective

Springer

Richard Crandall
Center for Advanced Computation
3754 S.E. Knight Street
Portland, OR 97202
USA
crandall@reed.edu

Carl Pomerance
Bell Labs
600 Mountain Avenue
Murray Hill, NJ 07974
USA
carlp@research.bell-labs.com

Cover illustration: In the cover's background is a blown up part of a manufactured poster of all (roughly 2 million) decimal digits of the Mersenne prime $2^{6972593}-1$. The prime was discovered using algorithms—some very old and some very new—detailed in this book. The printed poster was 1 meter tall, yet if digits were physically expanded as per the magnified view (actually digitized through a circular watchmaker's loupe), the poster would have been the size of a tennis court.

Library of Congress Cataloging-in-Publication Data
Crandall, Richard E. 1947-
Prime numbers: a computational perspective / Richard Crandall, Carl Pomerance.
p. cm.
Includes bibliographical references and index.
ISBN 0-387-94777-9 (alk. paper)
1. Numbers, Prime. I. Pomerance, Carl. II. Title.
QA246.C74 2000
512'.72—dc21 00-056270

ISBN 0-387-94777-9 Printed on acid-free paper.

Printed in the United States of America

9 8 7 6 5 4 3 2 (Corrected printing, 2002) SPIN 10866783

www.springer-ny.com

Springer-Verlag New York Berlin Heidelberg
A member of BertelsmannSpringer Science+Business Media GmbH

We dedicate this work to our parents Janice, Harold, Hilda, Leon, who—in unique and wonderful ways—taught their children how to think.

Preface

In this volume we have endeavored to provide a middle ground—hopefully even a bridge—between "theory" and "experiment" in the matter of prime numbers. Of course, we speak of number theory and computer experiment. There are great books on the abstract properties of prime numbers. Each of us working in the field enjoys his or her favorite classics. But the experimental side is relatively new. Even though it can be forcefully put that computer science is by no means young, as there have arguably been four or five computer "revolutions" by now, it is the case that the theoretical underpinnings of prime numbers go back centuries, even millennia. So, we believe that there is room for treatises based on the celebrated classical ideas, yet authored from a modern computational perspective.

Design and scope of this book

The book combines the essentially complementary areas of expertise of the two authors. (One author (RC) is more the computationalist, the other (CP) more the theorist.) The opening chapters are in a theoretical vein, even though some explicit algorithms are laid out therein, while heavier algorithmic concentration is evident as the reader moves well into the book. Whether in theoretical or computational writing mode, we have tried to provide the most up-to-date aspects of prime-number study. What we do not do is sound the very bottom of every aspect. Not only would that take orders of magnitude more writing, but, as we point out in the opening of the first chapter, it can be said that no mind harbors anything like a complete picture of prime numbers. We could perhaps also say that neither does any team of *two* investigators enjoy such omniscience. And this is definitely the case for the present team! What we have done is attempt to provide references to many further details about primes, which details we cannot hope to cover exhaustively. Then, too, it will undoubtedly be evident, by the time the book is available to the public, that various prime-number records we cite herein have been broken already. In fact, such are being broken as we write this very Preface. During the final stages of this book we were in some respects living in what electronics engineers call a "race condition," in that results on primes—via the Internet and personal word of mouth—were coming in as fast or faster than editing passes were carried out. So we had to decide on a cutoff point. (In compensation, we often give pointers to websites that do indeed provide up-to-the-minute results.) The race condition has become a natural part of the game, especially now that computers are on the team.

Exercises and research problems

The Exercises occur in roughly thematic order at the end of every chapter, and range from very easy to extremely difficult. Because it is one way of conveying the nature of the cutting edge in prime-number studies, we have endeavored to supply many exercises having a research flavor. These are set off after each chapter's "Exercises" section under the heading "Research problems". (But we still call both normal exercises and research problems "Exercises" during in-text reference.) We are not saying that all the normal exercises are easy; rather we flag a problem as a research problem if it can be imagined as part of a long-term, hopefully relevant investigation.

Algorithms and pseudocode

We put considerable effort—working at times on the threshold of frustration—into the manner of algorithm coding one sees presented herein. From one point of view, the modern art of proper "pseudocode" (meaning not machine-executable, but let us say human-readable code) is in a profound state of disrepair. In almost any book of today containing pseudocode, an incompatibility reigns between readability and symbolic economy. It is as if one cannot have both.

In seeking a balance we chose the C language style as a basis for our book pseudocode. The Appendix describes explicit examples of how to interpret various kinds of statements in our book algorithms. We feel that we shall have succeeded in our pseudocode design if two things occur:

(1) The programmer can readily create programs from our algorithms;

(2) All readers find the algorithm expositions clear.

We went as far as to ask some talented programmers to put our book algorithms into actual code, in this way verifying to some extent our goal (1). (Implementation code is available, in *Mathematica* form, at website http://www.perfsci.com.) Yet, as can be inferred from our remarks above, a completely satisfactory symbiosis of mathematics and pseudocode probably has to wait until an era when machines are more "human."

Acknowledgments

The authors express their profound thanks to a diverse population of colleagues, friends, supporters whom we name as follows: S. Arch, E. Bach, D. Bailey, A. Balog, M. Barnick, D. Bernstein, P. Bateman, O. Bonfim, D. Bleichenbacher, J. Borwein, D. Bradley, N. and P. Bragdon, R. Brent, D. Bressoud, D. Broadhurst, Y. Bugeaud, L. Buhler, G. Campbell, M. Campbell, D. Cao, P. Carmody, E. Catmull, H. Cohen, D. Copeland, D. Coppersmith, J. Cosgrave, T. Day, K. Dilcher, J. Doenias, G. Effinger, J. Essick, J. Fix, W. Galway, B. Garst, M. Gesley, G. Gong, A. Granville, D. Griffiths, E. Hasibar, D. Hayes, D. Hill, U. Hofmann, N. Howgrave-Graham, S. Jobs, A. Jones, B. Kaliski, W. Keller, M. Kida, J. Klivington, K. and S. Koblik, D. Kohel, D. Kramer, A. Kruppa, S. Kurowski, S. Landau, A. Lenstra,

H. Lenstra, M. Levich, D. Lichtblau, D. Lieman, I. Lindemann, M. Martin, E. Mayer, F. McGuckin, M. Mignotte, V. Miller, D. Mitchell, V. Mitchell, T. Mitra, P. Montgomery, W. Moore, G. Nebe, A. Odlyzko, F. Orem, J. Papadopoulos, N. Patson, A. Perez, J. Pollard, A. Powell, J. Powell, L. Powell, P. Ribenboim, B. Salzberg, A. Schinzel, T. Schulmeiss, J. Seamons, J. Shallit, M. Shokrollahi, J. Solinas, L. Somer, D. Symes, D. Terr, E. Teske, A. Tevanian, R. Thompson, M. Trott, S. Wagon, S. Wagstaff, M. Watkins, P. Wellin, N. Wheeler, M. Wiener, T. Wieting, S. Wolfram, G. Woltman, A. Wylde, A. Yerkes, A. Zaccagnini, and P. Zimmermann. These people contributed combinations of technical, theoretical, literary, moral, computational, debugging, and manuscript support. We would like to express a special gratitude to our long-time friend and colleague Joe Buhler who quite unselfishly, and in his inimitable, encyclopedic style aided us at many theoretical and computational junctures during the book project. Because of all of these spectacular colleagues, this book is immeasurably better than it would otherwise have been.

Portland, Oregon — Richard Crandall
Murray Hill, New Jersey — Carl Pomerance
December 2000
December 2001 (Second printing)

Contents

Chapter 1

PRIMES!

Prime numbers belong to an exclusive world of intellectual conceptions. We speak of those marvelous notions that enjoy simple, elegant description, yet lead to extreme—one might say unthinkable—complexity in the details. The basic notion of primality can be accessible to a child, yet no human mind harbors anything like a complete picture. In modern times, while theoreticians continue to grapple with the profundity of the prime numbers, vast toil and resources have been directed toward the computational aspect, the task of finding, characterizing, and applying the primes in other domains. It is this computational aspect on which we concentrate in the ensuing chapters. But we shall often digress into the theoretical domain in order to illuminate, justify, and underscore the practical import of the computational algorithms.

Simply put: A prime is a positive integer p having exactly two positive divisors, namely 1 and p. An integer n is composite if $n > 1$ and n is not prime. (The number 1 is considered neither prime nor composite.) Thus, an integer n is composite if and only if it admits a nontrivial factorization $n = ab$, where a, b are integers, each strictly between 1 and n. Though the definition of primality is exquisitely simple, the resulting sequence $2, 3, 5, 7, \dots$ of primes will be the highly nontrivial collective object of our attention. The wonderful properties, known results, and open conjectures pertaining to the primes are manifold. We shall cover some of what we believe to be theoretically interesting, aesthetic, and practical aspects of the primes. Along the way, we also address the essential problem of factorization of composites, a field inextricably entwined with the study of the primes themselves.

In the remainder of this chapter we shall introduce our cast of characters, the primes themselves, and some of the lore that surrounds them.

1.1 Problems and progress

1.1.1 Fundamental theorem and fundamental problem

The primes are the multiplicative building blocks of the natural numbers as is seen in the following theorem.

Theorem 1.1.1 (Fundamental theorem of arithmetic). *For each natural number n there is a unique factorization*

$$n = p_1^{a_1} p_2^{a_2} \cdots p_k^{a_k}$$

where exponents a_i *are positive integers and* $p_1 < p_2 < \cdots < p_k$ *are primes.*

(If n is itself prime, the representation of n in the theorem collapses to the special case $k = 1$ and $a_1 = 1$. If $n = 1$, sense is made of the statement by taking an empty product of primes, that is, $k = 0$.) The proof of Theorem 1.1.1 naturally falls into two parts, the existence of a prime factorization of n, and its uniqueness. Existence is very easy to prove (consider the first number that does not have a prime factorization, factor it into smaller numbers, and derive a contradiction). Uniqueness is a bit more subtle. It can be deduced from a simpler result, namely Euclid's "first theorem" (see Exercise 1.2).

The fundamental theorem of arithmetic gives rise to what might be called the "fundamental *problem* of arithmetic." Namely, given an integer $n > 1$, find its prime factorization. We turn now to the current state of computational affairs.

1.1.2 Technological and algorithmic progress

In a very real sense, there are no large numbers: Any explicit integer can be said to be "small." Indeed, no matter how many digits or towers of exponents you write down, there are only finitely many natural numbers smaller than your candidate, and infinitely many that are larger. Though condemned always to deal with small numbers, we can at least strive to handle numbers that are larger than those that could be handled before. And there has been remarkable progress. The number of digits of the numbers we can factor is about six times as large as just 25 years ago, and the number of digits of the numbers we can routinely prove prime is about 30 times larger.

It is important to observe that computational progress is two-pronged: There is progress in technology, but also progress in algorithm development. Surely, credit must be given to the progress in the quality and proliferation of computer hardware, but—just as surely—not all the credit. If we were forced to use the algorithms prior to 1970, even with the wonderful computing power available today, we might think that, say, 40 digits is about the limit of what can routinely be factored or proved prime.

So, what can we do these days? For factoring, 155 digits is the current record for worst-case numbers. A very famous factorization was of the 129-digit challenge number enunciated in M. Gardner's "Mathematical Games" column in *Scientific American* [Gardner 1977]. The number

$$\begin{aligned} \text{RSA129} = & 114381625757888867669235779976146612010218296721242362\backslash \\ & 25625618429357069352457338978305971235639587050589890\backslash \\ & 75147599290026879543541 \end{aligned}$$

had been laid as a test case for the then new RSA cryptosystem (see Chapter 8). Some projected that 40 quadrillion years would be required to factor RSA129. Nevertheless, in 1994 it was factored with the quadratic sieve (QS) algorithm (see Chapter 6) by D. Atkins, M. Graff, A. Lenstra, and P. Leyland. RSA129 was factored as

$$3490529510847650949147849619903898133417764638493387843990820577$$
$$\times$$
$$32769132993266709549961988190834461413177642967992942539798288533,$$

and the secret message was decrypted to reveal: "THE MAGIC WORDS ARE SQUEAMISH OSSIFRAGE."

Over the last decade, many other factoring and related milestones have been achieved. One interesting achievement has been the discovery of factors of various Fermat numbers $F_n = 2^{2^n} + 1$ discussed in Section 1.3.2. Some of the lower-lying Fermat numbers such as F_9, F_{10}, F_{11} have been completely factored, while impressive factors of some of the more gargantuan F_n have been uncovered. Depending on the size of a Fermat number, either the number field sieve (NFS) or the elliptic curve method (ECM) has been brought to bear on the problem (see Chapters 6 and 7). Factors having 30 or 40 or more decimal digits have been uncovered in this way. Using methods covered in various sections of the present book, it has been possible to perform a primality test on Fermat numbers as large as F_{24}, a number with more than five million decimal digits. Again, such achievements are due in part to advances in machinery and software, and in part to algorithmic advances. One possible future technology—quantum computation—may lead to such a tremendous machinery advance that factoring could conceivably end up being, in a few decades say, unthinkably faster than it is today. Quantum computation is discussed in Section 8.5.

We have indicated that prime numbers figure into modern cryptography—the science of encrypting and decrypting secret messages. Because many cryptographic systems depend on prime number studies, factoring, and related number-theoretical problems, technological and algorithmic advancement have become paramount. Our ability to uncover large primes and prove them prime has outstripped our ability to factor, a situation that gives some comfort to cryptographers. As of this writing, the largest number ever to have been proved prime is the gargantuan Mersenne prime $2^{13466917} - 1$, which can be thought of, roughly speaking, as a "book" full of decimal digits. The kinds of algorithms that make it possible to speedily do arithmetic with such giant numbers is discussed in Chapter 9. But again, alongside such algorithmic enhancements come machine improvements. To convey an idea of scale, the current hardware and algorithm marriage that found each of the most recent "largest known primes" performed thus: The primality proof/disproof for a single candidate $2^q - 1$ required in 2001 about one CPU-week, on a typical modern PC (see continually updating website [Woltman 2000]). By contrast, a number of order $2^{13000000}$ would have required, just a decade earlier, several years of a typical PC's CPU time! To convey again an idea of scale: At the start of the 21st century, a typical workstation equipped with the right software can multiply together two numbers, each with a million decimal digits, in about one second or less.

The special Mersenne form $2^q - 1$ of such numbers renders primality proofs feasible. For Mersenne numbers we have the very speedy Lucas–

Lehmer test, discussed in Chapter 4. What about primes of no special form—shall we say "random" primes? Primality proofs can be effected these days for such primes having a few thousand digits. Much of the implementation work has been pioneered by F. Morain, who applied ideas of A. Atkin and others to develop an efficient elliptic curve primality proving (ECPP) method, discussed in Chapter 7. A typically impressive recent ECPP result is that $(2^{7331} - 1)/458072843161$, possessed of 2196 decimal digits, is now a proven prime (by Mayer and Morain; see [Morain 2000]).

Alongside these modern factoring achievements and prime-number analyses there stand a great many record-breaking attempts geared to yet more specialized cases. From time to time we see new largest twin primes (pairs of primes $p, p+2$), an especially long arithmetic progression $\{p, p+d, \ldots, p+kd\}$ of primes, or spectacular cases of primes falling in other particular patterns. There are searches for primes we expect some day to find but have not yet found (such as new instances of the so-called Wieferich, Wilson, or Wall–Sun–Sun primes), which discoveries would have theoretical implications. In various sections of this book we refer to a few of these many endeavors, especially when the computational issues at hand lie within the scope of the book.

Details and special cases aside, the reader should be aware that there is a widespread "culture" of computational research. For a readable and entertaining account of prime number and factoring "records," see, for example, [Ribenboim 1996] as well as the popular and thorough newsletter of S. Wagstaff on state-of-the-art factorizations. Here at the dawn of the 21st century, vast distributed computations are not uncommon. A good lay reference is [Peterson 2000]. Another lay treatment about large-number achievements is [Crandall 1997a]. In the latter exposition appears an estimate that answers roughly the question, "How many computing operations have been performed by *all* machines across *all* of world history?" One is speaking of fundamental operations such as logical "and," or numerical "add," "multiply," and so on. The answer is relevant for various issues raised in the present book, and could be called the "mole rule." To put it roughly, right around the turn of the century (2000 A.D.), about one mole—that is, the Avogadro number $6 \cdot 10^{23}$ of chemistry, call it 10^{24}—is the total operation count for all machines for all of history. In spite of the usual mystery and awe that surrounds the notion of industrial and government supercomputing, it is the huge collection of personal computers that allows this 10^{24}, this mole. The relevance is that a task such as trial dividing an integer $N \approx 10^{50}$ directly for prime factors is hopeless in the sense that one would essentially have to replicate the machine effort of all time. To convey an idea of scale: A typical instance of the deepest factoring or primality-proving runs of the modern era involves perhaps 10^{16} to 10^{18} machine operations. Similarly, a full-length graphically rendered synthetic movie of today—for example, the 1999 Pixar/Disney movie *Toy Story 2*—involves operation counts in the same range. It is amusing that for this kind of Herculean machine effort one may either obtain a single answer (a factor, maybe even a single "prime/composite" decision bit) or create a full-length animated feature whose character is as culturally separate from a one-bit answer as can be. It is interesting that a computational task of say 10^{17} operations is one ten-millionth of the overall historical computing effort by all Earth-bound machinery.

1.1.3 The infinitude of primes

While modern technology and algorithms can uncover impressively large primes, it is an age-old observation that no single prime discovery can be the end of the story. Indeed, there exist infinitely many primes, as was proved by Euclid in 300 BC, while he was professor at the great university of Alexandria [Archibald 1949]. This achievement can be said to be the beginning of the abstract theory of prime numbers. The famous proof of the following theorem is essentially Euclid's.

Theorem 1.1.2 (Euclid). *There exist infinitely many primes.*

Proof. Assume that the primes are finite in number, and denote by p the largest. Consider one more than the product of all primes, namely:

$$n = 2 \cdot 3 \cdot 5 \cdots p + 1.$$

Now, n cannot be divisible by any of the primes 2 through p, because any such division leaves remainder 1. But we have assumed that the primes up through p comprise all of the primes. Therefore, n cannot be divisible by any prime, contradicting Theorem 1.1.1, so the assumed finitude of primes is contradicted. □

It might be pointed out that Theorem 1.1.1 was never explicitly stated by Euclid. However, the part of this theorem that asserts that every integer greater than 1 is divisible by some prime number was known to Euclid, and this is what is used in Theorem 1.1.2.

There are many variants of this classical theorem, both in the matter of its statement and its proofs (see Sections 1.3.2 and 1.4.1). Let us single out one particular variant, to underscore the notion that the fundamental Theorem 1.1.1 itself conveys information about the distribution of primes. Denote by $\mathcal{P}$ the set of all primes. We define the prime-counting function at real values of x by

$$\pi(x) = \#\{p \le x : p \in \mathcal{P}\},$$

that is, $\pi(x)$ is the number of primes not exceeding x. The fundamental Theorem 1.1.1 tells us that for positive integer x, the number of solutions to

$$\prod p_i^{a_i} \le x,$$

where now p_i denotes the i-th prime and the a_i are nonnegative, is precisely x itself. Each factor $p_i^{a_i}$ must not exceed x, so the number of possible choices of exponent a_i, including the choice zero, is bounded above by $\lfloor 1+(\ln x)/(\ln p_i) \rfloor$.

It follows that

$$x \le \prod_{p_i \le x} \left\lfloor 1 + \frac{\ln x}{\ln p_i} \right\rfloor \le \left(1 + \frac{\ln x}{\ln 2}\right)^{\pi(x)},$$

which leads immediately to the fact that for all $x \ge 8$,

$$\pi(x) > \frac{\ln x}{2 \ln \ln x}.$$

Though this bound is relatively poor, it does prove the infinitude of primes directly from the fundamental theorem of arithmetic.

The idea of Euclid in the proof of Theorem 1.1.2 is to generate new primes from old primes. Can we generate all of the primes this way? Here are a few possible interpretations of the question:

(1) Inductively define a sequence of primes $q_1, q_2, \ldots$, where $q_1 = 2$, and q_{k+1} is the least prime factor of $q_1 \cdots q_k + 1$. Does the sequence (q_i) contain every prime?

(2) Inductively define a sequence of primes $r_1, r_2, \ldots$, where $r_1 = 2$, and r_{k+1} is the least prime not already chosen that divides some $d+1$, where d runs over the divisors of the product $r_1 \cdots r_k$. Does the sequence (r_i) contain every prime?

(3) Inductively define a sequence of primes $s_1, s_2, \ldots$, where $s_1 = 2$, $s_2 = 3$, and s_{k+1} is the least prime not already chosen that divides some $s_i s_j + 1$, where $1 \le i < j \le k$. Does the sequence (s_i) contain every prime? Is the sequence (s_i) infinite?

The sequence (q_i) of problem (1) was considered by Guy and Nowakowski and later by Shanks. [Wagstaff 1993] computed the sequence through the 43rd term. The computational problem inherent in continuing the sequence further is the enormous size of the numbers that must be factored. Already, the number $q_1 \cdots q_{43} + 1$ has 180 digits.

The sequence (r_i) of problem (2) was recently shown in unpublished work of Pomerance to contain every prime. In fact, for $i \ge 5$, r_i is the i-th prime. The proof involved a direct computer search over the first (approximately) one million terms, followed with some explicit estimates from analytic number theory, about more of which theory we shall hear later in this chapter. This proof is just one of many examples that manifest the utility of the computational perspective.

The sequence (s_i) of problem (3) is not even known to be infinite, though it almost surely is, and almost surely contains every prime. We do not know whether anyone has attacked the problem computationally; perhaps you, the reader, would like to give it a try. The problem is due to M. Newman at the Australian National University.

Thus, even starting with the most fundamental and ancient ideas concerning prime numbers, one can quickly reach the fringe of modern research. Given the millennia that people have contemplated prime numbers, our continuing ignorance concerning the primes is stultifying.

1.1.4 Asymptotic relations and order nomenclature

At this juncture, in anticipation of many more asymptotic density results and computational complexity estimates, we establish asymptotic relation nomenclature for the rest of the book. When we intend

$$f(N) \sim g(N)$$

to be read "f is asymptotic to g as N goes to infinity," we mean that a certain limit exists and has value unity:

$$\lim_{N\to\infty} f(N)/g(N) = 1.$$

When we say

$$f(N) = O(g(N)),$$

to be read "f is big-O of g," we mean that f is bounded in this sense: There exists a positive number C such that, for all N, or for all N in a specified set,

$$|f(N)| \le C|g(N)|.$$

The "little-o" notation can be used when one function seriously dominates another; i.e., we say

$$f(N) = o(g(N))$$

to mean that

$$\lim_{N\to\infty} f(N)/g(N) = 0.$$

Some examples of the notation are in order. Since $\pi(x)$, the number of primes not exceeding x, is clearly less than x for any positive x, we can say

$$\pi(x) = O(x).$$

On the other hand, it is not so clear, and in fact takes some work to prove (see Exercises 1.11 and 1.13 for two approaches), that

$$\pi(x) = o(x). \tag{1.1}$$

The equation (1.1) can be interpreted as the assertion that at very high levels the primes are sparsely distributed, and get more sparsely distributed the higher one goes. If $\mathcal{A}$ is a subset of the natural numbers and $A(x)$ denotes the number of members of $\mathcal{A}$ that do not exceed x, then if $\lim_{x\to\infty} A(x)/x = d$, we call d the asymptotic density of the set $\mathcal{A}$. Thus equation (1.1) asserts that the set of primes has asymptotic density 0. Note that not all subsets of the natural numbers possess an asymptotic density; that is, the limit in the definition may not exist. As just one example, take the set of numbers with an even number of decimal digits.

Throughout the book, when we speak of computational complexity of algorithms we shall stay almost exclusively with "O" notation, even though some authors denote bit and operation complexity by such as $O_{\rm b}$, $O_{\rm op}$ respectively. So when an algorithm's complexity is cast in "O" form, we shall endeavor to specify in every case whether we mean bit or operation complexity. One should take care that these are not necessarily proportional, for it matters whether the "operations" are in a field, are additions or multiplies, or are comparisons (as occur within "if" statements). For example, we shall see in Chapter 9 that whereas a basic FFT multiplication method requires $O(D \ln D)$ floating-point *operations* when the operands possess D digits each (in some appropriate base), there exist methods having *bit* complexity $O(n \ln n \ln \ln n)$, where now n is the total operand bits. So in such a case there is no clear proportionality at work, the relationships between digit size, base, and bit size n are nontrivial (especially when floating-point errors figure into the computation), and so on. Another kind of nontrivial comparison might involve the Riemann zeta function, which for certain arguments can be evaluated to D good digits in $O(D)$ operations, but we mean *full-precision*, i.e., D-digit operations. In contrast, the bit complexity to obtain D good digits (or a proportional number of bits) grows faster than this. And of course we have a trivial comparison of the two complexities: The product of two large integers takes *one* (high-precision) operation, while a flurry of bit manipulations are generally required to effect this multiply! On the face of it, we are saying there is no obvious relation between these two complexity bounds. One might ask, "if these two types of bounds (bit- and operation-based bounds) are so different, isn't one superior, maybe more profound than the other?" The answer is that one is not necessarily better than the other. It might happen that the available machinery—hardware and software—is best suited for all operations to be full-precision; that is, every add and multiply is of the D-digit variety, in which case you are interested in the operation-complexity bound. If, on the other hand, you want to start from scratch and create special, optimal bit-complexity operations whose precision varies dynamically during the whole project, then you would be more interested in the bit-complexity bound. In general, the safe assumption to remember is that bit- versus operation-complexity comparisons can often be of the "apples and oranges" variety.

Because the phrase "running time" has achieved a certain vogue, we shall sometimes use this term as interchangeable with "bit complexity." This equivalence depends, of course, on the notion that the real, physical time a machine requires is proportional to the total number of relevant bit operations. Though this equivalence may well decay in the future—what with quantum computing, massive parallelism, advances in word-oriented arithmetic architecture, and so on—we shall throughout this book just assume running time and bit complexity are the same. Along the same lines, by "polynomial-time" complexity we mean that bit operations are bounded above by a fixed power of the number of bits in the input operands. So, for example, none of the dominant factoring algorithms of today (ECM, QS, NFS) is polynomial-time, but simple addition, multiplication, powering, and so on are polynomial-time. For example, powering, that is computing $x^y \bmod z$, using naive subroutines, has bit complexity $O(\ln^3 z)$ for positive integer operands x, y, z of comparable size, and so is polynomial time. Similarly, taking a gcd is polynomial-time, and so on.

1.1.5 How primes are distributed

In 1737, L. Euler achieved a new proof that there are infinitely many primes: He showed that the sum of the reciprocals of the primes is a divergent sum, and so must contain infinitely many terms (see Exercise 1.20).

In the mid-19th century, P. Chebyshev proved the following theorem, thus establishing the true order of magnitude for the prime-counting function.

Theorem 1.1.3 (Chebyshev). *There are positive numbers A, B such that for all $x \geq 3$,*

$$\frac{Ax}{\ln x} < \pi(x) < \frac{Bx}{\ln x}.$$

For example, Theorem 1.1.3 is true with $A = 1/2$ and $B = 2$. This was a spectacular result, because Gauss had conjectured in 1791 (at the age of fourteen!) the asymptotic behavior of $\pi(x)$, about which conjecture little had been done for half a century prior to Chebyshev. This conjecture of Gauss is now known as the celebrated "prime number theorem" (PNT):

Theorem 1.1.4 (Hadamard and de la Vallée Poussin). *As $x \to \infty$,*

$$\pi(x) \sim \frac{x}{\ln x}.$$

It would thus appear that Chebyshev was close to a resolution of the PNT. In fact, it was even known to Chebyshev that if $\pi(x)$ were asymptotic to some $Cx/\ln x$, then C would of necessity be 1. But the real difficulty in the PNT is showing that the $\lim_{x\to\infty} \pi(x)/(x/\ln x)$ exists at all; this final step was achieved a half-century later, by J. Hadamard and C. de la Vallée Poussin, independently, in 1896. What was actually established was that for some positive number C,

$$\pi(x) = \mathrm{li}\,(x) + O\left(xe^{-C\sqrt{\ln x}}\right), \tag{1.2}$$

where $\mathrm{li}\,(x)$, the logarithmic-integral function, is defined as follows (for a variant of this integral definition see Exercise 1.35):

$$\mathrm{li}\,(x) = \int_2^x \frac{1}{\ln t}\,dt. \tag{1.3}$$

Since $\mathrm{li}\,(x) \sim x/\ln x$, as can easily be shown via integration by parts (or even more easily by L'Hôpital's rule), this stronger form of the PNT implies the form in Theorem 1.1.4. The size of the "error" $\pi(x) - \mathrm{li}\,(x)$ has been a subject of intense study—and refined only a little—in the century following the proof of the PNT. In Section 1.4 we return to the subject of the PNT. But for the moment, we note that one useful, albeit heuristic, interpretation of the PNT is that for random large integers x the "probability" that x is prime is about $1/\ln x$.

It is interesting to ponder how Gauss arrived at his remarkable conjecture. The story goes that he came across the conjecture numerically, by studying a table of primes. Though it is clearly evident from tables that the primes thin out as one gets to larger numbers, locally the distribution appears to be quite erratic. So what Gauss did was to count the number of primes in blocks of length 1000. This smoothes out enough of the irregularities (at low levels) for a "law" to appear, and the law is that near x, the "probability" of a random integer being prime is about $1/\ln x$. This then suggested to Gauss that a reasonable estimate for $\pi(x)$ might be the logarithmic-integral function.

Though Gauss's thoughts on $\pi(x)$ date from the late 1700s, he did not publish them until decades later. Meanwhile, Legendre had independently conjectured the PNT, but in the form

$$\pi(x) \sim \frac{x}{\ln x - B} \tag{1.4}$$

with $B = 1.08366$. No matter what choice is made for the number B, we have $x/\ln x \sim x/(\ln x - B)$, so the only way it makes sense to include a number B in the result, or to use Gauss's approximation $\mathrm{li}\,(x)$, is to consider which option gives a *better* estimation. In fact, the Gauss estimate is by far the better one. Equation (1.2) implies that $|\pi(x) - \mathrm{li}\,(x)| = O(x/\ln^k x)$ for every $k > 0$ (where the big-O constant depends on the choice of k). Since

$$\mathrm{li}\,(x) = \frac{x}{\ln x} + \frac{x}{\ln^2 x} + O\left(\frac{x}{\ln^3 x}\right),$$

it follows that the best numerical choice for B in (1.4) is not Legendre's choice, but $B = 1$. The estimate

$$\pi(x) \approx \frac{x}{\ln x - 1}$$

is attractive for estimations with a pocket calculator.

One can gain insight into the sharpness of the li approximation by inspecting a table of prime counts as in Table 1.1.

For example, consider $x = 10^{21}$. We know from a recent computation of X. Gourdon (based on earlier work of M. Deléglise, J. Rivat and P. Zimmermann) that

$$\pi\left(10^{21}\right) = 21127269486018731928,$$

while on the other hand

$$\mathrm{li}\,\left(10^{21}\right) \approx 21127269486616126181.3$$

and

$$\frac{10^{21}}{\ln(10^{21}) - 1} \approx 21117412262909985552.2\ .$$

x	$\pi(x)$
10^2	25
10^3	168
10^4	1229
10^6	78498
10^8	5761455
10^{12}	37607912018
10^{16}	279238341033925
10^{17}	2623557157654233
10^{18}	24739954287740860
10^{19}	234057667276344607
10^{20}	2220819602560918840
10^{21}	21127269486018731928

Table 1.1 Values of the prime-counting function $\pi(x)$.

It is astounding how good the li (x) approximation really is! We will revisit this issue later in the present chapter, in connection with the Riemann hypothesis (RH) (see Conjecture 1.4.1 and the remarks thereafter).

Another question of historical import is this: What arithmetic progressions, i.e., sets of integers $\{a, a+d, a+2d, \ldots\}$, contain primes, and for those that do, how dense are the occurrences of primes along such progressions? If a and d have a common prime factor, then such a prime divides every term of the arithmetic progression, and so the progression cannot contain more than this one prime. The central classical result is that this is essentially the only obstruction for the arithmetic progression to contain infinitely many primes.

Theorem 1.1.5 (Dirichlet). *If a, d are coprime integers (that is, they have no common prime factor) and $d > 0$, then the arithmetic progression $\{a, a+d, a+2d, \ldots\}$ contains infinitely many primes. In fact, the sum of the reciprocals of the primes in this arithmetic progression is infinite.*

This marvelous (and nontrivial) theorem has been given modern refinement. It is now known that if $\pi(x; d, a)$ denotes the number of primes in the stated progression that do not exceed x, then for fixed coprime integers a, d with $d > 0$,

$$\pi(x; d, a) \sim \frac{1}{\varphi(d)}\pi(x) \sim \frac{1}{\varphi(d)}\frac{x}{\ln x} \sim \frac{1}{\varphi(d)}\,\mathrm{li}\,(x), \tag{1.5}$$

where φ is the Euler totient function $\varphi(d)$ being the number of integers in $[1, d]$ that are coprime to d. Consider that residue classes modulo d that are not coprime to d can contain at most one prime each, so all but finitely many primes are forced into the remaining $\varphi(d)$ residue classes modulo d, and so (1.5) says that each such residue class modulo d receives, asymptotically speaking, its fair parcel of primes. Thus (1.5) is intuitively reasonable. We shall later discuss some key refinements in the matter of the asymptotic error term. The result (1.5) is known as the "prime number theorem for arithmetic progressions."

Incidentally, the question of a set of primes *themselves* forming an arithmetic progression is also interesting. For example,

$$\{1466999,\ 1467209,\ 1467419,\ 1467629,\ 1467839\}$$

is an arithmetic progression of five primes, with common difference $d = 210$. A longer progression with smaller primes is $\{7, 37, 67, 97, 127, 157\}$. It is amusing that if negatives of primes are allowed, this last example may be extended to the left to include $\{-113, -83, -53, -23\}$. See Exercises 1.84, 1.39, 1.40, 1.43 for more on primes lying in arithmetic progression.

1.2 Celebrated conjectures and curiosities

We have indicated that the definition of the primes is so very simple, yet questions concerning primes can be so very hard. In this section we exhibit various celebrated problems of history. The more one studies these questions, the more one appreciates the profundity of the games that primes play.

1.2.1 Twin primes

Consider the case of twin primes, meaning two primes that differ by 2. It is easy to find such pairs, take $11, 13$ or $197, 199$, for example. It is not so easy, but still possible, to find relatively large pairs, recent largest findings being the pair

$$835335 \cdot 2^{39014} \pm 1,$$

found in 1998 by R. Ballinger and Y. Gallot, the pair

$$361700055 \cdot 2^{39020} \pm 1,$$

found in 1999 by H. Lifchitz, and (see [Caldwell 1999]) the gargantuan twin-prime pair

$$2409110779845 \cdot 2^{60000} \pm 1,$$

discovered in 2000 by H. Wassing, A. Járai, and K.-H. Indlekofer.

Are there infinitely many pairs of twin primes? Can we predict, asymptotically, how many such pairs there are up to a given bound? Let us try to think heuristically, like the young Gauss might have. He had guessed that the probability that a random number near x is prime is about $1/\ln x$, and thus came up with the conjecture that $\pi(x) \approx \int_2^x dt/\ln t$ (see Section 1.1.5). What if we choose *two* numbers near x. If they are "independent prime events," then the probability they are both prime should be about $1/\ln^2 x$. Thus, if we denote the twin-prime-pair counting function by

$$\pi_2(x) = \#\{p \le x \ :\ p, p+2 \in \mathcal{P}\},$$

where $\mathcal{P}$ is the set of all primes, then we might guess that

$$\pi_2(x) \sim \int_2^x \frac{1}{\ln^2 t}\,dt.$$

However, it is somewhat dishonest to consider p and $p+2$ as independent prime events. In fact, the chance of one being prime influences the chance that the other is prime. For example, since all primes $p > 2$ are odd, the number $p+2$ is also odd, and so has a "leg up" on being prime. Random odd numbers have twice the chance of being prime as a random number not stipulated beforehand as odd. But being odd is only the first in a series of "tests" a purported prime must pass. For a fixed prime q, a large prime must pass the "q-test" meaning "not divisible by q." If p is a random prime and $q > 2$, then the probability that $p+2$ passes the q-test is $(q-2)/(q-1)$. Indeed, from (1.5), there are $\varphi(q) = q-1$ equally likely residue classes modulo q for p to fall in, and for exactly $q-2$ of these residue classes we have $p+2$ not divisible by q. But the probability that a completely random number passes the q-test is $(q-1)/q$. So, let us revise the above heuristic with the "fudge factor" $2C_2$, where $C_2 = 0.6601618158\ldots$ is the so-called "twin-prime constant":

$$C_2 = \prod_{2<q\in\mathcal{P}} \frac{(q-2)/(q-1)}{(q-1)/q} = \prod_{2<q\in\mathcal{P}} \left(1 - \frac{1}{(q-1)^2}\right). \tag{1.6}$$

We might then conjecture that

$$\pi_2(x) \sim 2C_2 \int_2^x \frac{1}{\ln^2 t}\,dt, \tag{1.7}$$

or perhaps, more simply, that

$$\pi_2(x) \sim 2C_2 \frac{x}{\ln^2 x}.$$

The two asymptotic relations are equivalent, which can be seen by integrating by parts. But the reason we have written the more ungainly expression in (1.7) is that, like the estimate $\pi(x) \approx \mathrm{li}\,(x)$, it may be an extremely good approximation.

Let us try out the approximation (1.7) at $x = 2.75 \cdot 10^{15}$. It is reported, see [Nicely 1999], that

$$\pi_2(2.75 \cdot 10^{15}) = 3049989272575,$$

while

$$2C_2 \int_2^{2.75\cdot 10^{15}} \frac{1}{\ln^2 t}\,dt \approx 3049988592860$$

Let's hear it for heuristic reasoning!

As strong as the numerical evidence may be, we still do not even know whether there are infinitely many pairs of twin primes, that is, whether $\pi_2(x)$ is unbounded. This remains one of the great unsolved problems in mathematics. The closest we have come to proving this is the theorem of Chen Jing-run in 1966 [Halberstam and Richert 1974] that there are infinitely many primes p such that either $p+2$ is prime or the product of two primes.

A striking upper bound result on twin primes was achieved in 1915 by V. Brun, who proved that

$$\pi_2(x) = O\left(x\left(\frac{\ln\ln x}{\ln x}\right)^2\right), \tag{1.8}$$

and a year later he was able to replace the expression $\ln\ln x$ with 1 (see [Halberstam and Richert 1974]). Thus, in some sense, the twin prime conjecture (1.7) is partially established. From (1.8) one can deduce (see Exercise 1.48) the following:

Theorem 1.2.1 (Brun). *The sum of the reciprocals of all primes belonging to some pair of twin primes is finite, that is, if $\mathcal{P}_2$ denotes the set of all primes p such that $p+2$ is also prime, then*

$$\sum_{p\in P_2}\left(\frac{1}{p}+\frac{1}{p+2}\right) < \infty.$$

(Note that the prime 5 is unique in that it appears in *two* pairs of twins, and in its honor, it gets counted *twice* in the displayed sum; of course, this has nothing whatsoever to do with convergence or divergence.) The Brun theorem is remarkable, since we know that the sum of the reciprocals of *all* primes diverges, albeit slowly (see Section 1.1.5). The sum in the theorem, namely

$$B' = (1/3+1/5)+(1/5+1/7)+(1/11+1/13)+\cdots,$$

is known as the Brun constant. Thus, though the set of twin primes may well be infinite, we do know that they must be significantly less dense than the primes themselves.

An interesting sidelight on the issue of twin primes is the numerical calculation of the Brun constant B'. There is a long history on the subject, with the current computational champion being Nicely. According to [Nicely 1999], B' is *likely* to be about 1.9021605822 to an error within ± 0.0000000008. The estimate was made by computing the reciprocal sum very accurately for twin primes up to $2.75 \cdot 10^{15}$ and then extrapolating to the infinite sum using (1.7) to estimate the tail of the sum. (All that is actually proved rigorously about B' is that it is between a number slightly larger than 1.82 and a number slightly smaller than 2.15.) In his earlier (1995) computations concerning the Brun constant, Nicely discovered the now-famous floating-point flaw in the Pentium computer chip, a discovery that, by some estimates, cost the Pentium manufacturer Intel millions of dollars. It seems safe to assume that Brun had no idea in 1909 that his remarkable theorem would have such a technological consequence!

1.2.2 Prime k-tuples and hypothesis H

The twin prime conjecture is actually a special case of the "prime k-tuples" conjecture, which in turn is a special case of "hypothesis H." What are these mysterious-sounding conjectures?

The prime k-tuples conjecture begins with the question, what conditions on integers $a_1, b_1, \ldots, a_k, b_k$ ensure that the k linear expressions $a_1 n + b_1, \ldots, a_k n + b_k$ are simultaneously prime for infinitely many positive integers n? One can see embedded in this question the first part of the Dirichlet Theorem 1.1.5, which is the case $k = 1$. And we can also see embedded the twin prime conjecture, which is the case of two linear expressions $n, n+2$.

Let us begin to try to answer the question by giving necessary conditions on the numbers a_i, b_i. We rule out the cases when some $a_i = 0$, since such a case collapses to a smaller problem. Then, clearly, we must have each $a_i > 0$ and each $\gcd(a_i, b_i) = 1$. This is not enough, though, as the case $n, n+1$ quickly reveals: There are surely not infinitely many integers n for which n and $n+1$ are both prime! What is going on here is that the prime 2 destroys the chances for n and $n+1$, since one of them is always even, and even numbers are not often prime. Generalizing, we see that another necessary condition is that for each prime p there is some value of n such that none of $a_i n + b_i$ is divisible by p. This condition automatically holds for all primes $p > k$; it follows from the condition that each $\gcd(a_i, b_i) = 1$. The prime k-tuples conjecture [Dickson 1904] asserts that these conditions are sufficient:

Conjecture 1.2.1 (Prime k-tuples conjecture). *If $a_1, b_1, \ldots, a_k, b_k$ are integers with each $a_i > 0$, each $\gcd(a_i, b_i) = 1$, and for each prime $p \le k$, there is some integer n with no $a_i n + b_i$ divisible by p, then there are infinitely many positive integers n with each $a_i n + b_i$ prime.*

Whereas the prime k-tuples conjecture deals with linear polynomials, Schinzel's hypothesis H [Schinzel and Sierpiński 1958] deals with arbitrary irreducible polynomials with integer coefficients. It is a generalization of an older conjecture of Bouniakowski, who dealt with a single irreducible polynomial.

Conjecture 1.2.2 (Hypothesis H). *Let $f_1, \ldots, f_k$ be irreducible polynomials with integer coefficients such that the leading coefficient of each f_i is positive, and such that for each prime $p \le k$ there is some integer n with none of $f_1(n), \ldots, f_k(n)$ divisible by p. Then there are infinitely many positive integers n such that each $f_i(n)$ is prime.*

A famous special case of hypothesis H is the single polynomial $n^2 + 1$. As with twin primes, we still do not know whether there are infinitely many primes of the form $n^2 + 1$. In fact, the only special case of hypothesis H that has been proved is Theorem 1.1.5 of Dirichlet.

The Brun method for proving (1.8) can be generalized to get upper bounds of the roughly conjectured order of magnitude for the distribution of the integers n in hypothesis H that make the $f_i(n)$ simultaneously prime. See [Halberstam and Richert 1974] for much more on this subject.

For polynomials in two variables we can sometimes say more. For example, Gauss proved that there are infinitely many primes of the form a^2+b^2. It was shown only very recently in [Friedlander and Iwaniec 1998] that there are infinitely many primes of the form a^2+b^4.

1.2.3 The Goldbach conjecture

In 1742, C. Goldbach stated, in a letter to Euler, a belief that every integer exceeding 5 is a sum of three primes. (For example, $6 = 2+2+2$ and $21 = 13+5+3$.) Euler responded that this follows from what has become known as the Goldbach conjecture, that every even integer greater than two is a sum of two primes. This problem belongs properly to the field of additive number theory, the study of how integers can be partitioned into various sums. What is maddening about this conjecture, and many "additive" ones like it, is that the empirical evidence and heuristic arguments in favor become overwhelming. In fact, large even integers tend to have a great many representations as a sum of two primes.

Denote the number of Goldbach representations of an even n by

$$R_2(n) = \#\{(p,q) \ : \ n = p+q;\ p,q \in \mathcal{P}\}.$$

Thinking heuristically as before, one might guess that for even n,

$$R_2(n) \sim \sum_{p \le n-3} \frac{1}{\ln(n-p)},$$

since the "probability" that a random number near x is prime is about $1/\ln x$. But such a sum can be shown, via the Chebyshev Theorem 1.1.3 (see Exercise 1.38) to be $\sim n/\ln^2 n$. The frustrating aspect is that to settle the Goldbach conjecture, all one needs is that $R_2(n)$ be *positive* for even $n > 2$. One can tighten the heuristic argument above, along the lines of the argument for (1.7), to suggest that for even integers n,

$$R_2(n) \sim 2C_2 \frac{n}{\ln^2 n} \prod_{p|n,\, p>2} \frac{p-1}{p-2}, \tag{1.9}$$

where C_2 is the twin-prime constant of (1.6). The Brun method can be used to establish that $R_2(n)$ is big-O of the right side of (1.9) (see [Halberstam and Richert 1974).

Checking (1.9) numerically, we have $R_2(10^8) = 582800$, while the right side of (1.9) is approximately 518809. One gets better agreement using the asymptotically equivalent expression $\mathcal{R}_2(n)$ defined as

$$\mathcal{R}_2(n) = 2C_2 \int_2^{n-2} \frac{dt}{(\ln t)(\ln(n-t))} \prod_{p|n,\, p>2} \frac{p-1}{p-2}, \tag{1.10}$$

which at $n = 10^8$ evaluates to about 583157.

As with twin primes, [Chen 1966] also established a profound theorem on the Goldbach conjecture: Any sufficiently large even number is the sum of a prime and a number that is either a prime or the product of two primes.

It has been known since the late 1930s, see [Ribenboim 1996], that "almost all" even integers have a Goldbach representation $p + q$, the "almost all" meaning that the set of even natural numbers that *cannot* be represented as a sum of two primes has asymptotic density 0 (see Section 1.1.4 for the definition of asymptotic density). In fact, it is now known that the number of exceptional even numbers up to x that do not have a Goldbach representation is $O\left(x^{1-c}\right)$ for some $c > 0$ (see Exercise 1.39).

The Goldbach conjecture has been checked numerically up through 10^{14} by [Deshouillers et al. 1998] and through $4 \cdot 10^{14}$ by [Richstein 2001], and yes, every even number from 4 up to $4 \cdot 10^{14}$ inclusive is indeed a sum of two primes.

As Euler noted, a corollary of the assertion that every even number after 2 is a sum of two primes is the additional assertion that every odd number after 5 is a sum of three primes. This second assertion is known as the "ternary Goldbach conjecture." In spite of the difficulty of such problems of additive number theory, Vinogradov did in 1937 resolve the ternary Goldbach conjecture, in the asymptotic sense that all sufficiently large odd integers n admit a representation in three primes: $n = p+q+r$. It was shown in 1989 by Chen and Y. Wang, see [Ribenboim 1996], that "sufficiently large" here can be taken to be $n > 10^{43000}$. Vinogradov gave the asymptotic representation count of

$$R_3(n) = \#\{(p, q, r) \ : \ n = p+q+r; \ p, q, r \in \mathcal{P}\} \tag{1.11}$$

as

$$R_3(n) = \Theta(n)\frac{n^2}{2\ln^3 n}\left(1 + O\left(\frac{\ln\ln n}{\ln n}\right)\right), \tag{1.12}$$

where Θ is the so-called singular series for the ternary Goldbach problem, namely

$$\Theta(n) = \prod_{p\in\mathcal{P}}\left(1 + \frac{1}{(p-1)^3}\right)\prod_{p|n, p\in\mathcal{P}}\left(1 - \frac{1}{p^2 - 3p + 3}\right).$$

It is not hard to see that $\Theta(n)$ for odd n is bounded below by a positive constant. This singular series can be given interesting alternative forms (see Exercise 1.65). Vinogradov's effort is an example of analytic number theory *par excellence* (see Section 1.4.4 for a very brief overview of the core ideas).

[Zinoviev 1997] shows that if one assumes the extended Riemann hypothesis (ERH) (Conjecture 1.4.2), then the ternary Goldbach conjecture holds for all odd $n > 10^{20}$. Further, [Saouter 1998] "bootstrapped" the then current bound of $4 \cdot 10^{11}$ for the binary Goldbach problem to show that the ternary Goldbach conjecture holds *unconditionally* for all odd numbers up to 10^{20}. Thus, with the Zinoviev theorem, the ternary Goldbach problem is completely solved under the assumption of the ERH.

It follows from the Vinogradov theorem that there is a number k such that every integer starting with 2 is a sum of k or fewer primes. This corollary was actually proved earlier by G. Shnirel'man in a completely different way. Shnirel'man used the Brun sieve method to show that the set of even numbers representable as a sum of two primes contains a subset with positive asymptotic density (this predated the results that almost all even numbers were so representable), and using just this fact was able to prove there is such a number k. (See Exercise 1.42 for a tour of one proof method.) The least number k_0 such that every number starting with 2 is a sum of k_0 or fewer primes is now known as the Shnirel'man constant. If Goldbach's conjecture is true, then $k_0 = 3$. Since we now know that the ternary Goldbach conjecture is true, conditionally on the ERH, it follows that on this condition, $k_0 \leq 4$. The best unconditional estimate is due to O. Ramaré who showed that $k_0 \leq 7$ [Ramaré 1995]. Ramaré's proof used a great deal of computational analytic number theory, some of it joint with R. Rumely.

1.2.4 The convexity question

One spawning ground for curiosities about the primes is the theoretical issue of their density, either in special regions or under special constraints. Are there regions of integers in which primes are especially dense? Or especially sparse? Amusing dilemmas sometimes surface, such as the following one. There is an old conjecture of Hardy and Littlewood on the "convexity" of the distribution of primes:

Conjecture 1.2.3. *If $x \geq y \geq 2$, then $\pi(x+y) \leq \pi(x) + \pi(y)$.*

On the face of it, this conjecture seems reasonable: After all, since the primes tend to thin out, there ought to be fewer primes in the interval $[x, x+y]$ than in $[0, y]$. But amazingly, Conjecture 1.2.3 is known to be *incompatible* with the prime k-tuples Conjecture 1.2.1 [Hensley and Richards 1973].

So, which conjecture is true? Maybe neither is, but the current thinking is that the Hardy–Littlewood convexity Conjecture 1.2.3 is false, while the prime k-tuples conjecture is true. It would seem fairly easy to actually prove that the convexity conjecture is false; you just need to come up with numerical values of x and y where $\pi(x+y), \pi(x), \pi(y)$ can be computed and $\pi(x+y) > \pi(x)+\pi(y)$. It sounds straightforward enough, and perhaps it is, but it also may be that any value of x required to demolish the convexity conjecture is enormous. (See Exercise 1.89 for more on such issues.)

1.2.5 Prime-producing formulae

Prime-producing formulae have been a popular recreation, ever since the observation of Euler that the polynomial

$$x^2 + x + 41$$

attains prime values for each integer x from 0 to 39 inclusive. Armed with modern machinery, one can empirically analyze other polynomials that give, over certain ranges, primes with high probability (see Exercise 1.17). Here are some other curiosities, of the type that have dubious value for the computationalist (nevertheless, see Exercises 1.5, 1.74):

Theorem 1.2.2 (Examples of prime-producing formulae). *There exists a real number $\theta > 1$ such that for every positive integer n, the number*

$$\left\lfloor \theta^{3^n} \right\rfloor$$

is prime. There also exists a real number α such that the n-th prime is given by:

$$p_n = \left\lfloor 10^{2^{n+1}} \alpha \right\rfloor - 10^{2^n} \left\lfloor 10^{2^n} \alpha \right\rfloor .$$

This first result depends on a nontrivial theorem on the distribution of primes in "short" intervals [Mills 1947], while the second result is just a realization of the fact that there exists a well-defined decimal expansion $\alpha = \sum p_m 10^{-2^{m+1}}$.

Such formulae, even when trivial or almost trivial, can be picturesque. By appeal to the Wilson theorem and its converse (Theorem 1.3.6), one may show that

$$\pi(n) = \sum_{j=2}^{n} \left(\left\lfloor \frac{(j-1)!+1}{j} \right\rfloor - \left\lfloor \frac{(j-1)!}{j} \right\rfloor \right),$$

but this has no evident value in the theory of the prime-counting function $\pi(n)$. Yet more prime-producing and prime-counting formulae are exhibited in the exercises.

Prime-producing formulae are often amusing but, relatively speaking, useless. There is a famous counterexample though. In connection with the ultimate resolution of Hilbert's tenth problem, which problem asks for a deterministic algorithm that can decide whether a polynomial in several variables with integer coefficients has an all integral root, an attractive side result was the construction of a polynomial in several variables with integral coefficients, such that the set of its *positive* values at positive integral arguments is exactly the set of primes (see Section 8.4).

1.3 Primes of special form

By prime numbers of special form we mean primes p enjoying some interesting, often elegant, algebraic classification. For example, the Mersenne numbers M_q and the Fermat numbers F_n defined by

$$M_q = 2^q - 1, \quad F_n = 2^{2^n} + 1$$

are sometimes prime. These numbers are interesting for themselves and for their history, and their study has been a great impetus for the development of computational number theory.

1.3.1 Mersenne primes

Searching for Mersenne primes can be said to be a centuries-old research problem (or recreation, perhaps). There are various easily stated constraints on exponents q that aid one in searches for Mersenne primes $M_q = 2^q - 1$. An initial result is the following:

Theorem 1.3.1. *If $M_q = 2^q - 1$ is prime, then q is prime.*

Proof. A number $2^c - 1$ with c composite has a proper factor $2^d - 1$, where d is any proper divisor of c. □

This means that in the search for Mersenne primes one may restrict oneself to prime exponents q. Note the important fact that the converse of the theorem is false. For example, $2^{11} - 1$ is not prime even though 11 is. The practical import of the theorem is that one may rule out a great many exponents, only considering prime ones during searches for Mersenne primes.

Yet more weeding out of Mersenne candidates can be achieved via the following knowledge concerning possible prime factors of M_q:

Theorem 1.3.2 (Euler). *For prime $q > 2$, any prime factor of $M_q = 2^q - 1$ must be 1 (mod q) and furthermore must be ± 1 (mod 8).*

Proof. Let r be a prime factor of $2^q - 1$, with q a prime, $q > 2$. Then $2^q \equiv 1 \pmod{r}$, and since q is prime, the least positive exponent h with $2^h \equiv 1 \pmod{r}$ must be q itself. Thus, in the multiplicative group of nonzero residues modulo r (a group of order $r - 1$) the residue 2 has order q. This immediately implies that $r \equiv 1 \pmod{q}$, since the order of an element in a group divides the order of the group. Since q is an odd prime, we in fact have $q | \frac{r-1}{2}$, so $2^{\frac{r-1}{2}} \equiv 1 \pmod{r}$. By Euler's criterion (2.6), 2 is a square modulo r, which in turn implies via (2.10) that $r \equiv \pm 1 \pmod{8}$. □

A typical Mersenne prime search runs, then, as follows. For some set of prime exponents Q, remove candidates $q \in Q$ by checking whether

$$2^q \equiv 1 \pmod{r}$$

for various small primes $r \equiv 1 \pmod{q}$ and $r \equiv \pm 1 \pmod{8}$. For the survivors, one then invokes the celebrated Lucas–Lehmer test, which is a rigorous primality test (see Chapter 3).

As of this writing the known Mersenne primes are those displayed in Table 1.2.

Over the years 1979–96, D. Slowinski found seven Mersenne primes, all of the Mersenne primes from $2^{44497} - 1$ to $2^{1257787} - 1$, inclusive, except for $2^{110503} - 1$ (the first of the seven was found jointly with H. Nelson and the last three with P. Gage). The "missing" prime $2^{110503} - 1$ was found by W. N. Colquitt and L. Welsh, Jr., in 1988. The record for *consecutive* Mersenne primes is still held by R. Robinson, who found the five starting with $2^{521} - 1$ in 1952. The prime $2^{1398269} - 1$ was found in 1996 by J. Armengaud

2^2-1	2^3-1	2^5-1	2^7-1	$2^{13}-1$
$2^{17}-1$	$2^{19}-1$	$2^{31}-1$	$2^{61}-1$	$2^{89}-1$
$2^{107}-1$	$2^{127}-1$	$2^{521}-1$	$2^{607}-1$	$2^{1279}-1$
$2^{2203}-1$	$2^{2281}-1$	$2^{3217}-1$	$2^{4253}-1$	$2^{4423}-1$
$2^{9869}-1$	$2^{9941}-1$	$2^{11213}-1$	$2^{19937}-1$	$2^{21701}-1$
$2^{23209}-1$	$2^{44497}-1$	$2^{86243}-1$	$2^{110503}-1$	$2^{132049}-1$
$2^{216091}-1$	$2^{756839}-1$	$2^{859433}-1$	$2^{1257787}-1$	$2^{1398269}-1$
$2^{2976221}-1$	$2^{3021377}-1$	$2^{6972593}-1$	$2^{13466917}-1$	

Table 1.2 Known Mersenne primes (as of December 2001).

and G. Woltman, while $2^{2976221}-1$ was found in 1997 by G. Spence and Woltman. The prime $2^{3021377}-1$, was discovered in 1998 by R. Clarkson, Woltman, S. Kurowski et al. (further verified by D. Slowinski as prime in a separate machine/program run). Then in 1999 the prime $2^{6972593}-1$ was found by N. Hajratwala, Woltman, and Kurowski, then verified by E. Mayer and D. Willmore. The current record Mersenne prime, $2^{13466917}-1$, was discovered in November 2001 by M. Cameron, Woltman, and Kurowski, then verified by Mayer, P. Novarese, and G. Valor. These last five largest known primes were found using a fast multiplication method—the IBDWT—discussed in Chapter 9 (Theorem 9.5.17 and Algorithm 9.5.18). This method has at least doubled the search efficiency over previous methods.

It should be mentioned that modern Mersenne searching is sometimes of the "hit and miss" variety; that is, random prime exponents q are used to check accordingly random candidates 2^q-1. (In fact, some Mersenne primes were indeed found out of order, as indicated above). But much systematic testing has also occurred. As of this writing, exponents q have been checked for all $q \le 7760000$ and for all $q \le 3945000$ double-checked, meaning that the Lucas–Lehmer test was carried out twice (see [Woltman 2000], which website continually updates).

As mentioned in Section 1.1.2, the prime $M_{13466917}$ is the current record holder as not only the largest known Mersenne prime, but also the largest number that has ever been proved prime. With few exceptions, the record for largest proved prime in the modern era has always been a Mersenne prime. One of the exceptions occurred in 1989, when the "Amdahl Six" found the prime [Caldwell 1999]

$$391581 \cdot 2^{216193} - 1,$$

which is larger than $2^{216091}-1$, the record Mersenne prime of that time. However, this is not the largest known explicit non-Mersenne prime, for Young found, in 1997, the prime $5 \cdot 2^{240937}+1$, and in 2001 Cosgrave found the prime

$$3 \cdot 2^{916773} + 1.$$

Mersenne primes figure uniquely into the ancient subject of so-called perfect numbers. A perfect number is a positive integer equal to the sum of its divisors other than itself. For example, $6 = 1+2+3$ and $28 = 1+2+4+7+14$ are perfect numbers. An equivalent way to define "perfection" is to denote by $\sigma(n)$ the sum of the divisors of n, whence n is perfect if and only if $\sigma(n) = 2n$. The following famous theorem completely characterizes the even perfect numbers.

Theorem 1.3.3 (Euclid–Euler). *An even number n is perfect if and only if it is of the form*

$$n = 2^{q-1} M_q,$$

where $M_q = 2^q - 1$ is prime.

Proof. Suppose $n = 2^a m$ is an even number, where m is the largest odd divisor of n. The divisors of n are of the form $2^j d$, where $0 \le j \le a$ and $d|m$. Let D be the sum of the divisors of m excluding m, and let $M = 2^{a+1} - 1 = 2^0 + 2^1 + \cdots + 2^a$. Thus, the sum of all such divisors of n is $M(D + m)$. If M is prime and $M = m$, then $D = 1$, and the sum of all the divisors of n is $M(1 + m) = 2n$, so that n is perfect. This proves the first half of the assertion. For the second, assume that $n = 2^a m$ is perfect. Then $M(D+m) = 2n = 2^{a+1} m = (M + 1)m$. Subtracting Mm from this equation, we see that

$$m = MD.$$

If $D > 1$, then D and 1 are distinct divisors of m less than m, contradicting the definition of D. So $D = 1$, m is therefore prime, and $m = M = 2^{a+1} - 1$. □

The first half of this theorem was proved by Euclid, while the second half was proved some two millennia later by Euler. It is evident that every newly discovered Mersenne prime immediately generates a new (even) perfect number. On the other hand, it is still not known whether there are *any* odd perfect numbers, the conventional belief being that none exist. Much of the research in this area is manifestly computational: It is known that if an odd perfect number n exists, then $n > 10^{300}$, a result of [Brent et al. 1993], and that n has at least eight different prime factors, an independent result of E. Chein and P. Hagis [Ribenboim 1996]. For more on perfect numbers, see Exercise 1.29.

There are many interesting open problems concerning Mersenne primes. We do not know whether there are infinitely many such primes. We do not even know whether infinitely many Mersenne numbers M_q with q prime are *composite*. However, the latter assertion follows from the prime k-tuples Conjecture 1.2.1. Indeed, it is easy to see that if $q \equiv 3 \pmod 4$ is prime and $2q + 1$ is also prime, then $2q + 1$ divides M_q. For example, 23 divides M_{11}. Conjecture 1.2.1 implies there are infinitely many such primes q.

Various interesting conjectures have been made in regard to Mersenne numbers, for example the "new Mersenne conjecture" of P. Bateman, J. Selfridge, and S. Wagstaff. This stems from Mersenne's original assertion in

1644 that the exponents q for which $2^q - 1$ is prime and $29 \le q \le 257$ are 31, 67, 127, and 257. (The smaller exponents were known at that time, and it was also known that $2^{37}-1$ is composite.) Considering that the numerical evidence below 29 was that every prime except 11 and 23 works, it is rather amazing that Mersenne would assert such a sparse sequence for the exponents. He was right on the sparsity, and on the exponents 31 and 127, but he missed 61, 89, and 107. With just five mistakes, no one really knows how Mersenne effected such a claim. However, it was noticed that the odd Mersenne exponents below 29 are all either 1 away from a power of 2, or 3 away from a power of 4 (while the two missing primes, 11 and 23, do not have this property), and Mersenne's list just continues this pattern (perhaps with 61 being an "experimental error," since Mersenne left it out). In [Bateman et al. 1989] the authors suggest a new Mersenne conjecture, that any two of the following implies the third: (a) the prime q is either 1 away from a power of 2, or 3 away from a power of 4, (b) $2^q - 1$ is prime, (c) $(2^q + 1)/3$ is prime. Once one gets beyond small numbers it is very difficult to find any primes q that satisfy two of the statements, and probably there are none beyond 127. That is, probably the conjecture is true, but so far it is only an assertion based on a very small set of primes.

It has also been conjectured that every Mersenne number M_q, with q prime, is square-free (which means not divisible by a square greater than 1), but we cannot even show that this holds infinitely often. Let $\mathcal{M}$ denote the set of primes that divide some M_q with q prime. We know that the number of members of $\mathcal{M}$ up to x is $o(\pi(x))$, and it is known on the assumption of the generalized Riemann hypothesis that the sum of the reciprocals of the members of $\mathcal{M}$ converges [Pomerance 1986].

It is possible to give a heuristic argument that supports the assertion that there are $\sim c \ln x$ primes $q \le x$ with M_q prime, where $c = e^{\gamma}/\ln 2$ and γ is Euler's constant. For example, this formula suggests there should be, on average, about 23.7 values of q in an interval $[x, 10000x]$. Assuming that the machine checks of the Mersenne exponents up to 3000000 are exhaustive, the actual number of values of q with M_q prime in $[x, 10000x]$ is 23 for $x = 100$ and 200, and it is 24 for $x = 300$. Despite the good agreement with practice, some still think the "correct" value of c is $2/\ln 2$ or something else. Until a theorem can actually be proved, we shall not know for sure.

We begin the heuristic with the claim that the probability that a random number near 2^q-1 is prime is about $1/\ln(2^q-1)$, or about $1/(q\ln 2)$. However, we should also compare the chance of $M_q = 2^q-1$ being prime with a random number of the same size. It is likely not the same, as Theorem 1.3.2 already indicates. For the "r-test," meaning "not divisible by r," a random number passes it with probability $1 - \frac{1}{r}$. But if r is not both 1 (mod q) and ± 1 (mod 8), Theorem 1.3.2 says that M_q passes with probability 1. For a prime r with $r \equiv 1 \pmod q, r \equiv \pm 1 \pmod 8$ let us assume that M_q passes with probability $1 - \frac{2q}{r-1}$. Indeed, for such a prime r, we have $r|M_q$ if and only if 2 is a $((r-1)/q)$-th power modulo r. There are q such powers modulo r, and all of them are squares modulo r. The condition $r \equiv \pm 1 \pmod 8$ ensures that 2 is a square modulo r, so if we guess that 2 is like a random square residue, the "probability" that it is such a power is $(2q)/(r-1)$. This suggests that the "fudge factor" for M_q to adjust its "probability" of being prime from that of a random number of the same size is the number

$$f_q = \lim_{x\to\infty} \frac{\prod_{r\le x, r\equiv 1 \pmod q, r\equiv \pm 1 \pmod 8}\left(1-\frac{2q}{r-1}\right)}{\prod_{r\le x}\left(1-\frac{1}{r}\right)},$$

where in both products, r runs over primes. It is possible to prove that the limit exists, but it is not so easy to estimate what it is. The denominator at least is known asymptotically, this is the Mertens theorem (see Theorem 1.4.2):

$$\prod_{r\le x}\left(1-\frac{1}{r}\right) \sim \frac{1}{e^\gamma \ln x}.$$

For the numerator, let us try to estimate it for a "typical" q. If we say that a number $r = nq+1$ is prime and $\equiv \pm 1 \pmod 8$ with probability $1/2\ln(nq)$, then we might change the numerator product in the definition of f_q to

$$\prod_{n\le x/q}\left(1-\frac{2q}{2nq\ln(nq)}\right) = \prod_{n\le x/q}\left(1-\frac{1}{n\ln(nq)}\right),$$

where n runs over the positive integers. This product can be estimated: It is $\sim \ln q/\ln x$ for $x > q^2$ and $q\to\infty$. If we use this expression we might guess that on average f_q is about $e^\gamma \ln q$, and so the "probability" that M_q is prime is

$$\frac{e^\gamma \ln q}{q\ln 2}.$$

Summing this expression for primes $q\le x$, we get the heuristic asymptotic expression for the number of Mersenne prime exponents up to x, namely $c\ln x$ with $c = e^\gamma/\ln 2$.

1.3.2 Fermat numbers

The celebrated Fermat numbers $F_n = 2^{2^n}+1$, like the Mersenne numbers, have been the subject of much scrutiny for centuries. In 1637 Fermat claimed that the numbers F_n are always prime, and indeed the first five, up to $F_4 = 65537$ inclusive, are prime. However, this is one of the few cases where Fermat was wrong, perhaps *very* wrong. Every other single F_n for which we have been able to decide the question is composite! The first of these composites, F_5, was factored by Euler.

A very remarkable theorem on prime Fermat numbers was proved by Gauss, again from his teen years. He showed that a regular polygon with n sides is constructible with straightedge and compass if and only if the largest odd divisor of n is a product of distinct Fermat primes. If $F_0,\ldots,F_4$ turn out to be the *only* Fermat primes, then the only n-gons that are constructible are those with $n = 2^a m$ with $m|2^{32}-1$ (since the product of these five Fermat primes is $2^{32}-1$).

If one is looking for primes that are 1 more than a power of 2, then one need look no further than the Fermat numbers:

Theorem 1.3.4. *If $p = 2^m + 1$ is an odd prime, then m is a power of two.*

Proof. Assume that $m = ab$, where a is the largest odd divisor of m. Then p has the factor $2^b + 1$. Therefore, a necessary condition that p be prime is that $p = 2^b + 1$, that is, $a = 1$ and $m = b$ is a power of 2. □

Again, as with the Mersenne numbers, there is a useful result that restricts possible prime factors of a Fermat number.

Theorem 1.3.5 (Euler). *For $n \geq 2$, any prime factor p of $F_n = 2^{2^n} + 1$ must have $p \equiv 1 \pmod{2^{n+2}}$.*

Proof. Let r be a prime factor of F_n and let h be the least positive integer with $2^h \equiv 1 \pmod r$. Then, since $2^{2^n} \equiv -1 \pmod r$, we have $h = 2^{n+1}$. As in the proof of Theorem 1.3.1, 2^{n+1} divides $r - 1$. Since $n \geq 2$, we thus have that $r \equiv 1 \pmod 8$. This condition implies via (2.10) that 2 is a square modulo r, so that $h = 2^{n+1}$ divides $\frac{r-1}{2}$, from which the assertion of the theorem is evident. □

It was this kind of result that enabled Euler to find a factor of F_5, and thus be the first to "dent" the ill-fated conjecture of Fermat. To this day, Euler's result is useful in factor searches on gargantuan Fermat numbers.

As with Mersenne numbers, Fermat numbers allow a very efficient test that rigorously determines prime or composite character. This is the Pepin test, or the related Suyama test (for Fermat cofactors); see Theorem 4.1.2 and Exercises 4.5, 4.7, 4.8.

By combinations of various methods, including the Pepin/Suyama tests or in many cases the newest factoring algorithms available, various Fermat numbers have been factored, either partially or completely, or, barring that, have been assigned known character (i.e., determined composite). The current situation for all $F_n, n \leq 24$, is displayed in Table 1.3.

We give a summary of the theoretically interesting points concerning Table 1.3 (note that many of the factoring algorithms that have been successful on Fermat numbers are discussed in Chapters 5, 6, and 7).

(1) F_7 was factored via the continued fraction method [Morrison and Brillhart 1975], while F_8 was found by a variant of the Pollard-rho method [Brent and Pollard 1981].

(2) The spectacular 49-digit factor of F_9 was achieved via the number field sieve (NFS) [Lenstra et al. 1993a].

(3) Thanks to the recent demolition, via the elliptic curve method, of F_{10} [Brent 1999], and an earlier resolution of F_{11} also by Brent, the smallest Fermat number not yet completely factored is F_{12}.

(4) The two largest known prime factors of F_{13}, and the largest prime factors of both F_{15} and F_{16} were found in recent years, via modern, enhanced

$$
\begin{aligned}
F_0 &= 3 = P\\
F_1 &= 5 = P\\
F_2 &= 17 = P\\
F_3 &= 257 = P\\
F_4 &= 65537 = P\\
F_5 &= 641 \cdot 6700417\\
F_6 &= 274177 \cdot 67280421310721\\
F_7 &= 59649589127497217 \cdot 5704689200685129054721\\
F_8 &= 1238926361552897 \cdot P\\
F_9 &= 2424833 \cdot 7455602825647884208337395736200454918783366342657 \cdot P\\
F_{10} &= 45592577 \cdot 6487031809 \cdot 4659775785220018543264560743076778192897 \cdot P\\
F_{11} &= 319489 \cdot 974849 \cdot 167988556341760475137 \cdot 3560841906445833920513 \cdot P\\
F_{12} &= 114689 \cdot 26017793 \cdot 63766529 \cdot 190274191361 \cdot 1256132134125569 \cdot C\\
F_{13} &= 2710954639361 \cdot 2663848877152141313 \cdot 3603109844542291969 \cdot\\
&\qquad 319546020820551643220672513 \cdot C\\
F_{14} &= C\\
F_{15} &= 1214251009 \cdot 2327042503868417 \cdot 168768817029516972383024127016961 \cdot C\\
F_{16} &= 825753601 \cdot 188981757975021318420037633 \cdot C\\
F_{17} &= 31065037602817 \cdot C\\
F_{18} &= 13631489 \cdot 81274690703860512587777 \cdot C\\
F_{19} &= 70525124609 \cdot 646730219521 \cdot C\\
F_{20} &= C\\
F_{21} &= 4485296422913 \cdot C\\
F_{22} &= C\\
F_{23} &= 167772161 \cdot ?\\
F_{24} &= C
\end{aligned}
$$

Table 1.3 What is known about the first 25 Fermat numbers (as of December 2001); P = a proven prime; C = a proven composite, ? = unknown character (prime/composite?); all explicitly written factors are primes. The smallest Fermat number of unknown character is F_{33}.

variants of the elliptic curve method (ECM) [Crandall 1996a], [Brent et al. 2000], as we discuss in Section 7.4.1. The most recent factor found in this way is the 23-digit factor of F_{18} found by R. McIntosh and C. Tardif in 1999.

(5) The numbers $F_{14}, F_{20}, F_{22}, F_{24}$ are, as of this writing, "genuine" composites, meaning that we know the numbers *not* to be prime, but do not know a single prime factor of any of the numbers. However, see Exercise 1.79 for conceptual difficulties attendant on the notion of "genuine" in this context.

(6) The Pepin test proved that F_{14} is composite [Selfridge and Hurwitz 1964], while F_{20} was shown composite in the same way [Buell and Young 1988].

(7) The character of F_{22} was resolved [Crandall et al. 1995], but in this case an interesting verification occurred: A completely independent (in terms of hardware, software, and location) research team in South America [Trevisan and Carvalho 1993] performed the Pepin test, and obtained the same result for F_{22}. Actually, what they found were the same Selfridge–Hurwitz residues, taken to be the least nonnegative residue modulo F_n then taken again modulo the three coprime moduli 2^{36}, $2^{36}-1$, $2^{35}-1$ to forge a kind of "parity check" with probability of error being roughly 2^{-107}. Despite the threat of machine error in a single such extensive calculation, the agreement between the independent parties leaves little doubt as to the composite character of F_{22}.

(8) The character of F_{24} was resolved in 1999 by Crandall, Mayer, and Papadopoulos [Crandall et al. 1999]. In this case, rigor was achieved by having (a) two independent floating-point Pepin "wavefront" tests (by Mayer and Papadopoulos, finishing in that order in August 1999), but also (b) a pure-integer convolution method for deterministic checking of the Pepin squaring chain. Again the remaining doubt as to composite character must be regarded as minuscule. More details are discussed in Exercise 4.6.

(9) Beyond F_{24}, every F_n through $n = 32$ inclusive has yielded at least one proper factor, and all of those factors were found by trial division with the aid of Theorem 1.3.5. (Most recently, A. Kruppa and T. Forbes found in 2001 that 46931635677864055013377 divides F_{31}.) The first Fermat number of unresolved character is thus F_{33}. By conventional machinery and Pepin test, the resolution of F_{33} would take us well beyond the next ice age! So the need for new algorithms is as strong as can be for future work on giant Fermat numbers.

There are many other interesting facets of Fermat numbers. There is the challenge of finding very large composite F_n. For example, W. Keller showed that F_{23471} is divisible by $5 \cdot 2^{23473} + 1$, while more recently J. Young (see [Keller 1999]) found that F_{213319} is divisible by $3 \cdot 2^{213321} + 1$, and even more recent is the discovery by J. Cosgrave (who used remarkable software by Y. Gallot) that F_{382447} is divisible by $3 \cdot 2^{382449} + 1$ (see Exercise 4.9). To show how hard these investigators must have searched, the prime divisor Cosgrave found is *itself* currently one of the dozen or so largest known primes. Similar efforts reported recently by [Dubner and Gallot 2000] include K. Herranen's generalized Fermat prime

$$101830^{2^{14}} + 1$$

and S. Scott's gargantuan prime

$$48594^{2^{16}} + 1.$$

A compendium of numerical results on Fermat numbers is available at [Keller 1999].

It is amusing that Fermat numbers allow still another proof of Theorem 1.1.2 that there are infinitely many primes: Since the Fermat numbers are odd and the product of $F_0, F_1, \ldots, F_{n-1}$ is $F_n - 2$, we immediately see that each prime factor of F_n does not divide any earlier F_j, and so there are infinitely many primes.

What about heuristic arguments: Can we give a suggested asymptotic formula for the number of $n \le x$ with F_n prime? If the same kind of argument is done as with Mersenne primes, we get that the number of Fermat primes is finite. This comes from the convergence of the sum of $n/2^n$, which expression one finds is proportional to the supposed probability that F_n is prime. If this kind of heuristic is to be taken seriously, it suggests that there are no more Fermat primes after F_4, the point where Fermat stopped, confidently predicting that all larger Fermat numbers are prime! A heuristic suggested by H. Lenstra, similar in spirit to the previous estimate on the density of Mersenne primes, says that the "probability" that F_n is prime is (here and throughout the book, lg means $\log_2$) approximately

$$\text{Prob} \approx \frac{e^\gamma \lg b}{2^n}, \tag{1.13}$$

where b is the *current* limit on the possible prime factors of F_n. If nothing is known about possible factors, one might use the smallest possible prime factor $b = 3 \cdot 2^{n+2} + 1$ for the numerator calculation, giving a rough *a priori* probability of $n/2^n$ that F_n is prime. (Incidentally, a similar probability argument for generalized Fermat numbers $b^{2^n} + 1$ appears in [Dubner and Gallot 2000].) It is from such a probabilistic perspective that Fermat's guess looms as ill-fated as can be.

1.3.3 Certain presumably rare primes

There are interesting classes of presumably rare primes. We say "presumably" because little is known in the way of rigorous density bounds, yet empirical evidence and heuristic arguments suggest relative rarity. For any odd prime p, Fermat's "little theorem" tells us that $2^{p-1} \equiv 1 \pmod{p}$. One might wonder whether there are primes such that

$$2^{p-1} \equiv 1 \pmod{p^2}, \tag{1.14}$$

such primes being called Wieferich primes. These special primes figure strongly in the so-called first case of Fermat's "last theorem," as follows. [Wieferich 1909] proved that if

$$x^p + y^p = z^p$$

where p is a prime that does not divide xyz, then p satisfies relation (1.14). Equivalently, we say that p is a Wieferich prime if the Fermat quotient:

$$q_p(2) = \frac{2^{p-1} - 1}{p}$$

vanishes (mod p). One might guess that the "probability" that $q_p(2)$ so vanishes is about $1/p$. Since the sum of the reciprocals of the primes is divergent (see Exercise 1.20), one might guess that there are infinitely many Wieferich primes. Since the prime reciprocal sum diverges very slowly, one might also guess that they are very few and far between.

The Wieferich primes 1093 and 3511 have long been known. Crandall, Dilcher and Pomerance, with the computational aid of Bailey, established that there are no other Wieferich primes below $4 \cdot 10^{12}$ [Crandall et al. 1997]. McIntosh has pushed the limit further—to $16 \cdot 10^{12}$. It is not known whether there are any more Wieferich primes beyond 3511. It is also not known whether there are infinitely many primes that are *not* Wieferich primes! (But see Exercise 8.15.)

A second, presumably sparse class is conceived as follows. We first state a classical result and its converse:

Theorem 1.3.6 (Wilson–Lagrange). *Let p be an integer greater than one. Then p is prime if and only if*

$$(p-1)! \equiv -1 \pmod{p}.$$

This motivates us to ask whether there are any instances of

$$(p-1)! \equiv -1 \pmod{p^2}, \tag{1.15}$$

such primes being called Wilson primes. For any prime p we may assign a Wilson quotient

$$w_p = \frac{(p-1)!+1}{p},$$

whose vanishing (mod p) signifies a Wilson prime. Again the "probability" that p is a Wilson prime should be about $1/p$, and again the rarity is empirically manifest, in the sense that except for 5, 13, and 563, there are no Wilson primes less than $5 \cdot 10^8$.

A third presumably sparse class is that of Wall–Sun–Sun primes, namely those primes p satisfying

$$u_{p-\left(\frac{p}{5}\right)} \equiv 0 \pmod{p^2}, \tag{1.16}$$

where u_n denotes the n-th Fibonacci number (see Exercise 2.5 for definition) and where $\left(\frac{p}{5}\right)$ is 1 if $p \equiv \pm 1 \pmod 5$, is -1 if $p \equiv \pm 2 \pmod 5$, and is 0 if $p = 5$. As with the Wieferich and Wilson primes, the congruence (1.16) is always satisfied (mod p). R. McIntosh has shown that there are no Wall–Sun–Sun primes whatsoever below $3.2 \cdot 10^{12}$. The Wall–Sun–Sun primes also figure into the first case of Fermat's last theorem, in the sense that a prime exponent p for $x^p + y^p = z^p$, where p does not divide xyz, must also satisfy congruence (1.16) [Sun and Sun 1992].

Interesting computational issues arise in the search for Wieferich, Wilson, or Wall–Sun–Sun primes. Various such issues are covered in the exercises; for the moment we list a few salient points. First, computations (mod p^2) can be effected nicely by considering each congruence class to be a pair $(a, b) = a+bp$. Thus, for multiplication one may consider an operator $*$ defined by

$$(a, b) * (c, d) \equiv (ac, (bc + ad) \pmod p) \pmod {p^2},$$

and with this relation all the arithmetic necessary to search for the rare primes of this section can proceed with size-p arithmetic. Second, factorials in particular can be calculated using various enhancements, such as arithmetic progression-based products and polynomial evaluation, as discussed in Chapter 9. For example, it is known that for $p = 2^{40} + 5$,

$$(p-1)! \equiv -1 - 533091778023p \pmod {p^2},$$

as obtained by polynomial evaluation of the relevant factorial [Crandall et al. 1997]. This p is therefore not a Wilson prime, yet it is of interest that, in this day and age, machines can validate at least 12-digit primes via application of Lagrange's converse of the classical Wilson theorem.

In searches for these rare primes, some "close calls" have been encountered. Perhaps the only importance of a close call is to verify heuristic beliefs about the statistics of such as the Fermat and Wilson quotients. Examples of the near misses with their very small (but alas nonzero) quotients are

$$\begin{aligned} p &= 76843523891, & q_p(2) &\equiv -2 \pmod p, \\ p &= 12456646902457, & q_p(2) &\equiv \ 4 \pmod p, \\ p &= 56151923, & w_p &\equiv -1 \pmod p, \\ p &= 93559087, & w_p &\equiv -3 \pmod p, \end{aligned}$$

and we remind ourselves that the vanishing of any Fermat or Wilson quotient modulo p would have signaled a successful "strike."

1.4 Analytic number theory

Analytic number theory refers to the marriage of continuum analysis with the theory of the (patently discrete) integers. In this field, one can use integrals, complex domains, the tools of analysis to glean truths about the natural numbers. We speak of a beautiful and powerful subject that is both useful in the study of algorithms, and itself a source of many interesting algorithmic problems. In what follows we tour a few highlights of the analytic theory.

1.4.1 The Riemann zeta function

It was the brilliant leap of Riemann in the mid-19th century to ponder an entity so artfully employed by Euler,

$$\zeta(s) = \sum_{n=1}^{\infty} \frac{1}{n^s}, \tag{1.17}$$

but to ponder with powerful generality, namely, to allow s to attain complex values. The sum converges absolutely for $\mathrm{Re}(s) > 1$, and has an analytic continuation over the entire complex plane, regular except at the single point $s = 1$, where it has a simple pole with residue 1. (That is, $(s-1)\zeta(s)$ is analytic in the entire complex plane, and its value at $s = 1$ is 1.) It is fairly easy to see how $\zeta(s)$ can be continued to the half-plane $\mathrm{Re}(s) > 0$: For $\mathrm{Re}(s) > 1$ we have identities such as

$$\zeta(s) = \frac{s}{s-1} - s\int_1^\infty (x - \lfloor x \rfloor)x^{-s-1}\, dx.$$

But this formula continues to apply in the region $\mathrm{Re}(s) > 0, s \neq 1$, so we may take this integral representation as the definition of $\zeta(s)$ for the extended region. The equation also shows the claimed nature of the singularity at $s = 1$, and other phenomena, such as the fact that ζ has no zeros on the positive real axis. There are yet other analytic representations that give continuation to *all* complex values of s.

The connection with prime numbers was noticed earlier by Euler (with the variable s real), in the form of a beautiful relation that can be thought of as an analytic version of the fundamental Theorem 1.1.1:

Theorem 1.4.1 (Euler). *For* $\mathrm{Re}(s) > 1$ *and* $\mathcal{P}$ *the set of primes,*

$$\zeta(s) = \prod_{p\in\mathcal{P}} (1 - p^{-s})^{-1}. \tag{1.18}$$

Proof. The "Euler factor" $(1 - p^{-s})^{-1}$ may be rewritten as the sum of a geometric progression: $1 + p^{-s} + p^{-2s} + \cdots$. We consider the operation of multiplying together all of these separate progressions. The general term in the multiplied-out result will be $\prod_{p\in\mathcal{P}} p^{-a_p s}$, where each a_p is a positive integer or 0, and all but finitely many of these a_p are 0. Thus the general term is n^{-s} for some natural number n, and by Theorem 1.1.1, each such n occurs once and only once. Thus the right side of the equation is equal to the left side of the equation, which completes the proof. □

As was known to Euler, the zeta function admits various closed-form evaluations, such as

$$\zeta(2) = \pi^2/6,$$
$$\zeta(4) = \pi^4/90,$$

and in general, $\zeta(n)$ for even n is known; although not a single $\zeta(n)$ for odd $n > 2$ is known in closed form. But the real power of the Riemann zeta function, in regard to prime number studies, lies in the function's properties for $\mathrm{Re}(s) \le 1$. Closed-form evaluations such as

$$\zeta(0) = -1/2$$

are sometimes possible in this region. Here are some salient facts about theoretical applications of ζ:

(1) The fact that $\zeta(s) \to \infty$ as $s \to 1$ implies the infinitude of primes.

(2) The fact that $\zeta(s)$ has no zeros on the line $\mathrm{Re}(s) = 1$ leads to the prime number Theorem 1.1.4.

(3) The properties of ζ in the "critical strip" $0 < \mathrm{Re}(s) < 1$ lead to deep aspects of the distribution of primes, such as the essential error term in the PNT.

On the point (1), we can prove Theorem 1.1.2 as follows:

Another proof of the infinitude of primes. We consider $\zeta(s)$ for s real, $s > 1$. Clearly, from relation (1.17), $\zeta(s)$ diverges as $s \to 1^+$ because the harmonic sum $\sum 1/n$ is divergent. Indeed, for $s > 1$,

$$\begin{aligned}\zeta(s) &> \sum_{n \le 1/(s-1)} n^{-s} = \sum_{n \le 1/(s-1)} n^{-1} n^{-(s-1)} \\ &\ge e^{-1/e} \sum_{n \le 1/(s-1)} n^{-1} > e^{-1/e} |\ln(s-1)|.\end{aligned}$$

But if there were only finitely many primes, the product in (1.18) would tend to a finite limit as $s \to 1^+$, a contradiction. □

The above proof actually can be used to show that the sum of the reciprocals of the primes diverges. Indeed,

$$\ln\left(\prod_{p\in\mathcal{P}} (1-p^{-s})^{-1}\right) = -\sum_{p\in\mathcal{P}} \ln(1-p^{-s}) = \sum_{p\in\mathcal{P}} p^{-s} + O(1), \tag{1.19}$$

uniformly for $s > 1$. Since the left side of (1.19) goes to ∞ as $s \to 1^+$ and since $p^{-s} < p^{-1}$ when $s > 1$, the sum $\sum_{p\in\mathcal{P}} p^{-1}$ is divergent. (Compare with Exercise 1.20.) It is by a similar device that Dirichlet was able to prove Theorem 1.1.5; see Section 1.4.3.

Incidentally, one can derive much more concerning the partial sums of $1/p$ (henceforth we suppress the notation $p \in \mathcal{P}$, understanding that the index p is to be a prime variable unless otherwise specified):

Theorem 1.4.2 (Mertens). *As $x \to \infty$,*

$$\prod_{p \le x} \left(1 - \frac{1}{p}\right) \sim \frac{e^{-\gamma}}{\ln x}, \tag{1.20}$$

where γ is the Euler constant. Taking the logarithm of this relation, we have

$$\sum_{p \le x} \frac{1}{p} = \ln\ln x + B + o(1), \tag{1.21}$$

for the Mertens constant B defined as

$$B = \gamma + \sum_p \left(\ln\left(1 - \frac{1}{p}\right) + \frac{1}{p}\right).$$

This theorem is proved in [Hardy and Wright 1979]. The theorem is also a corollary of the prime number Theorem 1.1.4, but it is simpler than the PNT and predates it. The PNT still has something to offer, though; it gives smaller error terms in (1.20) and (1.21). Incidentally, the computation of the Mertens constant B is an interesting challenge (Exercise 1.87).

We have seen that certain facts about the primes can be thought of as facts about the Riemann zeta function. As one penetrates more deeply into the "critical strip," that is, into the region $0 < \mathrm{Re}(s) < 1$, one essentially gains more and more information about the detailed fluctuations in the distribution of primes. In fact it is possible to write down an explicit expression for $\pi(x)$ that depends on the zeros of $\zeta(s)$ in the critical strip. We illustrate this for a function that is related to $\pi(x)$, but is more natural in the analytic theory. Consider the function $\psi_0(x)$. This is the function $\psi(x)$ defined as

$$\psi(x) = \sum_{p^m \le x} \ln p = \sum_{p \le x} \ln p \left\lfloor \frac{\ln x}{\ln p} \right\rfloor, \tag{1.22}$$

except if $x = p^m$, in which case $\psi_0(x) = \psi(x) - \frac{1}{2}\ln p$. Then (see [Edwards 1974], [Davenport 1980], [Ivić 1985]) for $x > 1$,

$$\psi_0(x) = x - \sum_\rho \frac{x^\rho}{\rho} - \ln(2\pi) - \frac{1}{2}\ln\left(1 - x^{-2}\right), \tag{1.23}$$

where the sum is over the zeros ρ of $\zeta(s)$ with $\mathrm{Re}(\rho) > 0$. This sum is not absolutely convergent, and since the zeros ρ extend infinitely in both (vertical) directions in the critical strip, we understand the sum to be the limit as $T \to \infty$ of the finite sum over those zeros ρ with $|\rho| < T$. It is further understood that if a zero ρ is a multiple zero of $\zeta(s)$, it is counted with proper multiplicity in the sum. (It is widely conjectured that all of the zeros of $\zeta(s)$ are simple.)

Riemann posed what has become a central conjecture for all of number theory, if not for all of mathematics:

Conjecture 1.4.1 (Riemann hypothesis (RH)). *All the zeros of* $\zeta(s)$ *in the critical strip* $0 < \mathrm{Re}(s) < 1$ *lie on the line* $\mathrm{Re}(s) = 1/2$.

There are various equivalent formulations of the Riemann hypothesis. We have already mentioned one in Section 1.1.5. For another, consider the Mertens function

$$M(x) = \sum_{n \le x} \mu(n),$$

where $\mu(n)$ is the Möbius function defined to be 1 if n is square-free with an even number of prime factors, -1 if n is square-free with an odd number of prime factors, and 0 if n is not square-free. (For example, $\mu(1) = \mu(6) = 1$, $\mu(2) = \mu(105) = -1$, and $\mu(9) = \mu(50) = 0$.) The function $M(x)$ is related to the Riemann zeta function by

$$\frac{1}{\zeta(s)} = \sum_{n=1}^\infty \frac{\mu(n)}{n^s} = s\int_1^\infty \frac{M(x)}{x^{s+1}}\,dx, \tag{1.24}$$

valid certainly for $\mathrm{Re}(s) > 1$. It is interesting that the behavior of the Mertens function runs sufficiently deep that the following equivalences are known (in this and subsequent such uses of big-O notation, we mean that the implied constant depends on ϵ only):

Theorem 1.4.3. *The PNT is equivalent to the statement*

$$M(x) = o(x),$$

while the Riemann hypothesis is equivalent to the statement

$$M(x) = O\left(x^{\frac{1}{2}+\epsilon}\right)$$

for any fixed $\epsilon > 0$.

What a compelling notion, that the Mertens function, which one might envision as something like a random walk, with the Möbius μ contributing to the summation for M in something like the style of a random coin flip, should be so closely related to the great theorem (PNT) and the great conjecture (RH) in this way. The equivalences in Theorem 1.4.3 can be augmented with various alternative statements. One such is the elegant result that the PNT is equivalent to the statement

$$\sum_{n=1}^{\infty} \frac{\mu(n)}{n} = 0,$$

as shown by von Mangoldt. Incidentally, it is not hard to show that the sum in relation (1.24) converges absolutely for $\mathrm{Re}(s) > 1$; it is the rigorous sum evaluation at $s = 1$ that is difficult (see Exercise 1.19). In 1859, Riemann conjectured that for each fixed $\epsilon > 0$,

$$\pi(x) = \mathrm{li}\,(x) + O\left(x^{1/2+\epsilon}\right), \tag{1.25}$$

which conjecture is equivalent to the Riemann hypothesis, and perforce to the second statement of Theorem 1.4.3. In fact, the relation (1.25) is equivalent to the assertion that $\zeta(s)$ has no zeros in the region $\mathrm{Re}(s) > 1/2 + \epsilon$. The estimate (1.25) has not been proved for any $\epsilon < 1/2$.

In 1901, H. von Koch strengthened (1.25) slightly by showing that the Riemann hypothesis is true if and only if $|\pi(x) - \mathrm{li}\,(x)| = O(\sqrt{x}\ln x)$. In fact, for $x \geq 2.01$ we can take the big-O constant to be 1 in this assertion; see Exercise 1.36.

Let p_n denote the n-th prime. It follows from (1.25) that if the Riemann hypothesis is true, then

$$p_{n+1} - p_n = O\left(p_n^{1/2+\epsilon}\right)$$

holds for each fixed $\epsilon > 0$. Remarkably, we know rigorously that $p_{n+1} - p_n = O\left(p_n^{0.535+\epsilon}\right)$, a result of R. Baker and G. Harman. But much more is conjectured. The famous conjecture of H. Cramér asserts that

$$p_{n+1} - p_n \le (1 + o(1)) \ln^2 n$$

as $n \to \infty$. A. Granville has raised some doubt on this very precise conjecture, so it may be safer to weaken it slightly to $p_{n+1} - p_n = O\left(\ln^2 n\right)$.

1.4.2 Computational successes

The Riemann hypothesis (RH) remains open to this day. However, it is known after decades of technical development and a great deal of computer time that the first 1.5 billion zeros in the critical strip (ordered by increasing positive imaginary part) all lie precisely on the critical line $\mathrm{Re}(s) = 1/2$ [van de Lune et al. 1986]. It is highly intriguing—and such is possible due to a certain symmetry inherent in the zeta function—that one can numerically derive rigorous placement of the zeros with arithmetic of finite (yet perhaps high) precision. This is accomplished via rigorous counts of the number of zeros to various heights T (that is, the number of zeros with imaginary part in the interval $(0, T]$), and then an investigation of sign changes of a certain real function that is zero if and only if zeta is zero on the critical line. If the sign changes match the count, all of the zeros to that height T are accounted for in rigorous fashion [Brent 1979].

Another result along similar lines is the recent settling of the "Mertens conjecture," that

$$|M(x)| < \sqrt{x}. \tag{1.26}$$

Alas, the conjecture turns out to be ill-fated. An earlier conjecture that the right-hand side could be replaced by $\frac{1}{2}\sqrt{x}$ was first demolished in 1963 by Neubauer; later, H. Cohen found a minimal (least x) violation in the form

$$M(7725038629) = 43947.$$

But the Mertens conjecture (1.26) was finally demolished when [Odlyzko and te Riele 1985] showed that

$$\begin{aligned} \limsup x^{-1/2} M(x) &> 1.06, \\ \liminf x^{-1/2} M(x) &< -1.009. \end{aligned}$$

It has been shown by Pintz that for some x less than $10^{10^{65}}$ the ratio $M(x)/\sqrt{x}$ is greater than 1 [Ribenboim 1996]. Incidentally, it is known from statistical theory that the summatory function $m(x) = \sum_{n \le x} t_n$ of a random walk (with $t_n = \pm 1$, randomly and independently) enjoys (with probability 1) the relation

$$\limsup \frac{m(x)}{\sqrt{(x/2) \ln \ln x}} = 1,$$

so that on any notion of sufficient "randomness" of the Möbius μ function $M(x)/\sqrt{x}$ would be expected to be unbounded.

Yet another numerical application of the Riemann zeta function is in the assessment of the prime-counting function $\pi(x)$ for particular, hopefully large x. We address this computational problem later, in Section 3.6.2.

Analytic number theory is rife with big-O estimates. To the computationalist, every such estimate raises a question: What constant can stand in place of the big-O and in what range is the resulting inequality true? For example, it follows from a sharp form of the prime number theorem that for sufficiently large n, the n-th prime exceeds $n \ln n$. It is not hard to see that this is true for small n as well. Is it always true? To answer the question, one has to go through the analytic proof and put flesh on the various O-constants that appear, so as to get a grip on the "sufficiently large" aspect of the claim. In a wonderful manifestation of this type of analysis, [Rosser 1939] indeed showed that the n-th prime is always larger than $n \ln n$. Later, in joint work with Schoenfeld, many more explicit estimates involving primes were established. These collective investigations continue to be an interesting and extremely useful branch of computational analytic number theory.

1.4.3 Dirichlet L-functions

One can "twist" the Riemann zeta function by a Dirichlet character. To explain what this cryptic statement means, we begin at the end and explain what is a Dirichlet character.

Definition 1.4.4. Suppose D is a positive integer and χ is a function from the integers to the complex numbers such that

(1) For all integers m, n, $\chi(mn) = \chi(m)\chi(n)$.

(2) χ is periodic modulo D.

(3) $\chi(n) = 0$ if and only if $\gcd(n, D) > 1$.

Then χ is said to be a Dirichlet character to the modulus D.

For example, if $D > 1$ is an odd integer, then the Jacobi symbol $\left(\frac{n}{D}\right)$ is a Dirichlet character to the modulus D (see Definition 2.3.3).

It is a simple consequence of the definition that if χ is a Dirichlet character (mod D) and if $\gcd(n, D) = 1$, then $\chi(n)^{\varphi(D)} = 1$; that is, $\chi(n)$ is a root of unity. Indeed, $\chi(n)^{\varphi(D)} = \chi\left(n^{\varphi(D)}\right) = \chi(1)$, where the last equality follows from the Euler theorem (see (2.2)) that for $\gcd(n, D) = 1$ we have $n^{\varphi(D)} \equiv 1 \pmod{D}$. But $\chi(1) = 1$, since $\chi(1) = \chi(1)^2$ and $\chi(1) \neq 0$.

If χ_1 is a Dirichlet character to the modulus D_1 and χ_2 is one to the modulus D_2, then $\chi_1\chi_2$ is a Dirichlet character to the modulus $\operatorname{lcm}[D_1, D_2]$, where by $(\chi_1\chi_2)(n)$ we simply mean $\chi_1(n)\chi_2(n)$. Thus, the Dirichlet characters to the modulus D are closed under multiplication. In fact, they form a multiplicative group, where the identity is χ_0, the "principal character" to the modulus D. We have $\chi_0(n) = 1$ when $\gcd(n, D) = 1$, and 0 otherwise. The multiplicative inverse of a character χ to the modulus D is its complex conjugate, $\overline{\chi}$.

As with integers, characters can be uniquely factored. If D has the prime factorization $p_1^{a_1} \cdots p_k^{a_k}$, then a character $\chi \pmod{D}$ can be uniquely factored as $\chi_1 \cdots \chi_k$, where χ_j is a character $\pmod{p_j^{a_j}}$.

In addition, characters modulo prime powers are easy to construct and understand. Let $q = p^a$ be an odd prime power or 2 or 4. There are primitive roots $\pmod{q}$, say one of them is g. (A primitive root for a modulus D is a cyclic generator of the multiplicative group $\mathbf{Z}_D^*$ of residues modulo D that are coprime to D. This group is cyclic if and only if D is not properly divisible by 4 and not divisible by two different odd primes.) Then the powers of g $\pmod{q}$ run over all the residue classes $\pmod{q}$ coprime to q. So, if we pick a $\varphi(q)$-th root of 1, call it η, then we have picked the unique character χ $\pmod{q}$ with $\chi(g) = \eta$. We see there are $\varphi(q)$ different characters $\chi \pmod{q}$.

It is a touch more difficult in the case that $q = 2^a$ with $a > 2$, since then there is no primitive root. However, the order of 3 $\pmod{2^a}$ for $a > 2$ is always 2^{a-2}, and $2^{a-1}+1$, which has order 2, is not in the cyclic subgroup generated by 3. Thus these two residues, 3 and $2^{a-1}+1$, freely generate the multiplicative group of odd residues $\pmod{2^a}$. We can then construct the characters $\pmod{2^a}$ by choosing a 2^{a-2}-th root of 1, say η, and choosing $\varepsilon \in \{1, -1\}$, and then we have picked the unique character $\chi \pmod{2^a}$ with $\chi(3) = \eta, \chi(2^{a-1}+1) = \varepsilon$. Again there are $\varphi(q)$ characters $\chi \pmod{q}$.

Thus, there are exactly $\varphi(D)$ characters $\pmod{D}$, and the above proof not only lets us construct them, but it shows that the group of characters $\pmod{D}$ is isomorphic to the multiplicative group $\mathbf{Z}_D^*$ of residues $\pmod{D}$ coprime to D. To conclude our brief tour of Dirichlet characters we record the following two (dual) identities, which express a kind of orthogonality:

$$\sum_{\chi \pmod{D}} \chi(n) = \begin{cases} \varphi(D), & \text{if } n \equiv 1 \pmod{D}, \\ 0, & \text{if } n \not\equiv 1 \pmod{D}, \end{cases} \tag{1.27}$$

$$\sum_{n=1}^{D} \chi(n) = \begin{cases} \varphi(D), & \text{if } \chi \text{ is the principal character } \pmod{D} \\ 0, & \text{if } \chi \text{ is a nonprincipal character } \pmod{D}. \end{cases} \tag{1.28}$$

Now we can turn to the main topic of this section, Dirichlet L-functions. If χ is a Dirichlet character modulo D, let

$$L(s, \chi) = \sum_{n=1}^{\infty} \frac{\chi(n)}{n^s}.$$

The sum converges in the region $\mathrm{Re}(s) > 1$, and if χ is nonprincipal, then (1.28) implies that the domain of convergence is $\mathrm{Re}(s) > 0$. In analogy to (1.18) we have

$$L(s, \chi) = \prod_p \left(1 - \frac{\chi(p)}{p^s}\right)^{-1}. \tag{1.29}$$

It is easy to see from this formula that if $\chi = \chi_0$ is the principal character $\pmod{D}$, then $L(s, \chi_0) = \zeta(s) \prod_{p|D} (1 - p^{-s})$, that is, $L(s, \chi_0)$ is almost the same as $\zeta(s)$.

Dirichlet used his L-functions to prove Theorem 1.1.5 on primes in arithmetic progressions. The idea is to take the logarithm of (1.29) just as in (1.19), getting

$$\ln(L(s,\chi)) = \sum_p \frac{\chi(p)}{p^s} + O(1), \tag{1.30}$$

uniformly for $\mathrm{Re}(s) > 1$ and all Dirichlet characters χ. Then, if a is an integer coprime to D, we have

$$\begin{aligned}\sum_{\chi \pmod D} \overline{\chi}(a)\ln(L(s,\chi)) &= \sum_{\chi \pmod D}\sum_p \frac{\overline{\chi}(a)\chi(p)}{p^s} + O(\varphi(D)) \\ &= \varphi(D) \sum_{p\equiv a \pmod D} \frac{1}{p^s} + O(\varphi(D)),\end{aligned} \tag{1.31}$$

where the second equality follows from (1.27) and from the fact that $\overline{\chi}(a)\chi(p) = \chi(bp)$, where b is such that $ba \equiv 1 \pmod D$. Equation (1.31) thus contains the magic that is necessary to isolate the primes p in the residue class $a \pmod D$. If we can show the left side of (1.31) tends to infinity as $s \to 1^+$, then it will follow that there are infinitely many primes $p \equiv a \pmod D$, and in fact, they have an infinite reciprocal sum. We already know that the term on the left corresponding to the principal character χ_0 tends to infinity, but the other terms could cancel this. Thus, and this is the heart of the proof of Theorem 1.1.5, it remains to show that if χ is not a principal character $\pmod D$, then $L(1,\chi) \neq 0$. See [Davenport 1980] for a proof.

Just as the zeros of $\zeta(s)$ say much about the distribution of all of the primes, the zeros of the Dirichlet L-functions $L(s,\chi)$ say much about the distribution of primes in arithmetic progressions. In fact, the Riemann hypothesis has the following extension:

Conjecture 1.4.2 (The extended Riemann hypothesis (ERH)). *Let χ be an arbitrary Dirichlet character. Then the zeros of $L(s,\chi)$ in the region* $\mathrm{Re}(s) > 0$ *lie on the vertical line* $\mathrm{Re}(s) = \frac{1}{2}$.

We note that an even more general hypothesis, the generalized Riemann hypothesis (GRH) is relevant for more general algebraic domains, but we limit the scope of our discussion to the ERH above. (Note that one qualitative way to think of the ERH/GRH dichotomy is: The GRH says essentially that every general zeta-like function that should reasonably be expected not to have zeros in an interesting specifiable region indeed does not have any [Bach and Shallit 1996].) Conjecture 1.4.2 is of fundamental importance also in computational number theory. For example, one has the following conditional theorem.

Theorem 1.4.5. *Assume the ERH holds. For each positive integer D and each nonprincipal character $\chi \pmod D$, there is a positive integer $n < 2\ln^2 D$ with $\chi(n) \neq 1$ and a positive integer $m < 3\ln^2 D$ with $\chi(m) \neq 1$ and $\chi(m) \neq 0$.*

This result is in [Bach 1990]. That both estimates are $O\left(\ln^2 D\right)$, assuming the ERH, was originally due to N. Ankeny in 1952. Theorem 1.4.5 is what is behind ERH-conditional "polynomial time" primality tests, and it is also useful in other contexts.

The ERH has been checked computationally, but not as far as the Riemann hypothesis has. We know that it is true up to height 10000 for all characters χ with moduli up to 13, and up to height 2500 for all characters χ with moduli up to 72, and for various other moduli [Rumely 1993]. Using these calculations, [Ramaré and Rumely 1996] obtain explicit estimates for the distribution of primes in certain arithmetic progressions. (In recent unpublished calculations, Rumely has verified the ERH up to height 100000 for all characters with moduli up through 9.) Incidentally, the ERH implies an explicit estimate of the error in (1.5), the prime number theorem for arithmetic progressions; namely, for $x \ge 2,\ d \ge 2$,

$$\left|\pi(x;d,a) - \frac{1}{\varphi(d)}\,\mathrm{li}\,(x)\right| < x^{1/2}(\ln x + 2\ln d) \quad \text{(on the ERH).} \tag{1.32}$$

We note the important fact that there is here not only a tight error bound, but an *explicit* bounding constant (as opposed to the appearance of just an implied, nonspecific constant on the right-hand side). It is this sort of hard bounding that enables one to combine computations and theory, and settle conjectures in this way. Also on the ERH, if $d > 2$ and $\gcd(d,a) = 1$ there is a prime $p \equiv a \pmod d$ with $p < 2d^2 \ln^2 d$ (see [Bach and Shallit 1996] for these and related ERH-contingent results). As with the PNT itself, unconditional estimates (i.e., those not depending on the ERH) on $\pi(x;d,a)$ are less precise. For example, there is the following historically important (and unconditional) theorem:

Theorem 1.4.6 (Siegel–Walfisz). *For any number $\eta > 0$ there is a positive number $C(\eta)$ such that for all coprime positive integers a, d with $d < \ln^\eta x$,*

$$\pi(x;d,a) = \frac{1}{\varphi(d)}\,\mathrm{li}\,(x) + O\left(x\ \exp\left(-C(\eta)\sqrt{\ln x}\right)\right),$$

where the implied big-O constant is absolute.

Discussions of this and related theorems are found in [Davenport 1980]. It is interesting that the number $C(\eta)$ in Theorem 1.4.6 has not been computed for any $\eta \ge 1$. Furthermore it is *not computable* from the method of proof of the theorem. (It should be pointed out that numerically explicit error estimates for $\pi(x;d,a) - \frac{1}{\varphi(d)}\,\mathrm{li}\,(x)$ are possible in the range $1 \le \eta < 2$, though with an error bound not as sharp as in Theorem 1.4.6. For $\eta \ge 2$, no numerically explicit error estimate is known at all that is little-o of the main term.) Though error bounds of the Siegel–Walfisz type fall short of what is achievable on the ERH, such estimates nevertheless attain profound significance when combined with other analytic methods, as we discuss in Section 1.4.4.

1.4.4 Exponential sums

Beyond the Riemann zeta function and special arithmetic functions that arise in analytic number theory, there are other important entities, the exponential sums. These sums generally contain information—one might say "spectral" information—about special functions and sets of numbers. Thus, exponential sums provide a powerful bridge between complex Fourier analysis and number theory. For a real-valued function f, real t, and integers $a < b$, denote

$$E(f; a, b, t) = \sum_{a<n\le b} e^{2\pi i t f(n)}. \tag{1.33}$$

Each term in such an exponential sum has absolute value 1, but the terms can point in different directions in the complex plane. If the various directions are "random" or "decorrelated" in an appropriate sense, one would expect some cancellation of terms, reducing $|E|$ well below the trivial bound $b - a$. Thus, $E(f; a, b, t)$ measures in a certain sense the distribution of fractional parts for the sequence $(tf(n))$, $a < n \le b$. In fact, it is a celebrated theorem of [Weyl 1916] that the sequence $(f(n))$, $n = 1, 2, \ldots$ is equidistributed modulo 1 if and only if for every integer $h \ne 0$ we have $E(f; 0, N, h) = o(N)$. Though distribution of fractional parts is a constant undercurrent, the theory of exponential sums has wide application across many subfields of number theory. We give here a brief summary of the relevance of such sums to prime-number studies, ending with a brief, somewhat qualitative tour of Vinogradov's resolution of the ternary-Goldbach problem.

The theory of exponential sums began with Gauss and underwent a certain acceleration on the pivotal work of Weyl, who showed how to achieve rigorous upper bounds for specific classes of sums. In particular, Weyl discovered a simple but powerful estimation technique: Establish bounds on the absolute powers of a sum E. A fundamental observation is that

$$|E(f; a, b, t)|^2 = \sum_{n\in(a,b]} \sum_{k\in(a-n,b-n]} e^{2\pi i t(f(n+k)-f(n))}. \tag{1.34}$$

Now, something like a "derivative" of f appears in the exponent, allowing one to establish certain bounds on $|E|$ for *polynomial* f, by recursively applying a degree reduction. The manner in which one reduces the exponent degree can be instructive and gratifying; see, for example, Exercise 1.63 and other exercises referenced therein.

An important analytic problem one can address via exponential sums is that of the growth of the Riemann zeta function. The problem of bounding $\zeta(\sigma + it)$, for fixed real σ and varying real t, comes down to the bounding of sums

$$\sum_{N<n\le 2N} \frac{1}{n^{\sigma+it}},$$

which in turn can be bounded on the basis of estimates for the exponential sum

$$E(f;N,2N,t) = \sum_{N<n\le 2N} e^{-it\ln n},$$

where now the specific function is $f(n) = -(\ln n)/(2\pi)$. Expanding on Weyl's work, [van der Corput 1922] showed how to estimate such cases so that the bound on $\zeta(\sigma+it)$ could be given as a nontrivial power of t. For example, the Riemann zeta function can be bounded on the critical line $\sigma = 1/2$, as

$$\zeta(1/2+it) = O(t^{1/6}),$$

when $t \ge 1$; see [Graham and Kolesnik 1991]. The exponent has been successively reduced over the years; for example, [Bombieri and Iwaniec 1986] established the estimate $O\left(t^{9/56+\epsilon}\right)$ and [Watt 1989] obtained $O\left(t^{89/560+\epsilon}\right)$. The Lindelöf hypothesis is the conjecture that $\zeta(1/2+it) = O(t^\epsilon)$ for any $\epsilon > 0$. This conjecture also has consequences for the distribution of primes, such as the following result of [Yu 1996]: If p_n denotes the n-th prime, then on the Lindelöf hypothesis,

$$\sum_{p_n\le x} (p_{n+1}-p_n)^2 = x^{1+o(1)}.$$

The best that is known unconditionally is that the sum is $O\left(x^{23/18+\epsilon}\right)$ for any $\epsilon > 0$, a result of D. Heath-Brown. A consequence of Yu's conditional theorem is that for each $\epsilon > 0$, the number of integers $n \le x$ such that the interval $(n, n+n^\epsilon)$ contains a prime is $\sim x$. Incidentally, there is a connection between the Riemann hypothesis and the Lindelöf hypothesis: The former implies the latter.

It is not easy, but possible to get numerically explicit estimates via exponential sums. For example, [Cheng 1999] showed that

$$|\zeta(\sigma+it)| \le 175 t^{46(1-\sigma)^{3/2}} \ln^{2/3} t,$$

for $1/2 \le \sigma \le 1$ and $t \ge 2$. Such results can lead to explicit bounds relevant to various prime-number phenomena.

As for additive problems with primes, one may consider another important class of exponential sum, defined by

$$E_n(t) = \sum_{p\le n} e^{2\pi i t p}, \tag{1.35}$$

where p runs through primes. Certain integrals involving $E_n(t)$ over finite domains turn out to be associated with deep properties of the prime numbers. In fact, Vinogradov's proof that every sufficiently large odd integer is the sum of three primes starts essentially with the beautiful observation that the number of three-prime representations of n is precisely

$$R_3(n) = \int_0^1 \sum_{n\ge p,q,r\ \in\mathcal{P}} e^{2\pi i t(p+q+r-n)}\, dt \tag{1.36}$$

$$= \int_0^1 E_n^3(t) e^{-2\pi i t n}\, dt.$$

Vinogradov's proof was an extension of the earlier work of Hardy and Littlewood (see the monumental collection [Hardy 1966]), whose "circle method" was a *tour de force* of analytic number theory, essentially connecting exponential sums with general problems of additive number theory such as, but not limited to, the Goldbach problem.

Let us take a moment to give an overview of Vinogradov's method for estimating the integral (1.36). The guiding observation is that there is a strong correspondence between the distribution of primes and the *spectral* information embodied in $E_n(t)$. Assume that we have a general estimate on primes not exceeding n and belonging to an arithmetic progression $\{a, a+d, a+2d, \ldots\}$ with $\gcd(a,d) = 1$, in the form

$$\pi(n; d, a) = \frac{1}{\varphi(d)} \pi(n) + \epsilon(n; d, a),$$

which estimate, we assume, will be "good" in the sense that the error term ϵ will be suitably small for the problem at hand. (We have given a possible estimate in the form of the ERH relation (1.32) and the weaker, but unconditional Theorem 1.4.6.) Then for rational $t = a/q$ we develop an estimate for the sum (1.35) as

$$\begin{aligned} E_n(a/q) &= \sum_{f=0}^{q-1} \sum_{p \equiv f \pmod q,\ p \le n} e^{2\pi i p a/q} \\ &= \sum_{\gcd(f,q)=1} \pi(n; q, f) e^{2\pi i f a/q} + \sum_{p|q,\ p \le n} e^{2\pi i p a/q} \\ &= \sum_{\gcd(f,q)=1} \pi(n; q, f) e^{2\pi i f a/q} + O(q), \end{aligned}$$

where it is understood that the sums involving gcd run over the elements $f \in [1, q-1]$ that are coprime with q. It turns out that such estimates are of greatest value when the denominator q is relatively small. In such cases one may use the chosen estimate on primes in arithmetic progressions to arrive at

$$E_n(a/q) = \frac{c_q(a)}{\varphi(q)} \pi(n) + O(q + |\epsilon| \varphi(q)),$$

where $|\epsilon|$ denotes the maximum of $|\epsilon(n; q, f)|$ taken over all residues f coprime to q, and c_q is the well-studied Ramanujan sum

$$c_q(a) = \sum_{\gcd(f,q)=1} e^{2\pi i f a/q}. \tag{1.37}$$

We shall encounter this Ramanujan sum later, during our tour of discrete convolution methods, as in equation (9.23). For the moment, we observe that

[Hardy and Wright 1979]

$$c_q(a) = \frac{\mu(q/g)\varphi(q)}{\varphi(q/g)}, \qquad g = \gcd(a, q). \tag{1.38}$$

In particular, when a, q are coprime, we obtain a beautiful estimate of the form

$$E_n(a/q) = \sum_{p \le n} e^{2\pi i p a/q} = \frac{\mu(q)}{\varphi(q)}\pi(n) + \epsilon', \tag{1.39}$$

where the overall error ϵ' depends in complicated ways on a, q, n, and, of course, whatever is our theorem of choice on the distribution of primes in arithmetic progressions. We uncover thus a fundamental spectral property of primes: When q is small, the magnitude of the exponential sum is effectively reduced, by an explicit factor μ/φ, below the trivial estimate $\pi(n)$. Such reduction is due, of course, to cancellation among the oscillating summands; relation (1.39) quantifies this behavior.

Vinogradov was able to exploit the small-q estimate above in the following way. One chooses a cutoff $Q = \ln^B n$ for appropriately large B, thinking of q as "small" when $1 \le q \le Q$. (It turns out to be enough to consider only the range $Q < q < n/Q$ for "large" q.) Now, the integrand in (1.36) exhibits "resonances" when the integration variable t lies near to a rational a/q for the small $q \in [1, Q]$. These regions of t are traditionally called the "major arcs." The rest of the integral—over the "minor arcs" having $t \approx a/q$ with $q \in (Q, n/Q)$—can be thought of as "noise" that needs to be controlled (bounded). After some delicate manipulations, one achieves an integral estimate in the form

$$R_3(n) = \frac{n^2}{2\ln^3 n} \sum_{q=1}^{Q} \frac{\mu(q) c_q(n)}{\varphi^3(q)} + \epsilon'', \tag{1.40}$$

where we see a resonance sum from the major arcs, while ϵ'' now contains all previous arithmetic-progression errors plus the minor-arc noise. Already in the above summation over $q \in [1, Q]$ one can, with some additional algebraic effort, see how the final ternary-Goldbach estimate (1.12) results, as long as the error ϵ'' and the finitude of the cutoff Q and are not too troublesome (see Exercise 1.65).

It was the crowning achievement of Vinogradov to find an upper bound on the minor-arc component of the overall error ϵ''. The relevant theorem is this: If $\gcd(a, q) = 1$, $q \le n$, and a real t is near a/q in the sense $|t - a/q| \le 1/q^2$, then

$$|E_n(t)| < C\left(\frac{n}{q^{1/2}} + n^{4/5} + n^{1/2}q^{1/2}\right)\ln^3 n, \tag{1.41}$$

with absolute constant C. This result is profound, the proof difficult—involving intricate machinations with arithmetic functions—though having undergone some modern revision, notably by R. Vaughan (see references below). The bound is powerful because, for $q \in (Q, n/Q)$ and a real t of the theorem, the magnitude of $E_n(t)$ is reduced by a logarithmic-power factor below the total number $\pi(n)$ of summands. In this way the minor-arc noise has been bounded sufficiently to allow rigor in the ternary-Goldbach estimate. (Powerful as this approach may be, the *binary* Goldbach conjecture has so far been beyond reach, the analogous error term ϵ'', which includes yet noisier components, being so very difficult to bound.)

In summary: The estimate (1.39) is used for major-arc "resonances," yielding the main-term sum of (1.40), while the estimate (1.41) is used to bound the minor-arc "noise" and control the overall error ϵ'' . The relation (1.40) leads finally to the ternary-Goldbach estimate (1.12). Though this language has been qualitative, the reader may find the rigorous and compelling details—on this and related additive problems—in the references [Hardy 1966], [Davenport 1980], [Vaughan 1977, 1997], [Ellison and Ellison 1985, Theorem 9.4], [Nathanson 1996, Theorem 8.5], [Vinogradov 1985], [Estermann 1952].

Exponential-sum estimates can be, as we have just seen, incredibly powerful. The techniques enjoy application beyond just the Goldbach problem, even beyond the sphere of additive problems. Later, we shall witness the groundwork of Gauss on quadratic sums; e.g., Definition 2.3.6 involves variants of the form (1.33) with quadratic f. In Section 9.5.3 we take up the issue of discrete convolutions (as opposed to continuous integrals) and indicate through text and exercises how signal processing, especially discrete spectral analysis, connects with analytic number theory. What is more, exponential sums give rise to attractive and instructive computational experiments and research problems. For reader convenience, we list here some relevant Exercises: 1.34, 1.63, 1.65, 1.67, 2.23, 2.24, 9.42, 9.80.

1.4.5 Smooth numbers

Smooth numbers are extremely important for our computational interests, notably in factoring tasks. And there are some fascinating theoretical applications of smooth numbers, just one example being applications to a celebrated problem upon which we just touched, namely the Waring problem [Vaughan 1989]. We begin with a fundamental definition:

Definition 1.4.7. A positive integer is said to be y-smooth if it does not have any prime factor exceeding y.

What is behind the usefulness of smooth numbers? Basically, it is that for y not too large, the y-smooth numbers have a simple multiplicative structure, yet they are surprisingly numerous. For example, though only a vanishingly small fraction of the primes in $[1, x]$ are in the interval $[1, \sqrt{x}]$, nevertheless more than 30% of the numbers in $[1, x]$ are $\sqrt{x}$-smooth (for x sufficiently large). Another example illustrating this surprisingly high frequency of smooth numbers: The number of $(\ln^2 x)$-smooth numbers up to x exceeds $\sqrt{x}$ for all sufficiently large numbers x.

These examples suggest that it is interesting to study the counting function for smooth numbers. Let

$$\psi(x, y) = \#\{1 \le n \le x : n \text{ is } y\text{-smooth}\}. \tag{1.42}$$

Part of the basic landscape is the Dickman theorem from 1930:

Theorem 1.4.8 (Dickman). *For each fixed real number $u > 0$, there is a real number $\rho(u) > 0$ such that*

$$\psi(x, x^{1/u}) \sim \rho(u)x.$$

Moreover, Dickman described the function $\rho(u)$ as the solution of a certain differential equation: It is the unique continuous function on $[0, \infty)$ that satisfies (A) $\rho(u) = 1$ for $0 \le u \le 1$ and (B) for $u > 1$, $\rho'(u) = -\rho(u-1)/u$. In particular, $\rho(u) = 1 - \ln u$ for $1 \le u \le 2$, but there is no known closed form for $\rho(u)$ for $u > 2$. The function $\rho(u)$ can be approximated numerically, and it becomes quickly evident that it decays to zero rapidly. In fact, it decays somewhat faster than u^{-u}, though this simple expression can stand in as a reasonable estimate for $\rho(u)$ in various complexity studies. Indeed, we have

$$\ln \rho(u) \sim -u \ln u. \tag{1.43}$$

Theorem 1.4.8 is fine for estimating $\psi(x, y)$ when x, y tend to infinity with $u = \ln x / \ln y$ fixed or bounded. But how can we estimate $\psi\left(x, x^{1/\ln \ln x}\right)$ or $\psi\left(x, e^{\sqrt{\ln x}}\right)$ or $\psi\left(x, \ln^2 x\right)$? Estimates for these and similar expressions became crucial around 1980 when subexponential factoring algorithms were first being studied theoretically (see Chapter 6). Filling this gap, [Canfield et al. 1983] showed that

$$\psi\left(x, x^{1/u}\right) = xu^{-u+o(u)} \tag{1.44}$$

uniformly as $u \to \infty$ and $u < (1-\epsilon)\ln x / \ln \ln x$. Note that this is the expected estimate, since by (1.43) we have that $\rho(u) = u^{-u+o(u)}$. Thus we have a reasonable estimate for $\psi(x, y)$ when $y > \ln^{1+\epsilon} x$ and x is large. (We have reasonable estimates in smaller ranges for y as well, but we shall not need them in this book.)

It is also possible to prove explicit inequalities for $\psi(x, y)$. For example, in [Konyagin and Pomerance 1997] it is shown that for all $x \ge 4$ and $2 \le x^{1/u} \le x$,

$$\psi\left(x, x^{1/u}\right) \ge \frac{x}{\ln^u x}. \tag{1.45}$$

The implicit estimate here is reasonably good when $x^{1/u} = \ln^c x$, with $c > 1$ fixed (see Exercises 1.69 and 3.19).

1.5 Exercises

1.1. What is the largest integer N having the following property: All integers in $[2, \ldots, N-1]$ that have no common prime factor with N are themselves prime? What is the largest integer N divisible by every integer smaller than $\sqrt{N}$?

1.2. Prove Euclid's "first theorem": The product of two integers is divisible by a prime p if and only if one of them is divisible by p. Then show that Theorem 1.1.1 follows as a corollary.

1.3. Show that a positive integer n is prime if and only if

$$\sum_{m=1}^{\infty}\left(\left\lfloor\frac{n}{m}\right\rfloor - \left\lfloor\frac{n-1}{m}\right\rfloor\right) = 2.$$

1.4. Prove that for integer $x \geq 2$,

$$\pi(x) = \sum_{n=2}^{x}\left\lfloor\frac{1}{\sum_{k=2}^{n}\lfloor\lfloor n/k\rfloor k/n\rfloor}\right\rfloor.$$

1.5. Sometimes a prime-producing formula, even though computationally inefficient, has actual pedagogical value. Prove the Gandhi formula for the n-th prime:

$$p_n = \left\lfloor 1 - \log_2\left(-\frac{1}{2} + \sum_{d|p_{n-1}!}\frac{\mu(d)}{2^d - 1}\right)\right\rfloor.$$

One instructive way to proceed is to perform (symbolically) a sieve of Eratosthenes (see Chapter 3) on the binary expansion $1 = (0.11111\ldots)_2$.

1.6. By refining the method of proof for Theorem 1.1.2, one can achieve lower bounds (albeit relatively weak ones) on the prime-counting function $\pi(x)$. To this end, consider the "primorial of p," the number defined by

$$p\# = \prod_{q \leq p} q = 2 \cdot 3 \cdots p,$$

where the product is taken over primes q. Deduce, along the lines of Euclid's proof, that the n-th prime p_n satisfies:

$$p_n < p_{n-1}\#,$$

for $n \geq 3$. Then use induction to show that

$$p_n \leq 2^{2^{n-1}}.$$

Conclude that

$$\pi(x) > \frac{1}{\ln 2}\ln\ln x,$$

for $x \geq 2$.

Incidentally, the numerical study of primorial primes $p\# + 1$ is interesting in its own right. A modern example of a large primorial prime, discovered by C. Caldwell in 1999, is $42209\# + 1$, with more than eighteen thousand decimal digits.

1.7. By considering numbers of the form:

$$n = 2^2 \cdot 3 \cdot 5 \cdots p - 1,$$

prove that there exist infinitely many primes congruent to 3 modulo 4. Find a similar proof for primes that are congruent to 2 modulo 3. (Compare with Exercise 5.21.)

1.8. By considering numbers of the form:

$$(2 \cdot 3 \cdot \ldots \cdot p)^2 + 1,$$

prove that there are infinitely many primes $\equiv 1 \pmod 4$. Find a similar proof that there are infinitely many primes that are $\equiv 1 \pmod 3$.

1.9. Suppose a, n are natural numbers with $a \geq 2$. Let $N = a^n - 1$. Show that the order of $a \pmod N$ in the multiplicative group $\mathbf{Z}_N^*$ is n, and conclude that $n|\varphi(N)$. Use this to show that if n is prime, there are infinitely many primes congruent to 1 modulo n

1.10. Let $\mathcal{S}$ be a nonempty set of primes with sum of reciprocals $S < \infty$, and let $\mathcal{A}$ be the set of natural numbers that are not divisible by any member of $\mathcal{S}$. Show that $\mathcal{A}$ has asymptotic density less than e^{-S}. In particular show that if $\mathcal{S}$ has an infinite sum of reciprocals, then the density of $\mathcal{A}$ is zero. Using that the sum of reciprocals of the primes that are 3 (mod 4) is infinite, show that the set of numbers that can be written as a sum of two coprime squares has asymptotic density zero. (See Exercises 1.88 and 5.15.)

1.11. Starting from the fact that the sum of the reciprocals of the primes is infinite, use Exercise 1.10 to prove that the set of primes has asymptotic density zero, i.e., that $\pi(x) = o(x)$.

1.12. As we state in the text, the "probability" that a random positive integer x is prime is "about " $1/\ln x$. Assuming the PNT, cast this probability idea in rigorous language.

1.13. Using the definition

$$\phi(x, y) = \#\{1 \leq n \leq x : \text{ each prime dividing } n \text{ is greater than } y\}$$

(which appears later, Section 3.6.1, in connection with prime counting) argue that

$$\phi(x, \sqrt{x}) = \pi(x) - \pi(\sqrt{x}) + 1.$$

Then prove the classical Legendre relation

$$\pi(x) = \pi(\sqrt{x}) - 1 + \sum_{d|Q} \mu(d) \left\lfloor \frac{x}{d} \right\rfloor, \tag{1.46}$$

where Q is a certain product of primes, namely,

$$Q = \prod_{p \leq \sqrt{x}} p.$$

This kind of combinatorial reasoning can be used, as Legendre once did, to show $\pi(x) = o(x)$. To that end, show that

$$\phi(x,y) = x \prod_{p \le y} \left(1 - \frac{1}{p}\right) + E,$$

where the error term is $E = O(2^{\pi(y)})$. Now use this last relation and the fact that the sum of the reciprocals of the primes diverges to argue that $\pi(x)/x \to 0$ as $x \to \infty$. (Compare with Exercise 1.11.)

1.14. Starting with the fundamental Theorem 1.1.1, show that for any fixed $\epsilon > 0$, the number $d(n)$ of divisors of n (including always 1 and n) satisfies

$$d(n) = O(n^\epsilon).$$

How does the implied O-constant depend on the choice of ϵ? You might get started in this problem by first showing that, for fixed ϵ, there are finitely many prime powers q with $d(q) > q^\epsilon$.

1.15. Consider the sum of the reciprocals of all Mersenne numbers $M_n = 2^n - 1$ (for positive integers n), namely,

$$E = \sum_{n=1}^{\infty} \frac{1}{M_n}.$$

Prove the following alternative form involving the divisor function d (defined in Exercise 1.14):

$$E = \sum_{k=1}^{\infty} \frac{d(k)}{2^k}.$$

Actually, one can give this sum a faster-than-linear convergence. To that end show that we also have

$$E = \sum_{m=1}^{\infty} \frac{1}{2^{m^2}} \frac{2^m + 1}{2^m - 1}.$$

Incidentally, the number E has been well studied in some respects. For example, it is known [Erdős 1948], [Borwein 1991] that E is irrational, yet it has never been given a closed form. Possible approaches to establishing deeper properties of the number E are laid out in [Bailey and Crandall 2001b].

If we restrict such a sum to be over Mersenne *primes*, then on the basis of Table 1.2, and assuming said table is exhaustive up through its final entry (note that this is not rigorously known), to how many good decimal digits do we rigorously know

$$\sum_{M_q \in \mathcal{P}} \frac{1}{M_q} \quad ?$$

1.16. Euler's polynomial $x^2 + x + 41$ has prime values for each integer x with $-40 \le x \le 39$. Show that if $f(x)$ is a nonconstant polynomial with integer coefficients, then there are infinitely many integers x with $f(x)$ composite.

1.17. It can happen that a polynomial, while not *always* producing primes, is very likely to do so over certain domains. Show by computation that a polynomial found by [Dress and Olivier 1999],

$$f(x) = x^2 + x - 1354363,$$

has the astounding property that for a random integer $x \in [1, 10^4]$, the number $|f(x)|$ is prime with probability exceeding 1/2. An amusing implication is this: If you can remember the seven-digit "phone number" 1354363, then you have a mental mnemonic for generating thousands of primes.

1.18. Consider the sequence of primes $2, 3, 5, 11, 23, 47$. Each but the first is one away from the double of the prior prime. Show that there cannot be an infinite sequence of primes with this property, regardless of the starting prime.

1.19. As mentioned in the text, the relation

$$\frac{1}{\zeta(s)} = \sum_{n=1}^{\infty} \frac{\mu(n)}{n^s}$$

is valid (the sum converges absolutely) for $\mathrm{Re}(s) > 1$. Prove this. But the limit as $s \to 1$, for which we know the remarkable PNT equivalence

$$\sum_{n=1}^{\infty} \frac{\mu(n)}{n} = 0,$$

is not so easy. Two good exercises are these: First, via numerical experiments, furnish an estimate for the order of magnitude of

$$\sum_{n \le x} \frac{\mu(n)}{n}$$

as a function of x; and second, provide an at least heuristic argument as to why the sum should vanish as $x \to \infty$. For the first option, it is an interesting computational challenge to work out an efficient implementation of the μ function itself. As for the second option, you might consider the first few terms in the form

$$1 - \sum_{p \le x} \frac{1}{p} + \sum_{pq \le x} \frac{1}{pq} - \cdots$$

to see why the sum tends to zero for large x. It is of interest that even without recourse to the PNT, one can prove, as J. Gram did in 1884 [Ribenboim 1996], that the sum is bounded as $x \to \infty$.

1.20. Show that for all $x > 1$, we have

$$\sum_{p \le x} \frac{1}{p} > \ln \ln x - 1,$$

where p runs over primes. Conclude that there are infinitely many primes. One possible route is to establish the following intermediate steps:

(1) Show that $\sum_{n=1}^{\lfloor x \rfloor} \frac{1}{n} > \ln x$.

(2) Show that $\sum \frac{1}{n} = \prod_{p \le x} (1 - \frac{1}{p})^{-1}$, where the sum is over the natural numbers n not divisible by any prime exceeding x.

1.21. Use the multinomial (generalization of the binomial) theorem to show that for any positive integer u and any real number $x > 0$,

$$\frac{1}{u!}\left(\sum_{p \le x} \frac{1}{p}\right)^u \le \sum_{n \le x^u} \frac{1}{n},$$

where p runs over primes and n runs over natural numbers. Using this inequality with $u = \lfloor \ln \ln x \rfloor$ show that for $x \ge 3$,

$$\sum_{p \le x} \frac{1}{p} \le \ln \ln x + O(\ln \ln \ln x).$$

1.22. By considering the highest power of a given prime that divides a given factorial, prove that

$$N! = \prod_{p \le N} p^{\sum_{k=1}^{\infty} \lfloor N/p^k \rfloor},$$

where the product runs over primes p. Then use the inequality

$$N! > \left(\frac{N}{e}\right)^N$$

(which follows from $e^N = \sum_{k=0}^{\infty} N^k/k! > N^N/N!$), to prove that

$$\sum_{p \le N} \frac{\ln p}{p-1} > \ln N - 1.$$

Conclude that there are infinitely many primes.

1.23. Use the Stirling asymptotic formula

$$N! \sim \left(\frac{N}{e}\right)^N \sqrt{2\pi N}$$

and the method of Exercise 1.22 to show that

$$\sum_{p \le N} \frac{\ln p}{p} = \ln N + O(1).$$

Deduce that the prime-counting function $\pi(x)$ satisfies $\pi(x) = O(x/\ln x)$ and that if $\pi(x) \sim cx/\ln x$ for some number c, then $c = 1$.

1.24. Derive from the Chebyshev Theorem 1.1.3 the following bounds on the n-th prime number p_n for $n \geq 2$:

$$Cn \ln n < p_n < Dn \ln n,$$

where C, D are absolute constants.

1.25. As a teenager, P. Erdős proved that for each $x > 0$,

$$\prod_{p \leq x} p < 4^x.$$

Find a proof of this result, perhaps by first noting that it suffices to prove it for x an odd integer. Then you might proceed by induction, using

$$\prod_{n+1 < p \leq 2n+1} p \leq \binom{2n+1}{n} \leq 4^n.$$

1.26. Using Exercise 1.25 prove that $\pi(x) = O(x/\ln x)$. (Compare with Exercise 1.23.)

1.27. Prove the following theorem of Chebyshev, known as the Bertrand postulate: For a positive integer N there is at least one prime in the interval $(N, 2N]$. The following famous ditty places the Bertrand postulate as part of the lore of number theory:

Chebyshev said it,
we'll say it again:
There is always a prime
between N and $2N$.

Here is an outline of a possible strategy for the proof. Let P be the product of the primes p with $N < p \leq 2N$. We are to show $P > 1$. Show that P divides $\binom{2N}{N}$. Let Q be such that $\binom{2N}{N} = PQ$. Show that if q^a is the exact power of the prime q that divides Q, then $a \leq \ln(2N)/\ln q$. Show that the largest prime factor of Q does not exceed $2N/3$. Use Exercise 1.25 to show that

$$Q < 4^{\frac{2}{3}N} 4^{(2N)^{1/2}} 4^{(2N)^{1/3}} \cdots 4^{(2N)^{1/k}},$$

where $k = \lfloor \lg(2N) \rfloor$. Deduce that

$$P > \binom{2N}{N} 4^{-\frac{2}{3}N - (2N)^{1/2} - (2N)^{1/3} \lg(N/2)}.$$

Also show that $\binom{2N}{N} > 4^N/N$ for $N \geq 4$ (by induction) and deduce that $P > 1$ for $N \geq 250$. Handle the remaining cases of N by a direct argument.

1.28. Use the argument in Exercise 1.27 to show that there is a positive number c such that $\pi(x) > cx/\ln x$ for all $x \geq 2$, and give a suitable numerical value for c.

1.29. Here is an exercise involving a healthy mix of computation and theory. With $\sigma(n)$ denoting the sum of the divisors of n, and recalling from the discussion prior to Theorem 1.3.3 that n is deemed perfect if and only if $\sigma(n) = 2n$, do the following, wherein we adopt a unique prime factorization $n = p_1^{t_1} \cdots p_k^{t_k}$:

(1) Write a condition on the p_i, t_i alone that is equivalent to the condition $\sigma(n) = 2n$ of perfection.

(2) Use the relation from (1) to establish (by hand, with perhaps some minor machine computations) some lower bound on odd perfect numbers; e.g., show that any odd perfect number must exceed 10^6 (or an even larger bound).

(3) An "abundant number" is one with $\sigma(n) > 2n$, as in the instance $\sigma(12) = 28$. Find (by hand or by small computer search) an *odd* abundant number. Does an odd abundant number have to be divisible by 3?

(4) For odd n, investigate the possibility of "close calls" to perfection. For example, show (by machine perhaps) that every odd n with $10 < n < 10^6$ has $|\sigma(n) - 2n| > 5$.

(5) Explain why $\sigma(n)$ is almost always even. In fact, show that the number of $n \le x$ with $\sigma(n)$ odd is $\lfloor \sqrt{x} \rfloor + \lfloor \sqrt{x/2} \rfloor$.

(6) Show that for any fixed integer $k > 1$, the set of integers n with $k|\sigma(n)$ has asymptotic density 1. (Hint: Use the Dirichlet Theorem 1.1.5.) The case $k = 4$ is easier than the general case. Use this easier case to show that the set of odd perfect numbers has asymptotic density 0.

(7) Let $s(n) = \sigma(n) - n$ for natural numbers n, and let $s(0) = 0$. Thus, n is abundant if and only if $s(n) > n$. Let $s^{(k)}(n)$ be the function s iterated k times at n. Use the Dirichlet Theorem 1.1.5 to prove the following theorem of H. Lenstra: For each natural number k there is a number n with

$$n < s^{(1)}(n) < s^{(2)}(n) < \cdots < s^{(k)}(n). \tag{1.47}$$

It is not known whether there is any number n for which this inequality chain holds true for *every* k, nor is it known whether there is any number n for which the sequence $\left(s^{(k)}(n)\right)$ is unbounded. The smallest n for which the latter property is in doubt is 276. P. Erdős has shown that for each fixed k, the set of n for which $n < s(n)$, yet (1.47) fails, has asymptotic density 0.

1.30. [Vaughan] Prove, with $c_q(n)$ being the Ramanujan sum defined in relation (1.37), that n is a perfect number if and only if

$$\sum_{q=1}^{\infty} \frac{c_q(n)}{q^2} = \frac{12}{\pi^2}.$$

1.31. It is known [Copeland and Erdős 1946] that the number

$$0.235711131719\ldots,$$

where all the primes written in decimal are simply concatenated in order, is "normal to base 10," meaning that each finite string of k consecutive digits appears in this expansion with "fair" asymptotic frequency 10^{-k}. Argue a partial result: That each string of k digits appears infinitely often.

In fact, given two finite strings of decimal digits, show there are infinitely many primes that in base 10 begin with the first string and—regardless of what digits may appear in between—end with the second string, provided the last digit of the second string is $1, 3, 7$, or 9.

The relative density of primes having a given low-order decimal digit $1, 3, 7$ or 9 is $1/4$, as evident in relation (1.5). Does the set of all primes having a given *high*-order decimal digit have a similarly well-defined relative density?

1.32. Here we use the notion of normality of a number to a given base as enunciated in Exercise 1.31, and the notion of equidistribution enunciated in Exercise 1.34. Now think of the ordered, natural logarithms of the Fermat numbers as a pseudorandom sequence of real numbers. Prove this theorem: If said sequence is equidistributed modulo 1, then the number $\ln 2$ is normal to base 2. Is the converse of this theorem true?

Note that it remains unknown to this day whether $\ln 2$ is normal to any integer base. Unfortunately, the same can be said for any of the fundamental constants of history, such as π, e and so on. That is, except for instances of artificial digit construction as in Exercise 1.31, normality proofs remain elusive. A standard reference for rigorous descriptions of normality and equidistribution is [Kuipers and Niederreiter 1974]. A discussion of normality properties for specific fundamental constants such as $\ln 2$ is [Bailey and Crandall 2001a].

1.33. Using the PNT, or just Chebyshev's Theorem 1.1.3, prove that the set of rational numbers p/q with p, q prime is dense in the positive reals.

1.34. It is a theorem of Vinogradov that for any irrational number α, the sequence (αp_n), where the p_n are the primes in natural order, is equidistributed modulo 1. Equidistribution here means that if $\#(a, b, N)$ denotes the number of times any interval $[a, b) \subset [0, 1)$ is struck after N primes are used, then $\#(a, b, N)/N \sim (b - a)$ as $N \to \infty$. On the basis of this Vinogradov theorem, prove the following: For irrational $\alpha > 1$, and the set

$$S(\alpha) = \{\lfloor k\alpha \rfloor : k = 1, 2, 3, \ldots\},$$

the prime count defined by

$$\pi_\alpha(x) = \#\{p \le x \ : \ p \in \mathcal{P} \cap \mathcal{S}(\alpha)\}$$

behaves as

$$\pi_\alpha(x) \sim \frac{1}{\alpha} \frac{x}{\ln x}.$$

What is the behavior of π_α for α rational?

As an extension to this exercise, the Vinogradov equidistribution theorem itself can be established via the exponential sum ideas of Section 1.4.4. One uses the celebrated Weyl theorem on spectral properties of equidistributed sequences [Kuipers and Niederreiter 1974, Theorem 2.1] to bring the problem down to showing that for irrational α and any integer $h \neq 0$,

$$E_N(h\alpha) = \sum_{p \leq N} e^{2\pi i h \alpha p}$$

is $o(N)$. This, in turn, can be done by finding suitable rational approximants to α and providing bounds on the exponential sum, using essentially our book formula (1.39) for well-approximable values of $h\alpha$, while for other α using (1.41). The treatment of [Ellison and Ellison 1985] is pleasantly accessible on this matter.

As an extension, use exponential sums to study the count

$$\pi_c(x) = \#\left\{n \in [1, x] : \lfloor n^c \rfloor \in \mathcal{P}\right\}.$$

Heuristically, one might expect the asymptotic behavior

$$\pi_c(x) \sim \frac{1}{c} \frac{x}{\ln x}.$$

Show first, on the basis of the PNT, that for $c \leq 1$ this asymptotic relation indeed holds. Use exponential sum techniques to establish this asymptotic behavior for some $c > 1$; for example, there is the Piatetksi–Shapiro theorem [Graham and Kolesnik 1991] that the asymptotic relation holds for any c with $1 < c < 12/11$.

1.35. The study of primes can lead to considerations of truly stultifyingly large numbers, such as the Skewes numbers

$$10^{10^{10^{34}}}, \quad e^{e^{e^{e^{7.705}}}},$$

the second of these being a proven upper bound for the least x with $\pi(x) > \operatorname{li}_0(x)$, where $\operatorname{li}_0(x)$ is defined as $\int_0^x dt/\ln t$. (The first Skewes number is an earlier, celebrated bound that Skewes established conditionally on the Riemann hypothesis.) For $x > 1$ one takes the "principal value" for the singularity of the integrand at $t = 1$, namely,

$$\operatorname{li}_0(x) = \lim_{\epsilon \to 0} \left(\int_0^{1-\epsilon} \frac{1}{\ln t}\, dt + \int_{1+\epsilon}^{x} \frac{1}{\ln t}\, dt \right).$$

The function $\operatorname{li}_0(x)$ is $\operatorname{li}(x) + c$, where $c \approx 1.0451637801$. J. Littlewood has shown that $\pi(x) - \operatorname{li}_0(x)$ does change sign infinitely often.

An amusing first foray into the "Skewes world" is to express the *second* Skewes number above in decimal-exponential notation (in other words, replace the e's with 10's appropriately, as has been done already for the first Skewes number). Incidentally, a newer reference on the problem is [Kaczorowski 1984], while a modern estimate for the least x with $\pi(x) > \operatorname{li}_0(x)$ is $x <$

$1.4 \cdot 10^{316}$ [Bays and Hudson 2000a, 2000b]. In fact, these latter authors have recently demonstrated—using at one juncture 10^6 numerical zeros supplied by A. Odlyzko—that $\pi(x) > \mathrm{li}_0(x)$ for some $x \in (1.398201, 1.398244) \cdot 10^{316}$, and "probably" for very many x in said specific interval.

One interesting speculative exercise is to estimate roughly how many more years it will take researchers actually to *find and prove* an explicit case of $\pi(x) > \mathrm{li}_0(x)$. It is intriguing to guess how far calculations of $\pi(x)$ itself can be pushed in, say, 30 years. We discuss prime-counting algorithms in Section 3.6, although the state of the art is today $\pi\left(10^{21}\right)$ or somewhat higher than this (with new results emerging often).

Another speculative direction: Try to imagine numerical or even physical scenarios in which such huge numbers naturally arise. One reference for this recreation is [Crandall 1997a]. In that reference, what might be called preposterous physical scenarios—such as the annual probability of finding oneself accidentally quantum-tunneled bodily (and alive, all parts intact!) to planet Mars—are still not much smaller than A^{-A}, where A is the Avogadro number (a mole, or about $6 \cdot 10^{23}$). It is difficult to describe a statistical scenario relevant to the primes that begs of yet higher exponentiation as manifest in the Skewes number.

Incidentally, for various technical reasons the logarithmic-integral function li_0, on many modern numerical/symbolic systems, is best calculated in terms of $\mathrm{Ei}(\ln x)$, where we refer to the standard exponential-integral function

$$\mathrm{Ei}(z) = \int_{-\infty}^{z} t^{-1} e^t \, dt,$$

with principal value assumed for the singularity at $t = 0$. In addition, care must be taken to observe that some authors use the notation li for what we are calling li_0, rather than the integral from 2 in our defining equation (1.3) for li. Calling our book's function li, and the latter li_0, we can summarize this computational advice as follows:

$$\mathrm{li}\,(x) = \mathrm{li}_0(x) - \mathrm{li}_0(2) = \mathrm{Ei}(\ln x) - \mathrm{Ei}(\ln 2) \approx \mathrm{Ei}(\ln x) - 1.0451637801.$$

1.36. It is a result of [Schoenfeld 1976] that on the Riemann hypothesis we have the strict bound (for $x \geq 2657$)

$$|\pi(x) - \mathrm{li}_0(x)| < \frac{1}{8\pi}\sqrt{x}\,\ln x,$$

where $\mathrm{li}_0(x)$ is defined in Exercise 1.35. Show via computations that none of the data in Table 1.1 violates the Riemann hypothesis!

By the proof of Schoenfeld, it can be seen that the above estimate also holds in the stated range for the function $\mathrm{li}\,(x)$, again assuming the Riemann hypothesis. Show further, by direct examination of the relevant numbers x, that

$$|\pi(x) - \mathrm{li}\,(x)| < \frac{1}{8\pi}\sqrt{x}\,\ln x, \tag{1.48}$$

for $1451 \le x \le 2657$. Thus, assuming the Riemann hypothesis, (1.48) holds for all $x \ge 1451$.

Using the above, prove the assertion in the text that, assuming the Riemann hypothesis,

$$|\pi(x) - \mathrm{li}\,(x)| < \sqrt{x}\,\ln x$$

for $x \ge 2.01$.

1.37. Using the conjectured form of the PNT in (1.25), prove that there is a prime between every pair of sufficiently large cubes. Use (1.48) and any relevant computation to establish that (again, on the Riemann hypothesis) there is a prime between *every* two positive cubes. It was shown *unconditionally* by Ingham in 1937 that there is a prime between every pair of sufficiently large cubes, and it was shown, again unconditionally, by Cheng in 1999, that this is true for cubes greater than $e^{e^{15}}$.

1.38. Show that $\sum_{p\le n-2} 1/\ln(n-p) \sim n/\ln^2 n$, where the sum is over primes.

1.39. Using the known theorem that there is a positive number c such that the number of even numbers up to x that cannot be represented as a sum of two primes is $O(x^{1-c})$, show that there are infinitely many triples of primes in arithmetic progression. (For a different approach to the problem, see Exercise 1.40.)

1.40. It is known via the theory of exponential sums that

$$\sum_{n\le x} \left(R_2(2n) - \mathcal{R}_2(2n)\right)^2 = O\left(\frac{x^3}{\ln^5 x}\right), \tag{1.49}$$

where $R_2(2n)$ is, as in the text, the number of representations $p+q=2n$ with p, q prime, and where $\mathcal{R}_2(2n)$ is given by (1.10); see [Prachar 1978]. Further, we know from the Brun sieve method that

$$R_2(2n) = O\left(\frac{n \ln\ln n}{\ln^2 n}\right).$$

Show, too, that $\mathcal{R}_2(2n)$ enjoys the same big-O relation. Use these estimates to prove that the set of numbers $2p$ with p prime and with $2p$ *not* representable as a sum of two distinct primes has relative asymptotic density zero in the set of primes; that is, the number of these exceptional primes $p \le x$ is $o(\pi(x))$. In addition, let

$$A_3(x) = \#\left\{(p,q,r) \in \mathcal{P}^3 \ : \ 0 < q-p = r-q;\ q \le x\right\},$$

so that $A_3(x)$ is the number of 3-term arithmetic progressions $p<q<r$ of primes with $q \le x$. Prove that for $x \ge 2$,

$$A_3(x) = \frac{1}{2} \sum_{p\le x, p\in\mathcal{P}} \left(R_2(2p) - 1\right) \sim C_2 \frac{x^2}{\ln^3 x},$$

where C_2 is the twin-prime constant defined in (1.6).

In a computational vein, develop an efficient algorithm to compute $A_3(x)$ exactly for given values of x, and verify that $A_3(3000) = 15482$ (i.e., there are 15482 triples of distinct primes in arithmetic progressions with the middle prime not exceeding 3000), that $A_3(10^4) = 109700$, and that $A_3(10^6) = 297925965$. (The last value here was computed by R. Thompson.) There are at least two ways to proceed with such calculations: Use some variant of an Eratosthenes sieve, or employ Fourier transform methods (as intimated in Exercise 1.64). The above asymptotic formula for A_3 is about 16% too low at 10^6. Replacing $x^2/\ln^3 x$ with

$$\int_2^x \int_2^{2t-2} \frac{1}{(\ln t)(\ln s)(\ln(2t-s))}\, ds\, dt,$$

the changed formula is within 0.4% of the exact count at 10^6. Explain why the double integral should give a better estimation.

As a continuation, you might investigate the conjecture that there are infinitely many quadruplet of distinct primes that form an arithmetic progression. It is known that one can find infinitely many quadruplets $p < q < r < s$ in arithmetic progression where the first three are primes and the last is either a prime or a product of two primes, a result of D. Heath-Brown [Ribenboim 1996, p. 285]. One possible theoretical approach is to develop an exponential-sum criterion as in Exercise 1.64, except generalized to more than three primes in arithmetic progression. On the computational side, we note that R. Thompson also has calculated, for various x, the count $A_4(x)$ as the number of progressions $p < q < r < s$ with the third prime r not exceeding x. For aid in checking such computational effort, we give sample values: $A_4(10^5) = 554036$, $A_4(10^6) = 25302610$.

1.41. In [Saouter 1998], calculations are described to show how the validity of the binary Goldbach conjecture for even numbers up through $4 \cdot 10^{11}$ can be used to verify the validity of the ternary Goldbach conjecture for odd numbers greater than 7 and less then 10^{20}. We now know that the binary Goldbach conjecture is true for even numbers up to $4 \cdot 10^{14}$. Describe a calculation that could be followed to extend Saouter's bound for the ternary Goldbach conjecture to, say, 10^{23}.

Incidentally, research on the Goldbach conjecture can conceivably bring special rewards. In connection with the novel *Uncle Petros and Goldbach's Conjecture* by A. Doxiadis, the publisher has announced a \$1,000,000 prize for a proof of the (binary) Goldbach conjecture (see [Faber and Faber 2000]).

1.42. Here we prove (or at least finish the proof for) the result of Shnirel'man—as discussed in Section 1.2.3—that the set $S = \{p + q : p, q \in \mathcal{P}\}$ has "positive lower density" (the terminology to be clarified below). As in the text, denote by $R_2(n)$ the number of representations $n = p + q$ with p, q prime. Then:

(1) Argue from the Chebyshev Theorem 1.1.3 that

$$\sum_{n\le x} R_2(n) > A_1 \frac{x^2}{\ln^2 x},$$

for some positive constant A_1 and all sufficiently large values of x.

(2) Assume outright (here is where we circumvent a great deal of hard work!) the fact that

$$\sum_{n\le x} R_2(n)^2 < A_2 \frac{x^3}{\ln^4 x},$$

for $x > 1$, where A_2 is a constant. This result can be derived via such sophisticated techniques as the Selberg or Brun sieve [Nathanson 1996].

(3) Use (1), (2) and the Cauchy–Schwarz inequality

$$\left(\sum_{n=1}^{x} a_n b_n\right)^2 \le \left(\sum_{n=1}^{x} a_n^2\right)\left(\sum_{n=1}^{x} b_n^2\right)$$

(valid for arbitrary real numbers a_n, b_n) to prove that for some positive constant A_3 we have

$$\#\{n \le x \; : \; R_2(n) > 0\} > A_3 x,$$

for all sufficiently large values of x, this kind of estimate being what is meant by "positive lower density" for the set S. (Hint: Define $a_n = R_2(n)$ and (b_n) to be an appropriate *binary* sequence.)

As discussed in the text, Shnirel'man proved that this lower bound on density implies his celebrated result: That for some fixed s, every integer starting with 2 is the sum of at most s primes. It is intriguing that an *upper* bound on Goldbach representations—as in task (2)—is the key to this whole line of reasoning! That is because, of course, such an upper bound reveals that representation counts are kept "under control," meaning "spread around" such that a sufficient fraction of even n have representations. (See Exercise 9.80 for further applications of this basic bounding technique.)

1.43. Assuming the prime k-tuples Conjecture 1.2.1 show that for each k there is an arithmetic progression of k primes, and in fact an arithmetic progression of k consecutive primes.

1.44. Note that each of the Mersenne primes $2^2-1,\ 2^3-1,\ 2^5-1$ is a member of a pair of twin primes. Do any other of the known Mersenne primes from Table 1.2 enjoy this property?

1.45. Let q be a Sophie Germain prime, meaning that $s = 2q+1$ is likewise prime. Prove that if also $q \equiv 3 \pmod 4$ and $q > 3$, then the Mersenne number $M_q = 2^q - 1$ is composite, in fact divisible by s. A large Sophie Germain prime is Kerchner's and Gallot's

$$q = 18458709 \cdot 2^{32611} - 1,$$

with $2q+1$ also prime, so that the resulting Mersenne number M_q is a truly gargantuan composite of nearly 10^{10^4} decimal digits.

1.46. Prove the following relation between Mersenne numbers:

$$\gcd(2^a-1, 2^b-1) = 2^{\gcd(a,b)} - 1.$$

Conclude that for distinct primes q, r the Mersenne numbers M_q, M_r are coprime.

1.47. From W. Keller's lower bound on a factor p of F_{24}, namely,

$$p > 6 \cdot 10^{19},$$

estimate the *a priori* probability from relation (1.13) that F_{24} is prime (we now know it is *not* prime, but let us work in ignorance of that computational result here). Using what can be said about prime factors of arbitrary Fermat numbers, estimate the probability that there are *no more* Fermat primes beyond F_4 (that is, use the special form of possible factors and also the known character of some of the low-lying Fermat numbers).

1.48. Prove Theorem 1.2.1, assuming the Brun bound (1.8).

1.49. For the odd number $n = 3 \cdot 5 \cdots 101$ (consecutive odd-prime product) what is the approximate number of representations of n as a sum of three primes, on the basis of Vinogradov's estimate for $R_3(n)$? (See Exercise 1.65.)

1.50. Show by direct computation that 10^8 is *not* the sum of two base-2 pseudoprimes (see Section 3.3 for definitions). You might show in passing, however, that if p denotes prime and P_2 denotes odd base-2 pseudoprime, then

$$10^8 = p + P_2 \quad \text{or} \quad P_2 + p$$

in exactly 120 ways (this is a good check on any such programming effort). By the way, one fine representation is

$$10^8 = 99999439 + 561,$$

where 561 is well known as the smallest Carmichael number (see Section 3.3.2). Is 10^8 the sum of two pseudoprimes to some base other than 2? What is the statistical expectation of how many "pseudoreps" of various kinds $p + P_b$ should exist for a given n?

1.51. Prove: If the binary expansion of a prime p has all of its 1's lying in an arithmetic progression of positions, then p cannot be a Wieferich prime. Prove the corollary that neither a Mersenne prime nor a Fermat prime can be a Wieferich prime.

1.52. Show that if u^{-1} denotes a multiplicative inverse modulo p, then for each odd prime p,

$$\sum_{p/2<u<p} u^{-1} \equiv \frac{2^p-2}{p} \pmod{p}.$$

1.53. Use the Wilson–Lagrange Theorem 1.3.6 to prove that for any prime $p \equiv 1 \pmod 4$ the congruence $x^2 + 1 \equiv 0 \pmod p$ is solvable.

1.54. Prove the following theorem relevant to Wilson primes: if g is a primitive root of the prime p, then the Wilson quotient is given by:

$$w_p \equiv \sum_{j=1}^{p-1} \left\lfloor \frac{g^j}{p} \right\rfloor g^{p-1-j} \pmod p.$$

Then, using this result, give an algorithm that determines whether p with primitive root $g = 2$ is a Wilson prime, but using no multiplications; merely addition, subtraction, and comparison.

1.55. There is a way to connect the notion of twin-prime pairs with the Wilson–Lagrange theorem as follows. Let p be an odd integer greater than 1. Prove the theorem of Clement that $p, p+2$ is a twin-prime pair if and only if

$$(p-1)! \equiv -1 - p/4 \pmod{p(p+2)}.$$

1.56. How does one resolve the following "Mertens paradox"? Say x is a large integer and consider the "probability" that x is prime. As we know, primality can be determined by testing x for prime divisors not exceeding $\sqrt{x}$. But from Theorem 1.4.2, it would seem that when all the primes less than $\sqrt{x}$ are probabilistically sieved out, we end up with probability

$$\prod_{p \le \sqrt{x}} \left(1 - \frac{1}{p}\right) \sim \frac{2e^{-\gamma}}{\ln x}.$$

Arrive again at this same estimate by simply removing the floor functions in (1.46). However, the PNT says that the correct asymptotic probability that x is prime is $1/\ln x$. Note that $2e^{-\gamma} = 1.1229189\ldots$, so what is a resolution?

It has been said that the sieve of Eratosthenes is "more efficient than random," and that is one way to envision the "paradox." Actually, there has been some interesting work on ways to think of a resolution; for example, [Furry 1942] analyzed the action of the sieve of Eratosthenes on a prescribed interval $[x, x+d]$, and showed that there are some surprises in regard to how many composites are struck out of said interval; see [Bach and Shallit 1996, p. 365] for the historical summary.

1.57. By assuming that relation (1.24) is valid whenever the integral converges, prove that $M(x) = O(x^{1/2+\epsilon})$ implies the RH.

1.58. There is a compact way to quantify the relation between the PNT and the behavior of the Riemann zeta function. Using the relation

$$-\frac{\zeta'(s)}{\zeta(s)} = s \int_1^\infty \psi(x) x^{-s-1}\, dx,$$

show that the assumption

$$\psi(x) = x + O(x^\alpha)$$

implies that $\zeta(s)$ has no zeros in the half-plane $\mathrm{Re}(s) > \alpha$. This shows the connection between the essential error in the PNT estimate and the zeros of ζ.

For the other (harder) direction, assume that ζ has no zeros in the half-plane $\mathrm{Re}(s) > \alpha$. Looking at relation (1.23), prove:

$$\sum_{\mathrm{Im}(\rho)\,\leq\, T} \frac{x^\rho}{|\rho|} = O(x^\alpha \ln^2 T),$$

which proof is nontrivial and interesting in its own right [Davenport 1980]. Finally conclude that

$$\psi(x) = x + O\left(x^{\alpha+\epsilon}\right)$$

for any $\epsilon > 0$. These arguments reveal why the Riemann conjecture

$$\pi(x) = \mathrm{li}\,(x) + O(x^{1/2} \ln x)$$

is sometimes thought of as "the PNT form of the Riemann hypothesis."

1.59. Here we show how to evaluate the Riemann zeta function on the critical line, the exercise being to implement the formula and test against some high-precision values given below. We describe here, as compactly as we can, the celebrated Riemann–Siegel formula. This formula looms unwieldy on the face of it, but when one realizes the formula's power, the complications seem a small price to pay! In fact, the formula is precisely what has been used to verify that the first 1.5 billion zeros (of positive imaginary part) lie exactly on the critical line.

A first step is to define the Hardy function

$$Z(t) = e^{i\vartheta(t)}\zeta(1/2 + it),$$

where the assignment

$$\vartheta(t) = \mathrm{Im}\left(\ln\Gamma\left(\frac{1}{4} + \frac{it}{2}\right)\right) - \frac{1}{2} t \ln \pi$$

renders Z a *real-valued* function on the critical line (i.e., for t real). Moreover the sign changes in Z correspond to the zeros of ζ. Thus if $Z(a), Z(b)$ have opposite sign for reals $a < b$, there must be at least one zero in the interval (a, b). It is also convenient that

$$|Z(t)| = |\zeta(1/2 + it)|.$$

Note that one can either work entirely with the real Z, as in numerical studies of the Riemann hypothesis, or backtrack with appropriate evaluations of Γ and so on to get ζ itself on the critical line.

That having been said, the Riemann–Siegel prescription runs like so [Brent 1979]. Assign $\tau = t/(2\pi)$, $m = \lfloor\sqrt{\tau}\rfloor$, $z = 2(\sqrt{\tau} - m) - 1$. Then the computationally efficient formula is

$$\begin{aligned} Z(t) =& 2\sum_{n=1}^{m} n^{-1/2}\cos(t\ln n - \vartheta(t)) \\ &+ (-1)^{m+1}\tau^{-1/4}\sum_{j=0}^{M}(-1)^j\tau^{-j/2}\Phi_j(z) + R_M(t). \end{aligned}$$

Here, M is a cutoff integer of choice, the Φ_j are entire functions defined for $j \geq 0$ in terms of a function Φ_0 and its derivatives, and $R_M(t)$ is the error. A practical instance is the choice $M = 2$, for which we need

$$\Phi_0(z) = \frac{\cos(\frac{1}{2}\pi z^2 + \frac{3}{8}\pi)}{\cos(\pi z)},$$

$$\Phi_1(z) = \frac{1}{12\pi^2}\Phi_0^{(3)}(z),$$

$$\Phi_2(z) = \frac{1}{16\pi^2}\Phi_0^{(2)}(z) + \frac{1}{288\pi^4}\Phi_0^{(6)}(z).$$

In spite of the complexity here, it is to be stressed that the formula is immediately applicable in actual computation. In fact, the error R_2 can be rigorously bounded:

$$|R_2(t)| < 0.011t^{-7/4}, \quad \text{for all } t > 200.$$

Higher-order ($M > 2$) bounds, primarily due to [Gabcke 1979], are known, but just R_2 has served computationalists well for two decades.

Implement the Riemann–Siegel formula for $M = 2$, and test against some known values such as

$$\begin{aligned} \zeta(1/2 + 300i) &\approx 0.4774556718784825545360619 \\ &\qquad + 0.6079021332795530726590749\, i, \\ Z(1/2 + 300i) &\approx 0.7729870129923042272624525, \end{aligned}$$

which are accurate to the implied precision. Using your implementation, locate the nearest zero to this point $1/2+300i$, which zero should have $t \approx 299.84035$. You should also be able to find, still at the $M = 2$ approximation level and with very little machine time, the value

$$\zeta(1/2 + 10^6 i) \approx 0.0760890697382 + 2.805102101019\, i,$$

again correct to the implied precision.

When one is armed with a working Riemann–Siegel implementation, a beautiful world of computation in support of analytic number theory opens. For details on how actually to apply ζ evaluations away from the real axis, see [Brent 1979], [van de Lune et al. 1986], [Odlyzko 1994], [Borwein et al. 2000]. We should point out that in spite of the power and importance of the Riemann–Siegel formula, there are yet alternative means for efficient evaluation when imaginary parts are large. In fact it is possible to avoid the inherently asymptotic character of the Riemann–Siegel series, in favor of manifestly convergent expansions based on incomplete gamma function values, or on saddle points of certain integrals. Alternative schemes are discussed in [Galway 2000], [Borwein et al. 2000], and [Crandall 1999c].

1.60. Show that $\psi(x)$, defined in (1.22), is the logarithm of the least common multiple of all the positive integers not exceeding x. Show that the prime number theorem is equivalent to the assertion $\psi(x) \sim x$. Incidentally, [Deléglise and Rivat 1998] numerically evaluated $\psi\left(10^{15}\right)$, finding it to be $999999997476930.507683\ldots$, an attractive numerical instance of the relation $\psi(x) \sim x$. We see, in fact, that the error $|\psi(x) - x|$ is very roughly $\sqrt{x}$ for $x = 10^{15}$, such being the sort of error one expects on the basis of the Riemann hypothesis.

1.61. Perform computations that connect the distribution of primes with the Riemann critical zeros by way of the ψ function defined in (1.22). Starting with the classical exact relation (1.23), obtain a numerical table of the first $2K$ critical zeros (K of them having positive imaginary part), and evaluate the resulting numerical approximation to $\psi(x)$ for, say, noninteger $x \in (2, 1000)$. As a check on your computations, you should find, for $K = 200$ zeros and denoting by $\psi^{(K)}$ the approximation obtained via said $2K$ zeros, the amazing fact that

$$\left|\psi(x) - \psi^{(200)}(x)\right| < 5$$

throughout the possible x values. This means—heuristically speaking—that the first 200 critical zeros and their conjugates determine the prime occurrences in $(2, 1000)$ "up to a handful," if you will. Furthermore, a plot of the error vs. x is nicely noisy around zero, so the approximation is quite good in some sense of average. Try to answer this question: For a given range on x, about how many critical zeros are required to effect an approximation as good as $|\psi - \psi^{(K)}| < 1$ across the entire range? And here is another computational question: How good tends to be the approximation (based on the Riemann hypothesis)

$$\psi(x) = x + 2\sqrt{x}\sum_t \frac{\sin(t \ln x)}{t} + O\left(\sqrt{x}\right),$$

with t running over the imaginary parts of the critical zeros [Ellison and Ellison 1985]?

1.62. This, like Exercise 1.61, also requires a data base of critical zeros of the Riemann zeta function. There exist some useful tests of any computational scheme attendant on the critical line, and here is one such test. It is a consequence of the Riemann hypothesis that we would have an exact relation (see [Bach and Shallit 1996, p. 214]):

$$\sum_\rho \frac{1}{|\rho|^2} = 2 + \gamma - \ln(4\pi),$$

where ρ runs over all the zeros on the critical line. Verify this relation numerically, to as much accuracy as possible, by:

(1) Performing the sum for all zeros $\rho = 1/2+it$ for $|t| \le T$, some T of choice.

(2) Performing such a sum for $|t| \le T$ but appending an *estimate* of the remaining, infinite tail of the sum, using known formulae for the approximate distribution of zeros [Edwards 1974], [Titchmarsh 1986], [Ivić 1985].

Note in this connection Exercises 1.59 (for actual calculation of ζ values) and 8.30 (for more computations relating to the Riemann hypothesis).

1.63. There are attractive analyses possible for some of the simpler exponential sums. Often enough, estimates—particularly upper bounds—on such sums can be applied in interesting ways. Define, for odd prime p and integers a, b, c, the sum

$$S(a,b,c) = \sum_{x=0}^{p-1} e^{2\pi i(ax^2+bx+c)/p}.$$

Use the Weyl relation (1.34) to prove:

$$|S(a,b,c)| = 0,\ p, \text{ or } \sqrt{p},$$

and give conditions on a, b, c that determine precisely which of these three values for $|S|$ is attained. And here is an extension: Obtain results on $|S|$ when p is replaced by a composite integer N. With some care, you can handle even the cases when a, N are not coprime. Note that we are describing here a certain approach to the estimation of Gauss sums (see Exercises 2.23, 2.24).

Now, use the same basic approach on the following "cubic-exponential" sum (here for any prime p and any integer a):

$$T(a) = \sum_{x=0}^{p-1} e^{2\pi i a x^3/p}.$$

It is trivial that $0 \le |T(a)| \le p$. Describe choices of p, a such that equality (to 0 or p) occurs. Then prove: Whenever $a \not\equiv 0 \pmod p$ we always have an upper bound

$$|T(a)| < \sqrt{p^{3/2}+p} < 2p^{3/4}.$$

Note that one can do better, by going somewhat deeper than relation (1.34), to achieve a best-possible estimate $O\left(p^{1/2}\right)$ [Korobov 1992, Theorem 5], [Vaughan 1997, Lemma 4.3]. Yet, the 3/4 power already leads to some interesting results. In fact, just showing that $T(a) = o(p)$ establishes that as $p \to \infty$, the cubes mod p approach equidistribution (see Exercise 1.34). Note, too, that providing upper bounds on exponential sums can allow certain other sums to be given *lower* bounds. See Exercises 9.42 and 9.80 for additional variations on these themes.

1.64. The relation (1.36) is just one of many possible integral relations for interesting prime-related representations. With our nomenclature

$$E_N(t) = \sum_{p \le N} e^{2\pi i t p}$$

adopted, establish each of the following equivalences:

(1) The infinitude of twin primes is equivalent to the divergence as $N \to \infty$ of

$$\int_0^1 e^{4\pi i t} E_N(t) E_N(-t)\, dt.$$

(2) The infinitude of prime triples in arithmetic progression (a known result; see Exercises 1.39, 1.40) is equivalent to the divergence as $N \to \infty$ of

$$\int_0^1 E_N^2(t) E_N(-2t)\, dt.$$

(3) The (binary) Goldbach conjecture is equivalent to

$$\int_0^1 e^{-2\pi i t N} E_N^2(t)\, dt \ne 0$$

for even $N > 2$, and the ternary Goldbach conjecture is equivalent to

$$\int_0^1 e^{-2\pi i t N} E_N^3(t)\, dt \ne 0$$

for odd $N > 5$.

(4) The infinitude of Sophie Germain primes (i.e., primes p such that $2p + 1$ is likewise prime) is equivalent to the divergence as $N \to \infty$ of

$$\int_0^1 e^{2\pi i t} E_N(2t) E_N(-t)\, dt.$$

1.65. We mentioned in Section 1.4.4 that there is a connection between exponential sums and the singular series Θ arising in the Vinogradov resolution (1.12) for the ternary Goldbach problem. Prove that the Euler product form for $\Theta(n)$ converges (what about the case n even?), and is equal to an absolutely convergent sum, namely,

$$\Theta(n) = \sum_{q=1}^{\infty} \frac{\mu(q)}{\varphi^3(q)} c_q(n),$$

where the Ramanujan sum c_q is defined in (1.37). It is helpful to observe that μ, φ, c are all multiplicative, the latter function in the sense that if $\gcd(a, b) = 1$, then $c_a(n)c_b(n) = c_{ab}(n)$. Show also that for sufficiently large B in the assignment $Q = \ln^B n$, that the sum (1.40) being only to Q (and not to ∞) causes negligible error in the overall ternary-Goldbach estimate.

Next, derive the representation count, call it $R_s(n)$, for n the sum of s primes, in the following way. It is known that for $s > 2$, $n \equiv s \pmod 2$,

$$R_s(n) = \frac{\Theta_s(n)}{(s-1)!} \frac{n^{s-1}}{\ln^s n}\left(1 + O\left(\frac{\ln\ln n}{\ln n}\right)\right),$$

where now the general singular series is given from exponential-sum theory as

$$\Theta_s(n) = \sum_{q=1}^{\infty} \frac{\mu^s(q)}{\varphi^s(q)} c_q(n).$$

Cast this singular series into an Euler product form, which should agree with our text formula for $s = 3$. Verify that there are positive constants C_1, C_2 such that for all $s > 2$ and $n \equiv s \pmod 2$,

$$C_1 < \Theta_s(n) < C_2.$$

Do you obtain the (conjectured, unproven) singular series in (1.9) for the case $s = 2$? Of course, it is not that part but the error term in the theory that has for centuries been problematic. Analysis of such error terms has been a topic of fascination for much of the 20th century, with new bounds being established, it seems, every few years. For example, the work of [Languasco 2000] exemplifies a historical chain of results involving sharp error bounds for any $s \geq 3$, obtained conditionally on the generalized Riemann hypothesis.

As a computational option, give a good numerical value for the singular series in (1.12), say for $n = 10^8 - 1$, and compare the actual representation count $R_3(n)$ with the Vinogradov estimate (1.12). Might the expression $n^2/\ln^3 n$ be replaced by an integral so as to get a closer agreement? Compare with the text discussion of the exact value of $R_2(10^8)$.

1.66. Define a set

$$S = \{n\lfloor \ln n \rfloor : n = 1, 2, 3 \ldots\},$$

and prove that every sufficiently large integer is in $S+S$, that is, can be written as a sum of two numbers from S. (A proof can be effected either through combinatorics and the Chinese remainder theorem—see Section 2.1.3—or via convolution methods discussed elsewhere in this book.) Is every integer greater than 221 in $S + S$? For the set

$$T = \{\lfloor n \ln n \rfloor : n = 1, 2, 3, \ldots\},$$

is every integer greater than 25 in $T + T$?

Since the n-th prime is asymptotically $n \ln n$, these results indicate that the Goldbach conjecture has nothing to fear from just the sparseness of primes. Interesting questions abound in this area. For example, can you find a set of integers U such that the n-th member of U is asymptotically $n \ln n$, yet the set of numbers in $U + U$ has asymptotic density 0?

1.67. This exercise is a mix of theoretical and computational tasks pertaining to exponential sums. All of the tasks concern the sum we have denoted by E_N, for which we discussed the estimate

$$E_N(a/q) = \sum_{p \le N} e^{2\pi i p a/q} \approx \frac{\mu(q/g)}{\varphi(q/g)} \pi(N),$$

where $g = \gcd(a, q)$. We remind ourselves that the approximation here is useful mainly when $g = 1$ and q is small. Let us start with some theoretical tasks.

(1) Take $q = 2$ and explain why the above estimate on E_N is obvious for $a = 0, 1$.

(2) Let $q = 3$, and for $a = 1, 2$ explain using a vector diagram in the complex plane how the above estimate works.

(3) Let $q = 4$, and note that for some a values the right-hand side of the above estimate is actually zero. In such cases, use an error estimate (such as the conditional result (1.32)), to give sharp, nonzero estimates on $E_N(a/4)$ for $a = 1, 3$.

These theoretical examples reveal the basic behavior of the exponential sum for small q.

For a computational foray, test numerically the behavior of E_N by way of the following steps:

(1) Choose $N = 10^5$, $q = 31$, and by direct summation over primes $p \le N$, create a table of E values for $a \in [0, q-1]$. (Thus there will be q complex elements in the table.)

(2) Create a second table of values of $\pi(N)\frac{\mu(q/g)}{\varphi(q/g)}$, also for each $a \in [0, q-1]$.

(3) Compare, say graphically, the two tables. Though the former table is "noisy" compared to the latter, there should be fairly good average agreement. Is the discrepancy between the two tables consistent with theory?

(4) Explain why the latter table is so smooth (except for a glitch at the $(a = 0)$-th element). Finally, explain how the former table can be constructed via fast Fourier transform (FFT) on a binary signal (i.e., a certain signal consisting of only 0's and 1's).

Another interesting task is to perform direct numerical integration to verify (for small cases of N, say) some of the conjectural equivalences of Exercise 1.64.

1.68. Verify the following: There exist precisely 35084 numbers less than 10^{100} that are 4-smooth. Prove that for a certain constant c, the number of 4-smooth numbers not exceeding x is

$$\psi(x, 4) \sim c \ln^2 x,$$

giving the explicit c and also as sharp an error bound on this estimate as you can. Generalize by showing that for each $y \geq 2$ there is a positive number c_y such that

$$\psi(x, y) \sim c_y \ln^{\pi(y)} x, \text{ where } y \text{ is fixed and } x \to \infty.$$

1.69. Carry out some numerical experiments to verify the claim after equation (1.45) that the implicit lower bound is a "good" one.

1.70. Compute by empirical means the approximate probability that a random integer having 100 decimal digits has all of its prime factors less than 10^{10}. The method of [Bernstein 1998] might be used in such an endeavor. Note that the probability predicted by Theorem 1.4.8 is $\rho(10) \approx 2.77 \times 10^{-11}$.

1.71. What is the approximate probability that a random integer (but of size x, say) has all but one of its prime factors not exceeding B, with a single outlying prime in the interval $(B, C]$? This problem has importance for factoring methods that employ a "second stage," which, after a first stage exhausts (in a certain, algorithm-dependent sense) the first bound B, attempts to locate the outlying prime in $(B, C]$. It is typical in implementations of various factoring methods that C is substantially larger than B, for usually the operations of the second stage are much cheaper. See Exercise 3.5 for related concepts.

1.72. Here is a question that leads to interesting computational issues. Consider the number

$$c = 1/3 + \frac{1}{1/5 + \frac{1}{1/7 + \cdots}},$$

where the so-called elements of this continued fraction are the reciprocals of all the odd primes in natural order. It is not hard to show that c is well-defined. (In fact, a simple continued fraction—a construct having all 1's in the numerators—converges if the sum of the elements, in the present case $1/3 + 1/5 + \cdots$, diverges.) First, give an approximate numerical value for the constant c. Second, provide numerical (but rigorous) proof that c is not equal to 1. Third, investigate this peculiar idea: that using all primes, that is starting the fraction as $1/2 + \frac{1}{1/3+\cdots}$, results in nearly the same fraction value! Prove that if the two fractions in question were, in fact, equal, then we would have $c = \left(1 + \sqrt{17}\right)/4$. By invoking more refined numerical experiments, try to settle the issue of whether c is actually this exact algebraic value.

1.73. It is a corollary of an attractive theorem of [Bredihin 1963] that if n is a power of two, the number of solutions

$$N(n) = \#\{(x, y, p) \ : \ n = p + xy;\ p \in \mathcal{P};\ x, y \in \mathbf{Z}^+\}$$

enjoys the following asymptotic relation:

$$\frac{N(n)}{n} \sim \frac{105}{2\pi^4}\zeta(3) \approx 0.648\ldots.$$

From a computational perspective, consider the following tasks. First, attempt to verify this asymptotic relation by direct counting of solutions. Second, drop the restriction that n be a power of two, and try to verify experimentally, theoretically, or both that the constant 105 should in general be replaced by

$$315 \prod_{p\in\mathcal{P},\ p|n} \frac{(p-1)^2}{p^2-p+1}.$$

1.6 Research problems

1.74. In regard to the Mills theorem (the first part of Theorem 1.2.2), try to find an explicit number θ and a large number n such that $\left\lfloor \theta^{3^j} \right\rfloor$ is prime for $j = 1, 2, \ldots, n$. For example if one takes the specific rational $\theta = 165/92$, show that each of

$$\left\lfloor \theta^{3^1} \right\rfloor, \left\lfloor \theta^{3^2} \right\rfloor, \left\lfloor \theta^{3^3} \right\rfloor, \left\lfloor \theta^{3^4} \right\rfloor$$

is prime, yet the number $\left\lfloor \theta^{3^5} \right\rfloor$ is, alas, composite. Can you find a simple rational θ that has all cases up through $n = 5$ prime, or even further? Say a (finite or infinite) sequence of primes $q_1 < q_2 < \ldots$ is a "Mills sequence" if there is some number θ such that $q_j = \left\lfloor \theta^{3^j} \right\rfloor$ for $j = 1, 2, \ldots$. Is it true that any finite Mills sequence can be extended to an infinite Mills sequence (not necessarily with the same θ, but keeping the same initial sequence of primes)? If so, it would follow that for each prime p there is an infinite Mills sequence starting with p. It may be possible to settle the more general question for q_1 sufficiently large using the original method in [Mills 1947] (also see [Ellison and Ellison 1985, p. 31]). Of course, if the more general question is false, it may be possible to prove it so with a numerical example. In [Weisstein 2000] it is reported that a number θ slightly larger than 1.3 works in the Mills theorem. This has not yet been rigorously proved, so a research problem is to prove this conjecture.

1.75. Is there a real number $\theta > 1$ such that the sequence $(\lfloor \theta^n \rfloor)$ consists entirely of primes? The existence of such a θ seems unlikely, yet the authors are unaware of results along these lines. For $\theta = 1287/545$, the integer parts of the first 8 powers are $2, 5, 13, 31, 73, 173, 409, 967$, each of which is prime. Find a longer chain. If an infinite chain were to exist, there would be infinitely many triples of primes p, q, r for which there is some α with $p = \lfloor \alpha \rfloor, q = \lfloor \alpha^2 \rfloor, r = \lfloor \alpha^3 \rfloor$. Probably there are infinitely many such triples of primes p, q, r, and maybe this is not so hard to prove, but again the authors are unaware of such a result. It is known that there are infinitely many pairs of primes p, q of the form $p = \lfloor \alpha \rfloor, q = \lfloor \alpha^2 \rfloor$; this result is in [Balog 1989].

1.76. For a sequence $\mathcal{A} = (a_n)$, let $D(\mathcal{A})$ be the sequence $(|a_{n+1} - a_n|)$. For $\mathcal{P}$ the sequence of primes, consider $D(\mathcal{P})$, $D(D(\mathcal{P}))$, etc. Is it true that each of these sequences begins with the number 1? This has been verified by Odlyzko for the first $3 \cdot 10^{11}$ sequences [Ribenboim 1996], but has never been proved in general.

1.77. Find large primes of the form $(2^n+1)/3$, invoking possible theorems on allowed small factors, and so on. Three recent examples, due to R. McIntosh, are

$$p = (2^{42737} + 1)/3, \quad q = (2^{83339} + 1)/3, \quad r = (2^{95369} + 1)/3.$$

These numbers are "probable primes" (see Chapter 3). True primality proofs have not been achieved (and these examples may well be out of reach, for the foreseeable future!).

1.78. Candidates $M_p = 2^p - 1$ for Mersenne primes are often ruled out in practice by finding an actual nontrivial prime factor. Work out software for finding factors for Mersenne numbers, with a view to the very largest ones accessible today. You would use the known form of any factor of M_p and sequentially search over candidates. You should be able to ascertain, for example, that

$$460401322803353 \mid 2^{20295923} - 1.$$

On the issue of such large Mersenne numbers; see Exercise 1.79.

1.79. In the numerically accessible region of $2^{20000000}$ there has been at least one attempt at a compositeness proof, using not a search for factors but the Lucas–Lehmer primality test. The result (unverified as yet) by G. Spence is that $2^{20295631} - 1$ is composite. As of this writing, that would be a "genuine" composite, in that no explicit proper factor is known. One may notice that this giant Mersenne number is even larger than F_{24}, the latter recently having been shown composite. However, the F_{24} result was carefully verified with independent runs and so might be said still to be the largest "genuine" composite.

These ruminations bring us to a research problem. Note first a curious dilemma, that this "game of genuine composites" can lead one to trivial claims, as pointed out by L. Washington to [Lenstra 1991]. Indeed, if C be proven composite, then $2^C - 1$, $2^{2^C-1} - 1$ and so on are automatically composite. So in absence of new knowledge about factors of numbers in this chain, the idea of "largest genuine composite" is a dubious one. Second, observe that if $C \equiv 3 \pmod 4$ and $2C+1$ happens to be prime, then this prime is a factor of $2^C - 1$. Such a C could conceivably be a genuine composite (i.e., no factors known) yet the next member of the chain, namely $2^C - 1$, would have an explicit factor. Now for the research problem at hand: Find and prove composite some number $C \equiv 3 \pmod 4$ such that nobody knows any factors of C (nor is it easy to find them), you also have proof that $2C + 1$ is prime, so you *also* know thus an explicit factor of $2^C - 1$. The difficult part of this is to be able to prove primality of $2C+1$ without recourse to the factorization of C. This might be accomplished via the methods of Chapter 4 using a factorization of $C+1$.

1.80. Though it is unknown whether there are infinitely many Mersenne or Fermat primes, some results are known for other special number classes. Denote the n-th Cullen number by $C_n = n2^n + 1$. The Cullen and related numbers provide fertile ground for various research initiatives.

One research direction is computational: to attempt the discovery of prime Cullen numbers, perhaps by developing first a rigorous primality test for the Cullen numbers. Similar tasks pertain to the Sierpiński numbers described below.

A good, simple exercise is to prove that there are infinitely many composite Cullen numbers, by analyzing say C_{p-1} for odd primes p. In a different vein, C_n is divisible by 3 whenever $n \equiv 1, 2 \pmod 6$ and C_n is divisible by 5 whenever $n \equiv 3, 4, 6, 17 \pmod{20}$. In general show there are $p-1$ residue classes modulo $p(p-1)$ for n where C_n is divisible by the prime p. It can be shown via sieve methods that the set of integers n for which C_n is composite has asymptotic density 1 [Hooley 1976].

For another class where something, at least, is known, consider Sierpiński numbers, being numbers k such that $k2^n + 1$ is composite for every positive integer n. Sierpiński proved that there are infinitely many such k. Prove this Sierpiński theorem, and in fact show, as Sierpiński did, that there is an infinite arithmetic progression of integers k such that $k2^n + 1$ is composite for all positive integers n. Every Sierpiński number known is a member of such an infinite arithmetic progression. For example, the smallest known Sierpiński number, $k = 78557$, is in an infinite arithmetic progression of Sierpiński numbers; perhaps you would enjoy finding such a progression. It is an interesting open problem in computational number theory to decide whether 78557 actually is the smallest. (Erdős and Odlyzko have shown on the other side that there is a set of odd numbers k of positive asymptotic density such that for each k in the set, there is at least one number n with $k2^n + 1$ prime; see [Guy 1994].)

1.81. Initiate a machine search for a large prime of the form $n = k2^q \pm 1$, alternatively a twin-prime pair using both $+$ and $-$. Assume the exponent q is fixed and that k runs through small values. You wish to eliminate various k values for which n is clearly composite. First, describe precisely how various values of k could be eliminated by sieving, using a sieving base consisting of odd primes $p \le B$, where B is a fixed bound. Second, answer this important practical question: If k survives the sieve, what is now the conditional heuristic "probability" that n is prime?

Note that in Chapter 3 there is material useful for the practical task of optimizing such prime searching. One wants to find the best tradeoff between sieving out k values and actually invoking a primality test on the remaining candidates $k2^q \pm 1$. Note also that under certain conditions on the

q, k, there are relatively rapid, deterministic means for establishing primality (see Chapter 4).

1.82. The study of prime n-tuplets can be interesting and challenging. Prove the easy result that there exists only one prime triplet $\{p, p+2, p+4\}$. Then specify a pattern in the form $\{p, p+a, p+b\}$ for fixed a, b such that there should be infinitely many such triplets, and describe an algorithm for efficiently finding triplets. One possibility is the pattern $(a = 2, b = 6)$, for which the starting prime

$$p = 2^{3456} + 5661177712051$$

gives a prime triplet, as found by T. Forbes in 1995 with primalities proved in 1998 by F. Morain [Forbes 1999].

Next, as for quadruplets, argue heuristically that $\{p, p+2, p+6, p+8\}$ should be an allowed pattern. The current largest known quadruplet with this pattern has its four member primes of the "titanic" class, i.e., exceeding 1000 decimal digits [Forbes 1999].

Next, prove that there is just one prime sextuplet with pattern: $\{p, p+2, p+6, p+8, p+12, p+14\}$. Then observe that there is a prime septuplet with pattern $\{p, p+2, p+6, p+8, p+12, p+18, p+20\}$; namely for $p = 11$. Find a different septuplet having this same pattern.

To our knowledge the largest septuplet known with the above specific pattern was found in 1997 by Atkin, with first term

$$p = 4269551436942131978484635747263286365530029980299077\backslash \\ 593801111410036792376 91.$$

1.83. Study the Smarandache–Wellin numbers, being

$$w_n = (p_1)(p_2)\cdots(p_n),$$

by which notation we mean that w_n is constructed in decimal by simple concatenation of digits. For example, the first few w_n are 2, 23, 235, 2357, 235711, ...

First, prove the known result that infinitely many of the w_n are composite. Then, find an asymptotic estimate (it can be heuristic, unproven) for the number of Smarandache–Wellin primes not exceeding x.

Incidentally an example of such a prime is

$$w_{719} = 23571113171923\ldots5441\,.$$

How many decimal digits does this w_{719} have? Incidentally, large as this example is, yet even larger ones are known [Wellin 1998].

1.84. Show the easy result that if k primes each larger than k lie in arithmetic progression, then the common difference d is divisible by every prime not exceeding k. Find a long arithmetic progression of primes ($k = 22$ is the current record [Pritchard et al. 1998]).

Find some number j of *consecutive* primes in arithmetic progression. The current record is $j = 10$, found by M. Toplic [Dubner et al. 1998]. The progression is $\{P + 210m : m = 0, \ldots, 9\}$, with the first member being

$$P = 100996972469714247637786655587969840329509324689190041803603417758904341703348882159067229719.$$

An interesting claim has been made with respect to this $j = 10$ example. Here is the relevant quotation, from [Dubner et al. 1998]:

> Although a number of people have pointed out to us that $10 + 1 = 11$, we believe that a search for an arithmetic progression of eleven consecutive primes is far too difficult. The minimum gap between the primes is 2310 instead of 210 and the numbers involved in an optimal search would have hundreds of digits. We need a new idea, or a trillion-fold increase in computer speeds. So we expect the Ten Primes record to stand for a long time to come.

1.85. [Honaker 1998] Note that 61 divides $67 \cdot 71 + 1$. Are there three larger consecutive primes $p < q < r$ such that $p|qr + 1$?

1.86. Though the converse of Theorem 1.3.1 is false, it was once wondered whether q being a Mersenne prime implies $2^q - 1$ is likewise a Mersenne prime. Demolish this restricted converse by giving a Mersenne prime q such that $2^q - 1$ is composite. (You can inspect Table 1.2 to settle this, on the assumption that the table is exhaustive for all Mersenne primes up to the largest entry.) A related possibility, still open, is that the numbers:

$$c_1 = 2^2 - 1 = 3, \quad c_2 = 2^{c_1} - 1 = 7, \quad c_3 = 2^{c_2} - 1 = 127,$$

and so on, are all primes. The extremely rapid growth, evidenced by the fact that c_5 has more than 10^{37} decimal digits, would seem to indicate trial division as the only factoring recourse, yet even that humble technique may well be impossible on conventional machines. (To underscore this skepticism you might show that a factor of c_5 is $> c_4$, for example.)

Along such lines of aesthetic conjectures, and in relation to the "new Mersenne conjecture" discussed in the text, J. Selfridge offers prizes of \$1000 each, for resolution of the character (prime/composite) of the numbers

$$2^{B(31)} - 1, \quad 2^{B(61)} - 1, \quad 2^{B(127)} - 1,$$

where

$$B(p) = (2^p + 1)/3.$$

Before going ahead and writing a program to attack such Mersenne numbers, you might first ponder how huge they really are.

1.87. Here we obtain a numerical value for the Mertens constant B, from Theorem 1.4.2. First, establish the formula

$$B = 1 - \ln 2 + \sum_{n=2}^{\infty} \frac{\mu(n) \ln \zeta(n) + (-1)^n(\zeta(n) - 1)}{n}$$

(see [Bach 1997b]). Then, noting that a certain part of the infinite sum is essentially the Euler constant, in the sense that

$$\gamma + \ln 2 - 1 = \sum_{n=2}^{\infty} (-1)^n \frac{\zeta(n) - 1}{n},$$

use known methods for rapidly approximating $\zeta(n)$ (see [Borwein et al. 2000]) to obtain from this geometrically convergent series a numerical value such as

$$B \approx 0.26149721284764278375542683860869585905156664826120\ldots.$$

Estimate how many actual primes would be required to attain the implied accuracy for B if you were to use only the defining product formula for B directly. Incidentally, there are other constants that also admit of rapidly convergent expansions devoid of explicit reference to prime numbers. One of these "easy" constants is the twin prime constant C_2, as in estimate (1.6). Another such is the Artin constant

$$A = \prod_p \left(1 - \frac{1}{p(p-1)}\right) \approx 0.3739558136\ldots,$$

which is the conjectured, relative density of those primes admitting of 2 as primitive root (with more general conjectures found in [Bach and Shallit 1996]). Try to resolve C_2, A, or some other interesting constant such as the singular series value in relation (1.12) to some interesting precision but without recourse to explicit values of primes, just as we have done above for the Mertens constant. One notable exception to all of this, however, is the Brun constant, for which no polynomial-time evaluation algorithm is yet known. See [Borwein et al. 2000] for a comprehensive treatment of such applications of Riemann-zeta evaluations. See also [Lindqvist and Peetre 1997] for interesting ways to accelerate the Mertens series.

1.88. There is a theorem of Landau giving the asymptotic density of numbers n that can be represented $a^2 + b^2$, namely,

$$\#\{1 \le n \le x : \ r_2(n) > 0\} \sim L \frac{x}{\sqrt{\ln x}},$$

where the Landau constant is

$$L = \frac{1}{\sqrt{2}} \prod_{p \equiv 3 \pmod 4} \left(1 - \frac{1}{p^2}\right)^{-1/2}.$$

One question from a computational perspective is: Can one develop a fast algorithm for high-resolution computation of L, along the lines, say, of Exercise 1.87? A relevant reference is [Shanks and Schmid 1966].

1.89. By performing appropriate computations, prove the claim that the convexity Conjecture 1.2.3 is incompatible with the prime k-tuples Conjecture 1.2.1. A reference is [Hensley and Richards 1973]. Remarkably, those authors showed that on assumption of the prime k-tuples conjecture, there must exist some y for which

$$\pi(y + 20000) - \pi(y) > \pi(20000).$$

What will establish incompatibility is a proof that the interval $(0, 20000]$ contains an "admissible" set with more than $\pi(20000)$ elements. A set of integers is admissible if for each prime p there is at least one residue class modulo p that is not represented in the set. If a finite set S is admissible, the prime k-tuples conjecture implies that there are infinitely many integers n such that $n + s$ is prime for each $s \in S$. So, the Hensley and Richards result follows by showing that for each prime $p \leq 20000$ there is a residue class a_p such that if all of the numbers congruent to a_p modulo p are cast out of the interval $(0, 20000]$, the residual set (which is admissible) is large, larger than $\pi(20000)$. A more modern example is that of [Vehka 1979], who found an admissible set of 1412 elements in the interval $(0, 11763]$, while on the other hand, $\pi(11763) = 1409$. In his master's thesis at Brigham Young University in 1996, N. Jarvis was able to do this with the "20000" of the original Hensley-Richards calculation cut down to "4930." We still do not know the least integer y such that $(0, y]$ contains an admissible set with more than $\pi(y)$ elements, but in [Gordon and Rodemich 1998] it is shown that such a number y must be at least 1731. For guidance in actual computations, there is some interesting analysis of particular dense admissible sets in [Bressoud and Wagon 2000]. Recently, S. Wagon has reduced the "4930" of Jarvis yet further, to "4893," and that seems to be the current record.

It seems a very tough problem to convert such a large admissible set into an actual counterexample of the convexity conjecture. If there is any hope in actually disproving the convexity conjecture, short of proving the prime k-tuples conjecture itself, it may lie in a direct search for long and dense clumps of primes. But we should not underestimate computational analytic number theory in this regard. After all, as discussed elsewhere in this book (Section 3.6.2), estimates on $\pi(x)$ can be obtained, at least in principle, for very large x. Perhaps some day it will be possible to bound below, by machine, an appropriate difference $\pi(x + y) - \pi(x)$, say *without* knowing all the individual primes involved, to settle this fascinating compatibility issue.

1.90. Naively speaking, one can test whether p is a Wilson prime by direct multiplication of all integers $1, \ldots, p-1$, with continual reduction (mod p^2) along the way. However, there is a great deal of redundancy in this approach, to be seen as follows. If N is even, one can invoke the identity

$$N! = 2^{N/2}\,(N/2)!\,N!!,$$

where $N!!$ denotes the product of all odd integers in $[1, N-1]$. Argue that the (about) p multiply-mods to obtain $(p-1)!$ can be reduced to about $3p/4$ multiply-mods using the identity.

If one invokes a more delicate factorial identity, say by considering more equivalence classes for numbers less than N, beyond just even/odd classes, how far can the p multiplies be reduced in this way?

1.91. Investigate how the Granville identity, valid for $1 < m < p$ and p prime,

$$\prod_{j=1}^{m-1} \binom{p-1}{\lfloor jp/m \rfloor} \equiv (-1)^{(p-1)(m-1)/2}(m^p - m + 1) \pmod{p^2},$$

can be used to accelerate the testing of whether p is a Wilson prime. This and other acceleration identities are discussed in [Crandall et al. 1997].

1.92. Study the statistically expected value of $\omega(n)$, the number of distinct prime factors of n. There are beautiful elementary arguments that reveal statistical properties of $\omega(n)$. For example, we know from the celebrated Erdős–Kac theorem that the expression

$$\frac{\omega(n) - \ln \ln n}{\sqrt{\ln \ln n}}$$

is asymptotically Gaussian-normal distributed with zero mean and unit variance. That is, the set of natural numbers n with the displayed statistic not exceeding u has asymptotic density equal to $\frac{1}{\sqrt{2\pi}} \int_{-\infty}^{u} e^{-t^2/2}\, dt$. See [Ruzsa 1999] for some of the history of this theorem.

These observations, though profound, are based on elementary arguments. Investigate the possibility of an analytic approach, using the beautiful formal identity

$$\sum_{n=1}^{\infty} \frac{2^{\omega(n)}}{n^s} = \frac{\zeta^2(s)}{\zeta(2s)}.$$

Here is one amusing, instructive exercise in the analytic spirit: Prove directly from this zeta-function identity, by considering the limit $s \to 1$, that there exist infinitely many primes. What more can be gleaned about the ω function via such analytic forays?

Beyond this, study (in any way possible!) the fascinating conjecture of J. Selfridge that the number of distinct prime factors of a Fermat number, that is, $\omega(F_n)$, is *not* a monotonic (nondecreasing) function of n. Note from Table 1.3 that this conjecture is so far nonvacuous. (Selfridge suspects that F_{14}, if it ever be factored, may settle the conjecture by having a notable paucity of factors.) This conjecture is, so far, out of reach in one sense: We cannot factor enough Fermat numbers to thoroughly test it. On the other hand, one might be able to provide a heuristic argument indicating in some sense the "probability" of the truth of the Selfridge conjecture. On the face of it, one might expect said probability to be one, even given that each Fermat number is roughly the *square* of the previous one. Indeed, the Erdős–Kac theorem asserts that for two random integers a, b with $b \approx a^2$, it is a roughly even toss-up that $\omega(b) \geq \omega(a)$.

Chapter 2

NUMBER-THEORETICAL TOOLS

In this chapter we focus specifically on those fundamental tools and associated computational algorithms that apply to prime number and factorization studies. Enhanced integer algorithms, including various modern refinements of the classical ones of the present chapter, are detailed in Chapter 9. The reader may wish to refer to that chapter from time to time, especially when issues of computational complexity and optimization are paramount.

2.1 Modular arithmetic

Throughout prime-number and factorization studies the notion of modular arithmetic is a constant reminder that one of the great inventions of mathematics is to consider numbers modulo N, in so doing effectively contracting the infinitude of integers into a finite set of residues. Many theorems on prime numbers involve reductions modulo p, and most factorization efforts will use residues modulo N, where N is the number to be factored.

A word is in order on nomenclature. Here and elsewhere in the book, we denote by $x \bmod N$ the least nonnegative residue $x \pmod N$. The mod notation without parentheses is convenient when thought of as an algorithm step or a machine operation (more on this operator notion is said in Section 9.1.3). So, the notation $x^y \bmod N$ means the y-th power of x, reduced to the interval $[0, N-1]$ inclusive; and we allow negative values for exponents y when x is coprime to N, so that an operation $x^{-1} \bmod N$ yields a reduced inverse, and so on.

2.1.1 Greatest common divisor and inverse

In this section we exhibit algorithms for one of the very oldest operations in computational number theory, the evaluation of the greatest common divisor function $\gcd(x, y)$. Closely related is the problem of inversion, the evaluation of $x^{-1} \bmod N$, which operation yields (when it exists) the unique integer $y \in [1, N-1]$ with $xy \equiv 1 \pmod N$. The connection between the gcd and inversion operations is especially evident on the basis of the following fundamental result.

Theorem 2.1.1 (Linear relation for gcd). *If x, y are integers not both 0, then there are integers a, b with*

$$ax + by = \gcd(x, y). \tag{2.1}$$

Proof. Let g be the least positive integer in the form $ax + yb$, where a, b are integers. (There is at least one positive integer in this form, to wit, $x^2 + y^2$.) We claim that $g = \gcd(x, y)$. Clearly, any common divisor of x and y divides $g = ax + by$. So $\gcd(x, y)$ divides g. Suppose g does not divide x. Then $x = tg + r$, for some integer r with $0 < r < g$. We then observe that $r = (1 - ta)x - tby$, contradicting the definition of g. Thus, g divides x, and similarly, g divides y. We conclude that $g = \gcd(x, y)$. □

The connection of (2.1) to inversion is immediate: If x, y are positive integers and $\gcd(x, y) = 1$, then we can solve $ax + by = 1$, whence

$$b \bmod x, \quad a \bmod y$$

are the inverses $y^{-1} \bmod x$ and $x^{-1} \bmod y$, respectively.

However, what is clearly lacking from the proof of Theorem 2.1.1 from a computational perspective is any clue on how one might find a solution a, b to (2.1). We investigate here the fundamental, classical methods, beginning with the celebrated centerpiece of the classical approach: the Euclid algorithm. It is arguably one of the very oldest computational schemes, dating back to 300 B.C., if not the oldest of all. In this algorithm and those following, we indicate the updating of two variables x, y by

$$(x, y) = (f(x, y), g(x, y)),$$

which means that the pair (x, y) is to be replaced by the pair of evaluations (f, g) but with the evaluations using the original (x, y) pair. In similar fashion, longer vector relations $(a, b, c, \ldots) = \cdots$ update all components on the left, each using the original values on the right side of the equation. (This rule for updating of vector components is discussed in the Appendix.)

Algorithm 2.1.2 (Euclid algorithm for greatest common divisor). For integers x, y with $x \ge y \ge 0$ and $x > 0$, this algorithm returns $\gcd(x, y)$.

1. [Euclid loop]
   while($y > 0$) $(x, y) = (y, x \bmod y)$;
   return x;

It is intriguing that this algorithm, which is as simple and elegant as can be, is not so easy to analyze in complexity terms. Though there are still some interesting open questions as to detailed behavior of the algorithm, the basic complexity is given by the following theorem:

Theorem 2.1.3 (Lamé, Dixon, Heilbronn). *Let $x > y$ be integers from the interval $[1, N]$. Then the number of steps in the loop of the Euclid Algorithm 2.1.2 does not exceed*

$$\left\lceil \ln\left(N\sqrt{5}\right) / \ln\left(\left(1 + \sqrt{5}\right)/2\right) \right\rceil - 2,$$

and the average number of loop steps (over all choices x, y) is asymptotic to

$$\frac{12 \ln 2}{\pi^2} \ln N.$$

The first part of this theorem stems from an interesting connection between Euclid's algorithm and the theory of simple continued fractions (see Exercise 2.4). The second part involves the measure theory of continued fractions.

If x, y are each of order of magnitude N, and we employ the Euclid algorithm together with, say, a classical mod operation, it can be shown that the overall complexity of the gcd operation will then be

$$O\left(\ln^2 N\right)$$

bit operations, essentially the square of the number of digits in an operand (see Exercise 2.6). This complexity can be genuinely bested via modern approaches, and not merely by using a faster mod operation, as we discuss later.

The Euclid algorithm can be extended to the problem of inversion. In fact, the appropriate extension of the Euclid algorithm will provide a complete solution to the relation (2.1):

Algorithm 2.1.4 (Euclid's algorithm extended, for gcd and inverse). For integers x, y with $x \geq y \geq 0$ and $x > 0$, this algorithm returns an integer triple (a, b, g) such that $ax + by = g = \gcd(x, y)$. (Thus when $g = 1$ and $y > 0$, the residues $b \pmod x$, $a \pmod y$ are the inverses of $y \pmod x, x \pmod y$, respectively.)

1. [Initialize]
   $(a, b, g, u, v, w) = (1, 0, x, 0, 1, y)$;
2. [Extended Euclid loop]
   while($w > 0$) {
      $q = \lfloor g/w \rfloor$;
      $(a, b, g, u, v, w) = (u, v, w, a - qu, b - qv, g - qw)$;
   }
   return (a, b, g);

Because the algorithm simultaneously returns the relevant gcd and both inverses (when the input integers are coprime and positive), it is widely used as an integral part of practical computational packages. Interesting computational details of this and related algorithms are given in [Cohen 2000], [Knuth 1981]. Modern enhancements are covered in Chapter 9 including asymptotically faster gcd algorithms, faster inverse, inverses for special moduli, and so on. Finally, note that in Section 2.1.2 we give an "easy inverse" method (relation (2.3)) that might be considered as a candidate in computer implementations.

2.1.2 Powers

It is a celebrated theorem of Euler that

$$a^{\varphi(m)} \equiv 1 \pmod m \tag{2.2}$$

holds for any positive integer m as long as a, m are coprime. In particular, for prime p we have

$$a^{p-1} \equiv 1 \pmod{p},$$

which is used frequently as a straightforward initial (though not absolute) primality criterion. The point is that powering is an important operation in prime number studies, and we are especially interested in powering with modular reduction. Among the many applications of powering is this one: A straightforward method for finding inverses is to note that when $a^{-1} \pmod{m}$ exists, we always have the equality

$$a^{-1} \bmod m = a^{\varphi(m)-1} \bmod m, \tag{2.3}$$

and this inversion method might be compared with Algorithm 2.1.4 when machine implementation is contemplated.

It is a primary computational observation that one usually does not need to take an n-th power of some x by literally multiplying together n symbols as $x*x*\cdots*x$. We next give a radically more efficient (for large powers) recursive powering algorithm that is easily written out and also easy to understand. The objects that we raise to powers might be integers, members of a finite field, polynomials, or something else. We specify in the algorithm that the element x comes only from a semigroup, namely, a setting in which $x * x * \cdots * x$ is defined.

Algorithm 2.1.5 (Recursive powering algorithm). Given an element x in a semigroup and a positive integer n, the goal is to compute x^n.

1. [Recursive function *pow*]

```
pow(x, n) {
    if(n == 1) return x;
    if(n even) return pow(x, n/2)^2;              // Even branch.
    return x * pow(x, (n − 1)/2)^2;               // Odd branch.
}
```

This algorithm is recursive and compact, but for actual implementation one should consider the ladder methods of Section 9.3.1, which are essentially equivalent to the present one but are more appropriate for large, array-stored arguments. To exemplify the recursion in Algorithm 2.1.5, consider $3^{13} \pmod{15}$. Since $n = 13$, we can see that the order of operations will be

$$\begin{aligned} 3 * pow(3,6)^2 &= 3 * \left(pow(3,3)^2\right)^2 \\ &= 3 * \left(\left(3 * pow(3,1)^2\right)^2\right)^2. \end{aligned}$$

If one desires $x^n \bmod m$, then the required modular reductions are to occur for each branch (even, odd) of the algorithm. If the modulus is $m = 15$, say, casual inspection of the final power chain above shows that the answer is $3^{13} \bmod 15 = 3 \cdot \left((-3)^2\right)^2 \bmod 15 = 3 \cdot 6 \bmod 15 = 3$. The important observation, though, is that there are three squarings and two multiplications, and such operation counts depend on the binary expansion of the exponent n, with typical operation counts being dramatically less than the value of n itself. In fact, if x, n are integers the size of m, and we are to compute $x^n \bmod m$ via naive multiply/add arithmetic and Algorithm 2.1.5, then $O(\ln^3 m)$ bit operations suffice for the powering (see Exercise 2.15 and Section 9.3.1).

2.1.3 Chinese remainder theorem

The Chinese remainder theorem (CRT) is a clever, and very old, idea from which one may infer an integer value on the basis of its residues modulo an appropriate system of smaller moduli. The CRT was known to Sun-Zi in the first century A.D. [Hardy and Wright 1979], [Ding et al. 1996]; in fact a legendary ancient application is that of counting a troop of soldiers. If there are n soldiers, and one has them line up in justified rows of 7 soldiers each, one inspects the last row and infers $n \bmod 7$, while lining them up in rows of 11 will give $n \bmod 11$, and so on. If one does "enough" such small-modulus operations, one can infer the exact value of n. In fact, one does not need the small moduli to be primes; it is sufficient that the moduli be pairwise coprime.

Theorem 2.1.6 (Chinese remainder theorem (CRT)). *Let $m_0, \ldots, m_{r-1}$ be positive, pairwise coprime moduli with product $M = \Pi_{i=0}^{r-1} m_i$. Let r respective residues n_i also be given. Then the system comprising the r relations and inequality*

$$n \equiv n_i \pmod{m_i}, \quad 0 \le n < M$$

has a unique solution. Furthermore, this solution is given explicitly by the least nonnegative residue modulo M of

$$\sum_{i=0}^{r-1} n_i v_i M_i,$$

where $M_i = M/m_i$, and the v_i are inverses defined by $v_i M_i \equiv 1 \pmod{m_i}$.

A simple example should serve to help clarify the notation. Let $m_0 = 3$, $m_1 = 5$, $m_2 = 7$, for which the overall product is $M = 105$, and let $n_0 = 2$, $n_1 = 2$, $n_2 = 6$. We seek a solution $n < 105$ to

$$n \equiv 2 \pmod 3, \; n \equiv 2 \pmod 5, \; n \equiv 6 \pmod 7.$$

We first establish the M_i, as

$$M_0 = 35, \; M_1 = 21, \; M_2 = 15.$$

Then we compute the inverses

$$v_0 = 2 = 35^{-1} \bmod 3, \quad v_1 = 1 = 21^{-1} \bmod 5, \quad v_2 = 1 = 15^{-1} \bmod 7.$$

Then we compute

$$\begin{aligned} n &= (n_0 v_0 M_0 + n_1 v_1 M_1 + n_2 v_2 M_2) \bmod M \\ &= (140 + 42 + 90) \bmod 105 \\ &= 62. \end{aligned}$$

Indeed, 62 modulo $3, 5, 7$, respectively, gives the required residues $2, 2, 6$.

Though ancient, the CRT algorithm still finds many applications. Some of these are discussed in Chapter 9 and its exercises. For the moment, we observe that the CRT affords a certain "parallelism." A set of separate machines can perform arithmetic, each machine doing this with respect to a small modulus m_i, whence some final value may be reconstructed. For example, if each of x, y has fewer than 100 digits, then a set of prime moduli $\{m_i\}$ whose product is $M > 10^{200}$ can be used for multiplication: The i-th machine would find $((x \bmod m_i) * (y \bmod m_i)) \bmod m_i$, and the final value $x * y$ would be found via the CRT. Likewise, on one computer chip, separate multipliers can perform the small-modulus arithmetic.

All of this means that the reconstruction problem is paramount; indeed, the reconstruction of n tends to be the difficult phase of CRT computations. Note, however, that if the small moduli are fixed over many computations, a certain amount of one-time precomputation is called for. In Theorem 2.1.6, one may compute the M_i and the inverses v_i just once, expecting many future computations with different residue sets $\{n_i\}$. In fact, one may precompute the products $v_i M_i$. A computer with r parallel nodes can then reconstruct $\sum n_i v_i M_i$ in $O(\ln r)$ steps.

There are other ways to organize the CRT data, such as building up one partial modulus at a time. One such method is the Garner algorithm [Menezes et al. 1997], which can also be done with preconditioning.

Algorithm 2.1.7 (CRT reconstruction with preconditioning (Garner)). Using the nomenclature of Theorem 2.1.6, we assume $r \geq 2$ fixed, pairwise coprime moduli $m_0, \ldots, m_{r-1}$ whose product is M, and a set of given residues $\{n_i \pmod{m_i}\}$. This algorithm returns the unique $n \in [0, M-1]$ with the given residues. After the precomputation step, the algorithm may be reentered for future evaluations of such n (with the $\{m_i\}$ remaining fixed).

1. [Precomputation]
   for($1 \leq i < r$) {
      $\mu_i = \prod_{j=0}^{i-1} m_j$;
      $c_i = \mu_i^{-1} \bmod m_i$;
   }
   $M = \mu_{r-1} m_{r-1}$;
2. [Reentry point for given input residues $\{n_i\}$]
   $n = n_0$;
   for($1 \leq i < r$) {
      $u = ((n_i - n)c_i) \bmod m_i$;
      $n = n + u\mu_i$; // Now $n \equiv n_j \pmod{m_j}$ for $0 \leq j \leq i$;
   }
   $n = n \bmod M$;
   return n;

This algorithm can be shown to be more efficient than a naive application of Theorem 2.1.6 (see Exercise 2.8). Moreover, in case a fixed modulus M

is used for repeated CRT calculations, one can perform [Precomputation] for Algorithm 2.1.7 just once, store an appropriate set of $r-1$ integers, and allow efficient reentry.

In Section 9.5.9 we describe a CRT reconstruction algorithm that not only takes advantage of preconditioning, but of fast methods to multiply integers.

2.2 Polynomial arithmetic

Many of the algorithms for modular arithmetic have almost perfect analogues in the polynomial arena.

2.2.1 Greatest common divisor for polynomials

We next give algorithms for polynomials analogous to the Euclid forms in Section 2.1.1 for integer gcd and inverse. When we talk about polynomials, the first issue is where the coefficients come from. We may be dealing with $\mathbf{Q}[x]$, the polynomials with rational coefficients, or $\mathbf{Z}_p[x]$, polynomials with coefficients in the finite field $\mathbf{Z}_p$. Or from some other field. We may also be dealing with polynomials with coefficients drawn from a ring that is not a field, as we do when we consider $\mathbf{Z}[x]$ or $\mathbf{Z}_n[x]$ with n not a prime.

Because of the ambiguity of the arena in which we are to work, perhaps it is better to go back to first principles and begin with the more primitive concept of divide with remainder. If we are dealing with polynomials in $F[x]$, where F is a field, there is a division theorem completely analogous to the situation with ordinary integers. Namely, if $f(x), g(x)$ are in $F[x]$ with f not the zero polynomial, then there are (unique) polynomials $q(x), r(x)$ in $F[x]$ with

$$g(x) = q(x)f(x) + r(x) \text{ and either } r(x) = 0 \text{ or } \deg r(x) < \deg f(x). \quad (2.4)$$

Moreover, we can use the "grammar-school" method of building up the quotient $q(x)$ term by term to find $q(x)$ and $r(x)$. Thinking about this method, one sees that the only special property of fields that is used that is not enjoyed by a general commutative ring is that the leading coefficient of the divisor polynomial $f(x)$ is invertible. So if we are in the more general case of polynomials in $R[x]$ where R is a commutative ring with identity, we can perform a divide with remainder if the leading coefficient of the divisor polynomial is a unit, that is, it has a multiplicative inverse in the ring.

For example, say we wish to divide $3x+2$ into x^2 in the polynomial ring $\mathbf{Z}_{10}[x]$. The inverse of 3 in $\mathbf{Z}_{10}$ (which can be found by Algorithm 2.1.4) is 7. We get the quotient $7x+2$ and remainder 6.

In sum, if $f(x), g(x)$ are in $R[x]$, where R is a commutative ring with identity and the leading coefficient of f is a unit in R, then there are unique polynomials $q(x), r(x)$ in $R[x]$ such that (2.4) holds. We use the notation $r(x) = g(x) \bmod f(x)$. For much more on polynomial remaindering, see Section 9.6.2.

Though it is possible sometimes to define the gcd of two polynomials in the more general case of $R[x]$, in what follows we shall restrict the discussion to the much easier case of $F[x]$, where F is a field. In this setting the algorithms and theory are almost entirely the same as for integers. (For a discussion of gcd in the case where R is not necessarily a field, see Section 4.3.) We define the polynomial gcd of two polynomials, not both 0, as a polynomial of greatest degree that divides both polynomials. Any polynomial satisfying this definition of gcd, when multiplied by a nonzero element of the field F, again satisfies the definition. To standardize things, we take among all these polynomials the monic one, that is the polynomial with leading coefficient 1, and it is this particular polynomial that is indicated when we use the notation $\gcd(f(x), g(x))$. Thus, $\gcd(f(x), g(x))$ is the monic polynomial common divisor of $f(x)$ and $g(x)$ of greatest degree. To render any nonzero polynomial monic, one simply multiplies through by the inverse of the leading coefficient.

Algorithm 2.2.1 (gcd for polynomials). For given polynomials $f(x), g(x)$ in $F[x]$, not both zero, this algorithm returns $d(x) = \gcd(f(x), g(x))$.

1. [Initialize]
   Let $u(x), v(x)$ be $f(x), g(x)$ in some order so that either $\deg u(x) \geq \deg v(x)$ or $v(x)$ is 0;
2. [Euclid loop]
   while($v(x) \neq 0$) $(u(x), v(x)) = (v(x), u(x) \bmod v(x))$;
3. [Make monic]
   Set c as the leading coefficient of $u(x)$;
   $d(x) = c^{-1}u(x)$;
   return $d(x)$;

Thus, for example, if we take

$$f(x) = 7x^{11} + x^9 + 7x^2 + 1,$$
$$g(x) = -7x^7 - x^5 + 7x^2 + 1,$$

in $\mathbf{Q}[x]$, then the sequence in the Euclid loop is

$$\begin{aligned}
&(7x^{11} + x^9 + 7x^2 + 1,\ -7x^7 - x^5 + 7x^2 + 1)\\
&\to (-7x^7 - x^5 + 7x^2 + 1,\ 7x^6 + x^4 + 7x^2 + 1)\\
&\to (7x^6 + x^4 + 7x^2 + 1,\ 7x^3 + 7x^2 + x + 1)\\
&\to (7x^3 + 7x^2 + x + 1,\ 14x^2 + 2)\\
&\to (14x^2 + 2,\ 0),
\end{aligned}$$

so the final value of $u(x)$ is $14x^2 + 2$, and the gcd $d(x)$ is $x^2 + \frac{1}{7}$. It is, of course, understood that all calculations in the algorithm are to be performed in the polynomial ring $F[x]$. So in the above example, if $F = \mathbf{Z}_{13}$, then $d(x) = x^2 + 2$, if $F = \mathbf{Z}_7$, then $d(x) = 1$; and if $F = \mathbf{Z}_2$, then the loop stops one step earlier and $d(x) = x^3 + x^2 + x + 1$.

Along with the polynomial gcd we shall need a polynomial inverse. In keeping with the notion of integer inverse, we shall generate a solution to

$$s(x)f(x) + t(x)g(x) = d(x),$$

for given f, g, where $d(x) = \gcd(f(x), g(x))$.

Algorithm 2.2.2 (Extended gcd for polynomials). Let F be a field. For given polynomials $f(x), g(x)$ in $F[x]$, not both zero, with either $\deg f(x) \geq \deg g(x)$ or $g(x) = 0$, this algorithm returns $(s(x), t(x), d(x))$ in $F[x]$ such that $d = \gcd(f, g)$ and $sg + th = d$. (For ease of notation we shall drop the x argument in what follows.)

1. [Initialize]
   $(s, t, d, u, v, w) = (1, 0, f, 0, 1, g)$;
2. [Extended Euclid loop]
   while($w \neq 0$) {
      $q = (d - (d \bmod w))/w$; // q is the quotient of $d \div w$.
      $(s, t, d, u, v, w) = (u, v, w, s - qu, t - qv, d - qw)$;
   }
3. [Make monic]
   Set c as the leading coefficient of d;
   $(s, t, d) = (c^{-1}s, c^{-1}t, c^{-1}d)$;
   return (s, t, d);

If $d(x) = 1$ and neither of $f(x), g(x)$ is 0, then $s(x)$ is the inverse of $f(x)$ (mod $g(x)$) and $t(x)$ is the inverse of $g(x)$ (mod $f(x)$). It is clear that if naive polynomial remaindering is used, as described above, then the complexity of the algorithm is $O(D^2)$ field operations, where D is the larger of the degrees of the input polynomials; see [Menezes et al. 1997].

2.2.2 Finite fields

Examples of infinite fields are the rational numbers $\mathbf{Q}$, the real numbers $\mathbf{R}$, and the complex numbers $\mathbf{C}$. In this book, however, we are primarily concerned with finite fields. A common example: If p is prime, the field

$$\mathbf{F}_p = \mathbf{Z}_p$$

consists of all residues $0, 1, \ldots, p-1$ with arithmetic proceeding under the usual modular rules.

Given a field F and a polynomial $f(x)$ in $F[x]$ of positive degree, we may consider the quotient ring $F[x]/(f(x))$. The elements of $F[x]/(f(x))$ are subsets of $F[x]$ of the form $\{g(x) + f(x)h(x) : h(x) \in F[x]\}$; we denote this subset by $g(x) + (f(x))$. It is a coset of the ideal $(f(x))$ with coset representative $g(x)$. (Actually, *any* polynomial in a coset can stand in as a representative for the coset, so that $g(x) + (f(x)) = G(x) + (f(x))$ if and only if $G(x) \in g(x) + (f(x))$ if and only if $G(x) - g(x) = f(x)h(x)$ for some

$h(x) \in F[x]$ if and only if $G(x) \equiv g(x) \pmod{f(x)}$. Thus, working with cosets can be thought of as a fancy way of working with congruences.) Each coset has a canonical representative, that is, a unique and natural choice, which is either 0 or has degree smaller than $\deg f(x)$.

We can add and multiply cosets by doing the same with their representatives:

$$\begin{aligned}\big(g_1(x) + (f(x))\big) \ + \ \big(g_2(x) + (f(x))\big) &= g_1(x) + g_2(x) \ + \ (f(x)),\\ \big(g_1(x) + (f(x))\big) \ \cdot \ \big(g_2(x) + (f(x))\big) &= g_1(x)g_2(x) \ + \ (f(x)).\end{aligned}$$

With these rules for addition and multiplication, $F[x]/(f(x))$ is a ring that contains an isomorphic copy of the field F: An element $a \in F$ is identified with the coset $a + (f(x))$.

Theorem 2.2.3. *If F is a field and $f(x) \in F[x]$ has positive degree, then $F[x]/(f(x))$ is a field if and only if $f(x)$ is irreducible in $F[x]$.*

Via this theorem we can create new fields out of old fields. For example, starting with $\mathbf{Q}$, the field of rational numbers, consider the irreducible polynomial $x^2 - 2$ in $\mathbf{Q}[x]$. Let us denote the coset $a + bx + (f(x))$, where $a, b \in \mathbf{Q}$, more simply by $a + bx$. We have the addition and multiplication rules

$$\begin{aligned}(a_1 + b_1 x) + (a_2 + b_2 x) &= (a_1 + a_2) + (b_1 + b_2)x,\\ (a_1 + b_1 x) \cdot (a_2 + b_2 x) &= (a_1 a_2 + 2b_1 b_2) + (a_1 b_2 + a_2 b_1)x.\end{aligned}$$

That is, one performs ordinary addition and multiplication of polynomials, except that the relation $x^2 = 2$ is used for reduction. We have "created" the field

$$\mathbf{Q}\left[\sqrt{2}\right] = \left\{a + b\sqrt{2} : a, b \in \mathbf{Q}\right\}.$$

Let us try this idea starting from the finite field $\mathbf{F}_7$. Say we take $f(x) = x^2 + 1$. A degree-2 polynomial is irreducible over a field F if and only if it has no roots in F. A quick check shows that $x^2 + 1$ has no roots in $\mathbf{F}_7$, so it is irreducible over this field. Thus, by Theorem 2.2.3, $\mathbf{F}_7[x]/(x^2 + 1)$ is a field. We can abbreviate elements by $a + bi$, where $a, b \in \mathbf{F}_7$ and $i^2 = -1$. Our new field has 49 elements.

More generally, if p is prime and $f(x) \in \mathbf{F}_p[x]$ is irreducible and has degree $d \geq 1$, then $\mathbf{F}_p[x]/(f(x))$ is again a finite field, and it has p^d elements. Interestingly, *all* finite fields up to isomorphism can be constructed in this manner.

An important difference between finite fields and fields such as $\mathbf{Q}$ and $\mathbf{C}$ is that repeatedly adding 1 to itself in a finite field, you will eventually get 0. In fact, the number of times must be a prime, for otherwise, one can get the product of two nonzero elements being 0.

Definition 2.2.4. The characteristic of a field is the additive order of 1, unless said order is infinite, in which case the characteristic is 0.

As indicated above, the characteristic of a field, if it is positive, must be a prime number. Fields of characteristic 2 play a special role in applications, mainly because of the simplicity of doing arithmetic in such fields.

We collect some relevant classical results on finite fields as follows:

Theorem 2.2.5 (Basic results on finite fields).

(1) *A finite field F has nonzero characteristic, which must be a prime.*

(2) *Every finite field has p^k elements for some positive integer k, where p is the characteristic.*

(3) *For given prime p and exponent k, there is exactly one field with p^k elements (up to isomorphism), which field we denoted by $\mathbf{F}_{p^k}$.*

(4) *$\mathbf{F}_{p^k}$ contains as subfields unique copies of $\mathbf{F}_{p^j}$ for each $j|k$, and no other subfields.*

(5) *The multiplicative group $\mathbf{F}_{p^k}^*$ of nonzero elements in $\mathbf{F}_{p^k}$ is cyclic, that is, there is a single element whose powers constitute the whole group.*

The multiplicative group $\mathbf{F}_{p^k}^*$ is an important concept in studies of powers, roots, and cryptography.

Definition 2.2.6. A primitive root of a field $\mathbf{F}_{p^k}$ is an element whose powers constitute all of $\mathbf{F}_{p^k}^*$. That is, the root is a generator of the cyclic group $\mathbf{F}_{p^k}^*$.

For example, in the example above where we created a field with 49 elements, namely $\mathbf{F}_{7^2}$, the element $3 + i$ is a primitive root.

A cyclic group with n elements has $\varphi(n)$ generators in total, where φ is the Euler totient function. Thus, a finite field $\mathbf{F}_{p^k}$ has $\varphi(p^k - 1)$ primitive roots.

One way to detect primitive roots is to use the following result.

Theorem 2.2.7 (Test for primitive root). *An element g in $\mathbf{F}_{p^k}^*$ is a primitive root if and only if*

$$g^{(p^k-1)/q} \neq 1$$

for every prime q dividing $p^k - 1$.

As long as $p^k - 1$ can be factored, this test provides an efficient means of establishing a primitive root. A simple algorithm, then, for finding a primitive root is this: Choose random $g \in \mathbf{F}_{p^k}^*$, compute powers $g^{(p^k-1)/q} \bmod p$ for successive prime factors q of $p^k - 1$, and if any one of these powers is 1, choose another g. If g survives the chain of powers, it is a primitive root by Theorem 2.2.7.

Much of this book is concerned with arithmetic in $\mathbf{F}_p$, but at times we shall have occasion to consider higher prime-power fields. Though general $\mathbf{F}_{p^k}$ arithmetic can be complicated, it is intriguing that some algorithms can actually enjoy improved performance when we invoke such higher fields. As we saw above, we can "create" the finite field $\mathbf{F}_{p^k}$ by coming up with an irreducible polynomial $f(x)$ in $\mathbf{F}_p[x]$ of degree k. We thus say a little about how one might do this.

Every element a in $\mathbf{F}_{p^k}$ has the property that $a^{p^k} = a$, that is, a is a root of $x^{p^k} - x$. In fact this polynomial splits into linear factors over $\mathbf{F}_{p^k}$ with no repeated factors. We can use this idea to see that $x^{p^k} - x$ is the product of *all* monic irreducible polynomials in $\mathbf{F}_p[x]$ of degrees dividing k. From this we get a formula for the number $N_k(p)$ of monic irreducible polynomials in $\mathbf{F}_p[x]$ of exact degree k: One begins with the identity

$$\sum_{d|k} dN_d(p) = p^k,$$

on which we can use Möbius inversion to get

$$N_k(p) = \frac{1}{k}\sum_{d|k} p^d \mu(k/d). \tag{2.5}$$

Here, μ is the Möbius function discussed in Section 1.4.1. It is easy to see that the last sum is dominated by the term $d = k$, so that $N_k(p)$ is approximately p^k/k. That is, about 1 out of every k monic polynomials of degree k in $\mathbf{F}_p[x]$ is irreducible. Thus a random search for one of these should be successful in $O(k)$ trials. But how can we recognize an irreducible polynomial? An answer is afforded by the following result.

Theorem 2.2.8. *A polynomial $f(x)$ in $\mathbf{F}_p[x]$ of degree k is irreducible if and only if* $\gcd(f(x), x^{p^j} - x) = 1$ *for each* $j = 1, 2, \ldots, \lfloor k/2 \rfloor$.

This theorem is then what is behind the following irreducibility test.

Algorithm 2.2.9 (Irreducibility test). Given prime p and a polynomial $f(x) \in \mathbf{F}_p[x]$ of degree $k \geq 2$, this algorithm determines whether $f(x)$ is irreducible over $\mathbf{F}_p$.

1. [Initialize]
   $g(x) = x$;
2. [Testing loop]
   for($1 \le i \le \lfloor k/2 \rfloor$) {
      $g(x) = g(x)^p \bmod f(x)$; // Powering by Algorithm 2.1.5.
      $d(x) = \gcd(f(x), g(x) - x)$; // Polynomial gcd.
      if($d(x) \neq 1$) return NO;
   }
   return YES; // f is irreducible.

Let us now recapitulate the manner of field computations. Armed with a suitable irreducible polynomial f of degree k over $\mathbf{F}_p$, one represents any element $a \in \mathbf{F}_{p^k}$ as

$$a = a_0 + a_1 x + a_2 x^2 + \cdots + a_{k-1} x^{k-1},$$

with each $a_i \in \{0, \ldots, p-1\}$. That is, we represent a as a vector in $\mathbf{F}_p^k$. Note that there are clearly p^k such vectors. Addition is ordinary vector addition, but of course the arithmetic in each coordinate is modulo p. Multiplication is more complicated: We view it merely as multiplication of polynomials, but not only is the coordinate arithmetic modulo p, but we also reduce high-degree polynomials modulo $f(x)$. That is to say, to multiply $a * b$ in $\mathbf{F}_{p^k}$, we simply form a polynomial product $a(x)b(x)$, doing a mod p reduction when a coefficient during this process exceeds $p-1$, then taking this product mod $f(x)$ via polynomial mod, again reducing mod p whenever appropriate during that process. In principle, one could just form the unrestricted product $a(x)b(x)$, do a mod f reduction, then take a final mod p reduction, in which case the final result would be the same but the interior integer multiplies might run out of control, especially if there were many polynomials being multiplied. It is best to take a reduction modulo p at every meaningful juncture.

Here is an example for explicit construction of a field of characteristic 2, namely $\mathbf{F}_{16}$. According to our formula (2.5), there are exactly 3 irreducible degree-4 polynomials in $\mathbf{F}_2[x]$, and a quick check shows that they are x^4+x+1, x^4+x^3+1, and $x^4+x^3+x^2+x+1$. Though each of these can be used to create $\mathbf{F}_{16}$, the first has the pleasant property that reduction of high powers of x to lower powers is particularly simple: The mod $f(x)$ reduction is realized through the simple rule $x^4 = x+1$ (recall that we are in characteristic 2, so that $1 = -1$). We may abbreviate typical field elements $a_0+a_1x+a_2x^2+a_3x^3$, where each $a_i \in \{0, 1\}$ by the binary string $(a_0a_1a_2a_3)$. We add componentwise modulo 2, which amounts to an "exclusive-or" operation, for example

$$(0111) + (1011) = (1100).$$

To multiply $a * b = (0111) * (1011)$ we can simulate the polynomial multiplication by doing a convolution on the coordinates, first getting (0110001), a string of length 7. (Calling this $(c_0c_1c_2c_3c_4c_5c_6)$ we have $c_j = \sum_{i_1+i_2=j} a_{i_1}b_{i_2}$, where the sum is over pairs i_1, i_2 of integers in $\{0, 1, 2, 3\}$ with sum j.) To get the final answer, we take any 1 in places $6, 5, 4$, in this order, and replace them via the modulo $f(x)$ relation. In our case, the 1 in place 6 gets replaced with 1's in places 2 and 3, and doing the exclusive-or, we get (0101000). There are no more high-order 1's to replace, and our product is (0101); that is, we have

$$(0111) * (1011) = (0101).$$

Though this is only a small example, all the basic notions of general field arithmetic via polynomials are present.

2.3 Squares and roots

2.3.1 Quadratic residues

We start with some definitions.

Definition 2.3.1. For coprime integers m, a with m positive, we say that a is a quadratic residue (mod m) if and only if the congruence

$$x^2 \equiv a \pmod{m}$$

is solvable for integer x. If the congruence is not so solvable, a is said to be a quadratic nonresidue (mod m).

Note that quadratic residues and nonresidues are defined only when $\gcd(a, m) = 1$. So, for example, $0 \pmod{m}$ is always a square but is neither a quadratic residue nor a nonresidue. Another example is $3 \pmod{9}$. This residue is not a square, but it is *not* considered a quadratic nonresidue since 3 and 9 are not coprime. When the modulus is prime the only non-coprime case is the 0 residue, which is one of the choices in the next definition.

Definition 2.3.2. For odd prime p, the Legendre symbol $\left(\frac{a}{p}\right)$ is defined as

$$\left(\frac{a}{p}\right) = \begin{cases} 0, & \text{if } a \equiv 0 \pmod{p}, \\ 1, & \text{if } a \text{ is a quadratic residue} \pmod{p}, \\ -1, & \text{if } a \text{ is a quadratic nonresidue} \pmod{p}. \end{cases}$$

Thus, the Legendre symbol signifies whether or not $a \not\equiv 0 \pmod{p}$ is a square (mod p). Closely related, but differing in some important ways, is the Jacobi symbol:

Definition 2.3.3. For odd natural number m (whether prime or not), and for any integer a, the Jacobi symbol $\left(\frac{a}{m}\right)$ is defined in terms of the (unique) prime factorization

$$m = \prod p_i^{t_i}$$

as

$$\left(\frac{a}{m}\right) = \prod \left(\frac{a}{p_i}\right)^{t_i},$$

where $\left(\frac{a}{p_i}\right)$ are Legendre symbols, with $\left(\frac{a}{1}\right) = 1$ understood.

Note, then, that the function $\chi(a) = \left(\frac{a}{m}\right)$, defined for all integers a, is a character modulo m; see Section 1.4.3. It is important to note right off that for composite, odd m, a Jacobi symbol $\left(\frac{a}{m}\right)$ can sometimes be $+1$ when $x^2 \equiv a \pmod{m}$ is unsolvable. An example is

$$\left(\frac{2}{15}\right) = \left(\frac{2}{3}\right)\left(\frac{2}{5}\right) = (-1)(-1) = 1,$$

even though 2 is not, in fact, a square modulo 15. However, if $\left(\frac{a}{m}\right) = -1$, then a is coprime to m and the congruence $x^2 \equiv a \pmod{m}$ is not solvable. And $\left(\frac{a}{m}\right) = 0$ if and only if $\gcd(a, m) > 1$.

It is clear that in principle the symbol $\left(\frac{a}{m}\right)$ is computable: One factors m into primes, and then computes each underlying Legendre symbol by exhausting all possibilities to see whether the congruence $x^2 \equiv a \pmod p$ is solvable. What makes Legendre and Jacobi symbols so very useful, though, is that they are indeed very easy to compute, with no factorization or primality test necessary, and with no exhaustive search. The following theorem gives some of the beautiful properties of Legendre and Jacobi symbols, properties that make their evaluation a simple task, about as hard as taking a gcd.

Theorem 2.3.4 (Relations for Legendre and Jacobi symbols). *Let p denote an odd prime, let m, n denote arbitrary positive odd integers (including possibly primes), and let a, b denote integers. Then we have the Euler test for quadratic residues modulo primes, namely*

$$\left(\frac{a}{p}\right) \equiv a^{(p-1)/2} \pmod p. \tag{2.6}$$

We have the multiplicative relations

$$\left(\frac{ab}{m}\right) = \left(\frac{a}{m}\right)\left(\frac{b}{m}\right), \tag{2.7}$$

$$\left(\frac{a}{mn}\right) = \left(\frac{a}{m}\right)\left(\frac{a}{n}\right) \tag{2.8}$$

and special relations

$$\left(\frac{-1}{m}\right) = (-1)^{(m-1)/2}, \tag{2.9}$$

$$\left(\frac{2}{m}\right) = (-1)^{(m^2-1)/8}. \tag{2.10}$$

Furthermore, we have the law of quadratic reciprocity for coprime m, n:

$$\left(\frac{m}{n}\right)\left(\frac{n}{m}\right) = (-1)^{(m-1)(n-1)/4}. \tag{2.11}$$

Already (2.6) shows that when $|a| < p$, the Legendre symbol $\left(\frac{a}{p}\right)$ can be computed in $O\left(\ln^3 p\right)$ bit operations using naive arithmetic and Algorithm 2.1.5; see Exercise 2.15. But we can do better, and we do not even need to recognize primes.

Algorithm 2.3.5 (Calculation of Legendre/Jacobi symbol).
Given positive odd integer m, and integer a, this algorithm returns the Jacobi symbol $\left(\frac{a}{m}\right)$, which for m an odd prime is also the Legendre symbol.

1. [Reduction loops]
    $a = a \bmod m$;
    $t = 1$;
    while($a \neq 0$) {
        while(a even) {
            $a = a/2$;

```
            if(m mod 8 ∈ {3,5}) t = -t;
        }
        (a,m) = (m,a);                              // Swap variables.
        if(a ≡ m ≡ 3 (mod 4)) t = -t;
        a = a mod m;
    }
2. [Termination]
    if(m == 1) return t;
    return 0;
```

It is clear that this algorithm does not take materially longer than using Algorithm 2.1.2 to find $\gcd(a, m)$, and so runs in $O\left(\ln^2 m\right)$ bit operations when $|a| < m$.

In various other sections of this book we make use of a celebrated connection between the Legendre symbol and exponential sums. The study of this connection runs deep; for the moment we state one central, useful result, starting with the following definition:

Definition 2.3.6. The quadratic Gauss sum $G(a; m)$ is defined for integers a, N, with N positive, as

$$G(a; N) = \sum_{j=0}^{N-1} e^{2\pi i a j^2/N}.$$

This sum is—up to conjugation perhaps—a discrete Fourier transform (DFT) as used in various guises in Chapter 9. A more general form—a character sum—is used in primality proving (Section 4.4). The central result we wish to cite makes an important connection with the Legendre symbol:

Theorem 2.3.7 (Gauss). *For odd prime p and integer $a \not\equiv 0 \pmod p$,*

$$G(a; p) = \left(\frac{a}{p}\right) G(1; p),$$

and generally, for positive integer m,

$$G(1; m) = \frac{1}{2}\sqrt{m}(1+i)(1+(-i)^m).$$

The first assertion is really very easy, the reader might consider proving it without looking up references. The two assertions of the theorem together allow for Fourier inversion of the sum, so that one can actually express the Legendre symbol for $a \not\equiv 0 \pmod p$ by

$$\left(\frac{a}{p}\right) = \frac{c}{\sqrt{p}} \sum_{j=0}^{p-1} e^{2\pi i a j^2/p} = \frac{c}{\sqrt{p}} \sum_{j=0}^{p-1} \left(\frac{j}{p}\right) e^{2\pi i a j/p}, \tag{2.12}$$

where $c = 1, -i$ as $p \equiv 1, 3 \pmod 4$, respectively. This shows that the Legendre symbol is, essentially, its own discrete Fourier transform (DFT).

For practice in manipulating Gauss sums, see Exercises 1.63, 2.23, 2.24, and 9.42.

2.3.2 Square roots

Armed now with algorithms for gcd, inverse (actually the -1 power), and positive integer powers, we turn to the issue of square roots modulo a prime. As we shall see, the technique actually calls for raising residues to high integral powers, and so the task is not at all like taking square roots in the real numbers.

We have seen that for odd prime p, the solvability of a congruence

$$x^2 \equiv a \not\equiv 0 \pmod{p}$$

is signified by the value of the Legendre symbol $\left(\frac{a}{p}\right)$. When $\left(\frac{a}{p}\right) = 1$, an important problem is to find a "square root" x, of which there will be two, one the other's negative (mod p). We shall give two algorithms for extracting such square roots, both computationally efficient but raising different issues of implementation.

The first algorithm starts from Euler's test (2.6). If the prime p is 3 (mod 4) and $\left(\frac{a}{p}\right) = 1$, then Euler's test says that $a^t \equiv 1 \pmod{p}$, where $t = (p-1)/2$. Then $a^{t+1} \equiv a \pmod{p}$, and as $t+1$ is even in this case, we may take for our square root $x \equiv a^{(t+1)/2} \pmod{p}$. Surely, this delightfully simple solution to the square root problem can be generalized! Yes, but it is not so easy. In general, we may write $p - 1 = 2^s t$, with t odd. Euler's test (2.6) guarantees us that $a^{2^{s-1}t} \equiv 1 \pmod{p}$, but it does not appear to say anything about $A = a^t \pmod{p}$.

Well, it does say something; it says that the multiplicative order of A modulo p is a divisor of 2^{s-1}. Suppose that d is a quadratic *nonresidue* modulo p, and let $D = d^t \bmod p$. Then Euler's test (2.6) says that the multiplicative order of D modulo p is exactly 2^s, since $D^{2^{s-1}} \equiv -1 \pmod{p}$. The same is true about $D^{-1} \pmod{p}$, namely, its multiplicative order is 2^s. Since the multiplicative group $\mathbf{Z}_p^*$ is cyclic, it follows that A is in the cyclic subgroup generated by D^{-1}, and in fact, A is an even power of D^{-1}, that is, $A \equiv D^{-2\mu} \pmod{p}$ for some integer μ with $0 \le \mu < 2^{s-1}$. Substituting for A we have $a^t D^{2\mu} \equiv 1 \pmod{p}$. Then after multiplying this congruence by a, the left side has all even exponents, and we can extract the square root of a modulo p as $a^{(t+1)/2} D^{\mu} \pmod{p}$.

To make this idea into an algorithm, there are two problems that must be solved:

(1) Find a quadratic nonresidue $d \pmod{p}$.

(2) Find an integer μ with $A \equiv D^{-2\mu} \pmod{p}$.

It might seem that problem (1) is simple and that problem (2) is difficult, since there are many quadratic residues modulo p and we only need one of them, any one, while for problem (2) there is a specific integer μ that we are searching for. In some sense, these thoughts are correct. However, we know no rigorous, *deterministic* way to find a quadratic nonresidue quickly. We will get around this impasse by using a *random* algorithm. And though problem (2) is an instance of the notoriously difficult discrete logarithm problem (see Chapter 5), the particular instance we have in hand here is simple. The following algorithm is due to A. Tonelli in 1891, based on earlier work of Gauss.

Algorithm 2.3.8 (Square roots (mod p)). Given an odd prime p and an integer a with $\left(\frac{a}{p}\right) = 1$, this algorithm returns a solution x to $x^2 \equiv a \pmod p$.

1. [Check simplest cases: $p \equiv 3, 5, 7 \pmod 8$]
    $a = a \bmod p$;
    if($p \equiv 3, 7 \pmod 8$) {
        $x = a^{(p+1)/4} \bmod p$;
        return x;
    }
    if($p \equiv 5 \pmod 8$) {
        $x = a^{(p+3)/8} \bmod p$;
        $c = x^2 \bmod p$; // Then $c \equiv \pm a \pmod p$.
        if($c \not\equiv a \bmod p$) $x = x2^{(p-1)/4} \bmod p$;
        return x;
    }
2. [Case $p \equiv 1 \pmod 8$]
    Find a random integer $d \in [2, p-1]$ with $\left(\frac{d}{p}\right) = -1$;
    // Compute Jacobi symbols via Algorithm 2.3.5.
    Represent $p - 1 = 2^s t$, with t odd;
    $A = a^t \bmod p$;
    $D = d^t \bmod p$;
    $m = 0$; // m will be 2μ of text discussion.
    for($0 \le i < s$){ // One may start at $i = 1$; see text.
        if($(AD^m)^{2^{s-1-i}} \equiv -1 \pmod p$) $m = m + 2^i$;
    } // Now we have $AD^m \equiv 1 \pmod p$.
    $x = a^{(t+1)/2} D^{m/2} \bmod p$;
    return x;

Note the following interesting features of this algorithm. First, it turns out that the $p \equiv 1 \pmod 8$ branch—the hardest case—will actually handle all the cases. (We have essentially used in the $p \equiv 5 \pmod 8$ case that we may choose $d = 2$. And in the $p \equiv 3 \pmod 4$ cases, the exponent m is 0, so we do not need a value of d.) Second, notice that built into the algorithm is the check that $A^{2^{s-1}} \equiv 1 \pmod p$, which is what ensures that m is even. If this fails, then we do not have $\left(\frac{a}{p}\right) = 1$, and so the algorithm may be amended to leave out this requirement, with a break called for if the case $i = 0$ in the loop produces the residue -1. If one is taking many square roots of residues a for which it is unknown whether a is a quadratic residue or nonresidue, then one may be tempted to just let Algorithm 2.3.8 decide the issue for us. However, if nonresidues occur a positive fraction of the time, it will be faster on average to first run Algorithm 2.3.5 to check the quadratic character of a, and thus avoid running the more expensive Algorithm 2.3.8 on the nonresidues.

As we have mentioned, there is no known deterministic, polynomial time algorithm for finding a quadratic nonresidue d for the prime p. However, if one assumes the ERH, it can be shown there is a quadratic nonresidue $d < 2\ln^2 p$; see Theorem 1.4.5, and so an exhaustive search to this limit succeeds in finding a quadratic nonresidue in polynomial time. Thus, on the ERH, one can find square roots for quadratic residues modulo the prime p in deterministic, polynomial time. It is interesting, from a theoretical standpoint, that for a *fixed*, there is a rigorously proved, deterministic, polynomial time algorithm for square root extraction, which algorithm is due to [Schoof 1985]. (The bit complexity is polynomial in the length of p, but exponential in the length of a, so that for a fixed it is correct to say that the algorithm is polynomial time.) Still, in spite of this fascinating theoretical state of affairs, the fact that half of all nonzero residues $d \pmod p$ satisfy $\left(\frac{d}{p}\right) = -1$ leads to the expectation of only a few random attempts to find a suitable d. In fact, the expected number of random attempts is 2.

The complexity of Algorithm 2.3.8 is dominated by the various exponentiations called for, and so is $O(s^2 + \ln t)$ modular operations. Assuming naive arithmetic subroutines, this comes out to, in the worst case (when s is large), $O\left(\ln^4 p\right)$ bit operations. However, if one is applying Algorithm 2.3.8 to many prime moduli p, it is perhaps better to consider its average case, which is just $O\left(\ln^3 p\right)$ bit operations. This is because there are very few primes p with $p-1$ divisible by a large power of 2.

The following algorithm is asymptotically faster than the worst case of Algorithm 2.3.8. A beautiful application of arithmetic in the finite field $\mathbf{F}_{p^2}$, the method is a 1907 discovery of M. Cipolla.

Algorithm 2.3.9 (Square roots (mod p) via $\mathbf{F}_{p^2}$ arithmetic). Given an odd prime p and a quadratic residue a modulo p, this algorithm returns a solution x to $x^2 \equiv a \pmod p$.

1. [Find a certain quadratic nonresidue]
   Find a random integer $t \in [0, p-1]$ such that $\left(\frac{t^2-a}{p}\right) = -1$;
   // Compute Jacobi symbols via Algorithm 2.3.5.
2. [Find a square root in $\mathbf{F}_{p^2} = \mathbf{F}_p(\sqrt{t^2-a})$]
   $x = (t + \sqrt{t^2-a})^{(p+1)/2}$; // Use $\mathbf{F}_{p^2}$ arithmetic.
   return x;

The probability that a random value of t will be successful in step [Find a certain quadratic nonresidue] is $(p-1)/2p$. It is not hard to show that the element $x \in \mathbf{F}_{p^2}$ is actually an element of the subfield $\mathbf{F}_p$ of $\mathbf{F}_{p^2}$, and that $x^2 \equiv a \pmod p$. (In fact, the second assertion forces x to be in $\mathbf{F}_p$, since a has the same square roots in $\mathbf{F}_p$ as it has in the larger field $\mathbf{F}_{p^2}$.)

A word is in order on the field arithmetic, which for this case of $\mathbf{F}_{p^2}$ is especially simple, as might be expected on the basis of Section 2.2.2. Let

$\omega = \sqrt{t^2 - a}$. Representing this field by

$$\mathbf{F}_{p^2} = \{x + \omega y : x, y \in \mathbf{F}_p\} = \{(x, y)\},$$

all arithmetic may proceed using the rule

$$\begin{aligned}(x, y) * (u, v) &= (x + y\omega)(u + v\omega) \\ &= xu + yv\omega^2 + (xv + yu)\omega \\ &= (xu + yv(t^2 - a), xv + yu),\end{aligned}$$

noting that $\omega^2 = t^2 - a$ is in $\mathbf{F}_p$. Of course, we view x, y, u, v, t, a as residues modulo p and the above expressions are always reduced to this modulus. Any of the binary ladder powering algorithms in this book may be used for the computation of x in step [Find a square root ...]. An equivalent algorithm for square roots is given in [Menezes et al. 1997], in which one finds a quadratic nonresidue $b^2 - 4a$, defines the polynomial $f(x) = x^2 - bx + a$ in $\mathbf{F}_p[x]$, and simply computes the desired root $r = x^{(p+1)/2} \bmod f$ (using polynomial-mod operations). Note finally that the special cases $p \equiv 3, 5, 7 \pmod 8$ can also be ferreted out of any of these algorithms, as was done in Algorithm 2.3.8, to improve average performance.

The complexity of Algorithm 2.3.9 is $O(\ln^3 p)$ bit operations (assuming naive arithmetic), which is asymptotically better than the worst case of Algorithm 2.3.8. However, if one is loath to implement the modified powering ladder for the $\mathbf{F}_{p^2}$ arithmetic, the asymptotically slower algorithm will usually serve. Incidentally, there is yet another, equivalent, approach for square rooting by way of Lucas sequences (see Exercise 2.27).

It is very interesting to note at this juncture that there is no known fast method of computing square roots of quadratic residues for general composite moduli. In fact, as we shall see later, doing so is essentially equivalent to factoring the modulus (see Exercise 6.5).

2.3.3 Finding polynomial roots

Having discussed issues of existence and calculation of square roots, we now consider the calculation of roots of a polynomial of arbitrary degree over a finite field. We specify the finite field as $\mathbf{F}_p$, but much of what we say generalizes to an arbitrary finite field.

Let $g \in \mathbf{F}_p[x]$ be a polynomial, that is, it is a polynomial with integer coefficients reduced (mod p). We are looking for the roots of g in $\mathbf{F}_p$, and so we might begin by replacing $g(x)$ with the gcd of $g(x)$ and $x^p - x$, since as we have seen, the latter polynomial is the product of $x - a$ as a runs over all elements of $\mathbf{F}_p$. If $p > \deg g$, one should first compute $x^p \bmod g(x)$ via Algorithm 2.1.5. If the gcd has degree not exceeding 2, the prior methods we have learned settle the matter. If it has degree greater than 2, then we take a further gcd with $(x+a)^{(p-1)/2} - 1$ for a random $a \in \mathbf{F}_p$. Any particular $b \neq 0$ in $\mathbf{F}_p$ is a root of $(x + a)^{(p-1)/2} - 1$ with probability $1/2$, so that we have a positive probability of splitting $g(x)$ into two polynomials of smaller degree. This suggests a recursive algorithm, which is what we describe below.

Algorithm 2.3.10 (Roots of a polynomial over $\mathbf{F}_p$).
Given a nonzero polynomial $g \in \mathbf{F}_p[x]$, with p an odd prime, this algorithm returns the set r of the roots (without multiplicity) in $\mathbf{F}_p$ of g. The set r is assumed global, augmented as necessary during all recursive calls.

```
1. [Initial adjustments]
      r = { };                                   // Root list starts empty.
      g(x) = gcd(x^p − x, g(x));                 // Using Algorithm 2.2.1.
      if(g(0) == 0) {                            // Check for 0 root.
         r = r ∪ {0};
         g(x) = g(x)/x;
      }
2. [Call recursive procedure and return]
      r = r ∪ roots(g);
      return r;
3. [Recursive function roots()]
      roots(g) {
         If deg(g) ≤ 2, use quadratic (or lower) formula, via Algorithm 2.3.8, or
            2.3.9, to append to r all roots of g, and return;
         while(h == 1 or h == g) {               // Random splitting.
            Choose random a ∈ [0, p − 1];
            h(x) = gcd((x + a)^((p−1)/2) − 1, g(x));
         }
         r = r ∪ roots(h) ∪ roots(g/h);
         return;
      }
```

The computation of $h(x)$ in the random-splitting loop can be made easier by using Algorithm 2.1.5 to first compute $(x+a)^{(p-1)/2} \bmod g(x)$ (and of course, the coefficients are always reduced (mod p)). It can be shown that the probability that a random a will succeed in splitting $g(x)$ (where $\deg(g) \geq 3$) is at least about $3/4$ if p is large, and is always bounded above 0. Note that we can use the random splitting idea on degree-2 polynomials as well, and thus we have a third square root algorithm! (If $g(x)$ has degree 2, then the probability that a random choice for a in step [Recursive ...] will split g is at least $(p-1)/(2p)$.) Various implementation details of this algorithm are discussed in [Cohen 2000]. Note that the algorithm is not actually factoring the polynomial; for example, a polynomial f might be the product of two irreducible polynomials, each of which is devoid of roots in $\mathbf{F}_p$. For actual polynomial factoring, there is the Berlekamp algorithm [Menezes et al. 1997], [Cohen 2000], but many important algorithms require only the root finding we have exhibited.

We now discuss the problem of finding roots of a polynomial to a composite modulus. Suppose the modulus is $n = ab$, where a, b are coprime. If we have an integer r with $f(r) \equiv 0 \pmod{a}$ and an integer s with $f(s) \equiv 0 \pmod{b}$, we can find a root to $f(x) \equiv 0 \pmod{ab}$ that "corresponds" to r and s. Namely, if the integer t simultaneously satisfies $t \equiv r \pmod{a}$ and $t \equiv s \pmod{b}$, then $f(t) \equiv 0 \pmod{ab}$. And such an integer t may be found by the Chinese remainder theorem; see Theorem 2.1.6. Thus, if the modulus n can be factored into primes, and we can solve the case for prime power moduli, then we can solve the general case.

To this end, we now turn our attention to solving polynomial congruences modulo prime powers. Note that for any polynomial $f(x) \in \mathbf{Z}[x]$ and any integer r, there is a polynomial $g_r(x) \in \mathbf{Z}[x]$ with

$$f(x+r) = f(r) + xf'(r) + x^2 g_r(x). \tag{2.13}$$

This can be seen either through the Taylor expansion for $f(x+r)$ or through the binomial theorem in the form

$$(x+r)^d = r^d + dr^{d-1}x + x^2 \sum_{j=2}^{d} \binom{d}{j} r^{d-j} x^{j-2}.$$

We can use Algorithm 2.3.10 to find solutions to $f(x) \equiv 0 \pmod{p}$, if there are any. The question is how we might be able to "lift" a solution to one modulo p^k for various exponents k. Suppose we have been successful in finding a root modulo p^i, say $f(r) \equiv 0 \pmod{p^i}$, and we wish to find a solution to $f(t) \equiv 0 \pmod{p^{i+1}}$ with $t \equiv r \pmod{p^i}$. We write t as $r + p^i y$, and so we wish to solve for y. We let $x = p^i y$ in (2.13). Thus

$$f(t) = f(r + p^i y) \equiv f(r) + p^i y f'(r) \pmod{p^{2i}}.$$

If the integer $f'(r)$ is not divisible by p, then we can use the methods of Section 2.1.1 to solve the congruence

$$f(r) + p^i y f'(r) \equiv 0 \pmod{p^{2i}},$$

namely by dividing through by p^i (recall that $f(r)$ is divisible by p^i), finding an inverse z for $f'(r) \pmod{p^i}$, and letting $y = -zf(r)p^{-i} \bmod p^i$. Thus, we have done more than we asked for, having instantly gone from the modulus p^i to the modulus p^{2i}. But there was a requirement that the integer r satisfy $f'(r) \not\equiv 0 \pmod{p}$. In general, if $f(r) \equiv f'(r) \equiv 0 \pmod{p}$, then there may be no integer $t \equiv r \pmod{p}$ with $f(t) \equiv 0 \pmod{p^2}$. For example, take $f(x) = x^2 + 3$ and consider the prime $p = 3$. We have the root $x = 0$, that is, $f(0) \equiv 0 \pmod{3}$. But the congruence $f(x) \equiv 0 \pmod{9}$ has no solution. For more on criteria for when a polynomial solution lifts to higher powers of the modulus, see Section 3.5.3 in [Cohen 2000].

The method described above is known as Hensel lifting, after the German mathematician K. Hensel. The argument essentially gives a criterion for there to be a solution of $f(x) = 0$ in the "p-adic" numbers: There is a solution if there is an integer r with $f(r) \equiv 0 \pmod{p}$ and $f'(r) \not\equiv 0 \pmod{p}$. What is more important for us, though, is using this idea as an algorithm to solve polynomial congruences modulo high powers of a prime. We summarize the above discussion in the following.

Algorithm 2.3.11 (Hensel lifting). We are given a polynomial $f(x) \in \mathbf{Z}[x]$, a prime p, and an integer r that satisfies $f(r) \equiv 0 \pmod{p}$ (perhaps supplied by Algorithm 2.3.10) and $f'(r) \not\equiv 0 \pmod{p}$. This algorithm describes how one constructs a sequence of integers $r_0, r_1, \ldots$ such that for each $i < j$, $r_i \equiv r_j \pmod{p^{2^i}}$ and $f(r_i) \equiv 0 \pmod{p^{2^i}}$. The description is iterative, that is, we give r_0 and show how to find r_{i+1} as a function of an already known r_i.

1. [Initial term]
   $r_0 = r$;
2. [Function $newr()$ that gives r_{i+1} from r_i]
   $newr(r_i)$ {
     $x = f(r_i)p^{-2^i}$;
     $z = (f'(r))^{-1} \bmod p^{2^i}$; // Via Algorithm 2.1.4.
     $y = -xz \bmod p^{2^i}$;
     $r_{i+1} = r_i + yp^{2^i}$;
     return r_{i+1};
   }

Note that for $j \geq i$ we have $r_j \equiv r_i \left(\bmod\ p^{2^i}\right)$, so that the sequence (r_i) converges in the p-adic numbers to a root of $f(x)$. In fact, Hensel lifting may be regarded as a p-adic version of the Newton methods discussed in Section 9.2.2.

2.3.4 Representation by quadratic forms

We next turn to a problem important to such applications as elliptic curves and primality testing. This is the problem of finding quadratic Diophantine representations, for positive integer d and odd prime p, in the form

$$x^2 + dy^2 = p,$$

or, in studies of complex quadratic orders of discriminant $D < 0$, $D \equiv 0, 1 \pmod{4}$, the form [Cohen 2000]

$$x^2 + |D|y^2 = 4p.$$

There is a beautiful approach for these Diophantine problems. The next two algorithms are not only elegant, they are very efficient. Incidentally, the following algorithm was attributed usually to Cornacchia until recently, when it became known that H. Smith had discovered it earlier, in 1885 in fact.

Algorithm 2.3.12 (Represent p as $x^2 + dy^2$: Cornacchia–Smith method). Given an odd prime p and a positive integer $d \not\equiv 0 \pmod{p}$, this algorithm either reports that no solution exists, or returns a solution (x, y).

1. [Test for solvability]
   if($\left(\frac{-d}{p}\right) == -1$) return { }; // Return empty: no solution.
2. [Initial square root]

```
      x0 = √(−d) mod p;                           // Via Algorithm 2.3.8 or 2.3.9.
      if(2x0 < p) x0 = p − x0;                                 // Justify the root.
3. [Initialize Euclid chain]
      (a, b) = (p, x0);
      c = ⌊√p⌋;                                          // Via Algorithm 9.2.11.
4. [Euclid chain]
      while(b > c) (a, b) = (b, a mod b);
5. [Final report]
      t = p − b^2;
      if(t ≢ 0 (mod d)) return { };                              // Return empty.
      if(t/d not a square) return { };                           // Return empty.
      return (±b, ±√(t/d));                                 // Solution(s) found.
```

This completely solves the computational Diophantine problem at hand. Note that an integer square-root finding routine (Algorithm 9.2.11) is invoked at two junctures. The second invocation—the determination as to whether t/d is a perfect square—can be done along the lines discussed in the text following the Algorithm 9.2.11 description. Incidentally, the proof that Algorithm 2.3.12 works is, in the words of [Cohen 2000], "a little painful." There is an elegant argument, due to H. Lenstra, in [Schoof 1995], and a clear explanation from an algorist's point of view (for $d = 1$) in [Bressoud and Wagon 2000, p. 283].

The second case, namely for the Diophantine equation $x^2 + |D|y^2 = 4p$, for $D < 0$, can be handled in the following way [Cohen 2000]. First we observe that if $D \equiv 0 \pmod 4$, then x is even, whence the problem comes down to solving $(x/2)^2 + (|D|/4)y^2 = p$, which we have already done. If $D \equiv 1 \pmod 8$, we have $x^2 - y^2 \equiv 4 \pmod 8$, and so x, y are both even, and again we defer to the previous method. Given the above argument, one could use the next algorithm only for $D \equiv 5 \pmod 8$, but in fact, the following will work for what turn out to be convenient cases $D \equiv 0, 1 \pmod 4$:

Algorithm 2.3.13. (Represent $4p$ as $x^2 + |D|y^2$ (modified Cornacchia–Smith)) Given a prime p and $-4p < D < 0$ with $D \equiv 0, 1 \pmod 4$, this algorithm either reports that no solution exists, or returns a solution (x, y).

```
1. [Case p = 2]
      if(p == 2) {
         if(D + 8 is a square) return (√(D + 8), 1);
         return { };                              // Return empty: no solution.
      }
2. [Test for solvability]
      if((D/p) < 1) return { };                                  // Return empty.
3. [Initial square root]
      x0 = √D mod p;                              // Via Algorithm 2.3.8 or 2.3.9.
      if(x0 ≢ D (mod 2)) x0 = p − x0;                  // Ensure x0^2 ≡ D (mod 4p).
4. [Initialize Euclid chain]
```

```
    (a, b) = (2p, x_0);
    c = ⌊2√p⌋;                                        // Via Algorithm 9.2.11.
5. [Euclid chain]
    while(b > c) (a, b) = (b, a mod b);
6. [Final report]
    t = 4p − b^2;
    if(t ≢ 0 (mod |D|)) return { };                   // Return empty.
    if(t/|D| not a square) return { };                // Return empty.
    return (±b, ±√(t/|D|));                           // Found solution(s).
```

Again, the algorithm either says that there is no solution, or reports the essentially unique solution to $x^2 + |D|y^2 = 4p$.

2.4 Exercises

2.1. Prove that 16 is, modulo any odd number, an eighth power.

2.2. Show that the least common multiple $\operatorname{lcm}(a,b)$ satisfies

$$\operatorname{lcm}(a,b) = \frac{ab}{\gcd(a,b)},$$

and generalize this formula for more than two arguments. Then, using the prime number theorem (PNT), find a reasonable estimate for the lcm of all the integers from 1 through (a large) n.

2.3. Denote by $\omega(n)$ the number of prime divisors of n, Prove that for any square-free integer d,

$$\#\{(x,y)\ :\ \operatorname{lcm}(x,y) = d\} = 3^{\omega(d)}.$$

2.4. Study the relation between the Euclid algorithm and simple continued fractions, with a view to proving the Lamé theorem (the first part of Theorem 2.1.3).

2.5. Fibonacci numbers are defined $u_0 = 0$, $u_1 = 1$, and $u_{n+1} = u_n + u_{n-1}$ for $n \geq 1$. Prove the remarkable relation

$$\gcd(u_a, u_b) = u_{\gcd(a,b)},$$

which shows, among many other things, that u_n, u_{n+1} are coprime for $n > 1$, and that if u_n is prime, then n is prime. Find a counterexample to the converse (find a prime p such that u_p is composite). By analyzing numerically several Fibonacci numbers, guess—then prove—a simple, general formula for the inverse of $u_n \pmod{u_{n+1}}$.

Fibonacci numbers appear elsewhere in this book, e.g., in Sections 1.3.3, 3.5.1 and Exercises 3.25, 3.41, 9.51.

2.6. Show that for $x \approx y \approx N$, and assuming classical divide with remainder, the bit-complexity of the classical Euclid algorithm is $O\left(\ln^2 N\right)$. It is helpful to observe that to find the quotient–remainder pair q, r with $x = qy+r$ requires $O((1 + \ln q) \ln x)$ bit operations, and that the quotients are constrained in a certain way during the Euclid loop.

2.7. Prove that Algorithm 2.1.4 works; that is, the correct gcd and inverse pair are returned. Answer the following question: When, if ever, do the returned a, b have to be reduced further, to $a \bmod y$ and $b \bmod x$, to yield legitimate, unique inverses?

2.8. Argue that for a naive application of Theorem 2.1.6 the mod operations involved consume at least $O\left(\ln^2 M\right)$ bit operations if arithmetic be done in grammar-school fashion, but only $O\left(r \ln^2 m\right)$ via Algorithm 2.1.7, where m denotes the maximum of the m_i.

2.9. Write a program to effect the asymptotically fast, preconditioned CRT Algorithm 9.5.25, and use this to multiply two numbers each of, say, 100 decimal digits, using sufficiently many small prime moduli.

2.10. Following Exercise 1.46 one can use, for CRT moduli, Mersenne numbers having pairwise coprime exponents (the Mersenne numbers need not themselves be prime). What computational advantages might there be in choosing such a moduli set (see Section 9.2.3)? Is there an easy way to find inverses $(2^a - 1)^{-1} \pmod{2^b - 1}$?

2.11. Give the computational complexity of the "straightforward inverse" algorithm implied by relation (2.3). Is there ever a situation when one should use this, or use instead Algorithm 2.1.4 to obtain $a^{-1} \bmod m$?

2.12. Does formula (2.5) generalize to give the number of irreducible polynomials of degree k in $\mathbf{F}_{p^n}[x]$?

2.13. Show how Algorithm 2.2.2 plays a role in finite field arithmetic, namely in the process of finding a multiplicative inverse of an element in $\mathbf{F}_{p^n}$.

2.14. Show how Algorithms 2.3.8 and 2.3.9 may be appropriately generalized to find square roots of squares in the finite field $\mathbf{F}_{p^n}$.

2.15. By considering the binary expansion of the exponent n, show that the computational complexity of Algorithm 2.1.5 is $O(\ln n)$ operations. Argue that if x, n are each of size m and we are to compute $x^n \bmod m$, and classical multiply-mod is used, that the overall bit complexity of this powering grows as the cube of the number of bits in m.

2.16. Say we wish to compute a power $x^y \bmod N$, with $N = pq$, the product of two distinct primes. Describe an algorithm that combines a binary ladder and Chinese remainder theorem (CRT) ideas, and that yields the desired power more rapidly than does a standard, (mod N)-based ladder.

2.17. The "repunit" number $r_{1031} = (10^{1031} - 1)/9$, composed of 1031 decimal ones, is known to be prime. Determine, via reciprocity, which of

$7, -7$ is a quadratic residue of this repunit. Then give an explicit square root (mod r_{1031}) of the quadratic residue.

2.18. Using appropriate results of Theorem 2.3.4, prove that for prime $p > 3$,

$$\left(\frac{-3}{p}\right) = (-1)^{\frac{(p-1) \bmod 6}{4}}.$$

Find a similar closed form for $\left(\frac{5}{p}\right)$ when $p \neq 2, 5$.

2.19. Show that for prime $p \equiv 1 \pmod 4$, the sum of the quadratic residues in $[1, p-1]$ is $p(p-1)/4$.

2.20. Develop an algorithm for computing the Jacobi symbol $\left(\frac{a}{m}\right)$ along the lines of the binary *gcd* method of Algorithm 9.4.2.

2.21. Prove: For prime p with $p \equiv 3 \pmod 4$, given any pair of square roots of a given $x \not\equiv 0 \pmod p$, one root is itself a quadratic residue and the other is not. (The root that is the quadratic residue is known as the principal square root.) See Exercises 2.22 and 2.39 for applications of the principal root.

2.22. We denote by $\mathbf{Z}_n^*$ the multiplicative group of the elements in $\mathbf{Z}_n$ that are coprime to n.

(1) Suppose n is odd and has exactly k distinct prime factors. Let J denote the set of elements $x \in \mathbf{Z}_n^*$ with the Jacobi symbol $\left(\frac{x}{n}\right) = 1$ and let S denote the set of squares in $\mathbf{Z}_n^*$. Show that J is a subgroup of $\mathbf{Z}_n^*$ of $\varphi(n)/2$ elements, and that S is a subgroup of J.

(2) Show that squares in $\mathbf{Z}_n^*$ have exactly 2^k square roots in $\mathbf{Z}_n^*$ and conclude that $\#S = \varphi(n)/2^k$.

(3) Now suppose n is a Blum integer, that is, $n = pq$ is a product of two different primes $p, q \equiv 3 \pmod 4$. (Blum integers have importance in cryptography (see [Menezes et al. 1997] and our Section 8.2).) From parts (1) and (2), $\#S = \frac{1}{2}\#J$, so that half of J's elements are squares, and half are not. From part (2), an element of S has exactly 4 square roots. Show that exactly one of these square roots is itself in S.

(4) For a Blum integer $n = pq$, show that the squaring function $s(x) = x^2 \bmod n$ is a permutation on the set S, and that its inverse function is

$$s^{-1}(y) = y^{((p-1)(q-1)+4)/8} \bmod n.$$

2.23. Using Theorem 2.3.7 prove the two equalities in relations (2.12).

2.24. Here we prove the celebrated quadratic reciprocity relation (2.11) for two distinct odd primes p, q. Starting with Definition 2.3.6, show that G is multiplicative; that is, if $\gcd(m, n) = 1$, then

$$G(m; n)G(n; m) = G(1; mn).$$

(Hint: $mj^2/n + nk^2/m$ is similar—in a specific sense—to $(mj + nk)^2/(mn)$.) Infer from this and Theorem 2.3.7 the relation (now for primes p, q)

$$\left(\frac{p}{q}\right)\left(\frac{q}{p}\right) = (-1)^{(p-1)(q-1)/4}.$$

These are examples *par excellence* of the potential power of exponential sums; in fact, this approach is one of the more efficient ways to arrive at reciprocity. Extend the result to obtain the formula of Theorem 2.3.4 for $\left(\frac{2}{p}\right)$. Can this approach be extended to the more general reciprocity statement (i.e., for coprime m, n) in Theorem 2.3.4? Incidentally, Gauss sums for nonprime arguments m, n can be evaluated in closed form, using the techniques of Exercise 1.63 or the methods summarized in references such as [Graham and Kolesnik 1991].

2.25. This exercise is designed for honing one's skills in manipulating Gauss sums. The task is to count, among quadratic residues modulo a prime p, the exact number of arithmetic progressions of given length. The formal count of length-3 progressions is taken to be

$$A(p) = \#\left\{(r, s, t)\ :\ \left(\tfrac{r}{p}\right) = \left(\tfrac{s}{p}\right) = \left(\tfrac{t}{p}\right) = 1;\ r \neq s;\ s - r \equiv t - s \pmod p\right\}.$$

Note we are taking $0 \le r, s, t \le p-1$, we are ignoring trivial progressions (r, r, r), and that 0 is not a quadratic residue. So the prime $p = 11$, for which the quadratic residues are $\{1, 3, 4, 5, 9\}$, enjoys a total of $A(11) = 10$ arithmetic progressions of length three. (One of these is 4, 9, 3; i.e., we allow wraparound (mod 11); and also, descenders such as 5, 4, 3 are allowed.)

First, prove that

$$A(p) = -\frac{p-1}{2} + \frac{1}{p}\sum_{k=0}^{p-1}\sum_{r,s,t} e^{2\pi i k(r-2s+t)/p},$$

where each of r, s, t runs through the quadratic residues. Then, use relations (2.12) to prove that

$$A(p) = \frac{p-1}{8}\left(p - 6 - 2\left(\frac{2}{p}\right) - \left(\frac{-1}{p}\right)\right).$$

Finally, derive for the exact progression count the attractive expression

$$A(p) = (p-1)\left\lfloor \frac{p-2}{8} \right\rfloor.$$

An interesting extension to this exercise is to analyze progressions of longer length. Another direction: How many progressions of a given length would be expected to exist amongst a *random* half of all residues $\{1,\ 2,\ 3,\ \ldots,\ p-1\}$ (see Exercise 2.38)?

2.26. Prove that square-root Algorithms 2.3.8 and 2.3.9 work.

2.27. Prove that the following algorithm (certainly reminiscent of the text Algorithm 2.3.9) works for square roots (mod p), for p an odd prime. Let x be the quadratic residue for which we desire a square root. Define a particular Lucas sequence (V_k) by $V_0 = 2, V_1 = h$, and for $k > 1$

$$V_k = hV_{k-1} - xV_{k-2},$$

where h is such that $\left(\frac{h^2-4x}{p}\right) = -1$. Then compute a square root of x as

$$y = \frac{1}{2}V_{(p+1)/2} \pmod{p}.$$

Note that the Lucas numbers can be computed via a binary Lucas chain; see Algorithm 3.5.7.

2.28. Implement Algorithm 2.3.8 or 2.3.9 or some other variant to solve each of

$$x^2 \equiv 3615 \pmod{2^{16}+1},$$

$$x^2 \equiv 552512556430486016984082237 \pmod{2^{89}-1}.$$

2.29. Show how to enhance Algorithm 2.3.8 by avoiding some of the powerings called for, perhaps by a precomputation.

2.30. Prove that a primitive root of an odd prime p is a quadratic nonresidue.

2.31. Prove that Algorithm 2.3.12 (alternatively 2.3.13) works. As intimated in the text, the proof is not entirely easy. It may help to first prove a special-case algorithm, namely for finding representations $p = a^2 + b^2$ when $p \equiv 1 \pmod 4$. Such a representation always exists and is unique.

2.32. Since we have algorithms that extract square roots modulo primes, give an algorithm for extracting square roots (mod n), where $n = pq$ is the product of two explicitly given primes. (The Chinese remainder theorem (CRT) will be useful here.) How can one extract square roots of a prime power $n = p^k$? How can one extract square roots modulo n if the complete prime factorization of n is known?

Note that in ignorance of the factorization of n, square root extraction is extremely hard—essentially equivalent to factoring itself; see Exercise 6.5.

2.33. Prove that for odd prime p, the number of roots of $ax^2 + bx + c \equiv 0 \pmod p$, where $a \not\equiv 0 \pmod p$, is given by $1 + \left(\frac{D}{p}\right)$, where $D = b^2 - 4ac$ is the discriminant. For the case $1 + \left(\frac{D}{p}\right) > 0$, give an algorithm for calculation of all the roots.

2.34. Find a prime p such that the least primitive root of p exceeds the number of binary bits in p. Find an example of such a prime p that is also a Mersenne prime (i.e., some $p = M_q = 2^q - 1$ whose least primitive root exceeds q). These findings show that the least primitive root can exceed $\lg p$. For more exploration along these lines see Exercise 2.35.

2.5 Research problems

2.35. Implement a primitive root-finding algorithm, and study the statistical occurrence of least primitive roots.

The study of least primitive roots is highly interesting. It is known on the GRH that 2 is a primitive root of infinitely many primes, in fact for a positive proportion $\alpha = \prod(1 - 1/p(p-1)) \approx 0.3739558$, the product running over all primes (see Exercise 1.87). Again on the GRH, a positive proportion whose least primitive root is not 2, has 3 as a primitive root and so on; see [Hooley 1976]. It is conjectured that the least primitive root for prime p is $O((\ln p)(\ln\ln p))$; see [Bach 1997a]. It is known, on the GRH, that the least primitive root for prime p is $O\left(\ln^6 p\right)$; see [Shoup 1992]. It is known unconditionally that the least primitive root for prime p is $O(p^{1/4+\epsilon})$ for every $\epsilon > 0$, and for infinitely many primes p it exceeds $c \ln p \ln\ln\ln p$ for some positive constant c, the latter a result of S. Graham and C. Ringrosee. The study of the least primitive root is not unlike the study of the least quadratic nonresidue—in this regard see Exercise 2.38.

2.36. Investigate the fast-CRT ideas of [Bernstein 1997b], who describes means for multiprecision multiplication via the CRT.

2.37. Investigate the use of CRT in the seemingly remote domains of integer convolution, or fast Fourier transforms, or public-key cryptography. A good reference is [Ding et al. 1996].

2.38. Here we explore what might be called "statistical" features of the Legendre symbol. For odd prime p, denote by $N(a,b)$ the number of residues whose *successive* quadratic characters are (a,b); that is, we wish to count those integers $x \in [1, p-2]$ such that

$$\left(\left(\frac{x}{p}\right), \left(\frac{x+1}{p}\right)\right) = (a,b),$$

with each of a, b attaining possible values ± 1. Prove that

$$4N(a,b) = \sum_{x=1}^{p-2} \left(1 + a\left(\frac{x}{p}\right)\right)\left(1 + b\left(\frac{x+1}{p}\right)\right)$$

and therefore that

$$N(a,b) = \frac{1}{4}\left(p - 2 - b - ab - a\left(\frac{-1}{p}\right)\right).$$

Establish the corollary that the number of pairs of consecutive quadratic residues is $(p-5)/4, (p-3)/4$, respectively, as $p \equiv 1, 3 \pmod 4$. Using the formula for $N(a,b)$, prove that for every prime p the congruence

$$x^2 + y^2 \equiv -1 \pmod{p}$$

is solvable.

One satisfying aspect of the $N(a,b)$ formula is the statistical notion that sure enough, if the Legendre symbol is thought of as generated by a "random coin flip," there ought to be about $p/4$ occurrences of a given pair $(\pm 1, \pm 1)$.

All of this makes sense: The Legendre symbol is in some sense random. But in another sense, it is not quite so random. Let us estimate a sum:

$$s_{A,B} = \sum_{A \le x < B} \left(\frac{x}{p}\right),$$

which can be thought of, in some heuristic sense we suppose, as a random walk with $N = B - A$ steps. On the basis of remarks following Theorem 2.3.7, show that

$$|s_{A,B}| \le \frac{1}{\sqrt{p}} \sum_{b=0}^{p-1} \left| \frac{\sin(\pi N b/p)}{\sin(\pi b/p)} \right| \le \frac{1}{\sqrt{p}} \sum_{b=0}^{p-1} \frac{1}{|\sin(\pi b/p)|}.$$

Finally, arrive at the Pólya–Vinogradov inequality:

$$|s_{A,B}| < \sqrt{p} \ln p.$$

Actually, the inequality is often expressed more generally, where instead of the Legendre symbol as character, any nonprincipal character applies. This attractive inequality says that indeed, the "statistical fluctuation" of the quadratic residue/nonresidue count, starting from any initial $x = A$, is always bounded by a "variance factor" $\sqrt{p}$ (times a log term). One can prove more than this; for example, using an inequality of [Cochrane 1987] one can obtain

$$|s_{A,B}| < \frac{4}{\pi^2} \sqrt{p} \ln p + 0.41\sqrt{p} + 0.61,$$

and it is known that on the GRH, $s_{A,B} = O\left(\sqrt{p} \ln \ln p\right)$ [Davenport 1980]. In any case, we deduce that out of any N consecutive integers, $N/2 + O(p^{1/2} \ln p)$ are quadratic residues (mod p). We also conclude that the least quadratic nonresidue (mod p) is bounded above by, at worst, $\sqrt{p} \ln p$. Further results on this interesting inequality are discussed in [Hildebrand 1988a, 1988b].

The Pólya–Vinogradov inequality thus restricted to quadratic characters tells us that *not* just any coin-flip sequence can be a Legendre-symbol sequence. The inequality says that we cannot, for large p say, have a Legendre-symbol sequence such as $(1, 1, 1, \ldots, -1 - 1 - 1)$ (i.e., first half are 1's second half -1's). We cannot even build up more than an $O\left(\sqrt{p} \ln p\right)$ excess of one symbol over the other. But in a truly random coin-flip game, any pattern of 1's and -1's is allowed; and even if you constrain such a game to have equal numbers of 1's and -1's as does the Legendre-symbol game, there are still vast numbers of possible coin-flip sequences that cannot be symbol sequences. In some sense, however, the Pólya–Vinogradov inequality puts the Legendre symbol sequence smack in the middle of the distribution of possible sequences: It is what we might expect for a *random* sequence of coin flips. Incidentally, in view of the coin-flip analogy, what would be the expected value of the least quadratic nonresidue (mod p)? In this regard see Exercise 2.35. For a different kind of constraint on presumably random quadratic residues, see the remarks at the end of Exercise 2.25.

2.39. Here is a fascinating line of research: Using the age-old and glorious theory of the arithmetic–geometric mean (AGM), investigate the notion of what we might call a "discrete arithmetic–geometric mean (DAGM)." It was a *tour de force* of analysis, due to Gauss, Legendre, Jacobi, to conceive of the analytic AGM, which is the asymptotic fixed point of the elegant iteration

$$(a, b) \mapsto \left(\frac{a+b}{2}, \sqrt{ab} \right),$$

that is, one replaces the pair (a, b) of real numbers with the new pair of arithmetic and geometric means, respectively. The classical AGM, then, is the real number c to which the two numbers converge; sure enough, $(c, c) \mapsto (c, c)$ so the process tends to stabilize for appropriate initial choices of a and b. This scheme is connected with the theory of elliptic integrals, the calculation of π to (literally) billions of decimal places, and so on [Borwein and Borwein 1987].

But consider doing this procedure not on real numbers but on residues modulo a prime $p \equiv 3 \pmod 4$, in which case an $x \pmod p$ that has a square root always has a so-called principal root (and so an unambiguous choice of square root can be taken; see Exercise 2.21). Work out a theory of the DAGM modulo p. Perhaps you would want to cast $\sqrt{ab}$ as a principal root if said root exists, but something like a different principal root, say $\sqrt{gab}$, for some fixed nonresidue g when ab is a nonresidue. Interesting theoretical issues are these: Does the DAGM have an interesting cycle structure? Is there any relation between your DAGM and the classical, analytic AGM? If there *were* any fortuitous connection between the discrete and analytic means, one might have a new way to evaluate with high efficiency certain finite hypergeometric series, as appear in Exercise 7.26.

Chapter 3

RECOGNIZING PRIMES AND COMPOSITES

Given a large number, how might one quickly tell whether it is prime or composite? In this chapter we begin to answer this fundamental question.

3.1 Trial division

3.1.1 Divisibility tests

A divisibility test is a simple procedure to be applied to the decimal digits of a number n so as to determine whether n is divisible by a particular small number. For example, if the last digit of n is even, so is n. (In fact, nonmathematicians sometimes take this criterion as the *definition* of being even, rather than being divisible by two.) Similarly, if the last digit is 0 or 5, then n is a multiple of 5.

The simple nature of the divisibility tests for 2 and 5 are, of course, due to 2 and 5 being factors of the base 10 of our numeration system. Digital divisibility tests for other divisors get more complicated. Probably the next most well-known test is divisibility by 3 or 9: The sum of the digits of n is congruent to n (mod 9), so by adding up digits themselves and dividing by 3 or 9 respectively reveals divisibility by 3 or 9 for the original n. This follows from the fact that 10 is one more than 9; if we happened to write numbers in base 12, for example, then a number would be congruent (mod 11) to the sum of its base-12 "digits."

In general, divisibility tests based on digits get more and more complicated as the multiplicative order of the base modulo the test divisor grows. For example, the order of 10 (mod 11) is 2, so there is a simple divisibility test for 11: The alternating sum of the digits of n is congruent to n (mod 11). For 7, the order of 10 is 6, and there is no such neat and tidy divisibility test, though there are messy ones.

From a computational point of view, there is little difference between a special divisibility test for the prime p and dividing by p to get the quotient and the remainder. And with dividing there are no special formulae or rules peculiar to the trial divisor p. So when working on a computer, or even for extensive hand calculations, trial division by various primes p is simpler and just as efficient as using various divisibility tests.

3.1.2 Trial division

Trial division is the method of sequentially trying test divisors into a number n so as to partially or completely factor n. We start with the first prime, the number 2, and keep dividing n by 2 until it does not go, and then we try the next prime, 3, on the remaining unfactored portion, and so on. If we reach a trial divisor that is greater than the square root of the unfactored portion, we may stop, since the unfactored portion is prime.

Here is an example. We are given the number $n = 7399$. We trial divide by 2, 3, and 5 and find that they are not factors. The next choice is 7. It is a factor; the quotient is 1057. We next try 7 again, and find that again it goes, the quotient being 151. We try 7 one more time, but it is not a factor of 151. The next trial is 11, and it is not a factor. The next trial is 13, but this exceeds the square root of 151, so we find that 151 is prime. The prime factorization of 7399 is $7^2 \cdot 151$.

It is not necessary that the trial divisors all be primes, for if a composite trial divisor d is attempted, where all the prime factors of d have previously been factored out of n, then it will simply be the case that d is not a factor when it is tried. So though we waste a little time, we are not led astray in finding the prime factorization.

Let us consider the example $n = 492$. We trial divide by 2 and find that it is a divisor, the quotient being 246. We divide by 2 again and find that the quotient is 123. We divide by 2 and find that it does not go. We divide by 3, getting the quotient 41. We divide by 3, 4, 5 and 6 and find they do not go. The next trial is 7, which is greater than $\sqrt{41}$, so we have the prime factorization $492 = 2^2 \cdot 3 \cdot 41$.

Now let us consider the neighboring number $n = 491$. We trial divide by 2, 3, and so on up through 22 and find that none are divisors. The next trial is 23, and $23^2 > 491$, so we have shown that 491 is prime.

To speed things up somewhat, one may exploit the fact that after 2, the primes are odd. So 2 and the odd numbers may be used as trial divisors. With $n = 491$, such a procedure would have stopped us from trial dividing by the even numbers from 4 to 22. Here is a short description of trial division by 2 and the odd integers greater than 2.

Algorithm 3.1.1 (Trial division). We are given an integer $n > 1$. This algorithm produces the multiset $\mathcal{F}$ of the primes that divide n. (A "multiset" is a set where elements may be repeated; that is, a set with multiplicities.)

1. [Divide by 2]
    $\mathcal{F} = \{\ \}$; // The empty multiset.
    $N = n$;
    while($2|N$) {
        $N = N/2$;
        $\mathcal{F} = \mathcal{F} \cup \{2\}$;
    }
2. [Main division loop]

```
d = 3;
while(d^2 ≤ N) {
    while(d|N) {
        N = N/d;
        F = F ∪ {d};
    }
    d = d + 2;
}
if(N == 1) return F;
return F ∪ {N};
```

After 3, primes are either 1 or 5 (mod 6), and one may step through the sequence of numbers that are 1 or 5 (mod 6) by alternately adding 2 and 4 to the latest number. This is a special case of a "wheel," which is a finite sequence of addition instructions that may be repeated indefinitely. For example, after 5, all primes may be found in one of 8 residue classes (mod 30), and a wheel that traverses these classes (beginning from 7) is

$$4, 2, 4, 2, 4, 6, 2, 6.$$

Wheels grow more complicated at a rapid rate. For example, to have a wheel that traverses the numbers that are coprime to all the primes below 30, one needs to have a sequence of 1021870080 numbers. And in comparison with the simple $2, 4$ wheel based on just the two primes 2 and 3, we save only little more than 50% of the trial divisions. (Specifically, about 52.6% of all numbers coprime to 2 and 3 have a prime factor less than 30.) It is a bit ridiculous to use such an ungainly wheel. If one is concerned with wasting time because of trial division by composites, it is much easier and more efficient to first prepare a list of the primes that one will be using for the trial division. In the next section we shall see efficient ways to prepare this list.

3.1.3 Practical considerations

It is perfectly reasonable to use trial division as a primality test when n is not too large. Of course, "too large" is a subjective quality; such judgment depends on the speed of the computing equipment and how much time you are willing to allow a computer to run. It also makes a difference whether there is just the occasional number you are interested in, as opposed to the possibility of calling trial division repeatedly as a subroutine in another algorithm. On a modern workstation, and very roughly speaking, numbers that can be proved prime via trial division in one minute do not exceed 13 decimal digits. In one day of current workstation time, perhaps a 19-digit number can be resolved. (Although these sorts of rules of thumb scale, naturally, according to machine performance in any given era.)

Trial division may also be used as an efficient means of obtaining a partial factorization $n = FR$ as discussed above. In fact, for every fixed trial division bound $B \geq 2$, at least one quarter of all numbers have a divisor F that is greater than B and composed solely of primes not exceeding B; see Exercise 3.4.

Trial division is a simple and effective way to recognize smooth numbers, or numbers without large prime factors, see Definition 1.4.7.

It is sometimes useful to have a "smoothness test," where for some parameter B, one wishes to know whether a given number n is B-smooth, that is, n has no prime factor exceeding B. Trial division up to B not only tells us whether n is B-smooth, it also provides us with the prime factorization of the largest B-smooth divisor of n.

The emphasis in this chapter is on recognizing primes and composites, and not on factoring. So we leave a further discussion of smoothness tests to a later time.

3.1.4 Theoretical considerations

Suppose we wish to use trial division to completely factor a number n into primes. What is the worst case running time? This is easy, for the worst case is when n is prime and we must try as potential divisors the numbers up to $\sqrt{n}$. If we are using just primes as trial divisors, the number of divisions is about $2\sqrt{n}/\ln n$. If we use 2 and the odd numbers as trial divisors, the number of divisions is about $\frac{1}{2}\sqrt{n}$. If we use a wheel as discussed above, the constant $\frac{1}{2}$ is replaced by a smaller constant.

So this is the running time for trial division as a primality test. What is its complexity as an algorithm to obtain the complete factorization of n when n is composite? The worst case is still about $\sqrt{n}$, for just consider the numbers that are the double of a prime. We can also ask for the average case complexity for factoring composites. Again, it is almost $\sqrt{n}$, since the average is dominated by those composites that have a very large prime factor. But such numbers are rare. It may be interesting to throw out the 50% worst numbers and compute the average running time for trial division to completely factor the remaining numbers. This turns out to be n^c, where $c = 1/(2\sqrt{e}) \approx 0.30327$; see Exercise 3.5.

As we shall see later in this chapter and in the next chapter, the problem of recognizing primes is much easier than the general case of factorization. In particular, we have much better ways than trial division to recognize primes. Thus, if one uses trial division as a factorization method, one should augment it with a faster primality test whenever a new unfactored portion of n is discovered, so that the last bit of trial division may be skipped when the last part turns out to be prime. So augmenting trial division, the time to completely factor a composite n essentially is the square root of the second largest prime factor of n.

Again the average is dominated by a sparse set of numbers, in this case those numbers that are the product of two primes of the same order of magnitude; the average being about $\sqrt{n}$. But now throwing out the 50% worst numbers gives a smaller estimate for the average of the remaining numbers. It is n^c, where $c \approx 0.23044$.

3.2 Sieving

Sieving can be a highly efficient means of determining primality and factoring when one is interested in the results for every number in a large, regularly spaced set of integers. On average, the number of arithmetic operations spent per number in the set can be very small, essentially bounded.

3.2.1 Sieving to recognize primes

Most readers are likely to be familiar with the sieve of Eratosthenes. In its most common form it is a device for finding the primes up to some number N. Start with an array of $N-1$ "ones," corresponding to the numbers from 2 to N. The first one corresponds to "2," so the ones in locations 4, 6, 8, and so on, are all changed to zeros. The next one is in the position "3," and we read this as an instruction to change any ones in locations 6, 9, 12, and so on, into zeros. (Entries that are already zeros in these locations are left unchanged.) We continue in this fashion. If the next entry one corresponds to "p," we change to zero any entry one at locations $2p$, $3p$, $4p$, and so on. However, if p is so large that $p^2 > N$, we may stop this process. This exit point can be readily detected by noticing that when we attempt to sieve by p there are no changes of ones to zeros to be made. At this point the one entries in the list correspond to the primes not exceeding N, while the zero entries correspond to the composites.

In passing through the list $2p$, $3p$, $4p$, and so on, one starts from the initial number p and sequentially adds p until we arrive at a number exceeding N. Thus the arithmetic operations in the sieve are all additions. The number of steps in the sieve of Eratosthenes is proportional to $\sum_{p\le N} N/p$, where p runs over primes. But

$$\sum_{p\le N}\frac{N}{p} = N\ln\ln N + O(N); \tag{3.1}$$

see Theorem 427 in [Hardy and Wright 1979]. Thus, the number of steps needed per number up to N is proportional to $\ln\ln N$. It should be noted that $\ln\ln N$, though it does go to infinity, does so *very* slowly. For example, $\ln\ln N < 10$ for all $N \le 10^{9565}$.

The biggest computer limitation on sieves is the enormous space they can consume. Sometimes it is necessary to segment the array from 2 to N. However, if the length of a segment drops below $\sqrt{N}$, the efficiency of the sieve of Eratosthenes begins to deteriorate. The time it takes to sieve a segment of length M with the primes up to $\sqrt{N}$ is proportional to

$$M\ln\ln N + \pi\left(\sqrt{N}\right) + O(M),$$

where $\pi(x)$ denotes the number of primes up to x. Since $\pi\left(\sqrt{N}\right) \sim 2\sqrt{N}/\ln N$, by the prime number theorem, we see that this term can be much larger than the "main term" $M\ln\ln N$ when M is small. In fact, it is an unsolved problem to come up with a method of finding all the primes in the interval $\left[N, N + N^{1/4}\right]$ that is appreciably faster than individually examining each number. This problem is specified in Exercise 3.46.

3.2.2 Eratosthenes pseudocode

We now give practical pseudocode for implementing the ordinary Eratosthenes sieve to find primes in an interval.

Algorithm 3.2.1 (Practical Eratosthenes sieve). This algorithm finds all primes in an interval (L, R) by establishing Boolean primality bits for successive runs of B odd numbers. We assume L, R even, with $R > L$, $B \mid R - L$ and $L > P = \lceil\sqrt{R}\rceil$. We also assume the availability of a table of the $\pi(P)$ primes $p_k \le P$.

1. [Initialize the offsets]
   for($k \in [2, \pi(P)]$) $q_k = \left(-\frac{1}{2}(L + 1 + p_k)\right) \bmod p_k$;
2. [Process blocks]
   $T = L$;
   while($T < R$) {
     for($j \in [0, B-1]$) $b_j = 1$;
     for($k \in [2, \pi(P)]$) {
       for($j = q_k$; $j < B$; $j = j + p_k$) $b_j = 0$;
       $q_k = (q_k - B) \bmod p_k$;
     }
     for($j \in [0, B-1]$) {
       if($b_j == 1$) report $T + 2j + 1$; // Output the prime $p = T + 2j + 1$.
     }
     $T = T + 2B$;
   }

Note that this algorithm can be used either to find the primes in (L, R), or just to count said primes precisely, though more sophisticated prime counting methods are covered in Section 3.6. By use of a wheel, see Section 3.1, the basic sieve Algorithm 3.2.1 may be somewhat enhanced (see Exercise 3.6).

3.2.3 Sieving to construct a factor table

By a very small change, the sieve of Eratosthenes can be enhanced so that it not only identifies the primes up to N, but also gives the least prime factor of each composite up to N. This is done as follows. Instead of changing "one" to "zero" when the prime p hits a location, you change any ones to p, where entries that have already been changed into smaller primes are left unchanged.

The time for this sieve is the same as for the basic sieve of Eratosthenes, though more space is required.

A factor table can be used to get the complete prime factorization of numbers in it. For example, by the entry 12033 one would see 3, meaning that 3 is the least prime factor of 12033. Dividing 3 into 12033, we get 4011, and this number's entry is also 3. Dividing by 3 again, we get 1337, whose entry is 7. Dividing by 7, we get 191, whose entry is 1. Thus 191 is prime and we have the prime factorization

$$12033 = 3^2 \cdot 7 \cdot 191.$$

Factor tables predate, by far, the computer era. Extensive hand-computed factor tables were indispensable to researchers doing numerical work in number theory for many decades prior to the advent of electronic calculating engines.

3.2.4 Sieving to construct complete factorizations

Again, at the cost of somewhat more space, but very little more time, one may adapt the sieve of Eratosthenes so that next to entry m is the complete prime factorization of m. One does this by appending the prime p to lists at locations p, $2p$, $3p, \ldots,$ $p\lfloor N/p \rfloor$. One also needs to sieve with the powers p^a of primes $p \le \sqrt{N}$, where the power p^a does not exceed N. At each multiple of p^a another copy of the prime p is appended. To avoid sieving with the primes in the interval $\left(\sqrt{N}, N\right]$, one can divide to complete the factorization.

For example, say $N = 20000$; let us follow what happens to the entry $m = 12033$. Sieving by 3, we change the 1 at location 12033 to 3. Sieving by 9, we change the 3 at location 12033 to $3, 3$. Sieving by 7, we change the entry to $3, 3, 7$. At the end of sieving (which includes sieving with all primes up to 139 and higher powers of these primes up to 20000), we return to each location in the sieve and multiply the list there. At the location 12033, we multiply $3 \cdot 3 \cdot 7$, getting 63. Dividing 63 into 12033, the quotient is 191, which is also put on the list. So the final list for 12033 is $3, 3, 7, 191$, giving the complete prime factorization of 12033.

3.2.5 Sieving to recognize smooth numbers

Using the sieve of Eratosthenes to get complete factorizations may be simplified and turned into a device to recognize all of the B-smooth numbers (see Definition 1.4.7) in $[2, N]$. We suppose that $2 \le B \le \sqrt{N}$. Perform the factorization sieve as in the above subsection, but with two simplifications: (1) Do not sieve with any p^a where p exceeds B, and (2) if the product of the list at a location is not equal to that location number, then do not bother dividing to get the quotient. The B-smooth numbers are precisely those at locations that are equal to the product of the primes in the list at that location.

To simplify slightly, we might multiply a running product at each location by p whenever p^a hits there. There is no need to keep the lists around if we are interested only in picking out the B-smooth numbers. At the end of the sieve, those locations whose location numbers are equal to the entry in the location are the B-smooth numbers.

For example, say $B = 10$ and $N = 20000$. The entry corresponding to 12033 starts as 1, and gets changed sequentially to 3, to 9, and finally to

63. Thus 12033 is not 10-smooth. However, the entry at 12000 gets changed sequentially to 2, 4, 8, 16, 32, 96, 480, 2400, and finally 12000. Thus 12000 is 10-smooth.

One important way of speeding this sieve is to do the arithmetic at each location in the sieve with logarithms. Doing exact arithmetic with logarithms involves infinite precision, but there is no need to be exact. For example, say we use the closest integer to the base-2 logarithm. For 12000 this is 14. We also use the approximations $\lg 2 \approx 1$ (this one being exact), $\lg 3 \approx 2, \lg 5 \approx 2, \lg 7 \approx 3$. The entry now at location 12000 gets changed sequentially to 1, 2, 3, 4, 5, 7, 9, 11, 13. This is close enough to the target 14 for us to recognize that 12000 is smooth. In general, if we are searching for B-smooth numbers, then an error smaller than $\lg B$ is of no consequence.

One should see the great advantage of working with approximate logarithms, as above. First, the numbers one deals with are very small. Second, the arithmetic necessary is addition, an operation that is much faster to perform on most computers than multiplication or division. Also note that the logarithm function moves very slowly for large arguments, so that all nearby locations in the sieve have essentially the same target. For example, above we had 14 the target for 12000. This same number is used as the target for all locations between $2^{13.5}$ and $2^{14.5}$, namely, all integers between 11586 and 23170.

We shall find later an important application for this kind of sieve in factorization algorithms. And, as discussed in Section 6.4, sieving for smooth numbers is also crucial in some discrete logarithm algorithms. In these settings we are not so concerned with doing a *perfect* job sieving, but rather just recognizing most B-smooth numbers without falsely reporting too many numbers that are not B-smooth. This is a liberating thought that allows further speed-ups in sieving. The time spent sieving with a prime p in the sieve is proportional to the product of the length of the sieve and $1/p$. In particular, small primes are the most time-consuming. But their logarithms contribute very little to the sum, and so one might agree to forgo sieving with these small primes, allowing a little more error in the sieve. In the above example, say we forgo sieving with the moduli 2, 3, 4, 5. We will sieve by higher powers of 2, 3, and 5, as well as all powers of 7, to recognize our 10-smooth numbers. Then the running sum in location 12000 is 3, 4, 5, 9, 11. This total is close enough to 14 to cause a report, and the number 12000 is not overlooked. But we were able to avoid the most costly part of the sieve to find it.

3.2.6 Sieving a polynomial

Suppose $f(x)$ is a polynomial with integer coefficients. Consider the numbers $f(1), f(2), \ldots, f(N)$. Say we wish to find the prime numbers in this list, or to prepare a factor table for the list, or to find the B-smooth numbers in this list. All of these tasks can easily be accomplished with a sieve. In fact, we have already seen a special case of this for the polynomial $f(x) = 2x + 1$, when we noticed that it was essentially sufficient to sieve just the odd numbers up to N when searching for primes.

To sieve the sequence $f(1), f(2), \ldots, f(N)$, we initialize with ones an array corresponding to the numbers $1, 2, \ldots, N$. An important observation is that if p is prime and a satisfies $f(a) \equiv 0 \pmod{p}$, then $f(a+kp) \equiv 0 \pmod{p}$ for every integer k. Of course, there may be as many as $\deg f$ such solutions a, and hence just as many distinct arithmetic progressions $\{a + kp\}$ for each sieving prime p.

Let us illustrate with the polynomial $f(x) = x^2 + 1$. We wish to find the primes of the form x^2+1 for x an integer, $1 \le x \le N$. For each prime $p \le N$, solve the congruence $x^2 + 1 \equiv 0 \pmod{p}$ (see Section 2.3.2). When $p \equiv 1 \pmod{4}$, there are two solutions, when $p \equiv 3 \pmod{4}$, there are no solutions, and when $p = 2$ there is exactly one solution. For each prime p and solution a (that is, $a^2 + 1 \equiv 0 \pmod{p}$ and $1 \le a < p$), we sieve the residue class $a \pmod{p}$ up to N, changing any ones to zeros. However, the very first place a may correspond to the prime p itself, which may easily be detected by the criterion $a < \sqrt{p}$, or by computing a^2+1 and seeing whether it is p. Of course, if $p = a^2 + 1$, we should leave the entry at this location as a 1.

Again, this sieve works because $a^2 + 1 \equiv 0 \pmod{p}$ if and only if $(a+kp)^2 + 1 \equiv 0 \pmod{p}$ for every integer k (and we only need the values of k such that $1 \le a + kp \le N$).

An important difference with the ordinary sieve of Eratosthenes is how far one must go to detect the primes. The general principle is that one must sieve with the primes up to the square root of the largest number in the sequence $f(1), f(2), \ldots, f(N)$. (We assume here that these values are all positive.) In the case of x^2+1 this means that we must sieve with all the primes up to N, rather than stopping at $\sqrt{N}$ as with the ordinary sieve of Eratosthenes.

The time it takes to sieve $x^2 + 1$ for primes for x running up to N is, after finding the solutions to the congruences $x^2 + 1 \equiv 0 \pmod{p}$, about the same as the ordinary sieve of Eratosthenes. This may seem untrue, since there are now many primes for which we must sieve two residue classes, and we must consider all of the primes up to N, not just $\sqrt{N}$. The reply to the first objection is that yes, this is correct, but there are also many primes for which we sieve no residue classes at all. On the second objection, the key here is that the sum of the reciprocals of all of the primes between $\sqrt{N}$ and N is bounded as N grows (it is asymptotically equal to $\ln 2$), so the extra sieving time is only $O(N)$. That is, what we are asserting is that not only do we have

$$\sum_{p \le \sqrt{N}} \frac{1}{p} = \ln\ln N + O(1),$$

we also have

$$\frac{1}{2} + 2 \sum_{p \le N,\ p \equiv 1 \pmod{4}} \frac{1}{p} = \ln\ln N + O(1)$$

(see Chapter 7 in [Davenport 1980]).

It is important to be able to sieve the consecutive values of a polynomial for B-smooth numbers, as in Section 3.2.5. All of the ideas of that section port most naturally to the ideas of this section.

3.2.7 Theoretical considerations

The complexity $N \ln \ln N$ of the sieve of Eratosthenes may be reduced somewhat by several clever arguments. The following algorithm is based on ideas of Mairson and Pritchard (see [Pritchard 1981]). It requires only $O(N/\ln \ln N)$ steps, where each step is either for bookkeeping or an addition with integers at most N. (Note that an explicit pseudocode display for a rudimentary Eratosthenes sieve appears in Section 3.2.2.)

Algorithm 3.2.2 (Fancy Eratosthenes sieve). We are given a number $N \geq 4$. This algorithm finds the set of primes in $[1, N]$. Let p_l denote the l-th prime, let $M_l = p_1 p_2 \cdots p_l$, and let S_l denote the set of numbers in $[1, N]$ that are coprime to M_l. Note that if $p_{m+1} > \sqrt{N}$, then the set of primes in $[1, N]$ is $(S_m \setminus \{1\}) \cup \{p_1, p_2, \ldots, p_m\}$. The algorithm recursively finds $S_k, S_{k+1}, \ldots, S_m$ starting from a moderately sized initial value k and ending with $m = \pi(\sqrt{N})$.

1. [Set up]
   Set k as the integer with $M_k \leq N/\ln N < M_{k+1}$;
   $m = \pi(\sqrt{N})$;
   Use the ordinary sieve of Eratosthenes (Algorithm 3.2.1) to find the primes $p_1, p_2, \ldots, p_k$ and to find the set of integers in $[1, M_k]$ coprime to M_k;
2. [Roll wheel]
   Roll the M_k "wheel" (see Section 3.1) to find the set S_k;
   $S = S_k$;
3. [Find gaps]
   for($l \in [k+1, m]$) {
       $p = p_l =$ the least member of S that exceeds 1;
       // At this point, $S = S_{l-1}$.
       Find the *set* G of gaps between consecutive members of $S \cap [1, N/p]$;
       // Each number that is a gap is counted only once in G.
       Find the set $pG = \{pg : g \in G\}$;
       // Use "repeated doubling method" (see Algorithm 2.1.5).
4. [Find special set]
       Find the set $pS \cap [1, N] = \{ps : ps \leq N, s \in S\}$ as follows: If s and s' are consecutive members of S with $s'p \leq N$ and sp has already been computed, then $ps' = ps + p(s' - s)$;
       // Note that $s' - s$ is a member of G and the number $p(s' - s)$ has already been computed in step [Find gaps]. So ps' may be computed via a subtraction (to find $s' - s$), a look-up (to find $p(s' - s)$) and an addition. (Since the least member of S is 1, the first value of ps is p itself and does not need any computation.)
5. [Find next set S]

$S = S \setminus (pS \cap [1, N])$; // Now $S = S_l$.
$l = l + 1$;
}

6. [Return the set of primes in $[1, N]$]
return $(S \setminus \{1\}) \cup \{p_1, p_2, \ldots, p_m\}$;

Each set S_l consists of the numbers in $[1, N]$ that are coprime to $p_1, p_2, \ldots, p_l$. Thus the first member after 1 is the $(l+1)$-th prime, p_{l+1}. Let us count the number of operations in Algorithm 3.2.2. The number of operations for step [Set up] is $O((N/\ln N)\sum_{i\le k} 1/p_i) = O(N \ln\ln N/\ln N)$. (In fact, the expression $\ln\ln N$ may be replaced with $\ln\ln\ln N$, but it is not necessary for the argument.) For step [Roll wheel], the number of operations is $\#S_k \le \lceil N/M_k\rceil \varphi(M_k) = O(N\varphi(M_k)/M_k)$, where φ is Euler's function. The fraction $\varphi(M_k)/M_k$ is exactly equal to the product of the numbers $1 - 1/p_i$ for $i = 1$ up to k. By the Mertens Theorem 1.4.2, this product is asymptotically $e^{-\gamma}/\ln p_k$. Further, from the prime number theorem, we have that p_k is asymptotically equal to $\ln N$. Thus the number of operations for step [Roll wheel] is $O(N/\ln\ln N)$.

It remains to count the number of operations for steps [Find gaps] and [Find special set]. Suppose $S = S_{l-1}$, and let $G_l = G$. The number of members of $S \cap [1, N/p_l]$ is $O(N/(p_l \ln p_{l-1}))$, by Mertens. Thus, the total number of steps to find all sets G_l is bounded by a constant times

$$\sum_{l=k+1}^{m} N/(p_l \ln p_{l-1}) = O(N/\ln p_k) = O(N/\ln\ln N).$$

The number of additions required to compute gp_l for g in G_l by the repeated doubling method is $O(\ln g) = O(\ln N)$. The sum of all of the values of g in G_l is at most N/p_l, so that the number of members of G_l is $O\left(\sqrt{N/p_l}\right)$. Thus the total number of additions in step [Find gaps] is bounded by a constant times

$$\sum_{l=k+1}^{m} \sqrt{\frac{N}{p_l}} \ln N \le \sum_{i=2}^{\lfloor\sqrt{N}\rfloor} \sqrt{\frac{N}{i}} \ln N \le \int_1^{\sqrt{N}} \sqrt{\frac{N}{t}} \ln N \, dt = 2N^{3/4} \ln N.$$

We cannot be so crude in our estimation of the number of operations in step [Find special set]. Each of these operations is the simple bookkeeping step of deleting a member of a set. Since no entry is deleted more than once, it suffices to count the total number of deletions in all iterations of step [Find special set]. But this total number of deletions is just the size of the set S_k less the number of primes in $\left[\sqrt{N}, N\right]$. This is bounded by $\#S_k$, which we have already estimated to be $O(N/\ln\ln N)$.

3.3 Pseudoprimes

Suppose we have a theorem, *"If n is prime, then S is true about n,"* where "S" is some easily checkable arithmetic statement. If we are presented with a large number n, and we wish to decide whether n is prime or composite, we may very well try out the arithmetic statement S and see whether it actually holds for n. If the statement fails, we have proved the theorem that n is composite. If the statement holds, however, it may be that n is prime, and it also may be that n is composite. So we have the notion of S-pseudoprime, which is a composite integer for which S holds.

One example might be the theorem: *If n is prime, then n is 2 or n is odd.* Certainly this arithmetic property is easily checked for any given input n. However, as one can readily see, this test is virtually useless, since there are many more pseudoprimes around for this test than there are genuine primes. Thus, for the concept of "pseudoprime" to be useful, it will have to be the case that there are, in some appropriate sense, few of them.

3.3.1 Fermat pseudoprimes

The fact that the residue $a^b \pmod n$ may be rapidly computed (see Algorithm 2.1.5) is fundamental to many algorithms in number theory. Not least of these is the exploitation of Fermat's little theorem as a means to distinguish between primes and composites.

Theorem 3.3.1 (Fermat's little theorem). *If n is prime, then for any integer a, we have*

$$a^n \equiv a \pmod n. \tag{3.2}$$

Proofs of Fermat's little theorem may be found in any elementary number theory text. One particularly easy proof uses induction on a and the binomial theorem to expand $(a+1)^n$.

When a is coprime to n we may divide both sides of (3.2) by a to obtain

$$a^{n-1} \equiv 1 \pmod n. \tag{3.3}$$

Thus, (3.3) holds whenever n is prime and n does not divide a.

We say that a composite number n is a (Fermat) pseudoprime if (3.2) holds. For example, $n = 91$ is a pseudoprime base 3, since 91 is composite and $3^{91} \equiv 3 \pmod{91}$. Similarly, 341 is a pseudoprime base 2. The base $a = 1$ is uninteresting, since every composite number is a pseudoprime base 1. We suppose now that $a \ge 2$.

Theorem 3.3.2. *For each fixed integer $a \ge 2$, the number of Fermat pseudoprimes base a that are less than or equal to x is $o(\pi(x))$ as $x \to \infty$. That is, Fermat pseudoprimes are rare compared with primes.*

For pseudoprimes defined via the congruence (3.3), this theorem was first proved in [Erdős 1950]. For the possibly larger class of pseudoprimes defined via (3.2), the theorem was first proved in [Li 1997].

Theorem 3.3.2 tells us that using the Fermat congruence to distinguish between primes and composites is potentially very useful. However, this was known as a practical matter long before the Erdős proof.

Note that odd numbers n satisfy (3.3) for $a = n-1$, so that the congruence does not say very much about n in this case. If (3.3) holds for a pair n, a, where $1 < a < n-1$, we say that n is a probable prime base a. Thus, if n is a prime, then it is a probable prime base a for every integer a with $1 < a < n - 1$. Theorem 3.3.2 asserts that for a fixed choice of a, most probable primes base a are actually primes. We thus have a simple test to distinguish between members of a set that contains a sparse set of composite numbers and all of the primes exceeding $a+1$, and members of the set of the remaining composite numbers exceeding $a + 1$.

Algorithm 3.3.3 (Probable prime test). We are given an integer $n > 3$ and an integer a with $2 \le a \le n - 2$. This algorithm returns either "n is a probable prime base a" or "n is composite."

1. [Compute power residue]
   $b = a^{n-1} \bmod n$; // Use Algorithm 2.1.5.
2. [Return decision]
   if($b == 1$) return "n is a probable prime base a";
   return "n is composite";

We have seen that with respect to a fixed base a, pseudoprimes (that is, probable primes that are composite) are sparsely distributed. However, paucity notwithstanding, there are infinitely many.

Theorem 3.3.4. *For each integer $a \ge 2$ there are infinitely many Fermat pseudoprimes base a.*

Proof. We shall show that if p is any odd prime not dividing $a^2 - 1$, then $n = \left(a^{2p} - 1\right) / \left(a^2 - 1\right)$ is a pseudoprime base a. For example, if $a = 2$ and $p = 5$, then this formula gives $n = 341$. First note that

$$n = \frac{a^p - 1}{a - 1} \cdot \frac{a^p + 1}{a + 1},$$

so that n is composite. Using (3.2) for the prime p we get upon squaring both sides that $a^{2p} \equiv a^2 \pmod{p}$. So p divides $a^{2p} - a^2$. Since p does not divide $a^2 - 1$, by hypothesis, and since $n - 1 = \left(a^{2p} - a^2\right) / \left(a^2 - 1\right)$, we conclude that p divides $n-1$. We can conclude a second fact about $n-1$ as well: Using the identity

$$n - 1 \equiv a^{2p-2} + a^{2p-4} + \cdots + a^2,$$

we see that $n - 1$ is the sum of an even number of terms of the same parity, so $n - 1$ must be even. So far, we have learned that both 2 and p are divisors of $n - 1$, so that $2p$ must likewise be a divisor. Then $a^{2p} - 1$ is a divisor of $a^{n-1} - 1$. But $a^{2p} - 1$ is a multiple of n, so that (3.3) holds, as does (3.2). □

3.3.2 Carmichael numbers

In search of a simple and quick method of distinguishing prime numbers from composite numbers, we might consider combining Fermat tests for various bases a. For example, though 341 is a pseudoprime base 2, it is not a pseudoprime base 3. And 91 is a base-3, but not a base-2 pseudoprime. Perhaps there are no composites that are simultaneously pseudoprimes base 2 and 3, or if such composites exist, perhaps there is some finite set of bases such that there are no pseudoprimes to all the bases in the set. It would be nice if this were true, since then it would be a simple computational matter to test for primes.

However, the number $561 = 3 \cdot 11 \cdot 17$ is not only a Fermat pseudoprime to both bases 2 and 3, it is a pseudoprime to every base a. It may be a shock that such numbers exist, but indeed they do. They were first discovered by R. Carmichael in 1910, and it is after him that we name them.

Definition 3.3.5. A composite integer n for which $a^n \equiv a \pmod{n}$ for every integer a is a Carmichael number.

It is easy to recognize a Carmichael number from its prime factorization.

Theorem 3.3.6 (Korselt criterion). *An integer n is a Carmichael number if and only if n is positive, composite, square-free, and for each prime p dividing n we have $p - 1$ dividing $n - 1$.*

Remark. A. Korselt stated this criterion for Carmichael numbers in 1899, eleven years before Carmichael came up with the first example. Perhaps Korselt felt sure that no examples could possibly exist, and developed the criterion as a first step toward proving this.

Proof. First, suppose n is a Carmichael number. Then n is composite. Let p be a prime factor of n. From $p^n \equiv p \pmod{n}$, we see that p^2 does not divide n. Thus, n is square-free. Let a be a primitive root modulo p. Since $a^n \equiv a \pmod{n}$, we have $a^n \equiv a \pmod{p}$, from which we see that $a^{n-1} \equiv 1 \pmod{p}$. But $a \pmod{p}$ has order $p - 1$, so that $p - 1$ divides $n - 1$.

Now, conversely, assume that n is composite, square-free, and for each prime p dividing n, we have $p-1$ dividing $n-1$. We are to show that $a^n \equiv a \pmod{n}$ for every integer a. Since n is square-free, it suffices to show that $a^n \equiv a \pmod{p}$ for every integer a and for each prime p dividing n. So suppose that $p|n$ and a is an integer. If a is not divisible by p, we have $a^{p-1} \equiv 1 \pmod{p}$ (by (3.3)), and since $p - 1$ divides $n - 1$, we have $a^{n-1} \equiv 1 \pmod{p}$. Thus, $a^n \equiv a \pmod{p}$. But this congruence clearly holds when a is divisible by p, so it holds for all a. This completes the proof of the theorem. □

Are there infinitely many Carmichael numbers? Again, unfortunately for primality testing, the answer is yes. This was shown in [Alford et al. 1994a]. P. Erdős had given a heuristic argument in 1956 that not only are there infinitely many Carmichael numbers, but they are not as rare as one might expect. That is, if $C(x)$ denotes the number of Carmichael numbers up to the bound x, then Erdős conjectured that for each $\varepsilon > 0$, there is a number $x_0(\varepsilon)$ such that $C(x) > x^{1-\varepsilon}$ for all $x \geq x_0(\varepsilon)$. The proof of Alford, Granville, and Pomerance starts from the Erdős heuristic and adds some new ingredients.

Theorem 3.3.7. (Alford, Granville, Pomerance). *There are infinitely many Carmichael numbers. In particular, for x sufficiently large, the number $C(x)$ of Carmichael numbers not exceeding x satisfies $C(x) > x^{2/7}$.*

The proof is beyond the scope of this book; it may be found in [Alford et al. 1994a].

The "sufficiently large" in Theorem 3.3.7 has not been calculated, but probably it is the 96th Carmichael number, 8719309. From calculations of [Pinch 1993] it seems likely that $C(x) > x^{1/3}$ for all $x \geq 10^{15}$. Already at 10^{15}, there are 105212 Carmichael numbers. Though Erdős has conjectured that $C(x) > x^{1-\varepsilon}$ for $x \geq x_0(\varepsilon)$, we know no numerical value of x with $C(x) > x^{1/2}$.

Is there a "Carmichael number theorem," which like the prime number theorem would give an asymptotic formula for $C(x)$? So far there is not even a conjecture for what this formula may be. However, there is a somewhat weaker conjecture.

Conjecture 3.3.1 (Erdős, Pomerance). *The number $C(x)$ of Carmichael numbers not exceeding x satisfies*

$$C(x) = x^{1-(1+o(1)) \ln\ln\ln x / \ln\ln x}$$

as $x \to \infty$.

An identical formula is conjectured for $P_2(x)$, the number of base-2 pseudoprimes up to x. It has been proved, see [Pomerance 1981], that both

$$C(x) < x^{1-\ln\ln\ln x/\ln\ln x},$$
$$P_2(x) < x^{1-\ln\ln\ln x/(2\ln\ln x)},$$

for all sufficiently large values of x.

3.4 Probable primes and witnesses

The concept of Fermat pseudoprime, developed in the previous section, is a good one, since it is easy to check and for each base $a > 1$ there are few pseudoprimes compared with primes (Theorem 3.3.2). However, there are composites, the Carmichael numbers, for which (3.2) is useless as a means of recognizing them as composite. As we have seen, there are infinitely many Carmichael numbers. There are also infinitely many Carmichael numbers that have no small prime factor (see [Alford et al. 1994b]), so that for these numbers, even the slightly stronger test (3.3) is computationally poor.

We would ideally like an easy test for which there are no pseudoprimes. Failing this, we would like a family of tests, such that each composite is not a pseudoprime for a fixed, positive fraction of the tests in the family. The Fermat family does not meet this goal, since there are infinitely many Carmichael numbers. However, a slightly different version of Fermat's little theorem (Theorem 3.3.1) does meet this goal.

Theorem 3.4.1. *Suppose that n is an odd prime and $n-1=2^s t$, where t is odd. If a is not divisible by n then*

$$\begin{cases} \text{either } a^t \equiv 1 \pmod{n} \\ \text{or } a^{2^i t} \equiv -1 \pmod{n} \text{ for some } i \text{ with } 0 \le i \le s-1. \end{cases} \tag{3.4}$$

The proof of Theorem 3.4.1 uses only Fermat's little theorem in the form (3.3) and the fact that for n an odd prime, the only solutions to $x^2 \equiv 1 \pmod{n}$ in $\mathbf{Z}_n$ are $x \equiv \pm 1 \pmod{n}$. We leave the details to the reader.

In analogy to probable primes, we can now define a strong probable prime base a. This is an odd integer $n > 3$ for which (3.4) holds for a, where $1 < a < n-1$. Since every strong probable prime base a is automatically a probable prime base a, and since every prime greater than $a+1$ is a strong probable prime base a, the only difference between the two concepts is that possibly fewer composites pass the strong probable prime test.

Algorithm 3.4.2 (Strong probable prime test). We are given an odd number $n > 3$, represented as $n = 1 + 2^s t$, with t odd. We are also given an integer a with $1 < a < n-1$. This algorithm returns either "n is a strong probable prime base a" or "n is composite."

1. [Odd part of $n-1$]
    $b = a^t \bmod n$; // Use Algorithm 2.1.5.
    if($b == 1$ or $b == n-1$) return "n is a strong probable prime base a";
2. [Power of 2 in $n-1$]
    for($j \in [1, s-1]$) { // j is a dummy counter.
        $b = b^2 \bmod n$;
        if($b == n-1$) return "n is a strong probable prime base a";
    }
    return "n is composite";

This test was first suggested by [Artjuhov 1966/67], and a decade later, J. Selfridge rediscovered the test and popularized it.

We now consider the possibility of showing that an odd number n is composite by showing that (3.4) fails for a particular number a. For example, we saw in the previous section that 341 is pseudoprime base 2. But (3.4) does *not* hold for $n = 341$ and $a = 2$. Indeed, we have $340 = 2^2 \cdot 85$, $2^{85} \equiv 32 \pmod{341}$, and $2^{170} \equiv 1 \pmod{341}$. In fact, we see that 32 is a nontrivial square root of 1 (mod 341).

Now consider the pair $n = 91$ and $a = 10$. We have $90 = 2^1 \cdot 45$ and $10^{45} \equiv -1 \pmod{91}$. So (3.4) holds.

Definition 3.4.3. We say that n is a strong pseudoprime base a if n is an odd composite, $n-1 = 2^s t$, with t odd, and (3.4) holds.

Thus, 341 is not a strong pseudoprime base 2, while 91 is a strong pseudoprime base 10. J. Selfridge proposed using Theorem 3.4.1 as a pseudoprime test in the early 1970s, and it was he who coined the term "strong pseudoprime." It is clear that if n is a strong pseudoprime base a, then n is a pseudoprime base a. The example with $n = 341$ and $a = 2$ shows that the converse is false.

For an odd composite integer n we shall let

$$\mathcal{S}(n) = \{a \pmod n : n \text{ is a strong pseudoprime base } a\}, \qquad (3.5)$$

and let $S(n) = \#\mathcal{S}(n)$. The following theorem was proved independently in [Monier 1980] and [Rabin 1980].

Theorem 3.4.4. *For each odd composite integer $n > 9$ we have $S(n) \le \frac{1}{4}\varphi(n)$.*

Recall that $\varphi(n)$ is Euler's function evaluated at n. It is the number of integers in $[1, n]$ coprime to n, that is, the order of the group $\mathbf{Z}_n^*$. If we know the prime factorization of n, it is easy to compute $\varphi(n)$: We have $\varphi(n) = n\prod_{p|n}(1 - 1/p)$, where p runs over the prime factors of n.

Before we prove Theorem 3.4.4, we first indicate why it is a significant result. If we have an odd number n and we wish to determine whether it is prime or composite, we might try verifying (3.4) for some number a with $1 < a < n - 1$. If (3.4) fails, then we have proved that n is composite. Such a number a might be said to be a witness for the compositeness of n. In fact, we make a formal definition.

Definition 3.4.5. If n is an odd composite number and a is an integer in $[1, n-1]$ for which (3.4) fails, we say that a is a witness for n. Thus, for an odd composite number n, a witness is a base for which n is not a strong pseudoprime.

A witness for n is thus the key to a short proof that n is composite.

Theorem 3.4.4 implies that at least 3/4 of all integers in $[1, n-1]$ are witnesses for n, when n is an odd composite number. Since one can perform a strong pseudoprime test very rapidly, it is easy to decide whether a particular number a is a witness for n. All said, it would seem that it is quite an easy task to produce witnesses for odd composite numbers. Indeed, it is, if one uses a probabilistic algorithm. The following is often referred to as "the Miller–Rabin test," though as one can readily see, it is Algorithm 3.4.2 done with a random choice of the base a. (The original test of [Miller 1976] was somewhat more complicated and was a deterministic, ERH-based test. It was [Rabin 1976, 1980] who suggested a probabilistic algorithm as below.)

Algorithm 3.4.6 (Random compositeness test). We are given an odd number $n > 3$. This probabilistic algorithm attempts to find a witness for n and thus prove that n is composite. If a is a witness, (a, YES) is returned; otherwise, (a, NO) is returned.

1. [Choose possible witness]

Choose random integer $a \in [2, n-2]$;
Via Algorithm 3.4.2 decide whether n is a strong probable prime base a;

2. [Declaration]
if(n is a strong probable prime base a) return $(a$, NO$)$;
return $(a$, YES$)$;

One can see from Theorem 3.4.4 that if $n > 9$ is an odd composite, then the probability that Algorithm 3.4.6 fails to produce a witness for n is $< 1/4$. No one is stopping us from using Algorithm 3.4.6 repeatedly. The probability that we fail to find a witness for an odd composite number n with k (independent) iterations of Algorithm 3.4.6 is $< 1/4^k$. So clearly we can make this probability vanishingly small by choosing k large.

Algorithm 3.4.6 is a very effective method for recognizing composite numbers. But what does it do if we try it on an odd prime? Of course it will fail to produce a witness, since Theorem 3.4.1 asserts that primes have no witnesses.

Suppose n is a large odd number and we don't know whether n is prime or composite. Say we try 20 iterations of Algorithm 3.4.6 and fail each time to produce a witness. What should be concluded? Actually, nothing at all can be concluded concerning whether n is prime or composite. Of course, it is reasonable to strongly *conjecture* that n is prime. The probability that 20 iterations of Algorithm 3.4.6 fail to produce a witness for a given odd composite is less than 4^{-20}, which is less than one chance in a trillion. So yes, n is most likely prime. But it has not been proved prime and in fact might not be.

The reader should consult Chapter 4 for strategies on proving prime those numbers we strongly suspect to be prime. However, for practical applications, one may be perfectly happy to use a number that is almost certainly prime, but has not actually been proved to be prime. It is with this mindset that people refer to Algorithm 3.4.6 as a "primality test." It is perhaps more accurate to refer to a number produced by such a test as an "industrial-grade prime," to use a phrase of H. Cohen.

The following algorithm may be used for the generation of random numbers that are likely to be prime.

Algorithm 3.4.7 ("Industrial-grade prime" generation). We are given an integer $k \geq 3$ and an integer $T \geq 1$. This probabilistic algorithm produces a random k-bit number (that is, a number in the interval $[2^{k-1}, 2^k)$) that has not been recognized as composite by T iterations of Algorithm 3.4.6.

1. [Choose candidate]
Choose a random odd integer n in the interval $(2^{k-1}, 2^k)$;

2. [Perform strong probable prime tests]
for($1 \leq i \leq T$) { // i is a dummy counter.
  Via Algorithm 3.4.6 attempt to find a witness for n;
  if(a witness is found for n) goto [Choose candidate];
}

```
    return n;                    // n is an "industrial-grade prime."
```

An interesting question is this: What is the probability that a number produced by Algorithm 3.4.7 is composite? Let this probability be denoted by $P(k,T)$. One might think that Theorem 3.4.4 immediately speaks to this question, and that we have $P(k,T) \le 4^{-T}$. However, the reasoning is fallacious. Suppose $k = 500, T = 1$. We know from the prime number theorem (Theorem 1.1.4) that the probability that a random odd 500-bit number is prime is about 1 chance in 173. Since it is evidently more likely that one will witness an event with probability 1/4 occurring before an event with probability 1/173, it may seem that there are much better than even odds that Algorithm 3.4.7 will produce composites. In fact, though, Theorem 3.4.4 is a worst-case estimate, and for most odd composite numbers the fraction of witnesses is much larger than 3/4. It is shown in [Burthe 1996] that indeed we do have $P(k,T) \le 4^{-T}$.

If k is large, one gets good results even with $T = 1$ in Algorithm 3.4.7. It is shown in [Damgård et al. 1993] that $P(k,1) < k^2 4^{2-\sqrt{k}}$. For specific large values of k the paper has even better results, for example, $P(500,1) < 4^{-28}$. Thus, if a randomly chosen odd 500-bit number passes just one iteration of a random strong probable prime test, the number is composite with vanishingly small probability, and may be safely accepted as a "prime" in all but the most sensitive practical applications.

Before proving Theorem 3.4.4 we first establish some lemmas.

Lemma 3.4.8. *Say n is an odd composite with $n-1 = 2^s t$, t odd. Let $\nu(n)$ denote the largest integer such that $2^{\nu(n)}$ divides $p-1$ for each prime p dividing n. If n is a strong pseudoprime base a, where $n - 1 = 2^s t$ with t odd, then $a^{2^{\nu(n)-1}t} \equiv \pm 1 \pmod{n}$.*

Proof. If $a^t \equiv 1 \pmod{n}$, it is clear that the conclusion of the lemma holds. Suppose we have $a^{2^i t} \equiv -1 \pmod{n}$ and let p be a prime factor of n. Then $a^{2^i t} \equiv -1 \pmod{p}$. If k is the order of $a \pmod{p}$ (that is, k is the least positive integer with $a^k \equiv 1 \pmod{p}$), then k divides $2^{i+1}t$, but k does not divide $2^i t$. Thus the exact power of 2 in the prime factorization of k must be 2^{i+1}. But also k divides $p-1$, so that $2^{i+1} | p-1$. Since this holds for each prime p dividing n, we have $i+1 \le \nu(n)$. Thus, $a^{2^{\nu(n)-1}t} \equiv 1 \pmod{n}$ or $-1 \pmod{n}$ depending on whether $i+1 < \nu(n)$ or $i+1 = \nu(n)$. □

For the next lemma, let

$$\overline{\mathcal{S}}(n) = \left\{ a \pmod{n} : a^{2^{\nu(n)-1}t} \equiv \pm 1 \pmod{n} \right\}, \quad \overline{S}(n) = \#\overline{\mathcal{S}}(n). \tag{3.6}$$

Lemma 3.4.9. *Recall the notation in Lemma 3.4.8 and (3.6). Let $\omega(n)$ be the number of different prime factors of n. We have*

$$\overline{S}(n) = 2 \cdot 2^{(\nu(n)-1)\omega(n)} \prod_{p|n} \gcd(t, p-1).$$

Proof. Let $m = 2^{\nu(n)-1}t$. Suppose that the prime factorization of n is $p_1^{j_1}p_2^{j_2}\cdots p_k^{j_k}$, where $k = \omega(n)$. We have that $a^m \equiv 1 \pmod n$ if and only if $a^m \equiv 1 \pmod{p_i^{j_i}}$ for $i = 1, 2, \ldots, k$. For an odd prime p and positive integer j, the group $\mathbf{Z}_{p^j}^*$ of reduced residues modulo p^j is cyclic of order $p^{j-1}(p-1)$; that is, there is a primitive root modulo p^j. (This theorem is mentioned in Section 1.4.3 and can be found in most books on elementary number theory. Compare, too, to Theorem 2.2.5.) Thus, the number of solutions $a \pmod{p_i^{j_i}}$ to $a^m \equiv 1 \pmod{p_i^{j_i}}$ is

$$\gcd(m, p_i^{j_i-1}(p_i-1)) = \gcd(m, p_i - 1) = 2^{\nu(n)-1} \cdot \gcd(t, p_i - 1).$$

(Note that the first equality follows from the fact that m divides $n-1$, so is not divisible by p_i.) We conclude, via the Chinese remainder theorem, that the number of solutions $a \pmod n$ to $a^m \equiv 1 \pmod n$ is

$$\prod_{i=1}^{k}\left(2^{\nu(n)-1} \cdot \gcd(t, p_i - 1)\right) = 2^{(\nu(n)-1)\omega(n)} \prod_{p|n} \gcd(t, p-1).$$

To complete the proof we must show that there are exactly as many solutions to the congruence $a^m \equiv -1 \pmod n$. Note that $a^m \equiv -1 \pmod{p_i^{j_i}}$ if and only if $a^{2m} \equiv 1 \pmod{p_i^{j_i}}$ and $a^m \not\equiv 1 \pmod{p_i^{j_i}}$. Since $2^{\nu(n)}$ divides $p_i - 1$ it follows as above that the number of solutions to $a^m \equiv -1 \pmod{p_i^{j_i}}$ is

$$2^{\nu(n)} \cdot \gcd(t, p_i - 1) - 2^{\nu(n)-1} \cdot \gcd(t, p_i - 1) = 2^{\nu(n)-1} \cdot \gcd(t, p_i - 1).$$

Thus there are just as many solutions to $a^m \equiv 1 \pmod n$ as there are to $a^m \equiv -1 \pmod n$, and the lemma is proved. □

Proof of Theorem 3.4.4. From Lemma 3.4.8 and (3.6), it will suffice to show that $\overline{S}(n)/\varphi(n) \le 1/4$ whenever n is an odd composite that is greater than 9. From Lemma 3.4.9, we have

$$\frac{\varphi(n)}{\overline{S}(n)} = \frac{1}{2}\prod_{p^a \| n} p^{a-1}\frac{p-1}{2^{\nu(n)-1}\gcd(t, p-1)},$$

where the notation $p^a \| n$ means that p^a is the exact power of the prime p in the prime factorization of n. Each factor $(p-1)/(2^{\nu(n)-1}\gcd(t, p-1))$ is an even integer, so that $\varphi(n)/\overline{S}(n)$ is an integer. In addition, if $\omega(n) \ge 3$, it follows that $\varphi(n)/\overline{S}(n) \ge 4$. If $\omega(n) = 2$ and n is not square-free, the product of the various p^{a-1} is at least 3, so that $\varphi(n)/\overline{S}(n) \ge 6$.

Now suppose $n = pq$, where $p < q$ are primes. If $2^{\nu(n)+1}|q-1$, then $2^{\nu(n)-1}\gcd(t, q-1) \le (q-1)/4$ and $\varphi(n)/\overline{S}(n) \ge 4$. We may suppose then that $2^{\nu(n)}\|q-1$. Note that $n - 1 \equiv p - 1 \pmod{q-1}$, so that $q-1$ does not divide $n-1$. This implies there is an odd prime dividing $q-1$ to a higher power than it divides $n-1$; that is, $2^{\nu(n)-1}\gcd(t, q-1) \le (q-1)/6$. We conclude in this case that $\varphi(n)/\overline{S}(n) \ge 6$.

Finally, suppose that $n = p^a$, where $a \ge 2$. Then $\varphi(n)/\overline{S}(n) = p^{a-1}$, so that $\varphi(n)/\overline{S}(n) \ge 5$, except when $p^a = 9$. □

3.4.1 The least witness for n

We have seen in Theorem 3.4.4 that an odd composite number n has at least $3n/4$ witnesses in the interval $[1, n-1]$. Let $W(n)$ denote the least of the witnesses for n. Then $W(n) \geq 2$. In fact, for almost all odd composites, we have $W(n) = 2$. This is an immediate consequence of Theorem 3.3.2. The following theorem shows that $W(n) \geq 3$ for infinitely many odd composite numbers n.

Theorem 3.4.10. *If p is a prime larger than 5, then $n = (4^p + 1)/5$ is a strong pseudoprime base 2, so that $W(n) \geq 3$.*

Proof. We first show that n is a composite integer. Since $4^p \equiv (-1)^p \equiv -1 \pmod 5$, we see that n is an integer. That n is composite follows from the identity

$$4^p + 1 = (2^p - 2^{(p+1)/2} + 1)(2^p + 2^{(p+1)/2} + 1).$$

Note that $2^{2p} \equiv -1 \pmod n$, so that if m is odd, we have $2^{2pm} \equiv -1 \pmod n$. But $n - 1 = 2^2 t$, where t is odd and a multiple of p, the latter following from Fermat's little theorem (Theorem 3.3.1). Thus, $2^{2t} \equiv -1 \pmod n$, so that n is a strong pseudoprime base 2. □

It is natural to ask whether $W(n)$ can be arbitrarily large. In fact, this question is crucial. If there is a number B that is not too large such that every odd composite number n has $W(n) \leq B$, then the whole subject of testing primality becomes trivial. One would just try each number $a \leq B$ and if (3.4) holds for each such a, then n is prime. Unfortunately, there is no such number B. The following result is shown in [Alford et al. 1994b].

Theorem 3.4.11. *There are infinitely many odd composite numbers n with*

$$W(n) > (\ln n)^{1/(3 \ln\ln\ln n)}.$$

In fact, the number of such composite numbers n up to x is at least

$$x^{1/(35 \ln\ln\ln x)}$$

when x is sufficiently large.

Failing a universal bound B, perhaps there is a slowly growing function of n which is always greater than $W(n)$. In [Bach 1985] the following is proved.

Theorem 3.4.12. *On the ERH, $W(n) < 2\ln^2 n$ for all odd composite numbers n.*

Proof. Let n be an odd composite. Exercise 3.19 says that $W(n) < \ln^2 n$ if n is divisible by the square of a prime, and this result is not conditional on any unproved hypotheses. We thus may assume that n is square-free. Suppose p is a prime divisor of n with $p - 1 = 2^{s'} t'$, t' odd. Then the same considerations that were used in the proof of Lemma 3.4.8 imply that if (3.4) holds, then $(a/p) = -1$ if and only if $a^{2^{s'-1}t} \equiv -1 \pmod{n}$. Since n is odd, composite, and square-free, it must be that n is divisible by two different odd primes, say p_1, p_2. Let $p_i - 1 = 2^{s_i} t_i$, t_i odd, for $i = 1, 2$, with $s_1 \le s_2$. Let $\chi_1(m) = (m/p_1p_2), \chi_2(m) = (m/p_2)$, so that χ_1 is a character to the modulus p_1p_2 and χ_2 is a character to the modulus p_2. First, consider the case $s_1 = s_2$. Under the assumption of the extended Riemann hypothesis, Theorem 1.4.5 says that there is a positive number $m < 2\ln^2(p_1p_2) \le 2\ln^2 n$ with $\chi_1(m) \ne 1$. Then $\chi_1(m) = 0$ or -1. If $\chi_1(m) = 0$, then m is divisible by p_1 or p_2, which implies that m is a witness. Suppose $\chi_1(m) = -1$, so that either $(m/p_1) = 1, (m/p_2) = -1$ or vice versa. Without loss of generality, assume the first holds. Then, as noted above, if (3.4) holds then $m^{2^{s_2-1}t} \equiv -1 \pmod{n}$, which in turn implies that $(m/p_1) = -1$, since $s_1 = s_2$. This contradiction shows that m is a witness for n. Now assume that $s_1 < s_2$. Again, Theorem 1.4.5 implies that there is a natural number $m < 2\ln^2 p_2 < 2\ln^2 n$ with $(m/p_2) = \chi_2(m) \ne 1$. If $(m/p_2) = 0$, then m is divisible by p_2 and is a witness. If $(m/p_2) = -1$, then as above, m is not a witness implies $m^{2^{s_2-1}t} \equiv -1 \pmod{n}$. Then Lemma 3.4.8 implies that $2^{s_2} | p_1 - 1$, so that $s_2 \le s_1$, a contradiction. Thus, m is a witness for n, and the proof is complete. □

We might ask what can be proved unconditionally. It is obvious that $W(n) \le n^{1/2}$, since the least prime factor of an odd composite number n is a witness for n. [Burthe 1997] showed that $W(n) \le n^{c+o(1)}$ as $n \to \infty$ through the odd composites, where $c = 1/(6\sqrt{e})$. Heath-Brown (see [Balasubramanian and Nagaraj 1997]) has recently shown this with $c = 1/10.82$.

We close this section with the Miller primality test. It is based on Theorem 3.4.12 and shows that if the extended Riemann hypothesis holds, then primality can be decided in deterministic polynomial time.

Algorithm 3.4.13 (Miller primality test). We are given an odd number $n > 1$. This algorithm attempts to decide whether n is prime (YES) or composite (NO). If NO is returned, then n is definitely composite. If YES is returned, n is either prime or the extended Riemann hypothesis is false.

1. [Witness bound]
   $W = \min\{\lfloor 2\ln^2 n \rfloor, n-1\}$;
2. [Strong probable prime tests]
   for($2 \le a \le W$) {
      Decide via Algorithm 3.4.2 whether n is a strong probable prime base a;
      if(n is not a strong probable prime base a) return NO;
   }
   return YES;

3.5 Lucas pseudoprimes

We may generalize many of the ideas of the past two sections to incorporate finite fields. Traditionally the concept of Lucas pseudoprimes has been cast in the language of binary recurrent sequences. It is profitable to view this pseudoprime construct using the language of finite fields, not just to be fashionable, but because the ideas then seem less *ad hoc*, and one can generalize easily to higher order fields.

3.5.1 Fibonacci and Lucas pseudoprimes

The sequence $0, 1, 1, 2, 3, 5, \ldots$ of Fibonacci numbers, say u_j is the j-th one starting with $j = 0$, has an interesting rule for the appearance of prime factors.

Theorem 3.5.1. *If n is prime, then*

$$u_{n-\varepsilon_n} \equiv 0 \pmod{n}, \tag{3.7}$$

where $\varepsilon_n = 1$ when $n \equiv \pm 1 \pmod 5$, $\varepsilon_n = -1$ when $n \equiv \pm 2 \pmod 5$, and $\varepsilon_n = 0$ when $n \equiv 0 \pmod 5$.

Remark. The reader should recognize the function ε_n. It is the Legendre symbol $\left(\frac{n}{5}\right)$; see Definition 2.3.2.

Definition 3.5.2. We say that a composite number n is a Fibonacci pseudoprime if (3.7) holds.

For example, the smallest Fibonacci pseudoprime coprime to 10 is 323.

The Fibonacci pseudoprime test is not just a curiosity. As we shall see below, it can be implemented on very large numbers. In fact, it takes only about twice as long to run a Fibonacci pseudoprime test as a conventional pseudoprime test. And for those composites that are $\pm 2 \pmod 5$ it is, when combined with the ordinary base-2 pseudoprime test, *very* effective. In fact, we know no number $n \equiv \pm 2 \pmod 5$ that is simultaneously a base-2 pseudoprime and a Fibonacci pseudoprime; see Exercise 3.41.

In proving Theorem 3.5.1 it turns out that with no extra work we can establish a more general result. The Fibonacci sequence satisfies the recurrence $u_j = u_{j-1} + u_{j-2}$, with recurrence polynomial $x^2 - x - 1$. We shall consider the more general case of binary recurrent sequences with polynomial $f(x) = x^2 - ax + b$, where a, b are integers with $\Delta = a^2 - 4b$ not a square. Let

$$\begin{aligned} U_j = U_j(a,b) &= \frac{x^j - (a-x)^j}{x - (a-x)} \pmod{f(x)}, \\ V_j = V_j(a,b) &= x^j + (a-x)^j \pmod{f(x)}, \end{aligned} \tag{3.8}$$

where the notation means that we take the remainder in $\mathbf{Z}[x]$ upon division by $f(x)$. The sequences (U_j), (V_j) both satisfy the recurrence for the polynomial $x^2 - ax + b$, namely,

$$U_j = aU_{j-1} - bU_{j-2}, \quad V_j = aV_{j-1} - bV_{j-2},$$

and from (3.8) we may read off the initial values

$$U_0 = 0,\ U_1 = 1, \quad V_0 = 2,\ V_1 = a.$$

If it was not already evident from (3.8), it is now clear that (U_j), (V_j) are *integer* sequences.

In analogy to Theorem 3.5.1 we have the following result. In fact, we can read off Theorem 3.5.1 as the special case corresponding to $a = 1$, $b = -1$.

Theorem 3.5.3. *Let a, b, Δ be as above and define the sequences (U_j), (V_j) via (3.8). If p is a prime with $\gcd(p, 2b\Delta) = 1$, then*

$$U_{p-\left(\frac{\Delta}{p}\right)} \equiv 0 \pmod{p}. \tag{3.9}$$

Note that for $\Delta = 5$ and p odd, $\left(\frac{5}{p}\right) = \left(\frac{p}{5}\right)$, so the remark following Theorem 3.5.1 is justified. Since the Jacobi symbol $\left(\frac{\Delta}{n}\right)$ (see Definition 2.3.3) is equal to the Legendre symbol when n is an odd prime, we may turn Theorem 3.5.3 into a pseudoprime test.

Definition 3.5.4. We say that a composite number n with $\gcd(n, 2b\Delta) = 1$ is a Lucas pseudoprime with respect to $x^2 - ax + b$ if $U_{n-\left(\frac{\Delta}{n}\right)} \equiv 0 \pmod{n}$.

Since the sequence (U_j) is constructed by reducing polynomials modulo $x^2 - ax + b$, and since Theorem 3.5.3 and Definition 3.5.4 refer to this sequence reduced modulo n, we are really dealing with objects in the ring $R = \mathbf{Z}_n[x]/(x^2 - ax + b)$. To somewhat demystify this concept, we explicitly list a complete set of coset representatives:

$$\{i + jx : i, j \text{ are integers with } 0 \le i, j \le n - 1\}.$$

We add coset representatives as vectors (mod n), and we multiply them via $x^2 = ax - b$. Thus, we have

$$\begin{aligned}(i_1 + j_1 x) + (i_2 + j_2 x) &= i_3 + j_3 x\\ (i_1 + j_1 x)(i_2 + j_2 x) &= i_4 + j_4 x,\end{aligned}$$

where

$$\begin{aligned} i_3 &= i_1 + i_2 \pmod{n}, & j_3 &= j_1 + j_2 \pmod{n},\\ i_4 &= i_1 i_2 - b j_1 j_2 \pmod{n}, & j_4 &= i_1 j_2 + i_2 j_1 + a j_1 j_2 \pmod{n}.\end{aligned}$$

We now prove Theorem 3.5.3. Suppose p is an odd prime with $\left(\frac{\Delta}{p}\right) = -1$. Then Δ is not a square in $\mathbf{Z}_p$, so that the polynomial $x^2 - ax + b$, which has discriminant Δ, is irreducible over $\mathbf{Z}_p$. Thus, $R = \mathbf{Z}_p[x]/(x^2 - ax + b)$ is isomorphic to the finite field $\mathbf{F}_{p^2}$ with p^2 elements. The subfield $\mathbf{Z}_p$ $(= \mathbf{F}_p)$ is recognized as those coset representatives $i + jx$ with $j = 0$.

In $\mathbf{F}_{p^2}$ the function σ that takes an element to its p-th power (known as the Frobenius automorphism) has the following pleasant properties, which are easily derived from the binomial theorem and Fermat's little theorem (see (3.2)): $\sigma(u + v) = \sigma(u) + \sigma(v)$, $\sigma(uv) = \sigma(u)\sigma(v)$, and $\sigma(u) = u$ if and only if u is in the subfield $\mathbf{Z}_p$.

We have created the field $\mathbf{F}_{p^2}$ so as to provide roots for $x^2 - ax + b$, which were lacking in $\mathbf{Z}_p$. Which coset representatives $i + jx$ are the roots? They are x itself, and $a - x$ $(= a + (p-1)x)$. Since x and $a - x$ are not in $\mathbf{Z}_p$ and σ must permute the roots of $f(x) = x^2 - ax + b$, we have

$$\text{in the case } \left(\tfrac{\Delta}{p}\right) = -1 \ : \quad \begin{cases} x^p \equiv a - x \pmod{(f(x), p)}, \\ (a-x)^p \equiv x \pmod{(f(x), p)}. \end{cases} \tag{3.10}$$

Then $x^{p+1} - (a-x)^{p+1} \equiv x(a-x) - (a-x)x \equiv 0 \pmod{(f(x),p)}$, so that (3.8) implies $U_{p+1} \equiv 0 \pmod{p}$.

The proof of (3.9) in the case where p is a prime with $\left(\frac{\Delta}{p}\right) = 1$ is easier. In this case we have that $x^2 - ax + b$ has two roots in $\mathbf{Z}_p$, so that the ring $R = \mathbf{Z}_p[x]/(x^2 - ax + b)$ is not a finite field. Rather, it is isomorphic to $\mathbf{Z}_p \times \mathbf{Z}_p$, and every element to the p-th power is itself. Thus,

$$\text{in the case } \left(\tfrac{\Delta}{p}\right) = 1 \ : \quad \begin{cases} x^p \equiv x \pmod{(f(x), p)}, \\ (a-x)^p \equiv a - x \pmod{(f(x), p)}. \end{cases} \tag{3.11}$$

Note, too, that our assumption that $\gcd(p, b) = 1$ implies that x and $a - x$ are invertible in R, since $x(a-x) \equiv b \pmod{f(x)}$. Hence $x^{p-1} = (a-x)^{p-1} = 1$ in R. Thus, (3.8) implies $U_{p-1} \equiv 0 \pmod{p}$. This concludes the proof of Theorem 3.5.3.

Because of Exercise 3.26, it is convenient to rule out the polynomial $x^2 - x + 1$ when dealing with Lucas pseudoprimes. A similar problem occurs with $x^2 + x + 1$, and we rule out this polynomial, too. No other polynomials with nonsquare discriminants are ruled out, though. (Only $x^2 \pm x + 1$ are monic, irreducible over the rationals, and have their roots also being roots of 1.)

3.5.2 Grantham's Frobenius test

The key role of the Frobenius automorphism (raising to the p-th power) in the Lucas test has been put in center stage in a new test of J. Grantham. It allows for an arbitrary polynomial in the place of $x^2 - ax + b$, but even in the case of quadratic polynomials, it is stronger than the Lucas test. One of the advantages of Grantham's approach is that it cuts the tie to recurrent sequences. We describe below his test for quadratic polynomials. A little is said about the general test in Section 3.5.5. For more on Frobenius pseudoprimes see [Grantham 2001].

The argument that establishes Theorem 3.5.3 also establishes on the way (3.10) and (3.11). But Theorem 3.5.3 only extracts part of the information from these congruences. The Frobenius test maintains their full strength.

Definition 3.5.5. Let a, b be integers with $\Delta = a^2 - 4b$ not a square. We say that a composite number n with $\gcd(n, 2b\Delta) = 1$ is a Frobenius pseudoprime with respect to $f(x) = x^2 - ax + b$ if

$$x^n \equiv \begin{cases} a - x \pmod{(f(x), n)}, & \text{if } \left(\frac{\Delta}{n}\right) = -1, \\ x \pmod{(f(x), n)}, & \text{if } \left(\frac{\Delta}{n}\right) = 1. \end{cases} \tag{3.12}$$

At first glance it may seem that we are still throwing away half of (3.10), (3.11), but we are not; see Exercise 3.27.

It is easy to give a criterion for a Frobenius pseudoprime with respect to a quadratic polynomial, in terms of the Lucas sequences $(U_m), (V_m)$.

Theorem 3.5.6. *Let a, b be integers with $\Delta = a^2 - 4b$ not a square and let n be a composite number with $\gcd(n, 2b\Delta) = 1$. Then n is a Frobenius pseudoprime with respect to $x^2 - ax + b$ if and only if*

$$U_{n-\left(\frac{\Delta}{n}\right)} \equiv 0 \pmod{n} \text{ and } V_{n-\left(\frac{\Delta}{n}\right)} \equiv \begin{cases} 2b, & \text{when } \left(\frac{\Delta}{n}\right) = -1 \\ 2, & \text{when } \left(\frac{\Delta}{n}\right) = 1. \end{cases}$$

Proof. Let $f(x) = x^2 - ax + b$. We use the identity

$$2x^m \equiv (2x - a)U_m + V_m \pmod{(f(x), n)},$$

which is self-evident from (3.8). Then the congruences in the theorem lead to $x^{n+1} \equiv b \pmod{(f(x), n)}$ in the case $\left(\frac{\Delta}{n}\right) = -1$ and $x^{n-1} \equiv 1 \pmod{(f(x), n)}$ in the case $\left(\frac{\Delta}{n}\right) = 1$. The latter case immediately gives $x^n \equiv x \pmod{(f(x), n)}$, and the former, via $x(a - x) \equiv b \pmod{(f(x), n)}$, leads to $x^n \equiv a - x \pmod{(f(x), n)}$. Thus, n is a Frobenius pseudoprime with respect to $f(x)$.

Now suppose n is a Frobenius pseudoprime with respect to $f(x)$. Exercise 3.27 shows that n is a Lucas pseudoprime with respect to $f(x)$, namely that $U_{n-\left(\frac{\Delta}{n}\right)} \equiv 0 \pmod{n}$. Thus, from the identity above, $2x^{n-\left(\frac{\Delta}{n}\right)} \equiv V_{n-\left(\frac{\Delta}{n}\right)} \pmod{(f(x), n)}$. Suppose $\left(\frac{\Delta}{n}\right) = -1$. Then $x^{n+1} \equiv (a - x)x \equiv b \pmod{(f(x), n)}$, so that $V_{n+1} \equiv 2b \pmod{n}$. Finally, suppose $\left(\frac{\Delta}{n}\right) = 1$. Then since x is invertible modulo $(f(x), n)$, we have $x^{n-1} \equiv 1 \pmod{(f(x), n)}$, which gives $V_{n-1} \equiv 2 \pmod{n}$. □

The first Frobenius pseudoprime with respect to $x^2 - x - 1$ is 5777. We thus see that not every Lucas pseudoprime is a Frobenius pseudoprime, that is, the Frobenius test is more stringent. In fact, the Frobenius pseudoprime test can be very effective. For example, for $x^2 + 5x + 5$ we don't know any examples at all of a Frobenius pseudoprime n with $\left(\frac{5}{n}\right) = -1$, though such numbers are conjectured to exist; see Exercise 3.42.

3.5.3 Implementing the Lucas and quadratic Frobenius tests

It turns out that we can implement the Lucas test in about twice the time of an ordinary pseudoprime test, and we can implement the Frobenius test in about three times the time of an ordinary pseudoprime test. However, if we approach these tests naively, the running time is somewhat more than just claimed. To achieve the factors two and three mentioned, a little cleverness is required.

As before, we let a, b be integers with $\Delta = a^2 - 4b$ not a square, and we define the sequences (U_j), (V_j) as in (3.8). We first remark that it is easy to deal solely with the sequence (V_j). If we have V_m and V_{m+1}, we may immediately recover U_m via the identity

$$U_m = \Delta^{-1}(2V_{m+1} - aV_m). \tag{3.13}$$

We next remark that it is easy to compute V_m for large m from earlier values using the following simple rule: If $0 \le j \le k$, then

$$V_{j+k} = V_j V_k - b^j V_{k-j}. \tag{3.14}$$

Suppose now that $b = 1$. We record the formula (3.14) in the special cases $k = j$ and $k = j + 1$:

$$V_{2j} = V_j^2 - 2, \quad V_{2j+1} = V_j V_{j+1} - a \quad \text{(in the case } b = 1\text{)}. \tag{3.15}$$

Thus, if we have the residues $V_j \pmod n$, $V_{j+1} \pmod n$, then we may compute, via (3.15), either the pair $V_{2j} \pmod n$, $V_{2j+1} \pmod n$ or the pair $V_{2j+1} \pmod n$, $V_{2j+2} \pmod n$, with each choice taking 2 multiplications modulo n and an addition modulo n. Starting from V_0, V_1 we can recursively use (3.15) to arrive at any pair V_m, V_{m+1}. For example, say m is 97. We travel from 0, 1 to 97, 98 as follows:

$$0, 1 \to 1, 2 \to 3, 4 \to 6, 7 \to 12, 13 \to 24, 25 \to 48, 49 \to 97, 98.$$

There are two types of moves, one that sends the pair $a, a + 1$ to $2a, 2a + 1$ and one that sends it to $2a + 1, 2a + 2$. An easy way to find which sequence of moves to make is to start from the target pair $m, m + 1$ and work backwards. Another easy way is to write m in binary and read the binary digits from most significant bit to least significant bit. A zero signifies the first type of move and a one signifies the second. So in binary, 97 is 1100001, and we see above after the initial 0,1 that we have two moves of the second type, followed by four moves of the first type, followed by a move of the second type.

Such a chain is called a binary Lucas chain. For more on this subject, see [Montgomery 1992b] and [Bleichenbacher 1996]. Here is our pseudocode summarizing the above ideas:

Algorithm 3.5.7 (Lucas chain). For a sequence $x_0, x_1, \ldots$ with a rule for computing x_{2j} from x_j and a rule for computing x_{2j+1} from x_j, x_{j+1}, this algorithm computes the pair (x_n, x_{n+1}) for a given positive integer n. We have n in binary as $(n_0, n_1, \ldots, n_{B-1})$ with n_{B-1} being the high-order bit. We write the rules as follows: $x_{2j} = x_j * x_j$ and $x_{2j+1} = x_j \circ x_{j+1}$. At each step in the for() loop in the algorithm we have $u = x_j, v = x_{j+1}$ for some nonnegative integer j.

1. [Initialization]
   $(u, v) = (x_0, x_1)$;

```
2. [Loop]
      for(B > j ≥ 0) {
          if(n_j == 1) (u, v) = (u ∘ v, v * v);
              else (u, v) = (u * u, u ∘ v);
      }
      return (u, v);                         // Returning (x_n, x_{n+1}).
```

Let us see how we might relax the condition $b = 1$, that is, we are back in the general case of $x^2 - ax + b$. If $a = cd, b = d^2$ we can use the identity

$$V_m(cd, d^2) = d^m V(c, 1)$$

to quickly return to the case $b = 1$. More generally, if b is a square, say $b = d^2$ and $\gcd(n, b) = 1$, we have

$$V_m(a, d^2) \equiv d^m V_m(ad^{-1}, 1) \pmod{n},$$

where d^{-1} is a multiplicative inverse of d modulo n. So again we have returned to the case $b = 1$. In the completely general case that b is not necessarily a square, we note that if we run through the V_m sequence at double time, it is as if we were running through a new V_j sequence. In fact,

$$V_{2m}(a, b) = V_m(a^2 - 2b, b^2),$$

and the "b" number for the second sequence is a square! Thus, if $\gcd(n, b) = 1$ and we let A be an integer with $A \equiv b^{-1}V_2(a, b) \equiv a^2b^{-1} - 2 \pmod{n}$, then we have

$$V_{2m}(a, b) \equiv b^m V_m(A, 1) \pmod{n}. \tag{3.16}$$

Similarly, we have

$$U_{2m}(a, b) \equiv ab^{m-1}U_m(A, 1) \pmod{n},$$

so that using (3.13) (with $A, 1$ for a, b, so that "Δ" in (3.13) is $A^2 - 4$), we have

$$U_{2m}(a, b) \equiv (a\Delta)^{-1}b^{m+1}\big(2V_{m+1}(A, 1) - AV_m(A, 1)\big) \pmod{n}. \tag{3.17}$$

We may use the above method of binary Lucas chains to efficiently compute the pair $V_m(A, 1) \pmod{n}, V_{m+1}(A, 1) \pmod{n}$, where n is a number coprime to b and we view A as an integer modulo n. Thus, via (3.16), (3.17), we may find $V_{2m}(a, b), U_{2m}(a, b) \pmod{n}$. And from these, with $2m = n - \left(\frac{\Delta}{n}\right)$, we may see whether n is a Lucas pseudoprime or Frobenius pseudoprime with respect to $x^2 - ax + b$.

We summarize these notions in the following theorem.

Theorem 3.5.8. *Suppose that a, b, Δ, A are as above and that n is a composite number coprime to $2ab\Delta$. Then n is a Lucas pseudoprime with respect to $x^2 - ax + b$ if and only if*

$$AV_{\frac{1}{2}(n-(\frac{\Delta}{n}))}(A, 1) \equiv 2V_{\frac{1}{2}(n-(\frac{\Delta}{n}))+1}(A, 1) \pmod{n}. \tag{3.18}$$

Moreover, n is a Frobenius pseudoprime with respect to $x^2 - ax + b$ if and only if the above holds and also

$$b^{(n-1)/2} V_{\frac{1}{2}\left(n - \left(\frac{\Delta}{n}\right)\right)}(A, 1) \equiv 2 \pmod{n}. \tag{3.19}$$

As we have seen above, for $m = \frac{1}{2}\left(n - \left(\frac{\Delta}{n}\right)\right)$, the pair $V_m(A,1)$, $V_{m+1}(A,1)$ may be computed modulo n using fewer than $2 \lg n$ multiplications mod n and $\lg n$ additions mod n. Half of the multiplications mod n are squarings mod n. A Fermat test also involves $\lg n$ squarings mod n, and up to $\lg n$ additional multiplications mod n, if we use Algorithm 2.1.5 for the binary ladder. We conclude from (3.18) that the time to do a Lucas test is at most twice the time to do a Fermat test. To apply (3.19) we must also compute $b^{(n-1)/2}$ (mod n), so we conclude that the time to do a Frobenius test (for a quadratic polynomial) is at most three times the time to do a Fermat test.

As with the Fermat test and the strong Fermat test, we apply the Lucas test and the Frobenius test to numbers n that are not known to be prime or composite. Following is pseudocode for these tests along the lines of this section.

Algorithm 3.5.9 (Lucas probable prime test).
We are given integers n, a, b, Δ, with $\Delta = a^2 - 4b$, Δ not a square, $n > 1$, $\gcd(n, 2ab\Delta) = 1$. This algorithm returns "n is a Lucas probable prime with parameters a, b" if either n is prime or n is a Lucas pseudoprime with respect to $x^2 - ax + b$. Otherwise, it returns "n is composite."

1. [Auxiliary parameters]
   $A = a^2 b^{-1} - 2 \bmod n$;
   $m = \left(n - \left(\frac{\Delta}{n}\right)\right)/2$;
2. [Binary Lucas chain]
   Using Algorithm 3.5.7 calculate the last two terms of the sequence $(V_0, V_1, \ldots, V_m, V_{m+1})$, with initial values $(V_0, V_1) = (2, A)$ and specific rules $V_{2j} = V_j^2 - 2 \bmod n$ and $V_{2j+1} = V_j V_{j+1} - A \bmod n$;
3. [Declaration]
   if($AV_m \equiv 2V_{m+1} \pmod{n}$) return "$n$ is a Lucas probable prime with parameters a, b";
   return "n is composite";

The algorithm for the Frobenius probable prime test is the same except that step [Declaration] is changed to

3'. [Lucas test]
   if($AV_m \not\equiv 2V_{m+1}$) return "n is composite";

and a new step is added:

4. [Frobenius test]
   $B = b^{(n-1)/2} \bmod n$;
   if($BV_m \equiv 2 \pmod{n}$) return "n is a Frobenius probable prime with parameters a, b";

return "n is composite";

3.5.4 Theoretical considerations and stronger tests

If $x^2 - ax + b$ is irreducible over $\mathbf{Z}$ and is not $x^2 \pm x + 1$, then the Lucas pseudoprimes with respect to $x^2 - ax + b$ are rare compared with the primes (see Exercise 3.26 for why we exclude $x^2 \pm x + 1$). This result is in [Baillie and Wagstaff 1980]. The best result in this direction is in [Gordon and Pomerance 1991]. Since the Frobenius pseudoprimes with respect to $x^2 - ax + b$ are a subset of the Lucas pseudoprimes with respect to this polynomial, they are if anything rarer still.

It has been proved that for each irreducible polynomial $x^2 - ax + b$ there are infinitely many Lucas pseudoprimes, and in fact, infinitely many Frobenius pseudoprimes. This was done in the case of Fibonacci pseudoprimes in [Lehmer 1964], in the general case for Lucas pseudoprimes in [Erdős et al. 1988], and in the case of Frobenius pseudoprimes in [Grantham 2001]. Grantham's proof on the infinitude of Frobenius pseudoprimes works only in the case $\left(\frac{\Delta}{n}\right) = 1$. There are some specific quadratics, for example, the polynomial $x^2 - x - 1$ for the Fibonacci recurrence, for which we know that there are infinitely many Frobenius pseudoprimes with $\left(\frac{\Delta}{n}\right) = -1$ (see [Parberry 1970] and [Rotkiewicz 1973]). Very recently, Rotkiewicz proved that for any $x^2 - ax + b$ with $\Delta = a^2 - 4b$ not a square, there are infinitely many Lucas pseudoprimes n with $\left(\frac{\Delta}{n}\right) = -1$.

In analogy to strong pseudoprimes (see Section 3.4), we may have strong Lucas pseudoprimes and strong Frobenius pseudoprimes. Suppose n is an odd prime not dividing $b\Delta$. In the ring $R = \mathbf{Z}_n[x]/(f(x))$ it is possible (in the case $\left(\frac{\Delta}{n}\right) = 1$) to have $z^2 = 1$ and $z \neq \pm 1$. For example, take $f(x) = x^2 - x - 1$, $n = 11$, $z = 3 + 5x$. However, if $(x(a-x)^{-1})^{2m} = 1$, then a simple calculation (see Exercise 3.30) shows that we must have $(x(a-x)^{-1})^m = \pm 1$. We have from (3.10) and (3.11) that $(x(a-x)^{-1})^{n - \left(\frac{\Delta}{n}\right)} = 1$ in R. Thus, if we write $n - \left(\frac{\Delta}{n}\right) = 2^s t$, where t is odd, then

$$\begin{aligned} &\text{either } \ (x(a-x)^{-1})^t \equiv 1 \pmod{(f(x), n)} \\ &\text{or } \ (x(a-x)^{-1})^{2^i t} \equiv -1 \pmod{(f(x), n)} \ \text{ for some } i,\ 0 \le i \le s-1. \end{aligned}$$

This then implies that

$$\begin{aligned} &\text{either } \ U_t \equiv 0 \pmod{n} \\ &\text{or } \ V_{2^i t} \equiv 0 \pmod{n} \ \text{ for some } i,\ 0 \le i \le s-1. \end{aligned}$$

If this last statement holds for an odd composite number n coprime to $b\Delta$, we say that n is a strong Lucas pseudoprime with respect to $x^2 - ax + b$. It is easy to see that every strong Lucas pseudoprime with respect to $x^2 - ax + b$ is also a Lucas pseudoprime with respect to this polynomial.

In [Grantham 2001] a strong Frobenius pseudoprime test is developed, not only for quadratic polynomials, but for all polynomials. We describe the quadratic case for $\left(\frac{\Delta}{n}\right) = -1$. Say $n^2 - 1 = 2^S T$, where n is an odd prime not dividing $b\Delta$ and where $\left(\frac{\Delta}{n}\right) = -1$. From (3.10) and (3.11), we have $x^{n^2-1} \equiv 1 \pmod{n}$, so that

$$\begin{aligned} &\text{either } x^T \equiv 1 \pmod{n} \\ &\text{or } x^{2^i T} \equiv -1 \pmod{n} \text{ for some } i,\ 0 \le i \le S-1. \end{aligned}$$

If this holds for a Frobenius pseudoprime n with respect to $x^2 - ax + b$, we say that n is a strong Frobenius pseudoprime with respect to $x^2 - ax + b$. (That is, the above congruence does not appear to imply that n is a Frobenius pseudoprime, so this condition is put into the definition of a strong Frobenius pseudoprime.) It is shown in [Grantham 1998] that a strong Frobenius pseudoprime n with respect to $x^2 - ax + b$, with $\left(\frac{\Delta}{n}\right) = -1$, is also a strong Lucas pseudoprime with respect to this polynomial.

As with the ordinary Lucas test, the strong Lucas test may be accomplished in time bounded by the cost of two ordinary pseudoprime tests. It is shown in [Grantham 1998] that the strong Frobenius test may be accomplished in time bounded by the cost of three ordinary pseudoprime tests. The interest in strong Frobenius pseudoprimes comes from the following result from [Grantham 1998]:

Theorem 3.5.10. *Suppose n is a composite number that is not a square and not divisible by any prime up to 50000. Then n is a strong Frobenius pseudoprime with respect to at most 1/7710 of all polynomials $x^2 - ax + b$, where a, b run over the integers in $[1, n]$ with $\left(\frac{a^2-4b}{n}\right) = -1$ and $\left(\frac{b}{n}\right) = 1$.*

This result should be contrasted with the Monier–Rabin theorem (Theorem 3.4.4). If one does three random strong pseudoprime tests, that result implies that a composite number will fail to be recognized as such at most 1/64 of the time. Using Theorem 3.5.10, in about the same time, one has a test that recognizes composites with failure at most 1/7710 of the time.

3.5.5 The general Frobenius test

In the last few sections we have discussed Grantham's Frobenius test for quadratic polynomials. Here we briefly describe how the idea generalizes to arbitrary monic polynomials in $\mathbf{Z}[x]$.

Let $f(x)$ be a monic polynomial in $\mathbf{Z}[x]$ with degree $d \ge 1$. We do not necessarily assume that $f(x)$ is irreducible. Suppose p is an odd prime that does not divide the discriminant, $\operatorname{disc}(f)$, of $f(x)$. (The discriminant of a monic polynomial $f(x)$ of degree d may be computed as $(-1)^{d(d-1)/2}$ times the resultant of $f(x)$ and its derivative. This resultant is the determinant of the $(2d-1)\times(2d-1)$ matrix whose i, j entry is the coefficient of x^{j-i} in $f(x)$ for $i = 1, \dots, d-1$ and is the coefficient of $x^{j-(i-d+1)}$ in $f'(x)$ for $i = d, \dots, 2d-1$, where if the power of x does not actually appear, the matrix entry is 0.) Since $\operatorname{disc}(f) \ne 0$ if and only if $f(x)$ has no repeated irreducible factors of positive degree, the hypothesis that p does not divide $\operatorname{disc}(f)$ automatically implies that f has no repeated factors.

By reducing its coefficients modulo p, we may consider $f(x)$ in $\mathbf{F}_p[x]$. To avoid confusion, we shall denote this polynomial by $\overline{f}(x)$. Consider the polynomials $F_1(x), F_2(x), \ldots, F_d(x)$ in $\mathbf{F}_p[x]$ defined by

$$\begin{aligned} F_1(x) &= \gcd(x^p - x, \overline{f}(x)), \\ F_2(x) &= \gcd(x^{p^2} - x, \overline{f}(x)/F_1(x)), \\ &\vdots \\ F_d(x) &= \gcd(x^{p^d} - x, \overline{f}(x)/(F_1(x)\cdots F_{d-1}(x))). \end{aligned}$$

Then the following assertions hold:

(1) i divides $\deg(F_i(x))$ for $i = 1, \ldots, d$,

(2) $F_i(x)$ divides $F_i(x^p)$ for $i = 1, \ldots, d$,

(3) for

$$S = \sum_{i \text{ even}} \frac{1}{i} \deg(F_i(x)),$$

we have

$$(-1)^S = \left(\frac{\operatorname{disc}(f)}{p}\right).$$

Assertion (1) follows, since $F_i(x)$ is precisely the product of the degree-i irreducible factors of $\overline{f}(x)$, so its degree is a multiple of i. Assertion (2) holds for all polynomials in $\mathbf{F}_p[x]$. Assertion (3) is a little trickier to see. The idea is to consider the Galois group for the polynomial $\overline{f}(x)$ over $\mathbf{F}_p$. The Frobenius automorphism (which sends elements of the splitting field of $\overline{f}(x)$ to their p-th powers) of course permutes the roots of $\overline{f}(x)$ in the splitting field. It acts as a cyclic permutation of the roots of each irreducible factor, and hence the sign of the whole permutation is given by -1 to the number of even-degree irreducible factors. That is, the sign of the Frobenius automorphism is exactly $(-1)^S$. However, it follows from basic Galois theory that the Galois group of a polynomial with distinct roots consists solely of even permutations of the roots if and only if the discriminant of the polynomial is a square. Hence the sign of the Frobenius automorphism is identical to the Legendre symbol $\left(\frac{\operatorname{disc}(f)}{p}\right)$, which then establishes the third assertion.

The idea of Grantham is that the above assertions can actually be numerically checked and done so easily, even if we are not sure that p is prime. If one of the three assertions does not hold, then p is revealed as composite. This, then, is the core of the Frobenius test. One says that n is a Frobenius pseudoprime with respect to the polynomial $f(x)$ if n is composite, yet the test does not reveal this.

For many more details, the reader is referred to [Grantham 1998, 2001].

3.6 Counting primes

The prime number theorem (Theorem 1.1.4) predicts approximately the value of $\pi(x)$, the number of primes p with $p \le x$. It is interesting to compare these predictions with actual values, as we did in Section 1.1.5. The computation of

$$\pi\left(10^{21}\right) = 21127269486018731928$$

was certainly not performed by having a computer actually count each and every prime up to 10^{21}. There are far too many of them. So how then was the task actually accomplished? We give in the next sections two different ways to approach the interesting problem of prime counting, a combinatorial method and an analytic method.

3.6.1 Combinatorial method

We shall study here an elegant combinatorial method due to Lagarias, Miller, and Odlyzko, with roots in the work of Meissel and Lehmer; see [Lagarias et al. 1985], [Deléglise and Rivat 1996]. The method allows the calculation of $\pi(x)$ in bit complexity $O\left(x^{2/3+\epsilon}\right)$, using $O\left(x^{1/3+\epsilon}\right)$ bits of space (memory).

Label the consecutive primes $p_1, p_2, p_3, \ldots$, where $p_1 = 2$, $p_2 = 3$, $p_3 = 5$, etc. Let

$$\phi(x, y) = \#\{1 \le n \le x \,:\, \text{each prime dividing } n \text{ is greater than } y\}.$$

Thus $\phi(x, p_a)$ is the number of integers left unmarked in the sieve of Eratosthenes, applied to the interval $[1, x]$, after sieving with $p_1, p_2, \ldots, p_a$. Since sieving up to $\sqrt{x}$ leaves only the number 1 and the primes in $(\sqrt{x}, x]$, we have

$$\pi(x) - \pi\left(\sqrt{x}\right) + 1 = \phi\left(x, \sqrt{x}\right).$$

One could easily use this idea to compute $\pi(x)$, the time taking $O(x \ln\ln x)$ operations and, if the sieve is segmented, taking $O\left(x^{1/2}\ln x\right)$ space. (We shall begin suppressing $\ln x$ and $\ln\ln x$ factors for simplicity, sweeping them under a rather large rug of $O(x^\epsilon)$. It will be clear that each x^ϵ could be replaced, with a little more work, with a small power of logarithm and/or double logarithm.)

A key thought is that the sieve not only allows us to count the primes, it also identifies them. If it is only the count we are after, then perhaps we can be speedier.

We shall partition the numbers counted by $\phi(x, y)$ by the number of prime factors they have, counted with multiplicity. Let

$$\phi_k(x, y) = \#\{n \le x \,:\, n \text{ has exactly } k \text{ prime factors, each exceeding } y\}.$$

Thus, if $x \ge 1$, $\phi_0(x, y)$ is 1, $\phi_1(x, y)$ is the number of primes in $(y, x]$, $\phi_2(x, y)$ is the number of numbers $pq \le x$ where p, q are primes with $y < p \le q$, and so on. We evidently have

$$\phi(x, y) = \phi_0(x, y) + \phi_1(x, y) + \phi_2(x, y) + \cdots.$$

Further, note that $\phi_k(x,y) = 0$ if $y^k \geq x$. Thus,

$$\phi\left(x, x^{1/3}\right) = 1 + \pi(x) - \pi\left(x^{1/3}\right) + \phi_2\left(x, x^{1/3}\right). \tag{3.20}$$

One then can find $\pi(x)$ if one can compute $\phi\left(x, x^{1/3}\right)$, $\phi_2\left(x, x^{1/3}\right)$ and $\pi\left(x^{1/3}\right)$.

The computation of $\pi\left(x^{1/3}\right)$ can be accomplished, of course, using the Eratosthenes sieve and nothing fancy. The next easiest ingredient in (3.20) is the computation of $\phi_2\left(x, x^{1/3}\right)$, which we now describe. This quantity is found via the identity

$$\phi_2(x, x^{1/3}) = \binom{\pi(x^{1/3})}{2} - \binom{\pi(x^{1/2})}{2} + \sum_{x^{1/3} < p \leq x^{1/2}} \pi(x/p), \tag{3.21}$$

where in the sum the letter p runs over primes. To see why (3.21) holds, we begin by noting that $\phi_2\left(x, x^{1/3}\right)$ is the number of pairs of primes p, q with $x^{1/3} < p \leq q$ and $pq \leq x$. Then $p \leq x^{1/2}$. For each fixed p, the prime q is allowed to run over the interval $[p, x/p]$, and so the number of choices for q is $\pi(x/p) - \pi(p) + 1$. Thus,

$$\begin{aligned}\phi_2(x, x^{1/3}) &= \sum_{x^{1/3} < p \leq x^{1/2}} (\pi(x/p) - \pi(p) + 1) \\ &= \sum_{x^{1/3} < p \leq x^{1/2}} \pi(x/p) - \sum_{x^{1/3} < p \leq x^{1/2}} (\pi(p) - 1).\end{aligned}$$

The last sum is

$$\begin{aligned}\sum_{\pi(x^{1/3}) < j \leq \pi(x^{1/2})} (j-1) &= \sum_{j=1}^{\pi(x^{1/2})} (j-1) - \sum_{j=1}^{\pi(x^{1/3})} (j-1) \\ &= \binom{\pi(x^{1/2})}{2} - \binom{\pi(x^{1/3})}{2},\end{aligned}$$

which proves (3.21).

To use (3.21) to compute $\phi_2\left(x, x^{1/3}\right)$ we shall compute $\pi\left(x^{1/3}\right)$, $\pi\left(x^{1/2}\right)$, and the sum of the $\pi(x/p)$. We have already computed $\pi\left(x^{1/3}\right)$. The computation of $\pi\left(x^{1/2}\right)$ can again be done using the simple Eratosthenes sieve, except that the sieve is segmented into blocks of size about $x^{1/3}$ to preserve the space bound for the algorithm. Note that in the sum of $\pi(x/p)$ in (3.21), each $x/p < x^{2/3}$. Thus a simple sieve of Eratosthenes can likewise compute the sum of $\pi(x/p)$ in total time $O\left(x^{2/3+\epsilon}\right)$. We do this within the space allotment of $O\left(x^{1/3+\epsilon}\right)$ as follows. Let $N \approx x^{1/3}$ be a convenient number for segmenting the sieve, that is, we look at intervals of length N, beginning at $x^{1/2}$. Assuming that we have already computed $\pi(z)$, we use a sieve (with stored primes less than $x^{1/3}$) in the interval $[z, z+N)$ to compute the various

$\pi(x/p)$ for x/p landing in the interval, and we compute $\pi(z+N)$ to be used in computations for the next interval. The various $\pi(x/p)$'s computed are put into a running sum, and not stored individually. To find which p have x/p landing in the interval, we have to apply a second sieve, namely to the interval $(x/(z+N), x/z]$, which lies in $\left(x^{1/3}, x^{1/2}\right]$. The length of this interval is less than N so that space is not an issue, and the sieve may be accomplished using a stored list of primes not exceeding $x^{1/4}$ in time $O\left(x^{1/3+\epsilon}\right)$. When z is large, the intervals $(x/(z+N), x/z]$ become very short, and some time savings may be made (without altering the overall complexity), by sieving an interval of length N in this range, storing the results, and using these for several different intervals in the upper range.

To compute $\pi(x)$ with (3.20) we are left with the computation of $\phi\left(x, x^{1/3}\right)$. At first glance, this would appear to take about x steps, since it counts the number of uncanceled elements in the sieve of Eratosthenes applied to $[1, x]$ with the primes up to $x^{1/3}$. The idea is to reduce the calculation of $\phi\left(x, x^{1/3}\right)$ to that of many smaller problems. We begin with the recurrence

$$\phi(y, p_b) = \phi(y, p_{b-1}) - \phi(y/p_b, p_{b-1}), \tag{3.22}$$

for $b \geq 2$. We leave the simple proof for Exercise 3.33. Since $\phi(y, 2) = \lfloor (y+1)/2 \rfloor$, we can continue to use (3.22) to eventually come down to expressions $\phi(y, 2)$ for various choices of y. For example,

$$\begin{aligned}
\phi(1000, 7) &= \phi(1000, 5) - \phi(142, 5) \\
&= \phi(1000, 3) - \phi(200, 3) - \phi(142, 3) + \phi(28, 3) \\
&= \phi(1000, 2) - \phi(333, 2) - \phi(200, 2) + \phi(66, 2) \\
&\quad - \phi(142, 2) + \phi(47, 2) + \phi(28, 2) - \phi(9, 2) \\
&= 500 - 167 - 100 + 33 - 71 + 24 + 14 - 5 \\
&= 228.
\end{aligned}$$

Using this scheme, we may express any $\phi(x, p_a)$ as a sum of 2^{a-1} terms. In fact, this bottom-line expression is merely the inclusion–exclusion principle applied to the divisors of $p_2 p_3 \cdots p_a$, the product of the first $a-1$ odd primes. We have

$$\phi(x, p_a) = \sum_{n \mid p_2 p_3 \cdots p_a} \mu(n)\phi(x/n, 2) = \sum_{n \mid p_2 p_3 \cdots p_a} \mu(n) \left\lfloor \frac{x/n + 1}{2} \right\rfloor,$$

where μ is the Möbius function see Section 1.4.1.

For $a = \pi(x^{1/3})$, clearly 2^{a-1} terms is too many, and we would have been better off just sieving to x. However, we do not have to consider any n in the sum with $n > x$, since then $\phi(x/n, 2) = 0$. This "truncation rule" reduces the number of terms to $O(x)$, which is starting to be competitive with merely sieving. By fiddling with this idea, we can reduce the O-constant to a fairly small number. Since $2 \cdot 3 \cdot 5 \cdot 7 \cdot 11 = 2310$, by computing a table of values of $\phi(x, 11)$ for $x = 0, 1, \ldots, 2309$, one can quickly compute any $\phi(x, 11)$: It is

$\varphi(2310)\lfloor x/2310 \rfloor + \phi(x \bmod 2310, 11)$, where φ is the Euler totient function. By halting the recurrence (3.22) whenever a b value drops to 11 or a y/p_b value drops below 1, we get

$$\phi(x, p_a) = \sum_{\substack{n \mid p_6 p_7 \cdots p_a \\ n \le x}} \mu(n) \phi(x/n, 11).$$

If $a = \pi\left(x^{1/3}\right)$, the number of terms in this sum is asymptotic to cx with $c = \rho(3)\zeta(2)\prod_{i=1}^{5} p_i/(p_i+1)$, where ρ is the Dickman function (see Section 1.4.5), and ζ is the Riemann zeta function (so that $\zeta(2) = 6/\pi^2$). This expression for c captures the facts that n has no prime factors exceeding $x^{1/3}$, n is square-free, and n has no prime factor strictly below 11. Using $\rho(3) \approx 0.0486$, we get that $c \approx 0.00987$. By reducing a to $\pi\left(x^{1/4}\right)$ (and agreeing to compute $\phi_3\left(x, x^{1/4}\right)$ in addition to $\phi_2\left(x, x^{1/4}\right)$), we reduce the constant c to an expression where $\rho(4) \approx 0.00491$ replaces $\rho(3)$, so that $c \approx 0.000998$. These machinations amount, in essence, to the method of Meissel, as improved by Lehmer, see [Lagarias et al. 1985].

However, our present goal is to reduce the bit complexity to $O\left(x^{2/3+\epsilon}\right)$. We do this by using a different truncation rule. Namely, we stop using the recurrence (3.22) at any point $\phi(y, p_b)$ where either

(1) $p_b = 2$ and $y \ge x^{2/3}$, or

(2) $y < x^{2/3}$.

Here, y corresponds to some number x/n where $n \mid p_2 p_3 \cdots p_a$. The number of type-1 terms clearly does not exceed $x^{1/3}$, since such terms correspond to values $n < x^{1/3}$. To count the number of type-2 terms, note that a "parent" of $\phi(x/n, p_b)$ in the hierarchy is either the term $\phi(x/n, p_{b+1})$ or the term $\phi(x/(n/p_{b+1}), p_{b+1})$. The latter case occurs only when p_{b+1} is the least prime factor of n and $n/p_{b+1} \le x^{1/3}$, and the former case never occurs, since it would already have been subjected to a type-2 truncation. Thus, the number of type-2 terms is at most the number of pairs m, p_b, where $m \le x^{1/3}$ and p_b is smaller than the least prime factor of m. This count is at most $x^{1/3}\pi(x^{1/3})$, so the number of type-2 terms is less than $x^{2/3}$.

For an integer $m > 1$, let

$$P_{\min}(m) = \text{ the least prime factor of } m.$$

We thus have using the above truncation rule that

$$\phi(x, p_a) = \sum_{\substack{m \mid p_2 p_3 \cdots p_a \\ m \le x^{1/3}}} \mu(m) \left\lfloor \frac{x/m + 1}{2} \right\rfloor \tag{3.23}$$

$$- \sum_{\substack{m \mid p_2 p_3 \cdots p_a \\ 1 < m \le x^{1/3}}} \mu(m) \sum_{\substack{p_{b+1} < P_{\min}(m) \\ p_{b+1} m > x^{1/3}}} \phi\left(\frac{x}{m p_{b+1}}, p_b\right).$$

We apply (3.23) with $a = \pi(x^{1/3})$. The first sum in (3.23), corresponding to type-1 terms, is easy to compute. With a sieve, prepare a table $\mathcal{T}$ of the odd square-free numbers $m \le x^{1/3}$, together with their least prime factor (which will be of use in the double sum), and the value $\mu(m)$. (Each sieve location corresponds to an odd number not exceeding $x^{1/3}$ and starts with the number 1. The first time a location gets hit by a prime, we record this prime as the least prime factor of the number corresponding to the sieve location. Every time a prime hits at a location, we multiply the entry at the location by -1. We do this for all primes not exceeding $x^{1/6}$ and then mark remaining entries with the number they correspond to, and change the entry to -1. Finally, we sieve with the squares of primes p^2 for $p \le x^{1/6}$, and any location that gets hit gets its entry changed to 0. At the end, the numbers with nonzero entries are the square-free numbers, the entry is μ of the number, and the prime recorded there is the least prime factor of the number.) The time and space to prepare table $\mathcal{T}$ is $O(x^{1/3+\epsilon})$, and with it we may compute the first sum in (3.23) in time $O(x^{1/3+\epsilon})$.

The heart of the argument is the calculation of the double sum in (3.23). We first describe how to compute this sum using $O\left(x^{2/3+\epsilon}\right)$ space and time, and later show how segmentation can cut down the space to $O(x^{1/3+\epsilon})$. Prepare a table $\mathcal{T}'$ of triples $\mu(m), \lfloor x/(mp_{b+1}) \rfloor,\ b$, where m runs over numbers greater than 1 in the table $\mathcal{T}$ previously computed, and b runs over numbers such that $p_{b+1} < P_{\min}(m)$ and $mp_{b+1} > x^{1/3}$. Note that all of the numbers $\lfloor x/(mp_{b+1}) \rfloor$ are less than $x^{2/3}$. Sieve the interval $\left[1, x^{2/3}\right]$ with the primes not exceeding $x^{1/3}$. At stage b we have sieved with $p_1, p_2, \ldots, p_b$, and thus we can read off $\phi(y, b)$ for any $y \le x^{2/3}$. We are interested in the values $y = \lfloor x/(mp_{b+1}) \rfloor$.

However, just knowing which numbers are coprime to $p_1 p_2 \cdots p_b$ is not the same as knowing how many there are up to y, which requires an additional computation. Doing this for each b would increase the bit complexity to $O\left(x^{1+\epsilon}\right)$. This problem is solved via a binary data structure. For $i = 0, 1, \ldots, \lfloor \lg n \rfloor$, consider the intervals

$$I_{i,j} = \left((j-1)2^i, j2^i\right]$$

for j a positive integer and $I_{i,j} \subset \left[1, x^{2/3}\right]$. The total number of these intervals is $O\left(x^{2/3}\right)$. For each of the intervals $I_{i,j}$, let

$$A(i,j,b) = \#\{n \in I_{i,j} : \gcd(n, p_1 p_2 \ldots p_b) = 1\}.$$

The plan is to compute all of the numbers $A(i,j,b)$ for a fixed b. Once these are computed, we may use the binary representation of $\lfloor x/(mp_{b+1}) \rfloor$ and add up the appropriate choices of $A(i,j,b)$ to compute $\phi(\lfloor x/(mp_{b+1}) \rfloor, p_b)$.

So, we now show how the numbers $A(i,j,b)$ are to be computed from the previous values $A(i,j,b-1)$ (where the initial values $A(i,j,0)$ are set equal to 2^i). Note that in the case $i = 0$, the interval $I_{0,j}$ contains only the integer j, so that $A(0,j,b)$ is 1 if j is coprime to $p_1 p_2 \cdots p_b$, and is 0 otherwise. For integers $l \le x/p_b$, we update the numbers $A(i,j,b)$ corresponding to intervals $I_{i,j}$ containing lp_b. The number of such intervals for a given lp_b is $O(\ln x)$. If $A(0,j,b-1) = 0$, where $j = lp_b$, then no update is necessary in any interval. If $A(0,j,b-1) = 1$, where again $j = lp_b$, we set each relevant $A(i,j,b)$ equal to $A(i,j,b-1)-1$. The total number of updates is $O\left(x^{2/3}(\ln x)/p_b\right)$, so summing for $p_b \le x^{1/3}$, an estimate $O\left(x^{2/3+\epsilon}\right)$ accrues.

The space for the above argument is $O(x^{2/3+\epsilon})$. To reduce it to $O(x^{1/3+\epsilon})$, we let k be the integer with $x^{1/3} \le 2^k < 2x^{1/3}$, and then we segment the interval $\left[1, x^{2/3}\right]$ in blocks of size 2^k, where perhaps the last block is short, or we go a little beyond $x^{2/3}$. The r-th block is $\left((r-1)2^k, r2^k\right]$, namely, it is the interval $I_{r,k}$. When we reach it, we have stored the numbers $\phi\left((r-1)2^k, p_b\right)$ for all $b \le \pi\left(x^{1/3}\right)$ from the prior block. We next use the table $\mathcal{T}$ computed earlier to find the triples $\mu(m), \lfloor x/(mp_{b+1})\rfloor$, b where $\lfloor x/(mp_{b+1})\rfloor$ is in the r-th block. The intervals $I_{i,j}$ fit neatly in the r-th block for $i \le k$, and we do not need to consider larger values of i. Everything proceeds as before, and we compute each relevant $\phi(x/(mp_{b+1}), p_b)$ where $\lfloor x/(mp_{b+1})\rfloor$ is in the r-th block, and we also compute $\phi(r2^k, p_b)$ for each b, so as to use these for the next block. The computed values of $\phi(x/(mp_{b+1}), p_b)$ are not stored, but are multiplied by $\mu(m)$ and added into a running sum that represents the second term on the right of (3.23). The time and space required to do these tasks for all $p_b \le x^{1/3}$ in the r-th block is $O(x^{1/3+\epsilon})$. The values of $\phi\left(r2^k, p_b\right)$ are written over the prior values $\phi((r-1)2^k, p_b)$, so the total space used is $O\left(x^{1/3+\epsilon}\right)$. The total number of blocks does not exceed $x^{1/3}$, so the total time used in this computation is $O\left(x^{2/3+\epsilon}\right)$, as advertised.

There are various ideas for speeding up this algorithm in practice, see [Lagarias et al. 1985] and [Deléglise and Rivat 1996].

3.6.2 Analytic method

Here we describe an analytic method, highly efficient in principle, for counting primes. The idea is that of [Lagarias and Odlyzko 1987], with recent extensions that we shall investigate. The idea is to exploit the fact that the Riemann zeta function embodies in some sense the properties of primes. A certain formal manipulation of the Euler product relation (1.18) goes like so. Start by taking the logarithm

$$\ln \zeta(s) = \ln \prod_{p \in \mathcal{P}} (1-p^{-s})^{-1} = -\sum_{p \in \mathcal{P}} \ln(1-p^{-s}),$$

and then introduce a logarithmic series

$$\ln \zeta(s) = \sum_{p \in \mathcal{P}} \sum_{m=1}^{\infty} \frac{1}{mp^{sm}}, \tag{3.24}$$

where all manipulations are valid (and the double sum can be interchanged if need be) for $\mathrm{Re}(s) > 1$, with the caveat that $\ln \zeta$ is to be interpreted as a continuously changing argument. (By modern convention, one starts with the positive real $\ln\zeta(2)$ and tracks the logarithm as the angle argument of ζ, along a contour that moves vertically to $2 + i\,\mathrm{Im}(s)$ then over to s.)

In order to use relation (3.24) to count primes, we define a function reminiscent of—but not quite the same as—the prime-counting function $\pi(x)$. In particular, we consider a sum over prime *powers* not exceeding x, namely

$$\pi^*(x) = \sum_{p\in\mathcal{P},\ m>0} \frac{\theta(x-p^m)}{m}, \tag{3.25}$$

where $\theta(z)$ is the Heaviside function, equal to $1,\ 1/2,\ 0$, respectively, as its argument z is positive, zero, negative. The introduction of θ means that the sum involves only prime powers p^m not exceeding x, but that whenever the real x actually equals a power p^m, the summand is $1/(2m)$. The next step is to invoke the Perron formula, which says that for nonnegative real x, positive integer n, and a choice of contour $\mathcal{C} = \{s : \mathrm{Re}(s) = \sigma\}$, with fixed $\sigma > 0$ and $t = \mathrm{Im}(s)$ ranging, we have

$$\frac{1}{2\pi i}\int_{\mathcal{C}} \left(\frac{x}{n}\right)^s \frac{ds}{s} = \theta(x-n). \tag{3.26}$$

It follows immediately from these observations that for a given contour (but now with $\sigma > 1$ so as to avoid any $\ln\zeta$ singularity) we have:

$$\pi^*(x) = \frac{1}{2\pi i}\int_{\mathcal{C}} \left(\frac{x}{n}\right)^s \ln\zeta(s)\frac{ds}{s}. \tag{3.27}$$

This last formula provides analytic means for evaluation of $\pi(x)$, because if x is not a prime power, say, we have from relation (3.25) the identity:

$$\pi^*(x) = \pi(x) + \frac{1}{2}\pi\left(x^{1/2}\right) + \frac{1}{3}\pi\left(x^{1/3}\right) + \cdots,$$

which series terminates as soon as the term $\pi\left(x^{1/n}\right)/n$ has $2^n > x$.

It is evident that $\pi(x)$ may be, in principle at least, computed from a contour integral (3.27), and relatively easy side calculations of $\pi\left(x^{1/n}\right)$ starting with $\pi\left(\sqrt{x}\right)$. One could also simply apply the contour integral relation recursively, since the leading term of $\pi^*(x) - \pi(x)$ is $\pi^*\left(x^{1/2}\right)/2$, and so on. This analytic approach thus comes down to numerical integration, yet such integration is the problematic stage. First of all, one has to evaluate ζ with sufficient accuracy. Second, one needs a rigorous bound on the extent to which the integral is to be taken along the contour. Let us address the latter problem first. Say we have in hand a sharp computational scheme for ζ itself, and we take $x = 100, \sigma = 3/2$. Numerical integration reveals that for sample integration limits $T \in \{10, 30, 50, 70, 90\}$, respective values are

$$\begin{aligned}\pi^*(100) &\approx \mathrm{Re}\,\frac{100^{3/2}}{\pi}\int_0^T \frac{100^{it}}{3/2+it}\ln\zeta(3/2+it)\,dt \\ &\approx 30.14,\ 29.72,\ 27.89,\ 29.13,\ 28.3,\end{aligned}$$

which values exhibit poor convergence of the contour integral: The true value of $\pi^*(100)$ can be computed directly, by hand, to be $428/15 \approx 28.533\dots$. Furthermore, on inspection the value as a function of integration limit T is rather chaotic in the way it hovers around the true value, and rigorous error bounds are, as might be expected, nontrivial to achieve (see Exercise 3.37).

The suggestions of [Lagarias and Odlyzko 1987] address, and in principle repair, the above drawbacks of the analytic approach. As for evaluation of ζ itself, the Riemann–Siegel formula is often recommended for maximum speed; in fact, whenever s has a formidably large imaginary part t, said formula has been the exclusive historical workhorse (although there has been some modern work on interesting variants to Riemann–Siegel, as we touch upon at the end of Exercise 1.59). What is more, there is a scheme due to [Odlyzko and Schönhage 1988] for a kind of "parallel" evaluation of $\zeta(s)$ values, along, say, a progression of imaginary ordinates of the argument s. This sort of simultaneous evaluation is just what is needed for numerical integration. For a modern compendium including variants on the Riemann–Siegel formula and other computational approaches, see [Borwein et al. 2000] and references therein. In [Crandall 1998] can be found various fast algorithms for simultaneous evaluation at various argument sets. The essential idea for acceleration of ζ computations is to use FFT, polynomial evaluation, or Newton-method techniques to achieve simultaneous evaluations of $\zeta(s)$ for a given set of s values. In the present book we have provided enough instruction—via Exercise 1.59—for one at least to get started on single evaluations of $\zeta(s+it)$ that require only $O\left(t^{1/2+\epsilon}\right)$ bit operations.

As for the problem of poor convergence of contour integrals, the clever ploy is to invoke a smooth (one might say "adiabatic") turn-off function that renders a (modified) contour integral more convergent. The phenomenon is akin to that of reduced spectral bandwidth for smoother functions in Fourier analysis. The Lagarias–Odlyzko identity of interest is (henceforth we shall assume that x is not a prime power)

$$\pi^*(x) = \frac{1}{2\pi i}\int_C F(s,x)\ln\zeta(s)\frac{ds}{s} + \sum_{p\in\mathcal{P},\ m>0} \frac{\theta(x-p^m)-c(p^m,x)}{m}, \tag{3.28}$$

where c, F form a Mellin-transform pair:

$$c(u,x) = \frac{1}{2\pi i}\int_C F(s,x)u^{-s}\,ds,$$

$$F(s,x) = \int_0^\infty c(u,x)u^{s-1}\,du.$$

To understand the import of this scheme, take the turn-off function $c(u,x)$ to be $\theta(x-u)$. Then $F(s,x) = x^s/s$, the final sum in (3.28) is zero, and we recover the original analytic representation (3.27) for π^*. Now, however, let us contemplate the class of continuous turn-off functions $c(u,x)$ that stay at 1 over the interval $u \in [0, x-y)$, decay smoothly (to zero) over $u \in (x-y, x]$, and vanish for all $u > x$. For optimization of computational efficiency, y will eventually be chosen to be of order $\sqrt{x}$. In fact, we can combine various of the above relations to write

$$\pi(x) = \frac{1}{2\pi i}\int_C F(s,x)\ln\zeta(s)\,\frac{ds}{s} - \sum_{p\in\mathcal{P}\ m>1} \frac{\theta(x-p^m)}{m} + \sum_{p\in\mathcal{P},\ m>0} \frac{\theta(x-p^m) - c(p^m,x)}{m}. \tag{3.29}$$

Indeed, the last summation is rather easy, since it has just $O(\sqrt{x})$ terms. The next-to-last summation, which just records the difference between $\pi(x)$ and $\pi^*(x)$, also has just $O(\sqrt{x})$ terms.

Let us posit a specific smooth decay, i.e., for $u \in (x-y, x]$ we define

$$c(u,x) = 3\frac{(x-u)^2}{y^2} - 2\frac{(x-u)^3}{y^3}.$$

Observe that $c(x-y,x) = 1$ and $c(x,x) = 0$, as required for continuous c functions in the stated class. Mellin transformation of c gives

$$\frac{y^3}{6}F(s,x) = \frac{-2x^{s+3} + (s+3)x^{s+2}y + (x-y)^s(2x^3 + (s-3)x^2y - 2sxy^2 + (s+1)y^3)}{s(s+1)(s+2)(s+3)}. \tag{3.30}$$

This expression, though rather unwieldy, allows us to count primes more efficiently. For one thing, the denominator of the second fraction is $O(t^4)$, which is encouraging. As an example, performing numerical integration as in relation (3.29) with the choices $x = 100, y = 10$, we find for the same trial set of integration limits $T \in \{10, 30, 50, 70, 90\}$ the results

$$\pi(100) \approx 25.3,\ 26.1,\ 25.27,\ 24.9398,\ 24.9942,$$

which are quite satisfactory, since $\pi(100) = 25$. (Note, however, that there is still some chaotic behavior until T be sufficiently large.) It should be pointed out that Lagarias and Odlyzko suggest a much more general, parameterized form for the Mellin pair c, F, and indicate how to optimize the parameters. Their complexity result is that one can either compute $\pi(x)$ with bit operation count $O\left(x^{1/2+\epsilon}\right)$ and storage space of $O\left(x^{1/4+\epsilon}\right)$ bits, or on the notion of limited memory one may replace the powers with $3/5+\epsilon,\ \epsilon$, respectively.

As of this writing, there has been no practical result of the analytic method on a par with the greatest successes of the aforementioned combinatorial methods. However, this impasse apparently comes down to just a matter of calendar time. In fact, [Galway 1998] has reported that values of $\pi(10^n)$ for $n = 13$, and perhaps 14, are attainable for a certain turn-off function c and (only) standard, double-precision floating-point arithmetic for the numerical integration. Perhaps 100-bit or higher precision will be necessary to press the analytic method on toward modern limits, say $x \approx 10^{21}$ or more; the required precision depends on detailed error estimates for the contour integral. The Galway functions are a clever choice of Mellin pair, and work out to be more efficient than the turn-off functions that lead to F of the type (3.30). Take

$$c(u,x) = \frac{1}{2}\operatorname{erfc}\left(\frac{\ln \frac{u}{x}}{2a(x)}\right),$$

where erfc is the standard error function:

$$\operatorname{erfc}(z) = \frac{2}{\sqrt{\pi}} \int_z^\infty e^{-t^2}\, dt$$

and a is chosen later for efficiency. This c function turns off smoothly at $u \sim x$, but at a rate tunable by choice of a. The Mellin companion works out nicely to be

$$F(s) = \frac{x^s}{s} e^{s^2 a(x)^2}. \tag{3.31}$$

For $s = \sigma + it$ the wonderful (for computational purposes) decay in F is $e^{-t^2a^2}$. Now numerical experiments are even more satisfactory. Sure enough, we can use relation (3.29) to yield, for $x = 1000$, decay function $a(x) = (2x)^{-1/2}$, $\sigma = 3/2$, and integration limits $T \in \{20, 40, 60, 80, 100, 120\}$, the successive values

$$\pi(1000) \approx 170.6,\ 169.5,\ 170.1,\ 167.75,\ 167.97,\ 167.998,$$

in excellent agreement with the exact value $\pi(1000) = 168$; and furthermore, during such a run the chaotic manner of convergence is, qualitatively speaking, not so manifest.

3.7 Exercises

3.1. In the spirit of the opening observations to the present chapter, denote by $S_B(n)$ the sum of the base-B digits of n. Interesting phenomena accrue for specific B, such as $B = 7$. Find the smallest prime p such that $S_7(p)$ is itself composite. (The magnitude of this prime might surprise you!) Then, find all possible composite values of $S_7(p)$ for the primes $p < 16000000$ (there are *very* few such values!). Here are two natural questions, the answers to which are unknown to the authors: Given a base B, are there infinitely many primes p with $S_B(p)$ prime? (composite?) Obviously, the answer is "yes" for at least one of these questions!

3.2. Sometimes other fields of thought can feed back into the theory of prime numbers. Let us look at a beautiful gem by [Golomb 1956] that uses clever combinatorics—and even some "visual" highlights—to prove Fermat's little Theorem 3.3.1.

For a given prime p you are to build necklaces having p beads. In any one necklace the beads can be chosen from n possible different colors, but you have the constraint that no necklace can be all one color.

(1) Prove: For necklaces laid out first as linear strings (i.e., not yet circularized) there are $n^p - n$ possible such strings.

(2) Prove: When the necklace strings are all circularized, the number of distinguishable necklaces is $(n^p - n)/p$.

(3) Prove Fermat's little theorem, that $n^p \equiv n \pmod{p}$.

(4) Where have you used that p is prime?

3.3. Prove that if $n > 1$ and $\gcd(a^n - a, n) = 1$ for some integer a, then not only is n composite, it is not a prime power.

3.4. For each number $B \geq 2$, let d_B be the asymptotic density of the integers that have a divisor exceeding B with said divisor composed solely of primes not exceeding B. That is, if $N(x, B)$ denotes the number of positive integers up to x that have such a divisor, then we are defining $d_B = \lim_{x\to\infty} N(x, B)/x$.

(1) Show that

$$d_B = 1 - \prod_{p \leq B}\left(1 - \frac{1}{p}\right) \cdot \sum_{m=1}^{B} \frac{1}{m},$$

where the product is over primes.

(2) Find the smallest value of B with $d_B > d_7$.

(3) Using the Mertens Theorem 1.4.2 show that $\lim_{B\to\infty} d_B = 1 - e^{-\gamma} \approx 0.43854$, where γ is the Euler constant.

(4) It is shown in [Rosser and Schoenfeld 1962] that if $x \geq 285$, then $e^\gamma \ln x \prod_{p\leq x}(1 - 1/p)$ is between $1 - 1/(2\ln^2 x)$ and $1 + 1/(2\ln^2 x)$. Use this to show that $0.25 \leq d_B < e^{-\gamma}$ for all $B \geq 2$.

3.5. Let c be a real number and consider the set of those integers n whose largest prime factor does not exceed n^c. Let c be such that the asymptotic density of this set is $1/2$. Show that $c = 1/(2\sqrt{e})$. A pleasantly interdisciplinary reference is [Knuth and Trabb Pardo 1976].

Now, consider the set of those integers n whose *second*-largest prime factor (if there is one) does not exceed n^c. Let c be such that the asymptotic density of this set is $1/2$. Show that c is the solution to the equation

$$I(c) = \int_c^{1/2} \frac{\ln(1-u) - \ln u}{u}\, du = \frac{1}{2},$$

and solve this numerically for c. An interesting modern approach for the numerics is to show, first, that this integral is given exactly by

$$I(c) = \frac{1}{12}\left(-\pi^2 + 6\ln^2 c + 12\mathrm{Li}_2(c)\right),$$

in which the standard polylogarithm $\mathrm{Li}_2(c) = c/1^2 + c^2/2^2 + c^3/3^2 + \cdots$ appears. Second, using any of the modern packages that know how to evaluate Li_2 to high precision, implement a Newton-method solver, in this way circumventing the need for numerical integration *per se*. You ought to be able to obtain, for example,

$$c \approx 0.2304366013159997457147108570060465575080754\ldots,$$

presumed correct to the implied precision.

Another intriguing direction: Work out a fast algorithm—having a value of c as input—for counting the integers $n \in [1, x]$ whose second-largest prime factor exceeds n^c (when there are less than two prime factors let us simply not count that n). For the high-precision c value given above, there are 548 such $n \in [1, 1000]$, whereas the theory predicts 500. Give the count for some much higher value of x.

3.6. Rewrite the basic Eratosthenes sieve Algorithm 3.2.1 with improvements. For example, reduce memory requirements (and increase speed) by observing that any prime $p > 3$ satisfies $p \pm 1 \pmod 6$; or use a modulus greater than 6 in this fashion.

3.7. Use the Korselt criterion, Theorem 3.3.6, to find by hand or machine some explicit Carmichael numbers.

3.8. Prove that every composite Fermat number $F_n = 2^{2^n} + 1$ is a Fermat pseudoprime base 2. Can a composite Fermat number be a Fermat pseudoprime base 3? (The authors know of no example, nor do they know a proof that this cannot occur.)

3.9. This exercise is an exploration of rough mental estimates pertaining to the statistics attendant on certain pseudoprime calculations. The great computationalist/theorist team of D. Lehmer and spouse E. Lehmer together pioneered in the mid-20th century the notion of primality tests (and a great many other things) via hand-workable calculating machinery. For example, they proved the primality of such numbers as the repunit $(10^{23} - 1)/9$ with a mechanical calculator at home, they once explained, working a little every day over many months. They would trade off doing the dishes vs. working on the primality crunching. Later, of course, the Lehmers were able to handle much larger numbers via electronic computing machinery.

Now, the exercise is, comment on the statistics inherent in D. Lehmer's (1969) answer to a student's question, "Professor Lehmer, have you in all your lifetime researches into primes ever been tripped up by a pseudoprime you had thought was prime (a composite that passed the base-2 Fermat test)?" to which Lehmer's response was as terse as can be: "Just once." So the question is, does "just once" make statistical sense? How dense are the base-2 pseudoprimes in the region of 10^n? Presumably, too, one would not be fooled, say, by those base-2 pseudoprimes that are divisible by 3, so revise the question to those base-2 pseudoprimes not divisible by any "small" prime factors. A reference on this kind of question is [Damgård et al. 1993].

3.10. Note that applying the formula in the proof of Theorem 3.3.4 with $a = 2$, the first legal choice for p is 5, and as noted, the formula in the proof gives $n = 341$, the first pseudoprime base 2. Applying it with $a = 3$, the first legal choice for p is 3, and the formula gives $n = 91$, the first pseudoprime base 3. Show that this pattern breaks down for larger values of a and, in fact, never holds again.

3.11. Show that if n is a Carmichael number, then n is odd and has at least three prime factors.

3.12. Show that a composite number n is a Carmichael number if and only if $a^{n-1} \equiv 1 \pmod{n}$ for all integers a coprime to n.

3.13. [Beeger] Show that if p is a prime, then there are at most finitely many Carmichael numbers with second largest prime factor p.

3.14. For any positive integer n let

$$\mathcal{F}(n) = \left\{a \pmod{n} : a^{n-1} \equiv 1 \pmod{n}\right\}.$$

(1) Show that $\mathcal{F}(n)$ is a subgroup of $\mathbf{Z}_n^*$, the full group of reduced residues modulo n, and that it is a proper subgroup if and only if n is a composite that is not a Carmichael number.

(2) [Monier, Baillie–Wagstaff] Let $F(n) = \#\mathcal{F}(n)$. Show that

$$F(n) = \prod_{p|n} \gcd(p-1, n-1).$$

(3) Let $F_0(n)$ denote the number of residues $a \pmod{n}$ such that $a^n \equiv a \pmod{n}$. Find a formula, as in (2) above, for $F_0(n)$. Show that if $F_0(n) < n$, then $F_0(n) \le \frac{2}{3}n$. Show that if $n \ne 6$ and $F_0(n) < n$, then $F_0(n) \le \frac{3}{5}n$. (It is not known whether there are infinitely many numbers n with $F_0(n) = \frac{3}{5}n$, nor is it known whether there is some $\varepsilon > 0$ such that there are infinitely many n with $\varepsilon n < F_0(n) < n$.)

We remark that it is known that if $h(n)$ is any function that tends to infinity, then the set of numbers n with $F(n) < \ln^{h(n)} n$ has asymptotic density 1 [Erdős and Pomerance 1986].

3.15. [Monier] In the notation of Lemmas 3.4.8 and 3.4.9 and with $S(n)$ given in (3.5), show that

$$S(n) = \left(1 + \frac{2^{\nu(n)\omega(n)} - 1}{2^{\omega(n)} - 1}\right) \prod_{p|n} \gcd(t, p-1).$$

3.16. [Haglund] Let n be an odd composite. Show that $\overline{\mathcal{S}}(n)$ is the subgroup of $\mathbf{Z}_n^*$ generated by $\mathcal{S}(n)$.

3.17. [Gerlach] Let n be an odd composite. Show that $\mathcal{S}(n) = \overline{\mathcal{S}}(n)$ if and only if n is a prime power or n is divisible by a prime that is 3 (mod 4). Conclude that the set of odd composite numbers n for which $\mathcal{S}(n)$ is *not* a subgroup of $\mathbf{Z}_n^*$ is infinite, but has asymptotic density zero. (See Exercises 1.10, 1.88, and 5.15.)

3.18. Say you have an odd number n and an integer a not divisible by n such that n is a pseudoprime base a, but n is not a strong pseudoprime base a. Describe an algorithm that with these numbers as inputs gives a nontrivial factorization of n in polynomial time.

3.19. [Lenstra, Granville] Show that if an odd number n be divisible by the square of some prime, then $W(n)$, the least witness for n, is less than $\ln^2 n$. (Hint: Use (1.45).)

3.20. Describe a probabilistic algorithm that gives nontrivial factorizations of Carmichael numbers in expected polynomial time.

3.21. We say that an odd composite number n is an Euler pseudoprime base a if a is coprime to n and

$$a^{(n-1)/2} \equiv \left(\frac{a}{n}\right) \pmod{n}, \tag{3.32}$$

where $\left(\frac{a}{n}\right)$ is the Jacobi symbol (see Definition 2.3.3). Euler's criterion (see Theorem 2.3.4) asserts that odd primes n satisfy (3.32). Show that if n is a strong pseudoprime base a, then n is an Euler pseudoprime base a, and that if n is an Euler pseudoprime base a, then n is a pseudoprime base a.

3.22. [Lehmer, Solovay–Strassen] Let n be an odd composite. Show that the set of residues $a \pmod{n}$ for which n is an Euler pseudoprime is a proper subgroup of $\mathbf{Z}_n^*$. Conclude that the number of such bases a is at most $\varphi(n)/2$.

3.23. Along the lines of Algorithm 3.4.6 develop a probabilistic compositeness test using Exercise 3.22. (This test is often referred to as the Solovay–Strassen primality test.) Using Exercise 3.21 show that this algorithm is majorized by Algorithm 3.4.6.

3.24. [Lenstra, Robinson] Show that if n is odd and if there exists an integer b with $b^{(n-1)/2} \equiv -1 \pmod{n}$, then any integer a with $a^{(n-1)/2} \equiv \pm 1 \pmod{n}$ also satisfies $a^{(n-1)/2} \equiv \left(\frac{a}{n}\right) \pmod{n}$. Using this and Exercise 3.22, show that if n is an odd composite and $a^{(n-1)/2} \equiv \pm 1 \pmod{n}$ for all a coprime to n, then in fact $a^{(n-1)/2} \equiv 1 \pmod{n}$ for all a coprime to n. Such a number must be a Carmichael number; see Exercise 3.12. (It follows from the proof of the infinitude of the set of Carmichael numbers that there are infinitely many odd composite numbers n such that $a^{(n-1)/2} \equiv \pm 1 \pmod{n}$ for all a coprime to n. The first example is Ramanujan's "taxicab" number, 1729.)

3.25. Show that there are seven Fibonacci pseudoprimes smaller than 323.

3.26. Show that every composite number coprime to 6 is a Lucas pseudoprime with respect to $x^2 - x + 1$.

3.27. Show that if (3.12) holds, then so does

$$(a-x)^n \equiv \begin{cases} x \pmod{(f(x), n)}, & \text{if } \left(\frac{\Delta}{n}\right) = -1, \\ a - x \pmod{(f(x), n)}, & \text{if } \left(\frac{\Delta}{n}\right) = 1. \end{cases}$$

In particular, conclude that a Frobenius pseudoprime with respect to $f(x) = x^2 - ax + b$ is also a Lucas pseudoprime with respect to $f(x)$.

3.28. Show that the definition of Frobenius pseudoprime in Section 3.5.5 for a polynomial $f(x) = x^2 - ax + b$ reduces to the definition in Section 3.5.2.

3.29. Show that if a, n are positive integers with n odd and coprime to a, then n is a Fermat pseudoprime base a if and only if n is a Frobenius pseudoprime with respect to the polynomial $f(x) = x - a$.

3.30. Let a, b be integers with $\Delta = a^2 - 4b$ not a square, let $f(x) = x^2 - ax + b$, let n be an odd prime not dividing $b\Delta$, and let $R = \mathbf{Z}_n[x]/(f(x))$. Show that if $(x(a-x)^{-1})^{2m} = 1$ in R, then $(x(a-x)^{-1})^m = \pm 1$ in R.

3.31. Show that a Frobenius pseudoprime with respect to $x^2 - ax + b$ is also an Euler pseudoprime (see Exercise 3.21) with respect to b.

3.32. Prove that the various identities in Section 3.5.3 are correct.

3.33. Prove that the recurrence (3.22) is valid.

3.34. Show that if $a = \pi\left(x^{1/3}\right)$, then the number of terms in the double sum in (3.23) is $O\left(x^{2/3}/\ln^2 x\right)$.

3.35. Show that with M computers where $M < x^{1/3}$, each with the capacity for $O\left(x^{1/3+\epsilon}\right)$ space, the prime-counting algorithm of Section 3.6 may be speeded up by a factor M.

3.36. Show that instead of using analytic relation (3.25) to get the modified count $\pi^*(x)$, one could, if desired, use the "prime-zeta" function

$$\mathcal{P}(s) = \sum_{p \in \mathcal{P}} \frac{1}{p^s}$$

in place of $\ln \zeta$ within the integral, whence the result on the left-hand side of (3.25) is, for noninteger x, the π function itself. Then show that this observation is not entirely vacuous, and might even be practical, by deriving the relation

$$\mathcal{P}(s) = \sum_{n=1}^{\infty} \frac{\mu(n)}{n} \ln \zeta(ns),$$

for $\operatorname{Re} s > 1$, and describing quantitatively the relative ease with which one can calculate $\zeta(ns)$ for large integers n.

3.37. By establishing theoretical bounds on the magnitude of the real part of the integral

$$\int_T^\infty \frac{e^{it\alpha}}{\beta + it}\, dt,$$

where T, α, β are positive reals, determine a bound on that portion of the integral in relation (3.27) that comes from $\mathrm{Im}(s) > T$. Describe, then, how large T must be for $\pi^*(x)$ to be calculated to within some $\pm\epsilon$ of the true value. See Exercise 3.47 for the analogous estimate on a much more efficient method.

3.38. Consider a specific choice for the Lagarias–Odlyzko turn-off function $c(u, x)$, namely, a straight-line connection between the $1, 0$ values. Specifically, for $y = \sqrt{x}$, define $c = 1, (x - u)/y, 0$ as $u \le x - y, u \in (x - y, x], u > x$, respectively. Show that the Mellin companion function is

$$F(s, x) = \frac{1}{y} \frac{x^{s+1} - (x - y)^{s+1}}{s(s + 1)}.$$

Now derive a bound, as in Exercise 3.37, on proper values of T such that $\pi(x)$ will be calculated correctly on the basis of

$$\pi^*(x) \approx \mathrm{Re} \int_0^T F(s, x) \ln \zeta(s)\, dt.$$

Calculate numerically some correct values of $\pi(x)$ using this particular turn-off function c.

3.39. In regard to the Galway functions of which F is defined by (3.31), make rigorous the notion that even though the Riemann zeta function somehow embodies, if you will, "all the secrets of the primes," we need to know ζ only to an imaginary height of "about" $x^{1/2}$ to count all primes not exceeding x.

3.40. Using integration by parts, show that the F defined by (3.31) is indeed the Mellin transform of the given c.

3.8 Research problems

3.41. Find a number $n \equiv \pm 2 \pmod 5$ that is simultaneously a base-2 pseudoprime and a Fibonacci pseudoprime. Pomerance, Selfridge, and Wagstaff offer \$620 for the first example. (The prime factorization must also be supplied.) The prize money comes from the three, but not equally: Selfridge offers \$500, Wagstaff offers \$100 and Pomerance offers \$20. However, they also agree to pay \$620, with Pomerance and Selfridge reversing their roles, for a proof that no such number n exists.

3.42. Find a composite number n, together with its prime factorization, that is a Frobenius pseudoprime for x^2+5x+5 and satisfies $\left(\frac{5}{n}\right) = -1$. J. Grantham has offered a prize of \$6.20 for the first example.

3.43. Consider the least witness function $W(n)$ defined for odd composite numbers n. It is relatively easy to see that $W(n)$ is never a power; prove this. Are there any other forbidden numbers in the range of $W(n)$? If some n exists with $W(n) = k$, let n_k denote the smallest such n. We have

$$\begin{aligned}
n_2 &= 9 & n_{12} &> 10^{16}\\
n_3 &= 2047 & n_{13} &= 2152302898747\\
n_5 &= 1373653 & n_{14} &= 1478868544880821\\
n_6 &= 134670080641 & n_{17} &= 3474749660383\\
n_7 &= 25326001 & n_{19} &= 4498414682539051\\
n_{10} &= 307768373641 & n_{23} &= 341550071728321.\\
n_{11} &= 3215031751
\end{aligned}$$

(These values were computed by D. Bleichenbacher, also see [Jaeschke 1993], and Exercise 4.22.) S. Li has shown that $W(n) = 12$ for

$$n = 1502401849747176241,$$

so we know that n_{12} exists. Find n_{12} and extend the above table. Using Bleichenbacher's computations, we know that any other value of n_k that exists must exceed 10^{16}.

3.44. Study, as a possible alternative to the simple trial-division Algorithm 3.1.1, the notion of taking (perhaps extravagant) gcd operations with the N to be factored. For example, you could compute a factorial of some B and take $\gcd(B!, N)$, hoping for a factor. Describe how to make such an algorithm complete, with the full prime factorizations resulting. This completion task is nontrivial: For example, one must take note that a factor k^2 of N with $k < B$ might not be extracted from a single factorial.

Then there are complexity issues. Should one instead multiply together sets of consecutive primes, i.e., partial "primorials" (see Exercise 1.6), to form numbers $\{B_i\}$, and then test various $\gcd(B_i, N)$?

3.45. Let $f(N)$ be a worst-case bound on the time it takes to decide primality on any particular number between N and $N + N^{1/4}$. By sieving first with the primes below $N^{1/4}$ we are left with the numbers in the interval $[N, N + N^{1/4}]$ that have no prime factor up to $N^{1/4}$. The number of these remaining numbers is $O(N^{1/4}/\ln N)$. Thus one can find all the primes in the interval in a time bound of $O(N^{1/4}f(N)/\ln N) + O(N^{1/4}\ln\ln N)$. Is there a way of doing this either in time $o(N^{1/4}f(N)/\ln N)$ or in time $O(N^{1/4}\ln\ln N)$?

3.46. The ordinary sieve of Eratosthenes, as discussed above, may be segmented, so that but for the final list of primes collected, the space required along the way is $O(N^{1/2})$. And this can be accomplished without sacrificing on the time bound of $O(N \ln\ln N)$ bit operations. Can one prepare a table of primes up to N in $o(N)$ bit operations, and use only $O(N^{1/2})$ space along the way? Algorithm 3.2.2 meets the time bound goal, but not the space bound. (The paper [Atkin and Bernstein 1999] nearly solves this problem.)

3.47. Along the lines of the formalism of Section 3.6.2, derive an integral condition on x, Δ and involving the Riemann ζ function such that there exist *no* primes in the interval $[x, x+\Delta]$. Describe how such a criterion could be used for given x, Δ to show numerically, but rigorously, whether or not primes exist in such an interval. Of course, any new *theoretical* inroads into the analysis of these "gaps" would be spectacular.

3.48. Suppose T is a probabilistic test that takes composite numbers n and, with probability $p(n)$, provides a proof of compositeness for n. (For prime inputs, the test T reports only that it has not succeeded in finding a proof of compositeness.) Is there such a test T that has $p(n) \to 1$ as n runs to infinity through the composite numbers, and such that the time to run T on n is no longer than doing k pseudoprime tests on n, for some fixed k?

3.49. For a positive integer n coprime to 12 and square-free, define $K(n)$ depending on n mod 12 according to one of the following equations:

$$K(n) = \#\{(u,v) \ : \ u > v > 0; \ n = u^2 + v^2\}, \quad \text{for } n \equiv 1, 5 \pmod{12},$$
$$K(n) = \#\{(u,v) \ : \ u > 0, \ v > 0; \ n = 3u^2 + v^2\}, \quad \text{for } n \equiv 7 \pmod{12},$$
$$K(n) = \#\{(u,v) \ : \ u > v > 0; \ n = 3u^2 - v^2\}, \quad \text{for } n \equiv 11 \pmod{12}.$$

Then it is a theorem of [Atkin and Bernstein 1999] that n is prime if and only if $K(n)$ is odd. First, prove this theorem using perhaps the fact (or related facts) that the number of representations of (any) positive n as a sum of two squares is

$$r_2(n) = 4 \sum_{d|n, \ d \text{ odd}} (-1)^{(d-1)/2},$$

where we count all $n = u^2 + v^2$ including negative u or v representations; e.g. one has as a check the value $r_2(25) = 12$.

A research question is this: Using the Atkin–Bernstein theorem can one fashion an efficient sieve for primes in an interval, by assessing the parity of K for many n at once? (See [Galway 2000].)

Another question is, can one fashion an efficient sieve (or even a primality test) using alternative descriptions of $r_2(n)$, for example by invoking various connections with the Riemann zeta function? See [Titchmarsh 1986] for a relevant formula connecting r_2 with ζ.

Yet another research question runs like so: Just how hard is it to "count up" all lattice points (in the three implied lattice regions) within a given "radius" $\sqrt{n}$, and look for representation numbers $K(n)$ as numerical discontinuities at certain radii. This technique may seem on the face of it to belong in some class of brute-force methods, but there *are* efficient formulae—arising in analyses for the celebrated Gauss circle problem (how many lattice points lie inside a given radius?)—that provide exact counts of points in surprisingly rapid fashion. In this regard, show an alternative lattice theorem, that if $n \equiv 1 \pmod 4$ is square-free, then n is prime if and only if $r_2(n) = 8$. A simple starting experiment that shows $n = 13$ to be prime by lattice counting, via analytic Bessel formulae, can be found in [Crandall 1994b, p. 68].

Chapter 4

PRIMALITY PROVING

In Chapter 3 we discussed probabilistic methods for quickly recognizing composite numbers. If a number is not declared composite by such a test, it is either prime, or we have been unlucky in our attempt to prove the number composite. Since we do not expect to witness inordinate strings of bad luck, after a while we become convinced that the number is prime. We do not, however, have a *proof*; rather, we have a conjecture substantiated by numerical experiments. This chapter is devoted to the topic of how one might actually prove that a number is prime.

4.1 The $n-1$ test

Small numbers can be tested for primality by trial division, but for larger numbers there are better methods (10^{12} is a possible size threshold, but this depends on the specific computing machinery used). One of these better methods is based on the same theorem as the simplest of all of the pseudoprimality tests, namely, Fermat's little theorem (Theorem 3.3.1). Known as the $n-1$ test, the method somewhat surprisingly suggests that we try our hand at factoring not n, but $n-1$.

4.1.1 The Lucas theorem and Pepin test

We begin with an idea of E. Lucas, from 1876.

Theorem 4.1.1 (Lucas theorem). *If a, n are integers with $n > 1$, and*

$$a^{n-1} \equiv 1 \pmod{n}, \textit{ but } a^{(n-1)/q} \not\equiv 1 \pmod{n} \textit{ for every prime } q|n-1, \tag{4.1}$$

then n is prime.

Proof. The first condition in (4.1) implies that the order of a in $\mathbf{Z}_n^*$ is a divisor of $n-1$, while the second condition implies that the order of a is not a proper divisor of $n-1$, that is, it is equal to $n-1$. But the order of a is also a divisor of $\varphi(n)$, by the Euler theorem (see (2.2)), so $n-1 \le \varphi(n)$. But if n is composite and has the prime factor p, then both p and n are integers in $\{1, 2, \dots, n\}$ that are not coprime to n, so from the definition of Euler's function $\varphi(n)$ (below (1.5)), $\varphi(n) \le n-2$. This is incompatible with $n-1 \le \varphi(n)$, so it must be the case that n is prime. □

Remark. The version of Theorem 4.1.1 above is due to Lehmer. Lucas had such a result where q runs through all of the proper divisors of $n-1$.

The hypothesis (4.1) of the Lucas theorem is not vacuous for prime numbers; such a number a is called a primitive root, and all primes have them. That is, if n is prime, the multiplicative group $\mathbf{Z}_n^*$ is cyclic; see Theorem 2.2.5. In fact, each prime $n > 200560490131$ has more than $n/(2\ln\ln n)$ primitive roots in $\{1, 2, \ldots, n-1\}$; see Exercise 4.1. (Note: The prime 200560490131 is 1 greater than the product of the first 11 primes.)

A consequence is that if $n > 200560490131$ is prime, it is easy to find a number satisfying (4.1) via a probabilistic algorithm. Just choose random integers a in the range $1 \le a \le n-1$ until a successful one is found. The expected number of trials is less than $2\ln\ln n$.

Though we know no deterministic polynomial-time algorithm for finding a primitive root for a prime, the principal hindrance in implementing the Lucas theorem as a primality test is *not* the search for a primitive root a, but rather finding the complete prime factorization of $n-1$. Factorization is hard in practice for many numbers, but it is not hard for every number. For example, consider a search for primes that are 1 more than a power of 2. As seen in Theorem 1.3.4, such a prime must be of the form $F_k = 2^{2^k} + 1$. Numbers in this sequence are called Fermat numbers after Fermat, who thought they were all prime.

In 1877, Pepin gave a criterion similar to the following for the primality of a Fermat number.

Theorem 4.1.2 (Pepin test). *For $k \ge 1$, the number $F_k = 2^{2^k}+1$ is prime if and only if $3^{(F_k-1)/2} \equiv -1 \pmod{F_k}$.*

Proof. Suppose the congruence holds. Then (4.1) holds with $n = F_k$, $a = 3$, so F_k is prime by the Lucas Theorem 4.1.1. Conversely, assume F_k is prime. Since 2^k is even, it follows that $2^{2^k} \equiv 1 \pmod 3$, so that $F_k \equiv 2 \pmod 3$. But also $F_k \equiv 1 \pmod 4$, so the Legendre symbol $\left(\frac{3}{F_k}\right)$ is -1, that is, 3 is not a square $\pmod{F_k}$. The congruence in the theorem thus follows from Euler's criterion (2.6). □

Actually, Pepin gave his test with the number 5 in place of the number 3 (and with $k \ge 2$). It was noticed by Proth and Lucas that one can use 3. In this regard, see [Williams 1998] and Exercise 4.5.

As of this writing, the largest F_k for which the Pepin test has been used is F_{24}. As discussed in Section 1.3.2, this number is composite, and in fact, so is every other Fermat number beyond F_4 for which the character (prime or composite) has been resolved.

4.1.2 Partial factorization

Since the hardest step, in general, in implementing the Lucas Theorem 4.1.1 as a primality test is coming up with the complete prime factorization of $n-1$, one might wonder whether any use can be made of a partial factorization of $n-1$. In particular, say

$$n-1 = FR, \text{ and the complete prime factorization of } F \text{ is known.} \tag{4.2}$$

If F is fairly large as a function of n, we may fashion a primality proof for n along the lines of (4.1), if indeed n happens to be prime. Our first result on these lines allows us to deduce information on the prime factorization of n.

Theorem 4.1.3 (Pocklington). *Suppose (4.2) holds and a is such that*

$$a^{n-1} \equiv 1 \pmod{n} \text{ and } \gcd(a^{(n-1)/q}-1, n) = 1 \text{ for each prime } q|F. \tag{4.3}$$

Then every prime factor of n is congruent to 1 $\pmod{F}$.

Proof. Let p be a prime factor of n. From the first part of (4.3) we have that the order of a^R in $\mathbf{Z}_p^*$ is a divisor of $(n-1)/R = F$. From the second part of (4.3) it is not a proper divisor of F, so is equal to F. Hence F divides the order of $\mathbf{Z}_p^*$, which is $p-1$. This concludes the proof. □

Corollary 4.1.4. *If (4.2) and (4.3) hold and $F \ge \sqrt{n}$, then n is prime.*

Proof. Theorem 4.1.3 implies that each prime factor of n is congruent to 1 $\pmod{F}$, and so each prime factor of n exceeds F. But $F \ge \sqrt{n}$, so each prime factor of n exceeds $\sqrt{n}$, so n must be prime. □

The next result allows a still smaller value of F.

Theorem 4.1.5 (Brillhart, Lehmer, and Selfridge). *Suppose (4.2) and (4.3) both hold and suppose that $n^{1/3} \le F < n^{1/2}$. Consider the base F representation of n, namely $n = c_2F^2 + c_1F + 1$, where c_1, c_2 are integers in $[0, F-1]$. Then n is prime if and only if $c_1^2 - 4c_2$ is not a square.*

Proof. Since $n \equiv 1 \pmod{F}$, it follows that the base-F "units" digit of n is 1. Thus n has its base-F representation in the form $c_2F^2 + c_1F + 1$, as claimed. Suppose n is composite. From Theorem 4.1.3, all the prime factors of n are congruent to 1 $\pmod{F}$, so must exceed $n^{1/3}$. We conclude that n has exactly two prime factors:

$$n = pq, \quad p = aF+1, \quad q = bF+1, \quad a \le b.$$

We thus have

$$c_2F^2 + c_1F + 1 = n = (aF+1)(bF+1) = abF^2 + (a+b)F + 1.$$

Our goal is to show that we must have $c_2 = ab$ and $c_1 = a+b$, for then it will follow that $c_1^2 - 4c_2$ is a square.

First note that $F^3 \ge n > abF^2$, so that $ab \le F-1$. It follows that either $a+b \le F-1$ or $a=1, b=F-1$. In the latter case, $n = (F+1)((F-1)F+1) = F^3+1$, contradicting $F \ge n^{1/3}$. Hence both ab and $a+b$ are positive integers smaller than F. From the uniqueness of the base-F representation of a number it follows that $c_2 = ab$ and $c_1 = a + b$ as claimed.

Now suppose, conversely, that $c_1^2 - 4c_2$ is a square, say u^2. Then

$$n = \left(\frac{c_1 + u}{2}F + 1\right)\left(\frac{c_1 - u}{2}F + 1\right).$$

The two fractions are both integers, since $c_1 \equiv u \pmod 2$. It remains to note that this factorization is nontrivial, since $c_2 > 0$ implies $|u| < c_1$. Thus, n is composite. □

To apply Theorem 4.1.5 as a primality test one should have a fast method of verifying whether the integer $c_1^2 - 4c_2$ in the theorem is a square. This is afforded by Algorithm 9.2.11.

The next result allows F to be even smaller.

Theorem 4.1.6 (Konyagin and Pomerance). *Suppose that $n \geq 214$, both (4.2) and (4.3) hold, and $n^{3/10} \leq F < n^{1/3}$. Say the base-$F$ expansion of n is $c_3F^3 + c_2F^2 + c_1F + 1$, and let $c_4 = c_3F + c_2$. Then n is prime if and only if the following conditions hold:*

(1) *$(c_1 + tF)^2 + 4t - 4c_4$ is not a square for $t = 0, 1, 2, 3, 4, 5$.*

(2) *With u/v the continued fraction convergent to c_1/F such that v is maximal subject to $v < F^2/\sqrt{n}$ and with $d = \lfloor c_4v/F + 1/2 \rfloor$, the polynomial $vx^3 + (uF - c_1v)x^2 + (c_4v - dF + u)x - d \in \mathbf{Z}[x]$ has no integral root a such that $aF + 1$ is a nontrivial factor of n.*

Proof. Since every prime factor of n is congruent to 1 (mod F) (by Theorem 4.1.3), we have that n is composite if and only if there are positive integers $a_1 \leq a_2$ with $n = (a_1F + 1)(a_2F + 1)$. Suppose n is composite and (1) and (2) hold. We begin by establishing some identities and inequalities. We have

$$n = c_4F^2 + c_1F + 1 = a_1a_2F^2 + (a_1 + a_2)F + 1,$$

and there is some integer $t \geq 0$ with

$$a_1a_2 = c_4 - t, \quad a_1 + a_2 = c_1 + tF. \tag{4.4}$$

Since (1) holds, we have $t \geq 6$. Thus

$$a_2 \geq \frac{a_1 + a_2}{2} \geq \frac{c_1 + 6F}{2} \geq 3F$$

and

$$a_1 < \frac{n}{a_2F^2} \leq \frac{n}{3F^3}. \tag{4.5}$$

We have from (4.4) that

$$t \leq \frac{a_1 + a_2}{F} \leq \frac{a_1a_2 + 1}{F} < \frac{c_4}{F} < \frac{n}{F^3}. \tag{4.6}$$

Also, (4.4) implies that

$$a_1c_1 + a_1tF = a_1^2 + c_4 - t. \tag{4.7}$$

With the notation of condition (2), we have from (4.7) that

$$\begin{aligned} a_1u + a_1tv - \frac{c_4v}{F} &= a_1v\left(\frac{u}{v} - \frac{c_1}{F}\right) + (a_1c_1 + a_1tF)\frac{v}{F} - \frac{c_4v}{F} \\ &= a_1v\left(\frac{u}{v} - \frac{c_1}{F}\right) + (a_1^2 + c_4 - t)\frac{v}{F} - \frac{c_4v}{F} \\ &= a_1v\left(\frac{u}{v} - \frac{c_1}{F}\right) + (a_1^2 - t)\frac{v}{F}. \end{aligned} \tag{4.8}$$

Note that (4.5), (4.6), and $t \ge 6$ imply that

$$|a_1^2 - t| < \max\{a_1^2, t\} \le \max\left\{\frac{1}{9}\left(\frac{n}{F^3}\right)^2, \frac{n}{F^3}\right\} \le \frac{1}{6}\left(\frac{n}{F^3}\right)^2. \tag{4.9}$$

First suppose that $u/v = c_1/F$. Then (4.8) and (4.9) imply that

$$\left|a_1u + a_1tv - \frac{c_4v}{F}\right| = |a_1^2 - t|\frac{v}{F} < \frac{1}{6}\left(\frac{n}{F^3}\right)^2 \frac{v}{F} < \frac{n^2}{6F^7} \cdot \frac{F^2}{\sqrt{n}} = \frac{n^{3/2}}{6F^5} \le \frac{1}{6}. \tag{4.10}$$

If $u/v \ne c_1/F$, let u'/v' be the next continued fraction convergent to c_1/F after u/v, so that

$$v < \frac{F^2}{\sqrt{n}} \le v', \quad \left|\frac{u}{v} - \frac{c_1}{F}\right| \le \frac{1}{vv'} \le \frac{\sqrt{n}}{vF^2}.$$

Thus, from (4.5), (4.8), and the calculation in (4.10),

$$\left|a_1u + a_1tv - \frac{c_4v}{F}\right| \le a_1v\frac{\sqrt{n}}{vF^2} + \frac{1}{6} < \frac{n^{3/2}}{3F^5} + \frac{1}{6} \le \frac{1}{2}.$$

Let $d = a_1u + a_1tv$, so that $|d - c_4v/F| < 1/2$, which implies that $d = \lfloor c_4v/F + 1/2 \rfloor$. Multiplying (4.7) by a_1v, we have

$$va_1^3 - c_1va_1^2 - a_1^2tvF - a_1tv + c_4a_1v = 0,$$

and using $-a_1tv = ua_1 - d$, we get

$$va_1^3 + (uF - c_1v)a_1^2 + (c_4v - dF + u)a_1 - d = 0.$$

Hence (2) does not hold after all, which proves that if n is composite, then either (1) or (2) does not hold.

Now suppose n is prime. If $t \in \{0, 1, 2, 3, 4, 5\}$ and $(c_1+tF)^2 - 4c_4 + 4t = u^2$, with u integral, then

$$\begin{aligned} n &= (c_4 - t)F^2 + (c_1 + tF)F + 1 \\ &= \left(\frac{c_1 + tF + u}{2}F + 1\right)\left(\frac{c_1 + tF - u}{2}F + 1\right). \end{aligned}$$

Since n is prime, this must be a trivial factorization of n, that is,

$$c_1 + tF - |u| = 0,$$

which implies $c_4 = t$. But $c_4 \geq F \geq n^{3/10} \geq 214^{3/10} > 5 \geq t$, a contradiction. So if (1) fails, n must be composite. It is obvious that if n is prime, then (2) holds. □

As with Theorem 4.1.5, if Theorem 4.1.6 is to be used as a primality test, one should use Algorithm 9.2.11 as a subroutine to recognize squares. In addition, one should use Newton's method or a divide and conquer strategy to search for integral roots of the cubic polynomial in condition (2) of the theorem. We next embody Theorems 4.1.3-4.1.6 in one algorithm.

Algorithm 4.1.7 (The $n-1$ test). Suppose we have an integer $n \geq 214$ and that (4.2) holds with $F \geq n^{3/10}$. This probabilistic algorithm attempts to decide whether n is prime (YES) or composite (NO).

```
1. [Pocklington test]
      Choose random a ∈ [2, n − 2];
      if(a^{n−1} ≢ 1 (mod n)) return NO;                    // n is composite.
      for(prime q|F) {
            g = gcd((a^{(n−1)/q} mod n) − 1, n);
            if(1 < g < n) return NO;
            if(g == n) goto [Pocklington test]
      }                        // Exhausting the 'for' loop means relation (4.3) holds.
2. [First magnitude test]
      if(F ≥ n^{1/2}) return YES;
3. [Second magnitude test]
      if(n^{1/3} ≤ F < n^{1/2}) {
            Cast n in base F : n = c_2 F^2 + c_1 F + 1;
            if(c_1^2 − 4c_2 not a square) return YES;
            return NO;
      }
4. [Third magnitude test]
      if(n^{3/10} ≤ F < n^{1/3}) {
            If conditions (1) and (2) of Theorem 4.1.6 hold, return YES;
            return NO;
      }
```

Though Algorithm 4.1.7 is probabilistic, any returned value YES (n is prime) or NO (n is composite) is a rigorous declaration.

4.1.3 Succinct certificates

The goal in primality testing is to quickly find a short proof of primality for prime inputs p. But how do we know that a short proof exists? Any search will necessarily be in vain if p does not have a short primality proof. We now show that every prime p has a short proof of primality, or what V. Pratt has called a "succinct certificate."

In fact, there is always a short proof that is based on the Lucas Theorem 4.1.1. This might appear obvious, for once you have somehow found the complete prime factorization of $p-1$ and the primitive root a, the conditions (4.1) may be quickly verified.

However, for the proof to be complete, one needs a demonstration that we indeed have the complete factorization of $p-1$; that is, that the numbers q appearing in (4.1) really are prime. This suggests an iteration of the method, but then arises the possibility that there may be a proliferation of cases. The heart of the proof is to show in the worst case, not too much proliferation can occur.

It is convenient to make a small, and quite practical, modification in the Lucas Theorem 4.1.1. The idea is to treat the prime $q=2$ differently from the other primes q dividing $p-1$. In fact, we know what $a^{(p-1)/2}$ should be congruent to (mod p) if it is not 1, namely -1. And if $a^{(p-1)/2} \equiv -1 \pmod{p}$, we do not need to check that $a^{p-1} \equiv 1 \pmod{p}$. Further, if q is an odd prime factor of $p-1$, let $m = a^{(p-1)/2q}$. If $m^q \equiv -1 \pmod{p}$ and $m^2 \equiv 1 \pmod{p}$, then $m \equiv -1 \pmod{p}$ (regardless of whether p is prime or composite). Thus, to show that $a^{(p-1)/q} \not\equiv 1 \pmod{p}$ it suffices to show $a^{(p-1)/2q} \not\equiv -1 \pmod{p}$. Thus we have the following result.

Theorem 4.1.8. *Suppose $p>1$ is an odd integer and*

$$\begin{cases} a^{(p-1)/2} \equiv -1 \pmod{p}, \\ a^{(p-1)/2q} \not\equiv -1 \pmod{p} \text{ for every odd prime } q|p-1. \end{cases} \tag{4.11}$$

Then p is prime. Conversely, if p is an odd prime, then every primitive root a of p satisfies conditions (4.11).

We now describe what might be called a "Lucas tree." It is a rooted tree with odd primes at the vertices, p at the root (level 0), and for each positive level k, a prime r at level k is connected to a prime q at level $k-1$ if and only if $r|q-1$. For example, here is the Lucas tree for $p=1279$:

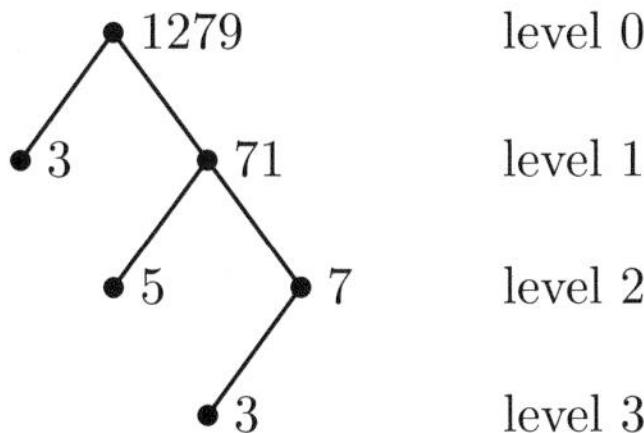

Let $M(p)$ be the number of modular multiplications (with integers not exceeding p) needed to prove p prime using Theorem 4.1.8 to traverse the Lucas tree for p, and using binary addition chains for the exponentiations (see Algorithm 2.1.5).

For example, consider $p = 1279$:

$$\begin{aligned}
3^{1278/2} &\equiv -1 \pmod{1279}, & 3^{1278/6} &\equiv 775 \pmod{1279},\\
& & 3^{1278/142} &\equiv 498 \pmod{1279},\\
2^{2/2} &\equiv -1 \pmod{3}, & &\\
7^{70/2} &\equiv -1 \pmod{71}, & 7^{70/10} &\equiv 14 \pmod{71},\\
& & 7^{70/14} &\equiv 51 \pmod{71},\\
2^{4/2} &\equiv -1 \pmod{5}, & &\\
3^{6/2} &\equiv -1 \pmod{7}, & 3^{6/6} &\equiv 3 \pmod{7},\\
2^{2/2} &\equiv -1 \pmod{3}. & &
\end{aligned}$$

If we use the binary addition chain for each exponentiation, we have the following number of modular multiplications:

$$\begin{aligned}
1278/2 &: 16\\
1278/6 &: 11\\
1278/142 &: 4\\
2/2 &: 0\\
70/2 &: 7\\
70/10 &: 4\\
70/14 &: 3\\
4/2 &: 1\\
6/2 &: 2\\
6/6 &: 0\\
2/2 &: 0.
\end{aligned}$$

Thus, using binary addition chains we have 48 modular multiplications, so $M(1279) = 48$.

The following result is essentially due to [Pratt 1975]:

Theorem 4.1.9. *For every odd prime p, $M(p) < 2\lg^2 p$.*

Proof. Let $N(p)$ be the number of (not necessarily distinct) odd primes in the Lucas tree for p. We first show that $N(p) < \lg p$. This is true for $p = 3$. Suppose it is true for every odd prime less than p. If $p - 1$ is a power of 2, then $N(p) = 1 < \lg p$. If $p - 1$ has the odd prime factors $q_1, \ldots, q_k$, then, by the induction hypothesis,

$$N(p) = 1 + \sum_{i=1}^{k} N(q_i) < 1 + \sum_{i=1}^{k} \lg q_i = 1 + \lg(q_1 \cdots q_k) \le 1 + \lg\left(\frac{p-1}{2}\right) < \lg p.$$

So $N(p) < \lg p$ always holds.

If r is one of the odd primes appearing in the Lucas tree for p, and $r < p$, then there is some other prime q also appearing in the Lucas tree with $r | q - 1$

and $q \le p$. We have to show at one point that for some a, $a^{(q-1)/2r} \not\equiv -1 \pmod{q}$, and, at another point, that for some b, $b^{(r-1)/2} \equiv -1 \pmod{r}$. Note that the number of modular multiplications in the binary addition chain for m does not exceed $2 \lg m$. Thus, the number of modular multiplications in the above two calculations does not exceed

$$2 \lg \left(\frac{q-1}{2r} \right) + 2 \lg \left(\frac{r-1}{2} \right) < 2 \lg q - 4 < 2 \lg p.$$

We conclude that

$$M(p) < 2 \lg \left(\frac{p-1}{2} \right) + (N(p) - 1) 2 \lg p < 2 \lg p + (\lg p - 1) 2 \lg p = 2 \lg^2 p.$$

This completes the proof. □

By using more efficient addition chains we may reduce the coefficient 2. We do not know whether there is some $c > 0$ such that for infinitely many primes p, the Lucas tree proof of primality for p actually requires at least $c \lg^2 p$ modular multiplications. We also do not know whether there are infinitely many primes p with $M(p) = o(\lg^2 p)$. It is known, however, that via Theorem 7.6.1 (see [Pomerance 1987a]), there exists in principle *some* primality proof for every prime p using only $O(\lg p)$ modular multiplications. As with the Lucas tree proof, existence is comforting to know, but the rub is in finding such a short proof.

4.2 The $n+1$ test

The principal difficulty in applying the $n-1$ test of the previous section to prove n prime is in finding a sufficiently large completely factored divisor of $n-1$. For some values of n, this is no problem, such as with Fermat numbers, for which we have the Pepin test. For other classes of numbers, such as the Mersenne numbers $M_p = 2^p - 1$, the prime factorization of 1 more than the number is readily apparent. Can we use this information in a primality test? Indeed, we can.

4.2.1 The Lucas–Lehmer test

With $a, b \in \mathbf{Z}$, let

$$f(x) = x^2 - ax + b, \ \Delta = a^2 - 4b. \tag{4.12}$$

We reintroduce the Lucas sequences $(U_k), (V_k)$, already discussed in Section 3.5.1:

$$U_k = \frac{x^k - (a-x)^k}{x - (a-x)} \pmod{f(x)}, \quad V_k = x^k + (a-x)^k \pmod{f(x)}. \tag{4.13}$$

Recall that the polynomials U_k, V_k do not have positive degree; that is, they are integers.

Definition 4.2.1. With the above notation, if n is a positive integer with $\gcd(n, 2b\Delta) = 1$, the rank of appearance of n, denoted by $r_f(n)$, is the least positive integer r with $U_r \equiv 0 \pmod{n}$.

This concept sometimes goes by the name "rank of apparition," but according to Ribenboim, this is due to a mistranslation of the French *apparition*. There is nothing ghostly about the rank of appearance!

It is apparent from the definition (4.13) that (U_k) is a "divisibility sequence," that is, if $k|j$ then $U_k|U_j$. (We allow the possibility that $U_k = U_j = 0$.) It follows that if $\gcd(n, 2b\Delta) = 1$, then $U_j \equiv 0 \pmod{n}$ if and only if $j \equiv 0 \pmod{r_f(n)}$. On the basis of Theorem 3.5.3 we thus have the following result:

Theorem 4.2.2. *With f, Δ as in (4.12) and p a prime not dividing $2b\Delta$, we have $r_f(p)|p - \left(\frac{\Delta}{p}\right)$.*

(Recall the Legendre symbol $\left(\frac{\cdot}{p}\right)$ from Definition 2.3.2.)

In analogy to Theorem 4.1.3, we have the following result:

Theorem 4.2.3 (Morrison). *Let f, Δ be as in (4.12) and let n be a positive integer with $\gcd(n, 2b) = 1$, $\left(\frac{\Delta}{n}\right) = -1$. If F is a divisor of $n + 1$ and*

$$U_{n+1} \equiv 0 \pmod{n}, \quad \gcd(U_{(n+1)/q}, n) = 1 \text{ for every prime } q|F, \tag{4.14}$$

then every prime p dividing n satisfies $p \equiv \left(\frac{\Delta}{p}\right) \pmod{F}$. In particular, if $F > \sqrt{n} + 1$ and (4.14) holds, then n is prime.

(Recall the Jacobi symbol $\left(\frac{\cdot}{n}\right)$ from Definition 2.3.3.)

Proof. Let p be a prime factor of n. Then (4.14) implies that F divides $r_f(p)$. So, by Theorem 4.2.2, $p \equiv \left(\frac{\Delta}{p}\right) \pmod{F}$. If, in addition, we have $F > \sqrt{n} + 1$, then every prime factor p of n has $p \geq F - 1 > \sqrt{n}$, so n is prime. □

If Theorem 4.2.3 is to be used in a primality test, we will need to find an appropriate f in (4.12). As with Algorithm 4.1.7 where a is chosen at random, we may choose a, b in (4.12) at random. When we start with a prime n, the expected number of choices until a successful pair is found is not large, as the following result indicates.

Theorem 4.2.4. *Let p be an odd prime and let N be the number of pairs $a, b \in \{0, 1, \ldots, p-1\}$ such that if f, Δ are given as in (4.12), then $\left(\frac{\Delta}{p}\right) = -1$ and $r_f(p) = p + 1$. Then $N = \frac{1}{2}(p-1)\varphi(p+1)$.*

We leave the proof as Exercise 4.11. A consequence of Theorem 4.2.4 is that if n is an odd prime and if a, b are chosen randomly in $\{0, 1, \ldots, n-1\}$ with not both 0, then the expected number of choices until one is found where the f in (4.12) satisfies $r_f(n) = n+1$ is $2(n+1)/\varphi(n+1)$. If $n > 892271479$, then this expected number of choices is less than $4 \ln\ln n$; see Exercise 4.15.

It is also possible to describe a primality test using the V sequence in (4.13).

Theorem 4.2.5. *Let f, Δ be as in (4.12) and let n be a positive integer with $\gcd(n, 2b) = 1$ and $\left(\frac{\Delta}{n}\right) = -1$. If F is an even divisor of $n+1$ and*

$$V_{F/2} \equiv 0 \pmod{n}, \quad \gcd(V_{F/2q}, n) = 1 \text{ for every odd prime } q|F, \tag{4.15}$$

then every prime p dividing n satisfies $p \equiv \left(\frac{\Delta}{p}\right) \pmod{F}$. In particular, if $F > \sqrt{n} + 1$, then n is prime.

Proof. Suppose p is an odd prime that divides both U_m, V_m. Then (4.13) implies $x^m \equiv (a-x)^m \pmod{(f(x), p)}$ and $x^m \equiv -(a-x)^m \pmod{f(x), p}$, so that $x^m \equiv 0 \pmod{(f(x), p)}$. Then $b^m \equiv (x(a-x))^m \equiv 0 \pmod{(f(x), p)}$, that is, p divides b. Since n is coprime to $2b$, and since $U_{2m} = U_m V_m$, we have

$$\gcd(U_{2m}, n) = \gcd(U_m, n) \cdot \gcd(V_m, n).$$

Thus, the first condition in (4.15) implies $U_F \equiv 0 \pmod{n}$ and $\gcd(U_{F/2}, n) = 1$. Now suppose q is an odd prime factor of F. We have $U_{F/q} = U_{F/2q} V_{F/2q}$ coprime to n. Indeed, $U_{F/2q}$ divides $U_{F/2}$, so that $\gcd(U_{F/2q}, n) = 1$, and so with the second condition in (4.15) we have that $\gcd(U_{F/q}, n) = 1$. Thus, $r_f(p) = F$, and as in the proof of Theorem 4.2.3, this is sufficient for the conclusion. □

Just as the $n-1$ is particularly well suited for Fermat numbers, the $n+1$ test is especially speedy for Mersenne numbers.

Theorem 4.2.6 (Lucas–Lehmer test for Mersenne primes). *Consider the sequence (v_k) for $k = 0, 1, \ldots$, recursively defined by $v_0 = 4$ and $v_{k+1} = v_k^2 - 2$. Let p be an odd prime. Then $M_p = 2^p - 1$ is prime if and only if $v_{p-2} \equiv 0 \pmod{M_p}$.*

Proof. Let $f(x) = x^2 - 4x + 1$, so that $\Delta = 12$. Since $M_p \equiv 3 \pmod 4$ and $M_p \equiv 1 \pmod 3$, we see that $\left(\frac{\Delta}{M_p}\right) = -1$. We apply Theorem 4.2.5 with $F = 2^{p-1} = (M_p + 1)/2$. The conditions (4.15) reduce to the single condition $V_{2^{p-2}} \equiv 0 \pmod{M_p}$. But

$$V_{2m} \equiv x^{2m} + (4-x)^{2m} = (x^m + (4-x)^m)^2 - 2x^m(4-x)^m \equiv V_m^2 - 2 \pmod{f(x)},$$

since $x(4-x) \equiv 1 \pmod{f(x)}$; see (3.15). Also, $V_1 = 4$. Thus, $V_{2^k} = v_k$, and it follows from Theorem 4.2.5 that if $v_{p-2} \equiv 0 \pmod{M_p}$, then M_p is prime.

Suppose, conversely, that $M = M_p$ is prime. Since $\left(\frac{\Delta}{M}\right) = -1$, $\mathbf{Z}[x]/(f(x), M)$ is isomorphic to the finite field $\mathbf{F}_{M^2}$. Thus, raising to the M power is an automorphism and $x^M \equiv 4 - x \pmod{(f(x), M)}$; see the proof of Theorem 3.5.3. We compute $(x-1)^{M+1}$ two ways. First, since $(x-1)^2 \equiv 2x \pmod{(f(x), M)}$ and by the Euler criterion we have $2^{(M-1)/2} \equiv (2/M) = 1 \pmod{M}$, so

$$\begin{aligned}(x-1)^{M+1} &\equiv (2x)^{(M+1)/2} = 2 \cdot 2^{(M-1)/2} x^{(M+1)/2} \\ &\equiv 2x^{(M+1)/2} \pmod{(f(x), M)}.\end{aligned}$$

Next,

$$(x-1)^{M+1} = (x-1)(x-1)^M \equiv (x-1)(x^M-1) \equiv (x-1)(3-x)$$
$$\equiv -2 \pmod{(f(x), M)}.$$

Thus, $x^{(M+1)/2} \equiv -1 \pmod{(f(x), M)}$, that is, $x^{2^{p-1}} \equiv -1 \pmod{(f(x), M)}$. Using our automorphism, we also have $(4-x)^{2^{p-1}} \equiv -1 \pmod{(f(x), M)}$, so that $U_{2^{p-1}} \equiv 0 \pmod{M}$. If $U_{2^{p-2}} \equiv 0 \pmod{M}$, then $x^{2^{p-2}} \equiv (4-x)^{2^{p-2}} \pmod{(f(x), M)}$, so that

$$-1 \equiv x^{2^{p-1}} \equiv x^{2^{p-2}}(4-x)^{2^{p-2}} \equiv (x(4-x))^{2^{p-2}} \equiv 1^{2^{p-2}} \equiv 1 \pmod{(f(x), M)},$$

a contradiction. Since $U_{2^{p-1}} = U_{2^{p-2}} V_{2^{p-2}}$, we have $V_{2^{p-2}} \equiv 0 \pmod{M}$. But we have seen that $V_{2^{p-2}} = v_{p-2}$, so the proof is complete. □

Algorithm 4.2.7 (Lucas–Lehmer test for Mersenne primes).
We are given an odd prime p. This algorithm decides whether $2^p - 1$ is prime (YES) or composite (NO).

1. [Initialize]
   $v = 4$;
2. [Compute Lucas–Lehmer sequence]
   for($k \in [1, p-2]$) $v = (v^2 - 2) \bmod (2^p - 1)$; // k is a dummy counter.
3. [Check residue]
   if($v == 0$) return YES; // $2^p - 1$ definitely prime.
   return NO; // $2^p - 1$ definitely composite.

The celebrated Lucas–Lehmer test for Mersenne primes has achieved some notable successes, as mentioned in Chapter 1 and in the discussion surrounding Algorithm 9.5.18. Not only is the test breathtakingly simple, there are ways to perform with high efficiency the $p-2$ repeated squarings in step [Compute Lucas–Lehmer sequence].

4.2.2 An improved $n+1$ test, and a combined n^2-1 test

As with the $n-1$ test, which is useful only in the case that we have a large, fully factored divisor of $n-1$, the principal hurdle in implementing the $n+1$ test for most numbers is coming up with a large, fully factored divisor of $n+1$. In this section we shall improve Theorem 4.2.3 to get a result similar to Theorem 4.1.5. That is, we shall only require the fully factored divisor of $n+1$ to exceed the cube root. (Using the ideas in Theorem 4.1.6, this can be improved to the 3/10 root.) Then we shall show how fully factored divisors of both $n-1$ and $n+1$, that is, a fully factored divisor of n^2-1, may be combined into one test.

Theorem 4.2.8. *Suppose f, Δ are as in (4.12) and n is a positive integer with $\gcd(n, 2b) = 1$ and $\left(\frac{\Delta}{n}\right) = -1$. Suppose $n+1 = FR$ with $F > n^{1/3}+1$ and (4.14) holds. Write R in base F, so that $R = r_1 F + r_0$, $0 \le r_i \le F-1$.*

Then n is prime if and only if neither $x^2+r_0x-r_1$ nor $x^2+(r_0+F)x-r_1-1$ has a positive integral root.

Note that in the case $R < F$ we have $r_1 = 0$, and so neither quadratic can have positive integral roots. Thus, Theorem 4.2.8 contains the final assertion of Theorem 4.2.3.

Proof. Theorem 4.2.3 implies that all prime factors p of n satisfy $p \equiv \left(\frac{\Delta}{p}\right) \pmod{F}$. So, if n is composite, it must be the product pq of just two prime factors. Indeed, if n has 3 or more prime factors, n exceeds $(F-1)$, a contradiction. Since $-1 = \left(\frac{\Delta}{n}\right) = \left(\frac{\Delta}{p}\right)\left(\frac{\Delta}{q}\right)$, we have, say, $\left(\frac{\Delta}{p}\right) = 1, \left(\frac{\Delta}{q}\right) = -1$. Thus, there are positive integers c, d with $p = cF+1$, $q = dF-1$. Since both $(F^2+1)(F-1) > n$, $(F+1)(F^2-1) > n$, we have $1 \le c, d \le F-1$. Note that

$$r_1F + r_0 = R = \frac{n+1}{F} = cdF + d - c,$$

so that $d - c \equiv r_0 \pmod{F}$. It follows that $d = c + r_0$ or $d = c + r_0 - F$, that is, $d = c + r_0 - iF$ for $i = 0$ or 1. Thus,

$$r_1F + r_0 = c(c + r_0 - iF)F + r_0 - iF,$$

so that $r_1 = c(c + r_0 - iF) - i$, which implies that

$$c^2 + (r_0 - iF)c - r_1 - i = 0.$$

But then $x^2 + (r_0 - iF) - r_1 - i$ has a positive integral root for one of $i = 0, 1$. This proves one direction.

Suppose now that $x^2 + (r_0 - iF) - r_1 - i$ has a positive integral root c for one of $i = 0, 1$. Undoing the above algebra we see that $cF+1$ is a divisor of n. But $n \equiv -1 \pmod{F}$, so n is composite, since the hypotheses imply $F > 2$. □

We can improve the $n+1$ test further, requiring only $F \ge n^{3/10}$. The proof is completely analogous to Theorem 4.1.6, and we leave said proof as Exercise 4.14.

Theorem 4.2.9. *Suppose $n \ge 214$ and the hypotheses of Theorem 4.2.8 hold, except that $n^{3/10} \le F \le n^{1/3}+1$. Say the base-$F$ expansion of $n+1$ is $c_3F^3 + c_2F^2 + c_1F$, and let $c_4 = c_3F + c_2$. Then n is prime if and only if the following conditions hold:*

1. *$(c_1 + tF)^2 - 4t + 4c_4$ is not a square for t integral, $|t| \le 5$,*
2. *with u/v the continued fraction convergent to c_1/F such that v is maximal subject to $v < F^2/\sqrt{n}$ and with $d = \lfloor c_4v/F + 1/2 \rfloor$, the polynomial $vx^3 - (uF - c_1v)x^2 - (c_4v - dF + u)x + d$ has no integral root a such that $aF+1$ is a nontrivial factor of n, and the polynomial $vx^3 + (uF - c_1v)x^2 - (c_4v + dF + u)x + d$ has no integral root b such that $bF-1$ is a nontrivial factor of n.*

The next result allows one to combine partial factorizations of both $n-1$ and $n+1$ in attempting to prove n prime.

Theorem 4.2.10 (Brillhart, Lehmer, and Selfridge). *Suppose that n is a positive integer, $F_1|n-1$, and that (4.3) holds for some integer a_1 and $F = F_1$. Suppose, too, that f, Δ are as in (4.12), $\gcd(n, 2b) = 1$, $\left(\frac{\Delta}{n}\right) = -1$, $F_2|n+1$, and that (4.14) holds for $F = F_2$. Let F be the least common multiple of F_1, F_2. Then each prime factor of n is congruent to either 1 or $n \pmod F$. In particular, if $F > \sqrt{n}$ and $n \bmod F$ is not a nontrivial factor of n, then n is prime.*

Note that if F_1, F_2 are both even, then $F = \frac{1}{2}F_1F_2$, otherwise $F = F_1F_2$.

Proof. Let p be a prime factor of n. Theorem 4.1.3 implies $p \equiv 1 \pmod{F_1}$, while Theorem 4.2.3 implies that $p \equiv \left(\frac{\Delta}{p}\right) \pmod{F_2}$. If $\left(\frac{\Delta}{p}\right) = 1$, then $p \equiv 1 \pmod F$, and if $\left(\frac{\Delta}{p}\right) = -1$, then $p \equiv n \pmod F$. The last assertion of the theorem is then immediate. □

4.2.3 Divisors in residue classes

What if in Theorem 4.2.10 we have $F < n^{1/2}$? The theorem would be useful if we had a quick way to search for prime factors of n that are either 1 or $n \pmod F$. The following algorithm of [Lenstra 1984] provides such a quick method when $F/n^{1/3}$ is not too small.

Algorithm 4.2.11 (Divisors in residue classes). We are given positive integers n, r, s with $r < s < n$ and $\gcd(r, s) = 1$. This algorithm creates a list of all divisors of n that are congruent to $r \pmod s$.

1. [Initialize]
    $r^* = r^{-1} \bmod s$;
    $r' = nr^* \bmod s$;
    $(a_0, a_1) = (s, r'r^* \bmod s)$;
    $(b_0, b_1) = (0, 1)$;
    $(c_0, c_1) = (0, (nr^* - ra_1)/s \bmod s)$;
2. [Euclidean chains]
    Develop the Euclidean sequences $(a_i), (q_i)$, where $a_i = a_{i-2} - q_i a_{i-1}$ and $0 \le a_i < a_{i-1}$ for i even, $0 < a_i \le a_{i-1}$ for i odd, terminating at $a_t = 0$ with t even;
    Develop the sequences $(b_i), (c_i)$ for $i = 0, 1, \ldots, t$ with the rules $b_i = b_{i-2} - q_i b_{i-1}$, $c_i = c_{i-2} - q_i c_{i-1}$;
3. [Loop]
    for($0 \le i \le t$) {
        For each integer $c \equiv c_i \pmod s$ with $|c| < s$ if i is even, $2a_ib_i < c < a_ib_i + n/s^2$ if i is odd, attempt to solve the following system for x, y:

$$xa_i + yb_i = c, \quad (xs + r)(ys + r') = n; \tag{4.16}$$

If a nonnegative integral solution (x, y) is found, report $xs+r$ as a divisor of n that is also $\equiv r \pmod{s}$;

}

The theoretical justification for this algorithm is as follows:

Theorem 4.2.12 (Lenstra). *Algorithm 4.2.11 creates the list of all divisors of n that are congruent to $r \pmod{s}$. Moreover, if $s \geq n^{1/3}$, then the running time is $O(\ln n)$ arithmetic operations on integers of size $O(n)$ and $O(\ln n)$ evaluations of the integer part of square root for arguments of size $O(n^7)$.*

Proof. We first note some simple properties of the sequences $(a_i), (b_i)$. We have

$$a_i > 0 \text{ for } 0 \leq i < t, \ a_t = 0. \tag{4.17}$$

In addition, we have

$$b_{i+1}a_i - a_{i+1}b_i = (-1)^i s \text{ for } 0 \leq i < t. \tag{4.18}$$

Indeed, the relation (4.18) holds for $i = 0$. If $0 < i < t$ and the relation holds for $i-1$, then

$$\begin{aligned} b_{i+1}a_i - a_{i+1}b_i &= (b_{i-1} - q_{i+1}b_i)a_i - (a_{i-1} - q_{i+1}a_i)b_i \\ &= b_{i-1}a_i - a_{i-1}b_i \\ &= (-1)^i s. \end{aligned}$$

Thus (4.18) follows from induction.

Finally, note that we have

$$b_0 = 0, \ b_i < 0 \text{ for } i \text{ even, and } i \neq 0, \ b_i > 0 \text{ for } i \text{ odd}. \tag{4.19}$$

Indeed, (4.19) holds for $i = 0, 1$, and from $b_i = b_{i-2} - q_i b_{i-1}$ and $q_i > 0$, we see that it holds for the general i if it holds for $i-1, i-2$. Thus (4.19) holds via induction.

Suppose now that $xs + r$ is a divisor of n with $x \geq 0$. We must show that the algorithm discovers it. There is an integer $y \geq 0$ with $n = (xs+r)(ys+r')$. We have

$$xa_i + yb_i \equiv c_i \pmod{s} \text{ for } 0 \leq i \leq t. \tag{4.20}$$

Indeed, (4.20) holds trivially for $i = 0$, it holds for $i = 1$ because of $n = (xs+r)(ys+r')$ and the definition of c_1, and it holds for larger values of i from the inductive definitions of the sequences $(a_i), (b_i), (c_i)$.

It thus suffices to show that there is some even value of i with $|xa_i + yb_i| < s$ or there is some odd value of i with $2a_ib_i < xa_i + yb_i < a_ib_i + n/s^2$. For if so, $xa_i + yb_i$ will be one of the numbers c computed in step [Loop] of Algorithm 4.2.11, because of (4.20). Thus, step [Loop] will successfully retrieve the numbers x, y.

We have $xa_0 + yb_0 = xa_0 \ge 0$ and $xa_t + yb_t = yb_t \le 0$, so there is some even index i with

$$xa_i + yb_i \ge 0, \quad xa_{i+2} + yb_{i+2} \le 0.$$

If one of these quantities is less than s in absolute value, we are done, so assume that the first quantity is $\ge s$ and the second is $\le -s$. Then from (4.17), (4.18), (4.19),

$$xa_i \ge xa_i + yb_i \ge s = b_{i+1}a_i - a_{i+1}b_i \ge b_{i+1}a_i,$$

from which we conclude that $x \ge b_{i+1}$. We also have

$$yb_{i+2} \le xa_{i+2} + yb_{i+2} \le -s = b_{i+2}a_{i+1} - a_{i+2}b_{i+1} < b_{i+2}a_{i+1},$$

so that $y > a_{i+1}$. Therefore,

$$xa_{i+1} + yb_{i+1} > 2a_{i+1}b_{i+1},$$

and from $(x - b_{i+1})(y - a_{i+1}) > 0$, we have

$$xa_{i+1} + yb_{i+1} \le xy + a_{i+1}b_{i+1} < a_{i+1}b_{i+1} + \frac{n}{s^2}.$$

This completes the proof of correctness.

The running-time assertion follows from Theorem 2.1.3 and Algorithm 2.1.4. These results imply that the calculation of r^* is within our time bound and that $t = O(\ln n)$. Moreover, if $s \ge n^{1/3}$, then for each i there are at most 2 values of c for which the system (4.16) must be solved. Solving such a system involves $O(1)$ arithmetic operations and a square root extraction, as we shall see. Thus, there are a total of $O(\ln n)$ arithmetic operations and square root extractions.

It remains to estimate the size of the integers for which we need to compute the integer part of the square root. Note that x, y are solutions to the system (4.16) if and only if $u = a_i(xs + r)$, $v = b_i(ys + r')$ are roots of the quadratic polynomial

$$T^2 - (cs + a_ir + b_ir')T + a_ib_in.$$

For this polynomial to have integral roots it is necessary and sufficient that

$$\Delta = (cs + a_ir + b_ir')^2 - 4a_ib_i$$

be a square. We now show that $\Delta = O(s^7) = O(n^7)$. Let $B = \max\{|b_i|\}$. We shall show that $B < s^{5/2}$. Then, since c, a_i, r, r' are all bounded in absolute value by $2s$, it follows that $\Delta = O(s^7)$. (To see that $|c| < 2s$, note that $|c| < s$ if i is even; and if i is odd, for the interval $(2a_ib_i, a_ib_i + n/s^2)$ to have any integers in it, then $0 < a_ib_i < n/s^2 \le s$.)

To see the bound on B note that

$$|b_i| = |b_{i-2}| + q_i|b_{i-1}| \text{ for } i = 2, \dots, t,$$

so that

$$B = |b_t| < \prod_{i=2}^{t}(1 + q_i) < 2^t \prod_{i=2}^{t} q_i.$$

But $a_{i-2} \ge q_i a_{i-1}$ for $i = 2, \ldots, t$, so that

$$s = a_0 \ge \prod_{i=2}^{t} q_i.$$

We conclude that $B < 2^t s$. From Theorem 2.1.3 we have that $t < \ln s / \ln((1+\sqrt{5})/2)$, so that $2^t < s^{3/2}$. Our estimate and the theorem follow. □

Remark. The integer square roots that are performed in the algorithm may be done via Algorithm 9.2.11. If $s < n^{1/3}$, Algorithm 4.2.11 still works, but the number of square root steps is then $O(n^{1/3}s^{-1}\ln n)$.

Note that if F in Theorem 4.2.10 is such that $F/n^{1/3}$ is not very small, we can use that theorem and Algorithm 4.2.11 as a speedy primality test. In general, we can use Algorithm 4.2.11 in a primality test if we have learned that each prime factor of n is congruent to $r_i \pmod{s}$ for some $i \in [1, k]$, where each $\gcd(r_i, s) = 1$, $0 < r_i < s$, and $s \ge n^{1/3}$. Then with k calls to Algorithm 4.2.11 we will either find a nontrivial factor of n, or failing this, prove that n is prime. However, if $s \ge \sqrt{n}$, there is no need to use Algorithm 4.2.11. Indeed, if none of the integers r_i are proper factors of n, then every prime dividing n exceeds $\sqrt{n}$, so n is prime.

One can use a recent result of [Coppersmith 1997] (also [Coppersmith et al. 2000]) to improve on Algorithm 4.2.11 and find all divisors of n that are congruent to $r \pmod{s}$ when r, s are coprime and $s > n^{1/4+\epsilon}$. The Coppersmith paper uses the fast lattice basis reduction method of A. Lenstra, H. Lenstra and L. Lovasz. This lattice basis reduction method is often useful in practice, and it may well be that Coppersmith's algorithm is practical. In fact, Howgrave-Graham informs us that it is indeed practical for moduli $s > n^{0.29}$, say. Theoretically, the method is deterministic and runs in polynomial time, but this running time depends on the choice of ϵ; the smaller the ϵ, the higher the running time.

It remains an open question whether an efficient algorithm can be found that finds divisors of n that are congruent to $r \pmod{s}$ when s is about $n^{1/4}$ or smaller.

Here is another attractive open question. Let $D(n, s, r)$ denote the number of divisors of n that are congruent to $r \pmod{s}$. Given $\alpha > 0$, is $D(n, s, r)$ bounded as n, s, r range over all triples with $\gcd(r, s) = 1$ and $s \ge n^\alpha$? This is known for every $\alpha > 1/4$, but it is open for $\alpha = 1/4$; see [Lenstra 1984].

4.3 The finite field primality test

This section is primarily theoretical and is not intended to supply a practical primality test. The algorithm described has a favorable complexity estimate, but there are other, more complicated algorithms that majorize it in practice. Some of these algorithms are discussed later in this chapter.

The preceding sections, and in particular Theorem 4.2.10 and Algorithm 4.2.11, show that if we have a completely factored divisor F of $n^2 - 1$ with $F \geq n^{1/3}$, then we can efficiently decide, with a rigorous proof, whether n is prime or composite. As an aside: If $F_1 = \gcd(F, n-1)$ and $F_2 = \gcd(F, n+1)$, then $\operatorname{lcm}(F_1, F_2) \geq \frac{1}{2}F$, so that the "$F$" of Theorem 4.2.10 is at least $\frac{1}{2}n^{1/3}$. In this section we shall discuss a method of [Lenstra 1985] that works if we have a fully factored divisor F of $n^I - 1$ for some positive integer I and that is efficient if $F \geq n^{1/3}$ and I is not too large.

Before we describe the algorithm, we discuss a subroutine that will be used. If $n > 1$ is an integer, consider the ring $\mathbf{Z}_n[x]$ of polynomials in the variable x with coefficients being integer residues modulo n. An ideal of $\mathbf{Z}_n[x]$ is a nonempty subset closed under addition and closed under multiplication by all elements of $\mathbf{Z}_n[x]$. For example, if $f, g \in \mathbf{Z}_n[x]$, the set of all af with $a \in \mathbf{Z}_n[x]$ is an ideal, and so is the set of all $af + bg$ with $a, b \in \mathbf{Z}_n[x]$. The first example is of a *principal ideal* (with generator f). The second example may or may not be principal. For example, say $n = 15, f(x) = 3x + 1, g(x) = x^2 + 4x$. Then the ideal generated by f and g is all of $\mathbf{Z}_{15}[x]$, and so is principally generated by 1. (To see that 1 is in the ideal, note that $f^2 - 9g = 1$.)

Definition 4.3.1. We shall say that $f, g \in \mathbf{Z}_n[x]$ are *coprime* if the ideal they generate is all of $\mathbf{Z}_n[x]$, that is, there are $a, b \in \mathbf{Z}_n[x]$ with $af + bg = 1$.

It is not so hard to prove that every ideal in $\mathbf{Z}_n[x]$ is principally generated if and only if n is prime (see Exercise 4.18). The following algorithm, which is merely a dressed-up version of the Euclid algorithm (Algorithm 2.2.1), either finds a monic principal generator for the ideal generated by two members $f, g \in \mathbf{Z}_n[x]$, or gives a nontrivial factorization of n. If the principal ideal generated by $h \in \mathbf{Z}_n[x]$ is the same ideal as that generated by f and g and if h is monic, we write $h = \gcd(f, g)$. Thus f, g are coprime in $\mathbf{Z}_n[x]$ if and only if $\gcd(f, g) = 1$.

Algorithm 4.3.2 (Finding principal generator). We are given an integer $n > 1$ and $f, g \in \mathbf{Z}_n[x]$, with g monic. This algorithm produces either a nontrivial factorization of n, or a monic element $h \in \mathbf{Z}_n[x]$ such that $h = \gcd(f, g)$, that is, the ideal generated by f and g is equal to the ideal generated by h. We assume that either $f = 0$ or $\deg f \leq \deg g$.

1. [Zero polynomial check]
   if($f == 0$) return g;
2. [Euclid step]
   Set c equal the leading coefficient of f;
   Attempt to find $c^* \equiv c^{-1} \pmod n$ by Algorithm 2.1.4, but if this attempt produces a nontrivial factorization of n, then return this factorization;
   $f = c^* f$; // Multiplication is modulo n; the polynomial f is now monic.
   $r = g \bmod f$; // Divide with remainder is possible since f is monic.

$(f, g) = (r, f)$;
goto [Zero polynomial check];

The next theorem is the basis of the finite field primality test.

Theorem 4.3.3 (Lenstra). *Suppose that n, I, F are positive integers with $n > 1$ and $F | n^I - 1$. Suppose $f, g \in \mathbf{Z}_n[x]$ are such that*

(1) *$g^{n^I - 1} - 1$ is a multiple of f in $\mathbf{Z}_n[x]$,*

(2) *$g^{(n^I-1)/q} - 1$ and f are coprime in $\mathbf{Z}_n[x]$ for all primes $q|F$,*

(3) *each of the I elementary symmetric polynomials in $g, g^n, \ldots, g^{n^{I-1}}$ is congruent $\pmod{f}$ to an element of $\mathbf{Z}_n$.*

Then each prime factor p of n is congruent to $n^j \pmod{F}$ for some $j \in [0, I-1]$.

We remark that if we show that the hypotheses of Theorem 4.3.3 hold and if we also show that n has no proper divisors in the residue classes $n^j \pmod{F}$ for $j = 0, 1, \ldots, I - 1$, then we have proved that n is prime. This idea is the basis of Lenstra's finite field primality test, which we shall describe after we prove the theorem.

Proof. Let p be a prime factor of n. Thinking of f now in $\mathbf{Z}_p[x]$, let $f_1 \in \mathbf{Z}_p[x]$ be an irreducible factor, so that $\mathbf{Z}_p[x]/(f_1) = \mathbf{F}$ is a finite field extension of $\mathbf{Z}_p$. Let $\bar{g}$ be the image of g in $\mathbf{F}$. The hypotheses (1), (2) imply that $\bar{g}^{n^I-1} = 1$ and $\bar{g}^{(n^I-1)/q} \neq 1$ for all primes $q|F$. So the order of $\bar{g}$ in $\mathbf{F}^*$ (the multiplicative group of the finite field $\mathbf{F}$) is a multiple of F. Hypothesis (3) implies that the polynomial $h(T) = (T - \bar{g})(T - \bar{g}^n) \cdots (T - \bar{g}^{n^{I-1}}) \in \mathbf{F}[T]$ is actually in $\mathbf{Z}_p[T]$. Now, for any polynomial in $\mathbf{Z}_p[T]$, if α is a root, so is α^p. Thus $h(\bar{g}^p) = 0$. But we have the factorization of $h(T)$, and we see that the only roots are $\bar{g}, \bar{g}^n, \ldots, \bar{g}^{n^{I-1}}$, so that we must have $\bar{g}^p \equiv \bar{g}^{n^j}$ for some $j = 0, 1, \ldots, I-1$. Since the order of $\bar{g}$ is a multiple of F, we have $p \equiv n^j \pmod{F}$. □

A number of questions naturally present themselves: If n is prime, will f, g as described in Theorem 4.3.3 exist? If f, g exist, is it easy to find examples? Can (1), (2), (3) in Theorem 4.3.3 be verified quickly?

The first question is easy. If n is prime, then any polynomial $f \in \mathbf{Z}_n[x]$ that is irreducible with $\deg f = I$, and any polynomial $g \in \mathbf{Z}_n[x]$ that is not a multiple of f will together satisfy (1) and (3). Indeed, if f is irreducible of degree I, then $\mathbf{F} = \mathbf{Z}_n[x]/(f)$ will be a finite field of order n^I, and so (1) just expresses the Lagrange theorem (a group element raised to the order of the group is the group identity) for the multiplicative group $\mathbf{F}^*$. To see (3) note that the Galois group of $\mathbf{F}$ is generated by the Frobenius automorphism: raising to the n-th power. That is, the Galois group consists of the I functions from $\mathbf{F}$ to $\mathbf{F}$, where the j-th function takes $\alpha \in \mathbf{F}$ and sends it to α^{n^j} for $j = 0, 1, \ldots, I-1$. Each of these functions fixes an expression that is symmetric in $g, g^n, \ldots, g^{n^{I-1}}$, so such an expression must be in the fixed field $\mathbf{Z}_n$. This is the assertion of (3).

It is not true that every choice for g with $g \not\equiv 0 \pmod f$ satisfies (2). But the group $\mathbf{F}^*$ is cyclic, and any cyclic generator satisfies (2). Moreover, there are quite a few cyclic generators, so a random search for g should not take long to find one. In particular, if g is chosen randomly as a nonzero polynomial in $\mathbf{Z}_n[x]$ of degree less than I, then the probability that g satisfies (2) is at least $\varphi(n^I - 1)/(n^I - 1)$ (given that n is prime and f is irreducible of degree I), so the expected number of choices before a valid g is found is $O(\ln\ln(n^I))$.

But what of f? Are there irreducible polynomials in $\mathbf{Z}_n[x]$ of degree I, can we quickly recognize one when we have it, and can we find one quickly? Yes, yes, yes. In fact (2.5) shows that not only are there irreducible polynomials of degree I, but that there are plenty of them, so that a random degree I polynomial has about a 1 in I chance of being irreducible. Further, Algorithm 2.2.9 provides an efficient way to test whether a polynomial is irreducible.

We now embody the above thoughts in the following explicit algorithm:

Algorithm 4.3.4 (Finite field primality test). We are given positive integers n, I, F with $F | n^I - 1$, $F \ge n^{1/3}$ and we are given the complete prime factorization of F. This probabilistic algorithm decides whether n is prime or composite, returning "n is prime" in the former case and "n is composite" in the latter case.

1. [Find irreducible polynomial of degree I]
   Via Algorithm 2.2.9, and using Algorithm 4.3.2 for the gcd steps, attempt to find a random monic polynomial f in $\mathbf{Z}_n[x]$ of degree I that is irreducible if n is prime. That is, continue testing random polynomials until Algorithm 2.2.9 either returns YES, or its gcd step finds a nontrivial factorization of n. In the latter case, return "n is composite";
   // The polynomial f is irreducible if n is prime.
2. [Find primitive element]
   Choose $g \in \mathbf{Z}_n[x]$ at random with g monic, $\deg g < I$;
   if($1 \ne g^{n^I-1} \bmod f$) return "n is composite";
   for(prime $q|F$) {
   Attempt to compute $\gcd(g^{(n^I-1)/q} - 1, f)$ via Algorithm 4.3.2, but if a nontrivial factorization of n is found in this attempt, return "n is composite";
   if($\gcd(g^{(n^I-1)/q} - 1, f) \ne 1$) goto [Find primitive element];
   }
3. [Symmetric expressions check]
   Form the polynomial $(T-g)(T-g^n)\cdots(T-g^{n^{I-1}}) = T^I + c_{I-1}T^{I-1} + \ldots + c_0$ in $\mathbf{Z}_n[x,T]/(f(x))$;
   // The coefficients c_j are in $\mathbf{Z}_n[x]$ and are reduced modulo f.
   for($0 \le j < I$) if($\deg c_j > 0$) return "n is composite";
4. [Divisor search]
   for($1 \le j < I$) {

Find the divisors of n that are $\equiv n^j \pmod{F}$, via Algorithm 4.2.11, and if a proper factor of n is found, return "n is composite";

}

return "n is prime";

If n is prime, the expected number of arithmetic operations with integers the size of n for Algorithm 4.3.4 to declare n prime is $O(I^c + \ln^c n)$ for some positive constant c. (We make no assertion on the expected running time for composite inputs.)

Given a prime n, the question remains on how one is supposed to come up with the numbers I, F. The criteria are as follows: that F is supposed to be large, namely, $F \geq n^{1/3}$; we are supposed to know the prime factorization of F, $F|n^I - 1$, with I not very large (since otherwise, the algorithm will not be very fast). For some numbers n we can choose $I = 1$ or 2; this was the subject of earlier sections in this chapter. The question remains for the general case whether we can find I, F that fit the above criteria.

An interesting observation is that we can pick up some small primes in $n^I - 1$ with very little work. For example, suppose $I = 12$. Then $n^I - 1$ is a multiple of $65520 = 2^4 \cdot 3^2 \cdot 5 \cdot 7 \cdot 13$, provided that n is coprime to 65520. In general, if q is a prime power that is coprime to n and $\varphi(q)|I$, then $q|n^I - 1$. This is just an assertion of the Euler theorem; see (2.2). (If q is a power of 2 higher than 4, then we need only $\frac{1}{2}\varphi(q)|I$.) Can such "cheap" divisors of $n^I - 1$ amount to much? Indeed they can. For example, say $I = 7! = 5040$. Then if n is not divisible by any prime up to 2521, then $n^{5040} - 1$ is divisible by

$$
\begin{aligned}
&15321986788854443284662612735663611380010431225771200 = \\
&2^6 \cdot 3^3 \cdot 5^2 \cdot 7^2 \cdot 11 \cdot 13 \cdot 17 \cdot 19 \cdot 29 \cdot 31 \cdot 37 \cdot 41 \cdot 43 \cdot 61 \cdot 71 \cdot 73 \cdot \\
&\quad 113 \cdot 127 \cdot 181 \cdot 211 \cdot 241 \cdot 281 \cdot 337 \cdot 421 \cdot 631 \cdot 1009 \cdot 2521.
\end{aligned}
$$

So $I = 5040$ can be used in Algorithm 4.3.4 for primes n up to $3.5 \cdot 10^{156}$ (and exceeding 2521).

From the above example with $I = 5040$ one might expect that in general a choice of I with enough "cheap" factors in $n^I - 1$ is a fairly small function of n. Indeed, we have the following theorem, which appeared in [Adleman et al. 1983]. The proof uses some deep tools in analytic number theory.

Theorem 4.3.5. *Let $I(x)$ be the least positive square-free integer I such that the product of the primes p with $p-1|I$ exceeds x. Then there is a number c such that $I(x) < (\ln x)^{c \ln\ln\ln x}$ for all $x > 16$.*

The reason for assuming $x > 16$ is to ensure that the triple-logarithm is positive. It is not necessary in the results so far that I be square-free, but because of an algorithm in the next section, this condition is included in the above result.

Corollary 4.3.6. *There is a positive number c' such that the expected running time for Algorithm 4.3.4 to declare a prime input n to be prime is less than* $(\ln n)^{c' \ln\ln\ln n}$.

Since the triple log function grows so slowly, this running-time bound is "almost" $\ln^{O(1)} n$, and so is "almost" polynomial time.

4.4 Gauss and Jacobi sums

In 1983, Adleman, Pomerance, and Rumely [Adleman et al. 1983] published a primality test with the running-time bound of $(\ln n)^{c \ln\ln\ln n}$ for prime inputs n and some positive constant c. The proof rested on Theorem 4.3.5 and on arithmetic properties of Jacobi sums. Two versions of the test were presented, a somewhat simpler and more practical version that was probabilistic, and a deterministic test. Both versions had the same complexity estimate. As with some of the other algorithms in this chapter, a declaration of primality in the probabilistic APR test definitely implies that the number is prime. The only thing in doubt is a prediction of the running time.

Shortly afterwards, there were two types of developments. In one direction, more practical versions of the test were found, and in the other, less practical, but simpler versions of the test were found. In the next section we shall discuss one of the second variety, the deterministic Gauss sums test of H. Lenstra [Lenstra 1981].

4.4.1 Gauss sums test

In Section 2.3.1 we introduced Gauss sums for quadratic characters. Here we consider Gauss sums for arbitrary Dirichlet characters. If q is a prime with primitive root g and if ζ is a complex number with $\zeta^{q-1} = 1$, then we can "construct" a character χ to the modulus q via $\chi(g^m) = \zeta^m$ for every integer m (and of course, $\chi(u) = 0$ if u is a multiple of q). (See Section 1.4.3 for a discussion of characters.) We may also "construct" the Gauss sum $\tau(\chi)$. With the notation $\zeta_n = e^{2\pi i/n}$ (which is a primitive n-th root of 1), we define $\tau(\chi) = \sum_{m=1}^{q-1} \chi(m)\zeta_q^m$.

To be even more specific, suppose $g = g_q$ is the least positive primitive root for q, suppose p is a prime dividing $q-1$, and let $\zeta = \zeta_p$. Then $\zeta^{q-1} = 1$. Call the resulting character $\chi_{p,q}$, so that $\chi_{p,q}(g_q^m) = \zeta_p^m$ for every integer m. Let

$$G(p,q) = \tau(\chi_{p,q}) = \sum_{m=1}^{q-1} \chi_{p,q}(m)\zeta_q^m = \sum_{k=1}^{q-1} \zeta_p^k \zeta_q^{g_q^k} = \sum_{k=1}^{q-1} \zeta_p^{k \bmod p} \zeta_q^{g_q^k \bmod q}.$$

(That this definition in the case $p = 2$ is equivalent to that in Definition 2.3.6 is the subject of Exercise 4.19.)

We are interested in the Gauss sums $G(p,q)$ for their arithmetic properties, though it may not be clear what a sum of lots of complex numbers has to do with arithmetic! The Gauss sum $G(p,q)$ is an element of the ring $\mathbf{Z}[\zeta_p, \zeta_q]$.

Elements of the ring can be expressed uniquely as sums $\sum_{j=0}^{p-2}\sum_{k=0}^{q-2} a_{j,k}\zeta_p^j\zeta_q^k$ where each $a_{j,k} \in \mathbf{Z}$. We thus can say what it means for two elements of $\mathbf{Z}[\zeta_p, \zeta_q]$ to be congruent modulo n; namely, the corresponding integer coefficients are congruent modulo n. Also note that if α is in $\mathbf{Z}[\zeta_p, \zeta_q]$, then so is its complex conjugate $\overline{\alpha}$.

It is very important in actual ring computations to treat ζ_p, ζ_q symbolically. As with Lucas sequences, where we work symbolically with the roots of quadratic polynomials, we treat ζ_p, ζ_q as symbols x, y, say, which obey the rules

$$x^{p-1} + x^{p-2} + \cdots + 1 = 0, \quad y^{q-1} + y^{q-2} + \cdots + 1 = 0.$$

In particular, one may avoid complex-floating-point methods.

We begin with a well-known result about Gauss sums.

Lemma 4.4.1. *If p, q are primes with $p \mid q-1$, then $G(p,q)\overline{G(p,q)} = q$.*

Proof. Let $\chi = \chi_{p,q}$. We have

$$G(p,q)\overline{G(p,q)} = \sum_{m_1=1}^{q-1}\sum_{m_2=1}^{q-1} \chi(m_1)\overline{\chi(m_2)}\zeta_q^{m_1-m_2}.$$

Let m_2^{-1} denote a multiplicative inverse of m_2 modulo q, so that $\overline{\chi(m_2)} = \chi(m_2^{-1})$. Note that if $m_1m_2^{-1} \equiv a \pmod q$, then $\chi(m_1)\overline{\chi(m_2)} = \chi(a)$ and $m_1 - m_2 \equiv (a-1)m_2 \pmod q$. Thus,

$$G(p,q)\overline{G(p,q)} = \sum_{a=1}^{q-1}\chi(a)\sum_{m=1}^{q-1}\zeta_q^{(a-1)m}.$$

The inner sum is $q-1$ in the case $a=1$ and is -1 in the cases $a>1$. Thus,

$$G(p,q)\overline{G(p,q)} = q - 1 - \sum_{a=2}^{q-1}\chi(a) = q - \sum_{a=1}^{q-1}\chi(a).$$

Finally, by (1.28), this last sum is 0, which proves the lemma. □

The next result begins to show a possible relevance of Gauss sums to primality testing. It may be viewed as an analogue to Fermat's little theorem.

Lemma 4.4.2. *Suppose p, q, n are primes with $p|q-1$ and* $\gcd(pq, n) = 1$. *Then*

$$G(p,q)^{n^{p-1}-1} \equiv \chi_{p,q}(n) \pmod n.$$

Proof. Let $\chi = \chi_{p,q}$. Since n is prime, the multinomial theorem implies that

$$G(p,q)^{n^{p-1}} = \left(\sum_{m=1}^{q-1}\chi(m)\zeta_q^m\right)^{n^{p-1}} \equiv \sum_{m=1}^{q-1}\chi(m)^{n^{p-1}}\zeta_q^{mn^{p-1}} \pmod n.$$

By Fermat's little theorem, $n^{p-1} \equiv 1 \pmod{p}$, so that $\chi(m)^{n^{p-1}} = \chi(m)$. Letting n^{-1} denote a multiplicative inverse of n modulo q, we have

$$\sum_{m=1}^{q-1} \chi(m)^{n^{p-1}} \zeta_q^{mn^{p-1}} = \sum_{m=1}^{q-1} \chi(m) \zeta_q^{mn^{p-1}} = \sum_{m=1}^{q-1} \chi(n^{-(p-1)}) \chi(mn^{p-1}) \zeta_q^{mn^{p-1}}$$
$$= \chi(n) \sum_{m=1}^{q-1} \chi(mn^{p-1}) \zeta_q^{mn^{p-1}} = \chi(n) G(p,q),$$

where the next to last equality uses that $\chi(n^p) = \chi(n)^p = 1$ and the last equality follows from the fact that mn^{p-1} traverses a reduced residue system $\pmod{q}$ as m does this. Thus,

$$G(p,q)^{n^{p-1}} \equiv \chi(n) G(p,q) \pmod{n}.$$

Let q^{-1} be a multiplicative inverse of q modulo n and multiply this last display by $q^{-1}\overline{G(p,q)}$. Lemma 4.4.1 then gives the desired result. □

The next lemma allows one to replace a congruence with an equality, in some cases.

Lemma 4.4.3. *If m, n are natural numbers with m not divisible by n and $\zeta_m^j \equiv \zeta_m^k \pmod{n}$, then $\zeta_m^j = \zeta_m^k$.*

Proof. By multiplying the congruence by ζ_m^{-k}, we may assume the given congruence is $\zeta_m^j \equiv 1 \pmod{n}$. Note that $\prod_{l=1}^{m-1}(x - \zeta_m^l) = (x^m - 1)/(x - 1)$, so that $\prod_{l=1}^{m-1}(1 - \zeta_m^l) = m$. Thus no factor in this last product is zero modulo n, which proves the result. □

We are now ready to describe the deterministic Gauss sums primality test.

Algorithm 4.4.4 (Gauss sums primality test). We are given an integer $n > 1$. This deterministic algorithm decides whether n is prime or composite, returning "n is prime" or "n is composite" in the appropriate case.

1. [Initialize]
   $I = -2$;
2. [Preparation]
   $I = I + 4$;
   Find the prime factors of I by trial division, but if I is not square-free, goto [Preparation];
   Set F equal to the product of the primes q with $q - 1|I$, but if $F^2 \leq n$ goto [Preparation];
   // Now I, F are square-free, and $F > \sqrt{n}$.
   If n is a prime factor of IF, return "n is prime";
   If $\gcd(n, IF) > 1$, return "n is composite";
   for(prime $q|F$) find the least positive primitive root g_q for q;
3. [Probable-prime computation]

for(prime $p|I$) factor $n^{p-1} - 1 = p^{s_p} u_p$ where p does not divide u_p;
for(primes p, q with $p|I,\ q|F,\ p|q-1$) {
Find the first positive integer $w(p,q) \le s_p$ with

$$G(p,q)^{p^{w(p,q)}u_p} \equiv \zeta_p^j \pmod{n} \text{ for some integer } j,$$

but if no such number $w(p,q)$ is found, return "n is composite";
} // Compute symbolically in the ring $\mathbf{Z}[\zeta_p, \zeta_q]$ (see text).

4. [Maximal order search]
for(prime $p|I$) set $w(p)$ equal to the maximum of $w(p,q)$ over all primes $q|F$ with $p|q-1$, and set $q_0(p)$ equal to the least such prime q with $w(p) = w(p,q)$;
for(primes p, q with $p|I,\ q|F,\ p|q-1$) find an integer $l(p,q) \in [0, p-1]$ with $G(p,q)^{p^{w(p)}u_p} \equiv \zeta_p^{l(p,q)} \pmod{n}$;

5. [Coprime check]
For those primes $p|I$ with $w(p) \ge 2$, and such that $l(p,q) = 0$ whenever prime $q|F$ with $p|q-1$ {
$h = G(p, q_0(p))^{p^{w(p)-1}u_p} \bmod n$;
// As in text, $h = \sum_{i=0}^{p-2} \sum_{k=0}^{q-2} a_{i,k} \zeta_p^i \zeta_q^k$ with $a_{i,k} \in [0, n-1]$.
if(h has some coefficient $a_{i,k} > 1$) {
Set a as the coefficient $a_{i,k}$ of h with $a_{i,k} > 1$ and (i,k) lexocographically minimal;
if($\gcd(a(a-1), n) > 1$) return "n is composite";
}
}

6. [Divisor search]
$l(2) = 0$;
for(odd prime $q|F$) use the Chinese remainder theorem (see Theorem 2.1.6) to construct an integer $l(q)$ with

$$l(q) \equiv l(p,q) \pmod{p} \text{ for each prime } p|q-1;$$

Use the Chinese remainder theorem to construct an integer l with

$$l \equiv g_q^{l(q)} \pmod{q} \text{ for each prime } q|F;$$

for($1 \le j < I$) if $l^j \bmod F$ is a nontrivial factor of n, return "n is composite";
return "n is prime";

Remark. We may omit the condition $F > \sqrt{n}$ and use Algorithm 4.2.11 for the divisor search as in the divisor search step of Algorithm 4.3.4. The algorithm will remain fast if $F > n^{1/3}$.

Theorem 4.4.5. *Algorithm 4.4.4 correctly identifies prime and composite inputs. The running time is bounded by $(\ln n)^{c \ln\ln\ln n}$ for some positive constant c.*

Proof. We first note that a declaration of prime or composite in step [Preparation] is certainly correct. That a declaration of composite in step [Probable-prime computation] is correct follows from Lemma 4.4.2. If the gcd calculation in step [Coprime check] is not 1, it reveals a proper factor of n, so it is correct to declare n composite. It is obvious that a declaration of composite is correct in step [Divisor search], so what remains to be shown is that composite numbers which have survived the prior steps must be factored in step [Divisor search] and so declared composite there.

Suppose n is composite with least prime factor r, and suppose n has survived steps 1–4. We first show that

$$p^{w(p)} | r^{p-1} - 1 \text{ for each prime } p|I. \tag{4.21}$$

This is clear if $w(p) = 1$, so assume $w(p) \geq 2$. Suppose some $l(p,q) \neq 0$. Then by Lemma 4.4.3

$$G(p,q)^{p^{w(p)}u_p} \equiv \zeta_p^{l(p,q)} \not\equiv 1 \pmod{n},$$

so the same is true (mod r). Let h be the multiplicative order of $G(p,q)$ modulo r, so that $p^{w(p)+1}|h$. But Lemma 4.4.2 implies that $h|p(r^{p-1}-1)$, so that $p^{w(p)}|r^{p-1}-1$, as claimed. So suppose that each $l(p,q) = 0$. Then from the calculation in step [Coprime check] we have

$$G(p,q_0)^{p^{w(p)}u_p} \equiv 1 \pmod{r}, \; G(p,q_0)^{p^{w(p)-1}u_p} \not\equiv \zeta_p^j \pmod{r}$$

for all j. Again with h the multiplicative order of $G(p,q_0)$ modulo r, we have $p^{w(p)}|h$. Also, $G(p,q_0)^m \equiv \zeta_p^j \pmod{r}$ for some integers m, j implies that $\zeta_p^j = 1$. Lemma 4.4.2 then implies that $G(p,q_0)^{r^{p-1}-1} \equiv 1 \pmod{r}$ so that $h|r^{p-1}-1$ and $p^{w(p)}|h$. This completes the proof of (4.21).

For each prime $p|I$, (4.21) implies there are integers a_p, b_p with

$$\frac{r^{p-1}-1}{p^{w(p)}u_p} = \frac{a_p}{b_p}, \quad b_p \equiv 1 \pmod{p}. \tag{4.22}$$

Let a be such that $a \equiv a_p \pmod{p}$ for each prime $p|I$. We now show that

$$r \equiv l^a \pmod{F}, \tag{4.23}$$

from which our assertion about step [Divisor search] follows. Indeed, since $F > \sqrt{n} \geq r$, we have r equal to the least positive residue of $l^a \pmod{F}$, so that the proper factor r of n will be discovered in step [Divisor search].

Note that the definition of $\chi_{p,q}$ and of l imply that

$$G(p,q)^{p^{w(p)}u_p} \equiv \zeta_p^{l(p,q)} = \chi_{p,q}(g_q^{l(p,q)}) = \chi_{p,q}(g_q^{l(q)}) = \chi_{p,q}(l) \pmod{r}$$

for every pair of primes p, q with $q|F, p|q-1$. Thus, from (4.22) and Lemma 4.4.2,

$$\begin{aligned}\chi_{p,q}(r) = \chi_{p,q}(r)^{b_p} \equiv G(p,q)^{(r^{p-1}-1)b_p} &= G(p,q)^{p^{w(p)}u_p a_p} \\ \equiv \chi_{p,q}(l)^{a_p} &= \chi_{p,q}(l^a) \pmod{r},\end{aligned}$$

and so by Lemma 4.4.3 we have

$$\chi_{p,q}(r) = \chi_{p,q}(l^a).$$

For each prime $q|F$ there is an integer ρ_q such that $r \equiv g_q^{\rho_q} \pmod q$, so that $\chi_{p,q}(r) = \zeta_p^{\rho_q}$. But $\chi_{p,q}(l^a) = \chi_{p,q}\left(g_q^{l(q)a}\right) = \zeta_p^{l(q)a}$, so that for each pair of primes p, q with $q|F$, $p|q-1$ we have $\rho_q \equiv l(q)a \pmod p$. For a given q the product of the various primes $p|q-1$ is in fact $q-1$ (since $q-1|I$ and I is square-free), so that $\rho_q \equiv l(q)a \pmod{q-1}$ and $r \equiv g_q^{l(q)a} \equiv l^a \pmod q$. Since this holds for each prime $q|F$ and F is square-free, (4.23) follows. This completes the proof of correctness of Algorithm 4.4.4.

It is clear that the running time is bounded by a fixed power of I, so the running time assertion follows immediately from Theorem 4.3.5. □

4.4.2 Jacobi sums test

There are a number of ways that the Gauss sums test can be improved in practice, and the principal way is not to use Gauss sums! Rather, as with the original test of Adleman, Pomerance and Rumely, Jacobi sums are used. The Gauss sums $G(p,q)$ are in the ring $\mathbf{Z}[\zeta_p, \zeta_q]$. Doing arithmetic in this ring modulo n requires dealing with vectors with $(p-1)(q-1)$ coordinates, with each coordinate being a residue modulo n. It is likely in practice that we can take the primes p to be very small, say less than $\ln n$. But the primes q can be somewhat larger, as large as $(\ln n)^{c \ln\ln\ln n}$. The Jacobi sums $J(p,q)$ that we shall presently introduce lie in the much smaller ring $\mathbf{Z}[\zeta_p]$, and so doing arithmetic with them is much speedier.

Recall the character $\chi_{p,q}$ from Section 4.4.1, where p, q are primes with $p|q-1$. We shall suppose that p is an odd prime. Let $b = b(p)$ be the least positive integer with $(b+1)^p \not\equiv b^p + 1 \pmod{p^2}$. (As shown in [Crandall et al. 1997] we may take $b = 2$ for every prime p up to 10^{12} except $p = 1093$ and $p = 3511$, for which we may take $b = 3$. It is probably true that $b(p) = 2$ or 3 for every prime p. We certainly have $b(p) < \ln^2 p$; see Exercise 3.19.)

We now define a Jacobi sum $J(p,q)$. This is

$$J(p,q) = \sum_{m=1}^{q-2} \chi_{p,q}\left(m^b(m-1)\right).$$

The connection to the supposed primality of n is made with the following more general result. Suppose n is an odd prime not divisible by p. Let f be the multiplicative order of n in $\mathbf{Z}_p^*$. Then the ideal (n) in $\mathbf{Z}[\zeta_p]$ factors into $(p-1)/f$ prime ideals $\mathcal{N}_1, \mathcal{N}_2, \ldots, \mathcal{N}_{(p-1)/f}$ each with norm n^f. If α is in $\mathbf{Z}[\zeta_p]$ but not in $\mathcal{N}_j$, then there is some integer a_j with $\alpha^{(n^f-1)/p} \equiv \zeta_p^{a_j} \pmod{\mathcal{N}_j}$. The Jacobi sums test tries this congruence with $\alpha = J(p,q)$ for the same pairs p, q (with $p > 2$) that appear in the Gauss sums test. To implement this, one also needs to find the ideals $\mathcal{N}_j$. This is accomplished by factoring the polynomial $x^{p-1} + x^{p-2} + \cdots + 1$ modulo n into $h_1(x)h_2(x)\cdots h_{(p-1)/f}(x)$, where each $h_j(x)$ is irreducible of degree f. Then we can take for $\mathcal{N}_j$ the ideal generated by n and $h_j(\zeta_p)$. These calculations can be attempted even if we don't know that n is prime, and if they should fail, then n is declared composite.

For a complete description of the test, the reader is referred to [Adleman et al. 1983]. For a practical version that represents the current state of the art, see [Bosma and van der Hulst 1990].

4.5 Exercises

4.1. Show that for n prime, $n > 200560490131$, the number of primitive roots modulo n is $> (n-1)/(2\ln\ln n)$. The following plan may be helpful:

(1) The number of primitive roots modulo n is $\varphi(n-1)$.

(2) If the product P of all the primes $p \le T$ is such that $P \ge m$, then

$$\frac{\varphi(m)}{m} \ge \prod_{p\le T}\left(1-\frac{1}{p}\right).$$

Use ideas such as this to show the inequality for $200560490131 < n < 5.6\cdot 10^{12}$.

(3) Complete the proof using the following estimate of [Rosser and Schoenfeld 1962]:

$$\frac{m}{\varphi(m)} < e^{\gamma}\ln\ln m + \frac{2.5}{\ln\ln m} \quad \text{for } m > 223092870.$$

4.2. Suppose (4.1) is replaced with "for each prime $q|n-1$ there is an integer a_q such that $a_q^{n-1} \equiv 1 \pmod{n}$ and $a_q^{(n-1)/q} \not\equiv 1 \pmod{n}$." Show that n must be prime.

4.3. Suppose we are given a prime n and the complete prime factorization of $n-1$ and we try to use Exercise 4.2 to prove n prime by choosing numbers a_q at random. However, if $q_1, q_2, \ldots, q_k$ are the distinct primes that divide $n-1$ and if $a_{q_1}, \ldots, a_{q_{i-1}}$ have already been found, we first check among these a_q's to see whether one of them works for q_i before we try random choices. Show that there is a number c, independent of n, such that the expected number of distinct a_q's does not exceed c.

4.4. Suppose elements $b_1, b_2, \ldots$ are chosen independently and uniformly at random from the multiplicative group $\mathbf{Z}_n^*$. Let $g(n)$ be the expected value for the least number g such that the subgroup generated by $b_1, \ldots, b_g$ is equal to $\mathbf{Z}_n^*$. In the spirit of Exercise 4.3 show that $g(n) < 3$ for all primes n. What can be said in general when n is not assumed to be prime?

4.5. Show that the Pepin test works with 5 instead of 3 for Fermat numbers larger than 5.

4.6. In 1999 a group of investigators (R. Crandall, E. Mayer, J. Papadopoulos) performed—and checked—a Pepin squaring chain for the twenty-fourth

Fermat number F_{24}. The number is composite. This could be called the deepest verified calculation ever performed prior to 2000 A.D. for a 1-bit (i.e., prime/composite) answer [Crandall et al. 1999]. (More recently, C. Percival has determined the quadrillionth bit of π's binary expansion to be 0; said calculation was somewhat more extensive than the F_{24} resolution.) F_{24} can also be said to be the current largest "genuine Fermat composite" (an F_n proven composite yet enjoying no known explicit proper factors). See Exercise 1.79 for more on the notion of genuine composites.

As of this writing, F_{33} is the smallest Fermat number of unknown character. Estimate how many total operations modulo F_{33} will be required for the Pepin test. How will this compare with the total number of machine operations performed for *all* purposes, worldwide, prior to 2000 A.D.? By what calendar year could F_{33} be resolved via the Pepin test? Note, in this connection, the itemized remarks pursuant to Table 1.3.

Analyze and discuss these issues:

(1) The possibility of parallelizing the Pepin squaring (nobody knows how to parallelize the squaring chain overall in an efficient manner, but indeed one can parallelize *within* one squaring operation by establishing each element of a convolution by way of parallel machinery and the CRT).

(2) The problem of *proving* the character of F_n is what the final Pepin residue says it is. This is an issue because, of course, a machine can sustain either cosmic-ray glitches (hardware) or bugs (software) that ruin the proof. Incidentally, hardware glitches *do* happen; after all, any computing machine, physics tells us, lives in an entropy bath; error probabilities are patently nonzero. As for checking software bugs, it is important to have different code on different machines that are supposed to be checking each other—one does not even want the same programmer responsible for all machines!

On this latter issue, consider the "wavefront" method, in which one, fastest available machine performs Pepin squaring, this continual squaring thought of as a wavefront, with other computations lagging behind in the following way. Using the wavefront machine's already deposited Pepin residues, a collection of (slower, say) machines verify the results of Pepin squarings at various intermediate junctures along the full Pepin squaring chain. For example, the fast, wavefront machine might deposit the millionth, two millionth, three millionth, and four millionth squares of 3; i.e., deposit powers

$$3^{2^{1000000}},\ 3^{2^{2000000}},\ 3^{2^{3000000}},\ 3^{2^{4000000}}$$

all modulo F_n, and each of the slow machines would grab a unique one of these residues, square it just one million times, and expect to find precisely the deterministic result (the next deposited power).

4.7. Prove the following theorems of Suyama (see [Williams 1998]):

(1) Suppose k is an odd number and $N = k2^n + 1$ divides the Fermat number F_m. Prove that if $N < (3 \cdot 2^{m+2} + 1)^2$, then N is prime.

(2) Suppose the Fermat number F_m is factored as FR, where we have the complete prime factorization of F, and R is the remaining unfactored portion. But perhaps R is prime and the factorization is complete. If R is composite, the following test often reveals this fact. Let $r_1 = 3^{F_m - 1} \bmod F_m$ and $r_2 = 3^{F-1} \bmod F_m$. If $r_1 \not\equiv r_2 \pmod R$ then R is composite. (This result is useful, since it replaces most of the mod R arithmetic with mod F_m arithmetic. The divisions by F_m are especially simple, as exemplified in Algorithm 9.2.13.)

4.8. Reminiscent of the Suyama results of Exercise 4.7 is the following scheme that has actually been used for some cofactors of large Fermat numbers [Crandall et al. 1999]. Say that F_n has been subjected to a Pepin test, and we have in hand the final Pepin residue, namely,

$$r = 3^{(F_n - 1)/2} \bmod F_n.$$

Say that someone discovers a factor f of F_n, so that we can write

$$F_n = fG.$$

Prove that if we assign

$$x = 3^{f-1} \bmod F_n,$$

then

$$\gcd(r^2 - x, G) = 1$$

implies that the cofactor G is neither a prime nor a prime power. As in Exercise 4.7, the relatively fast (mod F_n) operation is the reason why we interpose said operation prior to the implicit (mod G) operation in the gcd. All of this shows the importance of carefully squirreling away one's Pepin residues, to be used again in some future season!

4.9. There is an interesting way to find, rigorously, fairly large primes of the Proth form $p = k2^n + 1$. Prove this theorem of Suyama [Williams 1998], that if a p of this form divides some Fermat number F_m, and if $k2^{n-m-2} < 9{\cdot}2^{m+2}+6$, then p is prime.

4.10. Prove the following theorem of Proth: If $n > 1, 2^k | n-1, 2^k > \sqrt{n}$, and $a^{(n-1)/2} \equiv -1 \pmod n$ for some integer a, then n is prime.

4.11. Prove Theorem 4.2.4.

4.12. If the partial factorization (4.2) is found by trial division on $n - 1$ up to the bound B, then we have the additional information that R's prime factors are all $> B$. Show that if a satisfies (4.3) and also $\gcd(a^F - 1, n) = 1$, then every prime factor of n exceeds BF. In particular, if $BF \ge n^{1/2}$, then n is prime.

4.13. Suppose that in addition to the hypotheses of Theorem 4.2.10 we know that all of the prime factors of $R_1 R_2$ exceed B, where $n - 1 = F_1 R_1$,

$n+1 = F_2R_2$. Also suppose there is an integer a_1 such that $a_1^{n-1} \equiv 1 \pmod{n}$, $\gcd(a_1^{F_1} - 1, n) = 1$, and there are f, Δ as in (4.12) with $\gcd(n, 2b) = 1$, $\left(\frac{\Delta}{n}\right) = -1$, $U_{n+1} \equiv 0 \pmod{n}$, $\gcd(U_{F_2}, n) = 1$. Let F denote the least common multiple of F_1, F_2. Show that if the residue $n \bmod F$ is not a proper factor of n and $BF > \sqrt{n}$, then n is prime.

4.14. Prove Theorem 4.2.9.

4.15. By the methods of Exercise 4.1 show the following: If $n > 892271479$ is prime, let N denote the expected number of choices of random pairs $a, b \in \{0, 1, \ldots, n-1\}$, not both 0, until with f given in (4.12), we have $r_f(n) = n+1$. Then $N < 4 \ln \ln n$.

4.16. Prove that $n = 700001$ is prime, first using a factorization of $n-1$, and then again using a factorization of $n+1$.

4.17. Show how the algorithm of Coppersmith that is mentioned near the end of Section 4.2.3 can be used to improve the $n-1$ test, the $n+1$ test, the combined n^2-1 test, the finite field primality test, and the Gauss sums test.

4.18. Show that every ideal in $\mathbf{Z}_n[x]$ is principally generated (that is, is the set of multiples of one polynomial) if and only if n is prime.

4.19. Let q be an odd prime. With the notation of Section 4.4.1 and Definition 2.3.6 show that for integer m not divisible by q, we have $\chi_{2,q}(m) = \left(\frac{m}{q}\right)$ and that $G(2, q) = G(1, q)$.

4.20. Suppose that n survives steps [Preparation] and [Probable-prime computation] of Algorithm 4.4.4, and for each prime $p|I$ we either have $w(p) = 1$ or some $l(p, q) \neq 0$. Thus, no calculation is needed in step [Coprime check]. Show that l in step [Divisor search] may then be taken as n, so that the Chinese remainder theorem calculations in that step may be skipped.

4.21. In the algorithm based on Theorem 4.1.6, one is asked for the integral roots (if any) of a cubic polynomial with integer coefficients. As an initial foray, show how to do this efficiently using a Newton method or a divide-and-conquer strategy. Note the simple Algorithm 9.2.11 for design guidance. Consider the feasibility of rapidly solving even higher-order polynomials for possible integer roots.

4.6 Research problems

4.22. Design a practical algorithm that rigorously determines primality of an arbitrary integer $n \in [2, \ldots, x]$ for as large an x as possible, but carry out the design along the following lines.

Use a *probabilistic* primality test but create a (hopefully minuscule) table of *exceptions*. Or use a small combination of simple tests that has no exceptions up to the bound x. For example, in [Jaeschke 1993] it is shown that no composite below 341550071728321 simultaneously passes the strong probable prime test (Algorithm 3.4.2) for the prime bases below 20.

4.23. By consideration of the Diophantine equation

$$n^k - 4^m = 1,$$

prove that no Fermat number can be a power n^k, $k > 1$. That much is known. But unresolved to this day is this: Must a Fermat number be square-free? Show too that no Mersenne number M_p is a nontrivial power. Must a Mersenne number be square-free?

4.24. Recall the function $M(p)$ defined in Section 4.1.3 as the number of multiplications needed to prove p prime by traversing the Lucas tree for p. Prove or disprove: For all primes p, $M(p) = O(\lg p)$.

4.25. (Broadhurst). The Fibonacci series (u_n) as defined in Exercise 2.5 yields, for certain n, some impressive primes. Work out an efficient primality-testing scheme for Fibonacci numbers, perhaps using publicly available provers.

Incidentally, according to D. Broadhurst all indices are rigorously resolved, in regard to the primality question on u_n, for *all* n through $n = 35999$ inclusive (and, yes, u_{35999} is prime). Furthermore, u_{81839} is known to be prime, yet methods are still needed to resolve some suspected (probable) primes such as the u_n for $n \in \{37511, 50833, 104911\}$, and therefore to resolve the primality question through $n = 104911$.

Chapter 5

EXPONENTIAL FACTORING ALGORITHMS

For almost all of the multicentury history of factoring, the only algorithms available were exponential, namely, the running time was, in the worst case, a fixed positive power of the number being factored. But in the early 1970s, subexponential factoring algorithms began to come "on line." These methods, discussed in the next chapter, have their running time to factor n bounded by an expression of the form $n^{o(1)}$. One might wonder, then, why the current chapter exists in this book. We have several reasons for including it.

(1) An exponential factoring algorithm is often the algorithm of choice for small inputs. In particular, in some subexponential methods, smallish auxiliary numbers are factored in a subroutine, and such a subroutine might invoke an exponential factoring method.

(2) In some cases, an exponential algorithm is a direct ancestor of a subexponential algorithm. For example, the subexponential elliptic curve method grew out of the exponential $p-1$ method. One might think of the exponential algorithms as possible raw material for future developments, much as various wild strains of agricultural cash crops are valued for their possible future contributions to the plant gene pool.

(3) It is still the case that the fastest, rigorously analyzed, *deterministic* factoring algorithm is exponential.

(4) Some factoring algorithms, both exponential and subexponential, are the basis for analogous algorithms for discrete logarithm computations. For some groups the *only* discrete logarithm algorithms we have are exponential.

(5) Many of the exponential algorithms are pure delights.

We hope then that the reader is convinced that this chapter is worth it!

5.1 Squares

An old strategy to factor a number is to express it as the difference of two nonconsecutive squares. Let us now expand on this theme.

5.1.1 Fermat method

If one can write n in the form $a^2 - b^2$, where a, b are nonnegative integers, then one can immediately factor n as $(a+b)(a-b)$. If $a - b > 1$, then the factorization is nontrivial. Further, every factorization of every odd number n arises in this way. Indeed, if n is odd and $n = uv$, where u, v are positive integers, then $n = a^2 - b^2$ with $a = \frac{1}{2}(u+v)$ and $b = \frac{1}{2}|u-v|$.

For odd numbers n that are the product of two nearby integers, it is easy to find a valid choice for a, b and so to factor n. For example, consider $n = 8051$. The first square above n is $8100 = 90^2$, and the difference to n is $49 = 7^2$. So $8051 = (90+7)(90-7) = 97 \cdot 83$.

To formalize this as an algorithm, we take trial values of the number a from the sequence $\lceil \sqrt{n} \rceil, \lceil \sqrt{n} \rceil + 1, \ldots$ and check whether $a^2 - n$ is a square. If it is, say b^2, then we have $n = a^2 - b^2 = (a+b)(a-b)$. For n odd and composite, this procedure must terminate with a nontrivial factorization before we reach $a = \lfloor (n+9)/6 \rfloor$. The worst case occurs when $n = 3p$ with p prime, in which case the only choice for a that gives a nontrivial factorization is $(n+9)/6$ (and the corresponding b is $(n-9)/6$).

Algorithm 5.1.1 (Fermat method). We are given an odd integer $n > 1$. This algorithm either produces a nontrivial divisor of n or proves n prime.

1. [Main loop]
   for$\left(\lceil \sqrt{n} \rceil \le a \le (n+9)/6\right)$ {
   // Next, apply Algorithm 9.2.11.
   if$\left(b = \sqrt{a^2 - n}\text{ is an integer}\right)$ return $a - b$;
   }
   return "n is prime";

It is evident that in the worst case, Algorithm 5.1.1 is much more tedious than trial division. But the worst cases for Algorithm 5.1.1 are actually the easiest cases for trial division, and vice versa, so one might try to combine the two methods.

There are various tricks that can be used to speed up the Fermat method. For example, via congruences it may be discerned that various residue classes for a make it impossible for $a^2 - n$ to be a square. As an illustration, if $n \equiv 1 \pmod 4$, then a cannot be even, or if $n \equiv 2 \pmod 3$, then a must be a multiple of 3.

In addition, a multiplier might be used. As we have seen, if n is the product of two nearby integers, then Algorithm 5.1.1 finds this factorization quickly. Even if n does not have this product property, it may be possible for kn to be a product of two nearby integers, and $\gcd(kn, n)$ may be taken to obtain the factorization of n. For example, take $n = 2581$. Algorithm 5.1.1 has us start with $a = 51$ and does not terminate until the ninth choice, $a = 59$, where we find that $59^2 - 2581 = 900 = 30^2$ and $2581 = 89 \cdot 29$. (Noticing that $n \equiv 1 \pmod 4, n \equiv 1 \pmod 3$, we know that a is odd and not a multiple of 3, so 59 would be the third choice if we used this information.) But if we try Algorithm 5.1.1 on $3n = 7743$, we terminate on the first choice for a, namely $a = 88$, giving $b = 1$. Thus $3n = 89 \cdot 87$, and note that $89 = \gcd(89, n)$, $29 = \gcd(87, n)$.

5.1.2 Lehman method

But how do we know to try the multiplier 3 in the above example? The following method of R. Lehman formalizes the search for a multiplier.

Algorithm 5.1.2 (Lehman method). We are given an integer $n > 21$. This algorithm either provides a nontrivial factor of n or proves n prime.

1. [Trial division]
   Check whether n has a nontrivial divisor $d \le n^{1/3}$, and if so, return d;
2. [Loop]

```
for(1 ≤ k ≤ ⌈n^{1/3}⌉) {
    for(⌈2√(kn)⌉ ≤ a ≤ ⌊2√(kn) + n^{1/6}/(4√k)⌋) {
        if(b = √(a^2 − 4kn) is an integer) return gcd(a + b, n);
                                    // Via Algorithm 9.2.11.
    }
}
return "n is prime";
```

Assuming that this algorithm is correct, it is easy to estimate the running time. Step [Trial division] takes $O(n^{1/3})$ operations, and if step [Loop] is performed, it takes at most

$$\sum_{k=1}^{\lceil n^{1/3}\rceil}\left(\frac{n^{1/6}}{4\sqrt{k}}+1\right) = O(n^{1/3})$$

calls to Algorithm 9.2.11, each call taking $O(\ln\ln n)$ operations. Thus, in all, Algorithm 5.1.2 takes in the worst case $O(n^{1/3}\ln\ln n)$ arithmetic operations with integers the size of n. We now establish the integrity of the Lehman method.

Theorem 5.1.3. *The Lehman method (Algorithm 5.1.2) is correct.*

Proof. We may assume that n is not factored in step [Trial division]. If n is not prime, then it is the product of 2 primes both bigger than $n^{1/3}$. That is, $n = pq$, where p, q are primes and $n^{1/3} < p \le q$. We claim that there is a value of $k \le \lceil n^{1/3}\rceil$ such that k has the factorization uv, with u, v positive integers, and

$$|uq - vp| < n^{1/3}.$$

Indeed, by a standard result (see [Hardy and Wright 1979, Theorem 36]), for any bound $B > 1$, there are positive integers u, v with $v \le B$ and $|\frac{u}{v} - \frac{p}{q}| < \frac{1}{vB}$. We apply this with $B = n^{1/6}\sqrt{q/p}$. Then

$$|uq - vp| < \frac{q}{n^{1/6}\sqrt{q/p}} = n^{1/3}.$$

It remains to show that $k = uv \le \lceil n^{1/3}\rceil$. Since $\frac{u}{v} < \frac{p}{q} + \frac{1}{vB}$ and $v \le B$, we have

$$k = uv = \frac{u}{v}v^2 < \frac{p}{q}v^2 + \frac{v}{B} \le \frac{p}{q}\cdot\frac{q}{p}n^{1/3} + 1 = n^{1/3} + 1,$$

so the claim is proved.

With k, u, v as above, let $a = uq + vp$, $b = |uq - vp|$. Then $4kn = a^2 - b^2$. We show that $2\sqrt{kn} \le a < 2\sqrt{kn} + \frac{n^{1/6}}{4\sqrt{k}}$. Since $uq \cdot vp = kn$, we have $a = uq + vp \ge 2\sqrt{kn}$. Set $a = 2\sqrt{kn} + E$. Then

$$4kn + 4E\sqrt{kn} \le \left(2\sqrt{kn} + E\right)^2 = a^2 = 4kn + b^2 < 4kn + n^{2/3},$$

so that $4E\sqrt{kn} < n^{2/3}$, and $E < \frac{n^{1/6}}{4\sqrt{k}}$ as claimed.

Finally, we show that if a, b are returned in step [Loop], then $\gcd(a+b, n)$ is a nontrivial factor of n. Since n divides $(a+b)(a-b)$, it suffices to show that $a + b < n$. But

$$a + b < 2\sqrt{kn} + \frac{n^{1/6}}{4\sqrt{k}} + n^{1/3} < 2\sqrt{(n^{1/3}+1)n} + \frac{n^{1/6}}{4\sqrt{n^{1/3}+1}} + n^{1/3} < n,$$

the last inequality holding for $n \ge 21$. □

There are various ways to speed up the Lehman method, such as first trying values for k that have many divisors. We refer the reader to [Lehman 1974] for details.

5.1.3 Factor sieves

In the Fermat method we search for integers a such that $a^2 - n$ is a square. One path that has been followed is to try to make use of the many values of a for which $a^2 - n$ is not a square. For example, suppose $a^2 - n = 17$. Does this tell us anything useful about n? Indeed, it does. If p is a prime factor of n, then $a^2 \equiv 17 \pmod{p}$, so that if $p \ne 17$, then p is forced to lie in one of the residue classes $\pm 1, \pm 2, \pm 4, \pm 8 \pmod{17}$. That is, half of all the primes are ruled out as possible divisors of n in one fell swoop. With other values of a we similarly can rule out other residue classes for prime factors of n. It is then a hope that we can gain so much information about the residue classes that prime factors of n must lie in, that these primes are then completely determined and perhaps easily found.

The trouble with this kind of argument is the exponential growth in its complexity. Suppose we try this argument for k values of a, giving us k moduli $m_1, m_2, \cdots, m_k$, and for each we learn that prime factors p of n must lie in certain residue classes. For the sake of the argument, suppose the m_i's are different primes, and we have $\frac{1}{2}(m_i - 1)$ possible residue classes $\pmod{m_i}$ for the prime factors of n. Then modulo the product $M = m_1 m_2 \cdots m_k$, we have $2^{-k}(m_1 - 1)(m_2 - 1)\dots(m_k - 1) = 2^{-k}\varphi(M)$ possible residue classes $\pmod{M}$. On the one hand, this number is small, but on the other, it is large! That is, the probability that a random prime p is in one of these residue classes is 2^{-k}, so if k is large, this should greatly reduce the possibilities and pinpoint p. But we know no fast way of finding the small solutions that simultaneously satisfy all the required congruences, since listing the $2^{-k}\varphi(M)$ solutions to find the small ones is a prohibitive calculation. Early computational efforts at solving this problem involved ingenious apparatus with bicycle chains, cards, and photoelectric cells. There are also modern special purpose computers that have been built to solve this kind of problem. For much more on this approach, see [Williams and Shallit 1993].

5.2 Monte Carlo methods

There are several interesting heuristic methods that use certain deterministic sequences that are analyzed as if they were random sequences. Though the sequences may have a random seed, they are not truly random; we nevertheless refer to them as Monte Carlo methods. The methods in this section are all principally due to J. Pollard.

5.2.1 Pollard rho method for factoring

In 1975, J. Pollard introduced a most novel factorization algorithm, [Pollard 1975]. Consider a random function f from $\mathcal{S}$ to $\mathcal{S}$, where $\mathcal{S} = \{0, 1, \dots, l-1\}$. Let $s \in \mathcal{S}$ be a random element, and consider the sequence

$$s, f(s), f(f(s)), \dots .$$

Since f takes values in a finite set, it is clear that the sequence must eventually repeat a term, and then become cyclic. We might diagram this behavior with the letter ρ, indicating a precyclic part with the tail of the ρ, and the cyclic part with the oval of the ρ. How long do we expect the tail to be, and how long do we expect the cycle to be?

It should be immediately clear that the birthday paradox from elementary probability theory is involved here, and we expect the length of the tail and the oval together to be of order $\sqrt{l}$. But why is this of interest in factoring?

Suppose p is a prime, and we let $\mathcal{S} = \{0, 1, \dots, p-1\}$. Let us specify a particular function f from $\mathcal{S}$ to $\mathcal{S}$, namely $f(x) = x^2 + 1 \bmod p$. So if this function is "random enough," then we will expect that the sequence $(f^{(i)}(s))$, $i = 0, 1, \dots$, of iterates starting from a random $s \in \mathcal{S}$ begins repeating before $O(\sqrt{p})$ steps. That is, we expect there to be $0 \le j < k = O(\sqrt{p})$ steps with $f^{(j)}(s) = f^{(k)}(s)$.

Now suppose we are trying to factor a number n, and p is the least prime factor of n. Since we do not yet know what p is, we cannot compute the sequence in the above paragraph. However, we can compute values of the function F defined as $F(x) = x^2+1 \bmod n$. Clearly, $f(x) = F(x) \bmod p$. Thus, $F^{(j)}(s) \equiv F^{(k)}(s) \pmod{p}$. That is, $\gcd\left(F^{(j)}(s) - F^{(k)}(s), n\right)$ is divisible by p. With any luck, this gcd is not equal to n itself, so that we have a nontrivial divisor of n.

There is one further ingredient in the Pollard rho method. We surely should not be expected to search over all pairs j, k with $0 \le j < k$ and to compute $\gcd(F^{(j)}(s) - F^{(k)}(s), n)$ for each pair. This could easily take longer than a trial division search for the prime factor p, since if we search up to B, there are about $\frac{1}{2}B^2$ pairs j, k. And we do not expect to be successful until B is of order $\sqrt{p}$. So we need another way to search over pairs other than to examine all of them. This is afforded by a fabulous expedient, the Floyd cycle-finding method. Let $l = k - j$, so that for any $m \geq j$, $F^{(m)}(s) \equiv F^{(m+l)}(s) \equiv F^{(m+2l)}(s) \equiv \ldots \pmod{p}$. Consider this for $m = l\lceil j/l \rceil$, the first multiple of l that exceeds j. Then $F^{(m)}(s) \equiv F^{(2m)}(s) \pmod{p}$, and $m \leq k = O(\sqrt{p})$.

So the basic idea of the Pollard rho method is to compute the sequence $\gcd(F^{(i)}(s) - F^{(2i)}(s), n)$ for $i = 1, 2, \ldots$, and this should terminate with a nontrivial factorization of n in $O(\sqrt{p})$ steps, where p is the least prime factor of n.

Algorithm 5.2.1 (Pollard rho factorization method). We are given a composite number n. This algorithm attempts to find a nontrivial factor of n.

1. [Choose seeds]
   Choose random $a \in [1, n-3]$;
   Choose random $s \in [0, n-1]$;
   $U = V = s$;
   Define function $F(x) = (x^2 + a) \bmod n$;
2. [Factor search]
   $U = F(U)$;
   $V = F(V)$;
   $V = F(V)$; // $F(V)$ intentionally invoked twice.
   $g = \gcd(U - V, n)$;
   if($g == 1$) goto [Factor search];
3. [Bad seed]
   if($g == n$) goto [Choose seeds];
4. [Success]
   return g; // Nontrivial factor found.

A pleasant feature of the Pollard rho method is that very little space is required: Only the number n that is being factored and the current values of U, V need be kept in memory.

The main loop, step [Factor search], involves 3 modular multiplications (actually squarings) and a gcd computation. In fact, with the cost of one extra modular multiplication, one may put off the gcd calculation so that it is performed only rarely. Namely, the numbers $U - V$ may be accumulated (multiplied all together) modulo n for k iterations, and then the gcd of this modular product is taken with n. So if k is 100, say, the amortized cost of performing a gcd is made negligible, so that one generic loop consists of 3 modular squarings and one modular multiplication.

It is certainly possible for the gcd at step [Bad seed] to be n itself, and the chance for this is enhanced if one uses the above idea to put off performing gcd's. However, this defect can be mitigated by storing the values U, V at the last gcd. If the next gcd is n, one can return to the stored values U, V and proceed one step at a time, performing a gcd at each step.

There are actually many choices for the function $F(x)$. The key criterion is that the iterates of F modulo p should not have long ρ's, or as [Guy 1976] calls them, "epacts." The epact of a prime p with respect to a function F from $\mathbf{Z}_p$ to $\mathbf{Z}_p$ is the largest k for which there is an s with $F^{(0)}(s), F^{(1)}(s), \ldots, F^{(k)}(s)$ all distinct.

So a poor choice for a function $F(x)$ is $ax + b$, since the epact for a prime p is the multiplicative order of a modulo p (when $a \not\equiv 1 \pmod p$), usually a large divisor of $p - 1$. (When $a \equiv 1 \pmod p$ and $b \not\equiv 0 \pmod p$, the epact is p.)

Even among quadratic functions $x^2 + b$ there can be poor choices, for example $b = 0$. Another less evident, but nevertheless poor, choice is $x^2 - 2$. If x can be represented as $y + y^{-1}$ modulo p, then the k-th iterate is $y^{2^k} + y^{-2^k}$ modulo p.

It is not known whether the epact of $x^2 + 1$ for p is a suitably slow-growing function of p, but Guy conjectures it is $O\left(\sqrt{p \ln p}\right)$.

If we happen to know some information about the prime factors p of n, it may pay to use higher-degree polynomials. For example, since all prime factors of the Fermat number F_k are congruent to 1 (mod 2^{k+2}) when $k \geq 2$ (see Theorem 1.3.5), one might use $x^{2^{k+2}} + 1$ for the function F when attempting to factor F_k by the Pollard rho method. One might expect the epact for a prime factor p of F_k to be smaller than that of $x^2 + 1$ by a factor of $\sqrt{2^{k+1}}$. Indeed, iterating $x^2 + 1$ might be thought of as a random walk through the set of squares plus 1, a set of size $(p - 1)/2$, while using $x^{2^{k+2}} + 1$ we walk through the 2^{k+2} powers plus 1, a set of size $(p - 1)/2^{k+2}$. However, there is a penalty to using $x^{2^{k+2}} + 1$, since a typical loop now involves $3(k + 2)$ modular squarings and one modular multiplication. For large k the benefit is evident. In this connection see Exercise 5.23. Such acceleration was used successfully in [Brent and Pollard 1981] to factor F_8, historically the most spectacular factorization achieved with the Pollard rho method. The work of Brent and Pollard also discusses a somewhat faster cycle-finding method, which is to save certain iterate values and comparing future ones with those, as an alternative to the Floyd cycle-finding method.

5.2.2 Pollard rho method for discrete logarithms

Pollard has also suggested a rho method for discrete logarithm computations, but it does not involve iterating $x^2 + 1$, or any simple polynomial for that matter, [Pollard 1978]. If we are given a finite cyclic group G and a generator g of G, the discrete logarithm problem for G is to express given elements of G in the form g^l, where l is an integer. The rho method can be used for any group for which it is possible to perform the group operation and for which we can assign numerical labels to the group elements. However, we shall discuss it for the specific group $\mathbf{Z}_p^*$ of nonzero residues modulo p, where p is a prime greater than 3.

We view the elements of $\mathbf{Z}_p^*$ as integers in $\{1, 2, \ldots, p-1\}$. Let g be a generator and let t be an arbitrary element. Our goal is to find an integer l such that $g^l = t$, that is, $t = g^l \bmod p$. Since the order of g is $p-1$, it is really a residue class modulo $(p-1)$ that we are searching for, not a specific integer l, though of course, we might request the least nonnegative value.

Consider a sequence of pairs (a_i, b_i) of integers modulo $(p-1)$ and a sequence (x_i) of integers modulo p such that $x_i = t^{a_i} g^{b_i} \bmod p$, and we begin with the initial values $a_0 = b_0 = 0$, $x_0 = 1$. The rule for getting the $i+1$ terms from the i terms is as follows:

$$(a_{i+1}, b_{i+1}) = \begin{cases} ((a_i + 1) \bmod (p-1), b_i), & \text{if } 0 < x_i < \frac{1}{3}p, \\ (2a_i \bmod (p-1), 2b_i \bmod (p-1)), & \text{if } \frac{1}{3}p < x_i < \frac{2}{3}p, \\ (a_i, (b_i + 1) \bmod (p-1)), & \text{if } \frac{2}{3}p < x_i < p, \end{cases}$$

and so

$$x_{i+1} = \begin{cases} tx_i \bmod p, & \text{if } 0 < x_i < \frac{1}{3}p, \\ x_i^2 \bmod p, & \text{if } \frac{1}{3}p < x_i < \frac{2}{3}p, \\ gx_i \bmod p, & \text{if } \frac{2}{3}p < x_i < p. \end{cases}$$

Since which third of the interval $[0, p]$ an element is in has seemingly nothing to do with the group $\mathbf{Z}_p^*$, one may think of the sequence (x_i) as "random," and so it may be that there are numbers j, k with $j < k = O(\sqrt{p})$ with $x_j = x_k$. If we can find such a pair j, k, then we have $t^{a_j} g^{b_j} = t^{a_k} g^{b_k}$, so that if l is the discrete logarithm of t, we have

$$(a_j - a_k)l \equiv b_k - b_j \pmod{p-1}.$$

If $a_j - a_k$ is coprime to $p-1$, this congruence may be solved for the discrete logarithm l. If the gcd of $a_j - a_k$ with $p-1$ is $d > 1$, then we may solve for l modulo $(p-1)/d$, say $l \equiv l_0 \pmod{(p-1)/d}$. Then $l = l_0 + m(p-1)/d$ for some $m = 0, 1, \ldots, d-1$, so if d is small, these various possibilities may be checked.

As with the rho method for factoring, we use the Floyd cycle-finding algorithm. Thus, at the i-th stage of the algorithm we have at hand both x_i, a_i, b_i and x_{2i}, a_{2i}, b_{2i}. If $x_i = x_{2i}$, then we have our cycle match. If not, we go to the $(i+1)$-th stage, computing $x_{i+1}, a_{i+1}, b_{i+1}$ from x_i, a_i, b_i and computing $x_{2i+2}, a_{2i+2}, b_{2i+2}$ from x_{2i}, a_{2i}, b_{2i}. The principal work is in the calculation of the (x_i) and (x_{2i}) sequences, requiring 3 modular multiplications to travel from the i-th stage to the $(i+1)$-th stage. As with the Pollard rho method for factoring, space requirements are minimal.

[Teske 1998] describes a somewhat more complicated version of the rho method for discrete logs, with 20 branches for the iterating function at each point, rather than the 3 described above. Numerical experiments indicate that her random walk gives about a 20% improvement.

The rho method for discrete logarithms can be easily distributed to many processors, as described in connection with the lambda method below.

5.2.3 Pollard lambda method for discrete logarithms

In the same paper where the rho method for discrete logarithms is described, [Pollard 1978] also suggests a "lambda" method, so called because the "λ" shape evokes the image of two paths converging on one path. The idea is to take a walk from t, the group element whose discrete logarithm we are searching for, and another from T, an element whose discrete logarithm we know. If the two walks coincide, we can figure the discrete logarithm of t. Pollard views the steps in a walk as jumps of a kangaroo, and so the algorithm is sometimes referred to as the "kangaroo method." When we know that the discrete logarithm for which we are searching lies in a known short interval, the kangaroo method can be adapted to profit from this knowledge: We employ kangaroos with shorter strides.

One tremendous feature of the lambda method is that it is relatively easy to distribute the work over many computers. Each node in the network participating in the calculation chooses a random number r and begins a pseudorandom walk starting from t^r, where t is the group element whose discrete logarithm we are searching for. Each node uses the same easily computed pseudorandom function $f : G \to S$, where S is a relatively small set of integers whose mean value is comparable to the size of the group G. The powers g^s for $s \in S$ are precomputed. Then the "walk" starting at t^r is

$$w_0 = t^r, \; w_1 = w_0 g^{f(w_0)}, \; w_2 = w_1 g^{f(w_1)}, \; \ldots .$$

If another node, choosing r' initially and walking through the sequence $w_0', w_1', w_2', \ldots$, has a "collision" with the sequence $w_0, w_1, w_2, \ldots$, that is, $w_i' = w_j$ for some i, j, then

$$t^{r'} g^{f(w_0')+f(w_1')+\cdots+f(w_{i-1}')} = t^r g^{f(w_0)+f(w_1)+\cdots+f(w_{j-1})}.$$

So if $t = g^l$, then

$$(r' - r)l \equiv \sum_{\mu=0}^{j-1} f(w_\mu) - \sum_{\nu=0}^{i-1} f(w_\nu') \pmod{n},$$

where n is the order of the group.

The usual case where this method is applied is when the order n is prime, so as long as the various random numbers r chosen at the start by each node are all distinct modulo n, then the above congruence can be easily solved for the discrete logarithm l. (This is true unless we have the misfortune that the collision occurs on one of the nodes, that is, $r = r'$. However, if the number of nodes is large, an internodal collision is much more likely than an intranodal collision.)

It is also possible to use the pseudorandom function discussed in Section 5.2.2 in connection with the lambda method. In this case all collisions are useful: A collision occurring on one particular walk with itself can also be used to compute our discrete logarithm. That is, in this collision event, the lambda method has turned itself into the rho method. However, if one already knows that the discrete logarithm that one is searching for is in a small interval, the above method can be used, and the time spent should be about the square root of the interval length. However, the mean value of the set of integers in S needs to be smaller, so that the kangaroos are hopping only through the appropriate interval.

A central computer needs to keep track of all the sequences on all the nodes so that collisions may be detected. By the birthday paradox, we expect a collision when the number of terms of all the sequences is $O(\sqrt{n})$. It is clear that as described, this method has a formidable memory requirement for the central computer. The following idea, described in [van Oorschot and Wiener 1999] (and attributed to J.-J. Quisquater and J.-P. Delescaille, who in turn acknowledge R. Rivest) greatly mitigates the memory requirement, and so renders the method practical for large problems. It is to consider so-called distinguished points. We presume that the group elements are represented by integers (or perhaps tuples of integers). A particular field of length k of binary digits will be all zero about $1/2^k$ of the time. A random walk should pass through such a distinguished point about every 2^k steps on average. If two random walks ever collide, they will coincide thereafter, and both will hit the next distinguished point together. So the idea is to send only distinguished points to the central computer, which cuts the rather substantial space requirement down by a factor of 2^{-k}.

A notable success is the March 1998 calculation of a discrete logarithm in an elliptic-curve group whose order is a 97-bit prime n; see [Escott et al. 1998]. A group of 588 people in 16 countries used about 1200 computers over 53 days to complete the task. Roughly $2 \cdot 10^{14}$ elliptic-curve group additions were performed, with the number of distinguished points discovered being 186364. (The value of k in the definition of distinguished point was 30, so only about one out of each billion sequence steps was reported to the main computer.)

More recent advances in the world of parallel-rho methods include a cryptographic-DL treatment [van Oorschot and Wiener 1999] and an attempt at parallelization of actual Pollard-rho factoring (not DL) [Crandall 1999d]. In this latter regard, see Exercises 5.23 and 5.24. There is also a very accessible review article on the general DL problem [Odlyzko 2000a].

5.3 Baby-steps, giant-steps

Suppose $G = \langle g \rangle$ is a cyclic group of order not exceeding n, and suppose $t \in G$. We wish to find an integer l such that $g^l = t$. We may restrict our search for l to the interval $[0, n-1]$. Write l in base b, where $b = \lceil \sqrt{n} \rceil$. Then $l = l_0 + l_1 b$, where $0 \le l_0, l_1 \le b - 1$. Note that $g^{l_1 b} = t g^{-l_0} = t h^{l_0}$, where $h = g^{-1}$. Thus, we can search for l_0, l_1 by computing the lists $\{g^0, g^b, \ldots, g^{(b-1)b}\}$ and $\{th^0, th^1, \ldots, th^{b-1}\}$ and sorting them. Once they are sorted, one passes through one of the lists, finding where each element belongs in the sorted order of the second list, with a match then being readily apparent. (This idea is laid out in pseudocode in Algorithm 7.5.1.) If $g^{ib} = th^j$, then we may take $l = j + ib$, and we are through.

Here is a more formal description:

Algorithm 5.3.1 (Baby-steps, giant-steps for discrete logarithms). We are given a cyclic group G with generator g, an upper bound n for the order of G, and an element $t \in G$. This algorithm returns an integer l such that $g^l = t$. (It is understood that we may represent group elements in some numerical fashion that allows a list of them to be sorted.)

1. [Set limits]
   $b = \lceil \sqrt{n} \rceil$;
   $h = \left(g^{-1}\right)^b$; // Via Algorithm 2.1.5, for example.
2. [Construct lists]
   $A = \left\{g^i : i = 0, 1, \ldots, b-1\right\}$;
   $B = \left\{th^j : j = 0, 1, \ldots, b-1\right\}$;
3. [Sort and find intersection]
   Sort the lists A, B;
   Find an intersection, say $g^i = th^j$; // Via Algorithm 7.5.1.
   return $l = i + jb$;

Note that the hypothesis of the algorithm guarantees that the lists A, B will indeed have a common element. Note, too, that it is not necessary to sort both lists. Suppose, say, that A is generated and sorted. As the elements of B are sequentially generated, one can look for a match in A, provided that one has rapid means for content-searching in an ordered list. After the match is found, it is not necessary to continue to generate B, so that on average a savings of 50% can be gained.

The complexity for step [Construct lists] is $O(\sqrt{n})$ group operations, and for step [Sort and find intersection] is $O(\sqrt{n} \ln n)$ comparisons. The space required is what is needed to store $O(\sqrt{n})$ group elements. If one has no idea how large the group G is, one can let n run through the sequence 2^k for $k = 1, 2, \ldots$. If no match is found with one value of k, repeat the algorithm with $k + 1$. Of course, the sets from the previous run should be saved and enlarged for the next run. Thus if the group G has order m, we certainly will be successful in computing the logarithm of t in operation count $O(\sqrt{m} \ln m)$ and space $O(\sqrt{m})$ group elements.

A more elaborate version of this idea can be found in [Buchmann et al. 1997], [Terr 1999]. Also see [Blackburn and Teske 1999] for other baby-steps, giant-steps strategies.

We compare Algorithm 5.3.1 with the rho method for discrete logarithms in Section 5.2.2. There the running time is $O(\sqrt{m})$ and the space is negligible. However, the rho method is heuristic, while baby-steps, giant-steps is completely rigorous. In practice, there is no reason not to use a heuristic method for a discrete logarithm calculation just because a theoretician has not yet been clever enough to supply a proof that the method works and does so within the stated time bound. So in practice, the rho method majorizes the baby-steps, giant-steps method.

However, the simple and elegant idea behind baby-steps, giant-steps is useful in many contexts, as we shall see in Section 7.5. It also can be used for factoring, as shown in [Shanks 1971]. In fact, that paper introduced the baby-steps, giant-steps idea. The context here is the class group of binary quadratic forms with a given discriminant. We shall visit this method at the end of this chapter, in Section 5.6.4.

5.4 Pollard $p-1$ method

We know from Fermat's little theorem that if p is an odd prime, then $2^{p-1} \equiv 1 \pmod{p}$. Further, if $p-1|M$, then $2^M \equiv 1 \pmod{p}$. So if p is a prime factor of an integer n, then p divides $\gcd(2^M-1, n)$. The $p-1$ method of J. Pollard makes use of this idea as a tool to factor n. His idea is to choose numbers M with many divisors of the form $p-1$, and so search for many primes p as possible divisors of n in one fell swoop.

Let $M(k)$ be the least common multiple of the integers up to k. So, $M(1) = 1$, $M(2) = 2$, $M(3) = 6$, $M(4) = 12$, etc. The sequence $M(1), M(2), \ldots$ may be computed recursively as follows. Suppose $M(k)$ has already been computed. If $k+1$ is not a prime or a power of a prime, then $M(k+1) = M(k)$. If $k+1 = p^a$, where p is prime, then $M(k+1) = pM(k)$. A precomputation via a sieve, see Section 3.2, can locate all the primes up to some limit, and this may be easily augmented with the powers of the primes. Thus, the sequence $M(1), M(2), \ldots$ can be computed quite easily. In the following algorithm we arrive at $M(B)$ by using directly the primes up to B and their maximal powers up to B.

Algorithm 5.4.1 (Basic Pollard $p-1$ method). We are given a composite odd number n and a search bound B. This algorithm attempts to find a nontrivial factor of n.

1. [Establish prime-power base]
   Find, for example via Algorithm 3.2.1, the sequence of primes $p_1 < p_2 < \cdots < p_m \le B$, and for each such prime p_i, the maximum integer a_i such that $p_i^{a_i} \le B$;
2. [Perform power ladders]
```
   c = 2;                          // Actually, a random c can be tried.
   for(1 ≤ i ≤ m) {
      for(1 ≤ j ≤ a_i) c = c^{p_i} mod n;
   }
```
3. [Test gcd]
```
   g = gcd(c − 1, n);
   return g;                       // We hope for a success 1 < g < n.
```

There are two ways that the basic $p-1$ method can fail: (1) if $\gcd(c-1, n) = 1$, or (2) if this gcd is n itself. Here is an example to illustrate these problems. Suppose $n = 2047$ and $B = 10$. The prime powers are $2^3, 3^2, 5, 7$, and the final

g value is 1. However, we can increase the search bound. If we increase B to 12, there is one additional prime power, namely 11. Now, the final returned value is $g = n$ itself, and the algorithm still fails to yield a proper factor of n. Even taking more frequent gcd's in step [Test gcd] does not help for this n.

What is going on here is that $2047 = 2^{11} - 1 = 23 \cdot 89$. Thus, $\gcd\left(2^M - 1, n\right) = n$ if $11|M$ and is 1 otherwise. In the event of this type of failure, it is evident that increasing the search bound will not be of any help. However, one may replace the initial value $c = 2$ with $c = 3$ or some other number. With $c = 3$ one is computing $\gcd\left(3^{M(B)} - 1, n\right)$. However, this strategy does not work very well for $n = 2047$; the least initial value that splits n is $c = 12$. For this value we find $\gcd\left(12^{M(8)} - 1, n\right) = 89$.

There is a second alternative in case the algorithm fails with gcd equal to n. Choose a random integer for the initial value c, and reorganize the list of prime powers so that the 2 power comes at the end. Then take a gcd as in step [Test gcd] repeatedly, once before each factor of 2 is used. It is not hard to show that if n is divisible by at least 2 different odd primes, then the probability that a random c will cause a failure because the gcd is n is at most $1/2$.

It should be pointed out, though, that failing with gcd equal to n rarely occurs in practice. By far the more common form of failure occurs when the algorithm runs its course and the gcd is still 1 at the end. With this event, we may increase the search bound B, and/or apply the so-called *second stage*.

There are various versions of the second stage—we describe here the original one. Let us consider a second search bound B' that is somewhat larger than B. After searching through the exponents $M(1), M(2), \ldots, M(B)$, we next search through the exponents $QM(B)$, where Q runs over the primes in the interval $(B, B']$. This then has the chance of uncovering those primes $p|n$ with $p - 1 = Qu$, where Q is a prime in $(B, B']$ and $u|M(B)$. It is particularly easy to traverse the various exponents $QM(B)$. Suppose the sequence of primes in $(B, B']$ is $Q_1 < Q_2 < \cdots$. Note that $2^{Q_1 M(B)} \bmod n$ may be computed from $2^{M(B)} \bmod n$ in $O(\ln Q_1)$ steps. For $2^{Q_2 M(B)} \bmod n$, we multiply $2^{Q_1 M(B)} \bmod n$ by $2^{(Q_2 - Q_1)M(B)} \bmod n$, then by $2^{(Q_3 - Q_2)M(B)} \bmod n$ to get $2^{Q_3 M(B)} \bmod n$, and so on. The differences $Q_{i+1} - Q_i$ are all much smaller than the Q_i's themselves, and for various values d of these differences, the residues $2^{dM(B)} \bmod n$ can be precomputed. Thus, if $B' > 2B$, say, the amortized cost of computing all of the $2^{Q_i M(B)} \bmod n$ is just one modular multiplication per Q_i. If we agree to spend just as much time doing the second stage as the basic $p - 1$ method, then we may take B' much larger than B, perhaps as big as $B \ln B$.

There are many interesting issues pertaining to the second stage, such as means for further acceleration, birthday paradox manifestations, and so on. See [Montgomery 1987, 1992a], [Crandall 1996a], and Exercise 5.8 for some of these issues.

We shall see that the basic idea of the Pollard $p-1$ method is revisited with the Lenstra elliptic curve method (ECM) for factoring integers (see Section 7.4).

5.5 Polynomial evaluation method

Suppose the function $F(k, n) = k! \bmod n$ were easy to evaluate. Then a great deal of factoring and primality testing would also be easy. For example, the Wilson–Lagrange theorem (Theorem 1.3.6) says that an integer $n > 1$ is prime if and only if $F(n-1, n) = n-1$. Alternatively, $n > 1$ is prime if and only if $F(\lceil\sqrt{n}\rceil, n)$ is coprime to n. Further, we could factor almost as easily: Carry out a binary search for the least positive integer k with $\gcd(F(k, n), n) > 1$—this k, of course, will be the least prime factor of n.

As outlandish as this idea may seem, there is actually a fairly fast theoretical factoring algorithm based on it, an algorithm that stands as the fastest deterministic rigorously analyzed factoring algorithm of which we know. This is the polynomial evaluation method of [Pollard 1974] and [Strassen 1976].

The idea is as follows. Let $B = \lceil n^{1/4}\rceil$ and let $f(x)$ be the polynomial $x(x-1)\cdots(x-B+1)$. Then $f(jB) = (jB)!/((j-1)B)!$ for every positive integer j, so that the least j with $\gcd(f(jB), n) > 1$ isolates the least prime factor of n in the interval $((j-1)B, jB]$. Once we know this, if the gcd is in the stated interval, it is the least prime factor of n, and if the gcd is larger than jB, we may sequentially try the members of the interval as divisors of n, the first divisor found being the least prime divisor of n. Clearly, this last calculation takes at most B arithmetic operations with integers the size of n; that is, it is $O(n^{1/4})$. But what of the earlier steps? If we could compute each $f(jB) \bmod n$ for $j = 1, 2, \ldots, B$, then we would be in business to check each gcd and find the first that exceeds 1.

Algorithm 9.6.7 provides the computation of $f(x)$ as a polynomial in $\mathbf{Z}_n[x]$ (that is, the coefficients are reduced modulo n) and the evaluation of each $f(jB)$ modulo n for $j = 1, 2, \ldots, B$ in $O\left(B \ln^2 B\right) = O\left(n^{1/4} \ln^2 n\right)$ arithmetic operations with integers the size of n. This latter big-O expression then stands as the complexity of the Pollard–Strassen polynomial evaluation method for factoring n.

5.6 Binary quadratic forms

There is a rich theory of binary quadratic forms, as developed by Lagrange, Legendre, and Gauss in the late 1700s, a theory that played, and still plays, an important role in computational number theory.

5.6.1 Quadratic form fundamentals

For integers a, b, c we may consider the quadratic form $ax^2 + bxy + cy^2$. It is a polynomial in the variables x, y, but often we suppress the variables, and just refer to a quadratic form as an ordered triple (a, b, c) of integers.

We say that a quadratic form (a, b, c) represents an integer n if there are integers x, y with $ax^2 + bxy + cy^2 = n$. So attached to a quadratic form (a, b, c) is a certain subset of the integers, namely those numbers that (a, b, c)

represents. We note that certain changes of variables can change the quadratic form (a, b, c) to another form (a', b', c'), but keep fixed the set of numbers that are represented. In particular, suppose

$$x = \alpha X + \beta Y, \quad y = \gamma X + \delta Y,$$

where $\alpha, \beta, \gamma, \delta$ are integers. Making this substitution, we have

$$\begin{aligned} ax^2 + bxy + cy^2 &= a(\alpha X + \beta Y)^2 + b(\alpha X + \beta Y)(\gamma X + \delta Y) + c(\gamma X + \delta Y)^2 \\ &= a'X^2 + b'XY + c'Y^2, \end{aligned} \tag{5.1}$$

say. Thus every number represented by the quadratic form (a', b', c') is also represented by the quadratic form (a, b, c). We may assert the converse statement if there are integers $\alpha', \beta', \gamma', \delta'$ with

$$X = \alpha' x + \beta' y, \quad Y = \gamma' x + \delta' y.$$

That is, the matrices

$$\begin{pmatrix} \alpha & \beta \\ \gamma & \delta \end{pmatrix}, \quad \begin{pmatrix} \alpha' & \beta' \\ \gamma' & \delta' \end{pmatrix}$$

are inverses of each other. A square matrix with integer entries has an inverse with integer entries if and only if its determinant is ± 1. We conclude that if the quadratic forms (a, b, c) and (a', b', c') are related by a change of variables as in (5.1), then they represent the same set of integers if $\alpha\delta - \beta\gamma = \pm 1$.

Allowing both $+1$ and -1 for the determinant does not give much more leeway than restricting to just $+1$. (For example, one can go from (a, b, c) to $(a, -b, c)$ and to (c, b, a) via changes of variables with determinants -1, but these are easily recognized, and may be tacked on to a more complicated change of variables with determinant $+1$, so there is little loss of generality in just considering $+1$.) We shall say that two quadratic forms are *equivalent* if there is a change of variables as in (5.1) with determinant $+1$. Such a change of variables is called unimodular, and so two quadratic forms are called equivalent if you can go from one to the other by a unimodular change of variables.

Equivalence of quadratic forms is an "equivalence relation." That is, each form (a, b, c) is equivalent to itself; if (a, b, c) is equivalent to (a', b', c'), then the reverse is true, and two forms equivalent to the same form are equivalent to each other. We leave the proofs of these simple facts as Exercise 5.9.

There remains the computational problem of deciding whether two given quadratic forms are equivalent. The discriminant of a form (a, b, c) is the integer $b^2 - 4ac$. Equivalent forms have the same discriminant (see Exercise 5.11), so it is sometimes easy to see when two quadratic forms are *not* equivalent, namely this is so when their discriminants are unequal. However, the converse is not true. Witness the two forms $x^2+xy+4y^2$ and $2x^2+xy+2y^2$. They both have discriminant -15, but the first can have the value 1 (when $x = 1$ and $y = 0$), while the second cannot. So the two forms are not equivalent.

If it is the case that in each equivalence class of binary quadratic forms there is one distinguished form, and if it is the case that it is easy to find this distinguished form, then it will be easy to tell whether two given forms are equivalent. Namely, find the distinguished forms equivalent to each, and if these distinguished forms are the same form, then the two given forms are equivalent, and conversely.

This is particularly easy to do in the case of binary quadratic forms of negative discriminant. In fact, the whole theory of binary quadratic forms bifurcates on the issue of the sign of the discriminant. Forms of positive discriminant can represent both positive and negative values, but this is not the case for forms of negative discriminant. (Forms with discriminant zero are trivial objects—studying them is essentially studying the sequence of squares.)

The theory of binary quadratic forms of positive discriminant is somewhat more difficult than the corresponding theory of negative-discriminant forms. There are interesting factorization algorithms connected with the positive-discriminant case, and also with the negative-discriminant case. In the interests of brevity, we shall mainly consider the easier case of negative discriminants, and refer the reader to [Cohen 2000] for a description of algorithms involving quadratic forms of positive discriminant.

We make a further restriction. Since a binary quadratic form of negative discriminant does not represent both positive and negative numbers, we shall restrict attention to those forms that never represent negative numbers. If (a, b, c) is such a form, then $(-a, -b, -c)$ never represents positive numbers, so our restriction is not so severe. Another way of putting these restrictions is to say we are only considering forms (a, b, c) with $b^2 - 4ac < 0$ and $a > 0$. Note that these conditions then force $c > 0$.

We say that a form (a, b, c) of negative discriminant is reduced if

$$-a < b \le a < c \quad \text{or} \quad 0 \le b \le a = c. \tag{5.2}$$

Theorem 5.6.1 (Gauss). *No two different reduced forms of negative discriminant are equivalent, and every form (a, b, c) of negative discriminant with $a > 0$ is equivalent to some reduced form.*

Thus, Theorem 5.6.1 provides the mechanism for establishing a distinguished form in each equivalence class; namely, the reduced forms serve this purpose. For a proof of the theorem, see, for example, [Rose 1988].

We now discuss how to find the reduced form equivalent to a given form, and for this task there is a very simple algorithm due to Gauss.

Algorithm 5.6.2 (Reduction for negative discriminant). We are given a quadratic form (A, B, C), where A, B, C are integers with $B^2 - 4AC < 0, A > 0$. This algorithm constructs a reduced quadratic form equivalent to (A, B, C).

1. [Replacement loop]

```
    while(A > C or B > A or B ≤ −A) {
        if(A > C) (A, B, C) = (C, −B, A);                 // 'Type (1)' move.
        if(A ≤ C and (B > A or B ≤ −A)) {
```

```
            Find B*, C* such that the three conditions:
                    -A < B* ≤ A,
                    B* ≡ B (mod 2A),
                    B*^2 - 4AC* = B^2 - 4AC
            hold;
            (A, B, C) = (A, B*, C*);                    // 'Type (2)' move.
        }
    }
2. [Final adjustment]
    if(A == C and -A < B < 0) (A, B, C) = (A, -B, C);
    return (A, B, C);
```

Moves of type (2) leave the initial coordinate A unchanged, while a move of type (1) reduces it. So there can be at most finitely many type (1) moves. Further, we never do two type (2) moves in a row. Thus the algorithm terminates for each input. We leave it for Exercise 5.12 to show that the output is equivalent to the initial form. (This then shows that every form with negative discriminant and positive initial coordinate is equivalent to a reduced form, which is half of Theorem 5.6.1.)

5.6.2 Factoring with quadratic form representations

An old factoring strategy going back to Fermat is to try to represent n in two intrinsically different ways by the quadratic form $(1,0,1)$. That is, one tries to find two different ways to write n as a sum of two squares. For example, we have $65 = 8^2 + 1^2 = 7^2 + 4^2$. Then the gcd of $(8 \cdot 4 - 1 \cdot 7)$ and 65 is the proper factor 5. In general, if

$$n = x_1^2 + y_1^2 = x_2^2 + y_2^2, \quad x_1 \ge y_1 \ge 0, \quad x_2 \ge y_2 \ge 0, \quad x_1 > x_2,$$

then $1 < \gcd(x_1y_2 - y_1x_2, n) < n$. Indeed, let $A = x_1y_2 - y_1x_2$, $B = x_1y_2 + y_1x_2$. It will suffice to show that

$$AB \equiv 0 \pmod n, \quad 1 < A \le B < n.$$

The first follows from $y_i^2 \equiv -x_i^2 \pmod n$ for $i = 1, 2$, since $AB = x_1^2y_2^2 - y_1^2x_2^2 \equiv -x_1^2x_2^2 + x_1^2x_2^2 \equiv 0 \pmod n$. It is obvious that $A \le B$. To see that $A > 1$, note that $y_1x_2 < y_2x_2 < y_2x_1$. To see that $B < n$, note that $uv \le \frac{1}{2}u^2 + \frac{1}{2}v^2$ for positive numbers u, v, with equality if and only if $u = v$. Then, since $x_1 > y_2$, we have

$$B = x_1y_2 + y_1x_2 < \tfrac{1}{2}x_1^2 + \tfrac{1}{2}y_2^2 + \tfrac{1}{2}y_1^2 + \tfrac{1}{2}x_2^2 = \tfrac{1}{2}n + \tfrac{1}{2}n = n,$$

which completes the proof.

Two questions arise. Should we expect a composite number n to have two different representations as a sum of two squares? And if n does have two representations as a sum of two squares, should we expect to be able to find them easily? Unfortunately, the answer to both questions is in the negative. For the first question, it is a theorem that the set of numbers that can be represented as a sum of two squares in at least one way has asymptotic density zero. In fact, any number divisible by a prime $p \equiv 3 \pmod 4$ to an odd exponent has no representation as a sum of two squares, and these numbers constitute almost all natural numbers (see Exercise 5.15). However, there still are plenty of numbers that can be represented as a sum of two squares; in fact, any number pq where p, q are primes that are $1 \pmod 4$ can indeed be represented as a sum of two squares in two ways. But we know no way to easily find these representations.

Despite these obstacles, people have tried to work with this idea to come up with a factorization strategy. We now describe an algorithm in [McKee 1996] that can factor n in $O(n^{1/3+\epsilon})$ operations, for each fixed $\epsilon > 0$.

Observe that if (a, b, c) represents the positive integer n, say $ax^2 + bxy + cy^2 = n$, and if $D = b^2 - 4ac$ is the discriminant of (a, b, c), then $(2ax + by)^2 - Dy^2 = 4an$. That is, we have a solution u, v to $u^2 - Dv^2 \equiv 0 \pmod{4n}$. Let

$$\mathcal{S}(D, n) = \left\{(u, v) : u^2 - Dv^2 \equiv 0 \pmod{4n}\right\},$$

so that the above observation gives a mapping from representations of n by forms of discriminant D into $\mathcal{S}(D, n)$. It is straightforward to show that equivalent representations of n via (5.1) give pairs $(u, v), (u', v')$ in $\mathcal{S}(D, n)$ with the property that $uv' \equiv u'v \pmod{2n}$ (see Exercise 5.17).

Fix now the numbers D, n with $D < 0$ and n not divisible by any prime up to $\sqrt{|D|}$. If h is a solution to $h^2 \equiv D \pmod{4n}$, then the form (A, h, n), where $h^2 = D + 4An$, represents n via $x = 0, y = 1$. This maps to the pair $(h, 1)$ in $\mathcal{S}(D, n)$. Suppose now we reduce (A, h, n), and (a, b, c) is the reduced form equivalent to it. Say the corresponding representation of n is given by x, y, and this maps to the pair (u, v) in $\mathcal{S}(D, n)$. Then from the above paragraph, we have $u \equiv vh \pmod{2n}$. Moreover, v is coprime to n. Indeed, if p is a prime that divides both $v\ (= y)$ and n, then p also divides $u = 2ax + by$, so that p divides $2ax$. But $\gcd(x, y) = 1$, since a unimodular change of variables changed $0, 1$ to x, y. So p divides $2a$. But the form (a, b, c) is reduced, so that $0 < a \le \sqrt{|D|/3}$ (see Exercise 5.13). The assumption on n implies that $p > \sqrt{|D|} \ge 2$, so that p cannot divide $2a$ after all.

Now suppose we have two solutions h_1, h_2 to $h^2 \equiv D \pmod{4n}$ with $h_1 \not\equiv \pm h_2 \pmod n$. As in the above paragraph, these solutions give rise respectively to pairs (u_i, v_i) in $\mathcal{S}(D, n)$ with $u_i \equiv v_i h_i \pmod{2n}$ and $v_1 v_2$ coprime to n. We claim, then, that

$$1 < \gcd(u_1 v_2 - u_2 v_1, n) < n.$$

Indeed, we have $u_1^2 v_2^2 - u_2^2 v_1^2 \equiv D v_1^2 v_2^2 - D v_2^2 v_1^2 \equiv 0 \pmod{4n}$, so it will suffice to show that $u_1 v_2 \not\equiv \pm u_2 v_1 \pmod n$. If $u_1 v_2 \equiv u_2 v_1 \pmod n$, then

$$0 \equiv u_1 v_2 - u_2 v_1 \equiv v_1 h_1 v_2 - v_2 h_2 v_1 = v_1 v_2 (h_1 - h_2) \pmod n,$$

so that $h_1 \equiv h_2 \pmod{n}$, a contradiction. Similarly, if $u_1v_2 \equiv -u_2v_1 \pmod{n}$, then we get $h_1 \equiv -h_2 \pmod{n}$, again a contradiction.

We conclude that if there are two square roots h_1, h_2 of D modulo $4n$ such that $h_1 \not\equiv \pm h_2 \pmod{n}$, then there are two pairs $(u_1, v_1), (u_2, v_2)$ as above, where $\gcd(u_1v_2 - u_2v_1, n)$ is a nontrivial factor of n.

McKee thus proposes to search for pairs (u, v) in $\mathcal{S}(D, n)$ to come up with two pairs $(u_1, v_1), (u_2, v_2)$ as above. It is clear that we may restrict the search to pairs (u, v) with $u \geq 0, v \geq 0$.

Note that if (a, b, c) has negative discriminant D and if $ax^2+bxy+cy^2 = n$, then the corresponding pair (u, v) in $\mathcal{S}(D, n)$ satisfies $u^2 - Dv^2 = 4an$, so that $|u| \leq 2\sqrt{an}$. Further, if (a, b, c) is reduced, then $1 \leq a \leq \sqrt{|D|/3}$. McKee suggests we fix a choice for a with $1 \leq a \leq \sqrt{|D|/3}$ and then search for integers u with $0 \leq u \leq 2\sqrt{an}$ and $u^2 \equiv 4an \pmod{|D|}$. For each such u, check whether $(u^2 - 4an)/D$ is a square. If we know the prime factorization of D, then we may quickly solve for the residue classes modulo $|D|$ that u must lie in; there are fewer than $|D|^\epsilon$ of such classes. For each such residue class, our search for u is in an arithmetic progression of at most $\lceil 1 + 2\sqrt{an}/|D| \rceil$ terms. So, for a given a, we must search over at most $|D|^\epsilon + 2\sqrt{an}/|D|^{1-\epsilon}$ choices for u. Summing this expression for a up to $\sqrt{|D|/3}$ gives $O(|D|^{1/2+\epsilon} + \sqrt{n}/|D|^{1/4-\epsilon})$. So if we can find a suitable D with $|D|$ about $n^{2/3}$, we will have an algorithm that takes at most $O(n^{1/3+\epsilon})$ steps to factor n.

Such a suitable D is found very easily. Take $x_0 = \left\lfloor \sqrt{n - n^{2/3}} \right\rfloor$, so that if $d = n - x_0^2$, then $n^{2/3} \leq d < n^{2/3} + 2n^{1/2}$. We let $D = -4d$. Note that the quadratic form $(1, 0, d)$ is already reduced, it represents n with $x = x_0, y = 1$, and it gives rise to the pair $(2x_0, 1)$ in $\mathcal{S}(D, n)$. Thus, we get for free one of the two pairs we are looking for. Moreover, if n is divisible by at least 2 odd primes not dividing d, then there are two solutions h_1, h_2 to $h^2 \equiv D \pmod{4n}$ with $h_1 \not\equiv \pm h_2 \pmod{n}$. So the above search will be successful in finding a second pair in $\mathcal{S}(D, n)$, which, together with the pair $(2x_0, 1)$, will be successful in splitting n.

The following algorithm summarizes the above discussion.

Algorithm 5.6.3 (McKee test). We are given an integer $n > 1$ that has no prime factors below $3n^{1/3}$. This algorithm decides whether n is prime or composite, the algorithm giving in the composite case the prime factorization of n. (Note that any nontrivial factorization must be the prime factorization, since each prime factor of n exceeds the cube root of n.)

1. [Square test]

    If n is a square, say p^2, return the factorization $p \cdot p$;

    // A number may be tested for squareness via Algorithm 9.2.11.

2. [Side factorization]

    $d = n - \left\lfloor \sqrt{n - n^{2/3}} \right\rfloor^2$; // Thus, each prime factor of n is $> 2\sqrt{d}$.

    if $(\gcd(n, d) > 1)$ return the factorization $\gcd(n, d) \cdot (n/\gcd(n, d))$;

    By trial division, find the complete prime factorization of d;

3. [Congruences]
    for($1 \le a \le \lfloor 2\sqrt{d/3} \rfloor$) {
        Using the prime factorization of d and a method from Section 2.3.2 find the solutions $u_1, \ldots, u_t$ of the congruence $u^2 \equiv 4an \pmod{4d}$;
        for($1 \le i \le t$) { // If $t = 0$ this loop is not executed.
            For all integers u with $0 \le u \le 2\sqrt{an}$, $u \equiv u_i \pmod{4d}$, use Algorithm 9.2.11 to see whether $(4an - u^2)/4d$ is a square;
            If such a square is found, say v^2, and $u \not\equiv \pm 2x_0v \pmod{2n}$, goto [gcd computation];
        }
    }
    return "n is prime";
4. [gcd computation]
    $g = \gcd(2x_0v - u, n)$;
    return the factorization $g \cdot (n/g)$;
        // The factorization is nontrivial and the factors are primes.

Theorem 5.6.4. *Consider a procedure that on input of an integer $n > 1$ first removes from n any prime factor up to $3n^{1/3}$ (via trial division), and if this does not completely factor n, the unfactored portion is used as the input in Algorithm 5.6.3. In this way, the complete prime factorization of n is assembled. For each fixed $\epsilon > 0$, the running time of this procedure to find the complete prime factorization of n is $O(n^{1/3+\epsilon})$.*

For another McKee method of different complexity, see Exercise 5.20.

5.6.3 Composition and the class group

Suppose D is a nonsquare integer, (a_1, b, c_1), (a_2, b, c_2) are quadratic forms of discriminant D, and suppose c_1/a_2 is an integer. Since the middle coefficients are equal, we have $a_1c_1 = a_2c_2$, so that $c_1/a_2 = c_2/a_1$. We claim that the product of a number represented by the first form and a number represented by the second form is a number represented by the form $(a_1a_2, b, c_1/a_2)$. To see this assertion, it is sufficient to verify the identity

$$\left(a_1x_1^2 + bx_1y_1 + c_1y_1^2\right)\left(a_2x_2^2 + bx_2y_2 + c_2y_2^2\right) = a_1a_2x_3^2 + bx_3y_3 + (c_1/a_2)y_3^2,$$

where

$$x_3 = x_1x_2 - (c_1/a_2)y_1y_2, \quad y_3 = a_1x_1y_2 + a_2x_2y_1 + by_1y_2.$$

So in some sense, we can combine the two forms (a_1, b, c_1), (a_2, b, c_2) of discriminant D to get a third form $(a_1a_2, b, c_1/a_2)$. Note that this third form is also of discriminant D. This is the start of the definition of *composition* of forms.

We say that a binary quadratic form (a, b, c) is *primitive* if $\gcd(a, b, c) = 1$. Given an integer D that is not a square, but is 0 or 1 (mod 4), let $\mathcal{C}(D)$

denote the set of equivalence classes of primitive binary quadratic forms of discriminant D; where each class is the set of those forms equivalent to a given form. We shall use the notation $\langle a, b, c\rangle$ for the equivalence class containing the form (a, b, c).

Lemma 5.6.5. *Suppose* $\langle a_1, b, c_1\rangle = \langle A_1, B, C_1\rangle \in \mathcal{C}(D)$, $\langle a_2, b, c_2\rangle = \langle A_2, B, C_2\rangle \in \mathcal{C}(D)$, *and suppose that* $c_1/a_2, C_1/A_2$ *are integers. Then* $\langle a_1a_2, b, c_1/a_2\rangle = \langle A_1A_2, B, C_1/A_2\rangle$.

See [Rose 1988], for example.

Lemma 5.6.6. *Suppose* $(a_1, b_1, c_1), (a_2, b_2, c_2)$ *are primitive quadratic forms of discriminant* D. *Then there is a form* (A_1, B, C_1) *equivalent to* (a_1, b_1, c_1) *and a form* (A_2, B, C_2) *equivalent to* (a_2, b_2, c_2) *such that* $\gcd(A_1, A_2) = 1$.

Proof. We first show that there are coprime integers x_1, y_1 such that $a_1x_1^2 + b_1x_1y_1 + c_1y_1^2$ is coprime to a_2. Write $a_2 = m_1m_2m_3$, where every prime that divides m_1 also divides a_1, but does not divide c_1; every prime that divides m_2 also divides c_1, but does not divide a_1; and every prime that divides m_3 also divides $\gcd(a_1, c_1)$. Find integers u_1, v_1 such that $u_1m_1 + v_1m_2m_3 = 1$, and let $x_1 = u_1m_1$. Find integers u_2, v_2 such that $u_2m_2 + v_2m_3x_1 = 1$, and let $y_1 = u_2m_2$. Then x_1, y_1 have the desired properties.

Make the unimodular change of variables $x = x_1X - Y$, $y = y_1X + v_2m_3Y$. This changes (a_1, b_1, c_1) to an equivalent form (A_1, B_1, C_1'), where $A_1 = ax_1^2 + b_1x_1y_1 + c_1y_1^2$ is coprime to a_2. To bring B_1 and b_2 into agreement, find integers r, s such that $rA_1 + sa_2 = 1$, and let $k = r(b_2 - B_1)/2$. (Note that b_2 and B_1 have the same parity as D.) Set $B = B_1 + 2kA_1$, so that $B \equiv b_2 \pmod{2a_2}$. Then (see Exercise 5.17) (A_1, B_1, C_1') is equivalent to (A_1, B, C_1) for some integer C_1, and (a_2, b_2, c_2) is equivalent to (a_2, B, C_2) for some integer C_2. Let $A_2 = a_2$, and we are done. □

Given two primitive quadratic forms $(a_1, b_1, c_1), (a_2, b_2, c_2)$ of discriminant D, let $(A_1, B, C_1), (A_2, B, C_2)$ be the respectively equivalent forms given in Lemma 5.6.6. We define a certain operation like so:

$$\langle a_1, b_1, c_1\rangle * \langle a_2, b_2, c_2\rangle = \langle a_3, b_3, c_3\rangle,$$

where $a_3 = A_1A_2$, $b_3 = B$, $c_3 = C_1/A_2$. (Note that $A_1C_1 = A_2C_2$ and $\gcd(A_1, A_2) = 1$ imply that C_1/A_2 is an integer.) Then Lemma 5.6.5 asserts that "$*$" is a well-defined binary operation on $\mathcal{C}(D)$. This is the composition operation that we alluded to above. It is clearly commutative, and the proof that it is associative is completely straightforward. If D is even, then $\langle 1, 0, D/4\rangle$ acts as an identity for $*$, while if D is odd, then $\langle 1, 1, (1-D)/4\rangle$ acts as an identity. We denote this identity by 1_D. Finally, if $\langle a, b, c\rangle$ is in $\mathcal{C}(D)$, then $\langle a, b, c\rangle * \langle c, b, a\rangle = 1_D$ (see Exercise 5.19). We thus have that $\mathcal{C}(D)$ is an abelian group under $*$. This is called the class group of primitive binary quadratic forms of discriminant D.

It is possible to trace through the above argument and come up with an algorithm for the composition of forms. Here is a relatively compact procedure: it may be found in [Shanks 1971] and in [Schoof 1982].

Algorithm 5.6.7 (Composition of forms). We are given two primitive quadratic forms $(a_1, b_1, c_1), (a_2, b_2, c_2)$ of the same negative discriminant. This algorithm computes integers a_3, b_3, c_3 such that $\langle a_1, b_1, c_1\rangle * \langle a_2, b_2, c_2\rangle = \langle a_3, b_3, c_3\rangle$.

1. [Extended Euclid operation]
   $g = \gcd(a_1, a_2, (b_1 + b_2)/2)$;
   Find u, v, w such that $ua_1 + va_2 + w(b_1 + b_2)/2 = g$;
2. [Final assignment]
   Return the values:

$$a_3 = \frac{a_1 a_2}{g^2}, \quad b_3 = b_2 + 2\frac{a_2}{g}\left(\frac{b_1 - b_2}{2}v - c_2 w\right), \quad c_3 = \frac{b_3^2 - g}{4a_3}.$$

(To find the numbers g, u, v, w in step [Extended Euclid operation] first use Algorithm 2.1.4 to find integers U, V with $h = \gcd(a_1, a_2) = Ua_1 + Va_2$, and then to find integers U', V' with $g = \gcd(h, (b_1 + b_2)/2)) = U'h + V'(b_1 + b_2)/2$. Then $u = U'U$, $v = U'V$, $w = V'$.) We remark that even if $(a_1, b_1, c_1), (a_2, b_2, c_2)$ are reduced, the form (a_3, b_3, c_3) that is generated by the algorithm need not be reduced. One can follow Algorithm 5.6.7 with Algorithm 5.6.2 to get the reduced form in the class $\langle a_3, b_3, c_3\rangle$.

In the case that $D < 0$, Theorem 5.6.1 immediately implies that $\mathcal{C}(D)$ is a finite group. Indeed, each member of $\mathcal{C}(D)$ corresponds to a unique reduced form (a, b, c) satisfying (5.2). Thus $h(D)$, the order of $\mathcal{C}(D)$, is equal to the number of coprime triples a, b, c satisfying (5.2) and $b^2 - 4ac = D$. Using $|b| \le a$, we have $-D = 4ac - b^2 \ge 4ac - a^2$, and using $a \le c$, we have $-D \ge 3a^2$. Thus, $0 < a \le \sqrt{|D|/3}$. Since c is determined once a, b are chosen, we thus have $h(D) \le \sum 2a < 2|D|/3$.

But we can do better. Given an integer b with $|b| \le \sqrt{|D|/3}$ and $b \equiv D \pmod 2$, the number of choices of a that correspond to b is at most the number of divisors of $b^2 - D$. But the number of divisors of n is $n^{o(1)}$ as $n \to \infty$, so $h(D) \le |D|^{1/2+o(1)}$ as $D \to -\infty$.

And we can do better still. The famous Dirichlet class number formula (see [Davenport 1980]) asserts that for $D < 0$ and $D \equiv 0$ or $1 \pmod 4$,

$$h(D) = \frac{w}{\pi} L(1, \chi_D)\sqrt{|D|}, \tag{5.3}$$

where $w = 3$ if $D = -3$, $w = 2$ if $D = -4$, and $w = 1$ otherwise. The character χ_D is the Kronecker symbol $(D/\cdot)$. This is defined as follows: χ_D is completely multiplicative, $\chi_D(p)$ is the Legendre symbol (D/p) for p an odd prime, and $\chi_D(2)$ is 0 if D is even, is 1 if $D \equiv 1 \pmod 8$, and is -1 if $D \equiv 5 \pmod 8$. The L-function $L(s, \chi_D)$ is discussed in Section 1.4.3; $L(1, \chi_D)$ is the value of the infinite series $\sum \chi_D(n)/n$. In 1918, I. Schur showed that $L(1, \chi_D) < \frac{1}{2}\ln|D| + \ln\ln|D| + 1$, so that $\frac{w}{\pi}L(1, \chi_D) < \ln|D|$ for

$D \le -4$. Hence $h(D) < \sqrt{|D|}\ln|D|$ for these values of D. Since $h(-3) = 1$, the inequality holds for $D = -3$ as well, that is, it holds for all negative discriminants.

C. Siegel has shown that $h(D) = |D|^{1/2+o(1)}$ as $D \to -\infty$, but the proof is ineffective. That is, it is impossible to use the proof to give a bound, say, for the largest $|D|$ with $h(D) < 1000$, though the theorem says such a bound exists. After work of D. Goldfeld, B. Gross, and D. Zagier, [Oesterlé 1985] (also, see [Watkins 2000]) established the explicit inequality

$$h(D) > \frac{1}{7000}\ln|D|\prod_p\left(1 - \frac{\lfloor 2\sqrt{p}\rfloor}{p+1}\right),$$

where the product is over the primes that divide D and are smaller than $\sqrt{|D|/4}$. Combining this with the result $2^{k-1}|h(D)$, where k is the number of distinct odd prime factors of D (see Lemma 5.6.8), we get, for example, that $h(D) > 1000$ for $-D > 10^{1.3\cdot 10^{10}}$. Though almost surely very far from the truth, at least it is an explicit bound, something that cannot be obtained just with the Siegel theorem. Under an assumption of an unproved hypothesis that is weaker than the ERH, namely that the L-functions $L(s,\chi)$ never have a *real* zero greater than 1/2, [Tatuzawa 1951] gave an inequality that would imply that $h(D) > 1000$ for $-D > 1.9 \cdot 10^{11}$. Probably even this greatly lowered bound is about 100 times too high. It may well be possible to establish this remaining factor of 100 or so conditionally on the ERH.

In a computational (and theoretical) *tour de force*, [Watkins 2000] shows unconditionally that $h(D) > 16$ for $-D > 31243$.

The following formula for $h(D)$ is attractive (but admittedly not very efficient when $|D|$ is large) in that it replaces the infinite sum implicit in $L(1,\chi_D)$ with a finite sum. The formula is due to Dirichlet, see [Narkiewicz 1986]. For $D < 0$, D a fundamental discriminant (this means that either $D \equiv 1 \pmod 4$ and D is square-free or $D \equiv 8$ or $12 \pmod{16}$ and $D/4$ is square-free), we have

$$h(D) = \frac{w}{D}\sum_{n=1}^{|D|}\chi_D(n)n.$$

Though an appealing formula, such a summation with its $|D|$ terms is suitable for the exact computation of $h(D)$ only for small $|D|$, say $|D| < 10^8$. There are various ways to accelerate such a series; for example, in [Cohen 2000] one can find error-function summations of only $O(|D|^{1/2})$ summands, and such formulae allow one easily to handle $|D| \approx 10^{16}$. Moreover, it can be shown that directly counting the primitive reduced forms (a,b,c) of negative discriminant D computes $h(D)$ in $O\left(|D|^{1/2+\epsilon}\right)$ operations. And the Shanks baby-steps, giant-steps method reduces the exponent from 1/2 to 1/4. We revisit the complexity of computing $h(D)$ in the next section.

5.6.4 Ambiguous forms and factorization

It is not very hard to list all of the elements of the class group $\mathcal{C}(D)$ that are their own inverse. When $D < 0$, the reduced member of such a class is called an "ambiguous" form. They come in three types: $(a, 0, c), (a, a, c), (a, b, a)$. These forms have an intimate relationship with factorizations of the discriminant into two coprime factors.

We state the classification, and leave the simple verification to the reader.

Lemma 5.6.8. *Suppose D is a negative discriminant. If D is even, then the ambiguous forms of discriminant D include the forms $(u, 0, v)$, where $0 < u \le v$, $\gcd(u, v) = 1$, and $uv = -D/4$. In addition, if $uv = -D/4$, with $\gcd(u, v) = 1$ or 2 and $\frac{1}{2}(u + v)$ odd, we have the forms $\left(\frac{1}{2}(u+v), v-u, \frac{1}{2}(u+v)\right)$ when $\frac{1}{3}v \le u < v$ and the forms $\left(2u, 2u, \frac{1}{2}(u+v)\right)$ when $0 < u < \frac{1}{3}v$. If D is odd, then the ambiguous forms of discriminant D are the forms $\left(\frac{1}{4}(u+v), \frac{1}{2}(v-u), \frac{1}{4}(u+v)\right)$, where $-D = uv$ with $0 < \frac{1}{3}v \le u \le v$, $\gcd(u, v) = 1$, and the forms $\left(u, u, \frac{1}{4}(u+v)\right)$, where $-D = uv$, $0 < u \le \frac{1}{3}v$, $\gcd(u, v) = 1$.*

Note that the form $(1, 0, |D|/4)$ in the case that D is even, and the form $(1, 1, (1 - D)/4)$ in the case that D is odd, are ambiguous. As we have seen in the previous section, each is, in its respective case, the reduced form in the class 1_D. They correspond to the trivial factorization of $D/4$ or D where one factor is 1. Also, if $D \equiv 12 \pmod{16}$ and $D \le -20$, then the ambiguous form $(2, 2, (4 - D)/8)$ corresponds to the trivial factorization of $D/4$. We also have the ambiguous forms $(4, 4, 1 - D/16)$ corresponding to the trivial factorization of $D/4$ when $D \equiv 0 \pmod{32}$ and $D \le -64$, and the form $(3, 2, 3)$ with discriminant -32. However, every other ambiguous form gives rise, and arises from, a nontrivial factorization of $D/4$ or D. Suppose that D has k distinct odd prime factors. It follows from Lemma 5.6.8 that there are 2^{k-1} ambiguous forms of discriminant D, except for the cases $D \equiv 12 \pmod{16}$ and the cases $D \equiv 0 \pmod{32}$, when there are 2^k and 2^{k+1} ambiguous forms, respectively.

Suppose now that n is a positive odd integer divisible by at least two distinct primes. If $n \equiv 3 \pmod 4$, then $D = -n$ is a discriminant, while if $n \equiv 1 \pmod 4$, then $D = -4n$ is a discriminant. If we can find any ambiguous form in the first case, other than $(1, 1, (1 + n)/4)$, we will have a nontrivial factorization of n. And if we can find any ambiguous form in the second case, other than $(1, 0, n)$ and $(2, 2, (1 + n)/2)$, then we will have a nontrivial factorization of n. And in either case, if we find all of the ambiguous forms, we can use these to construct the complete prime factorization of n.

Thus, one can say that the search for nontrivial factorizations is really a search for ambiguous forms.

So, let us see how one might find an ambiguous form, given a negative discriminant D. Let $h = h(D)$ denote the class number, that is, the order of the group $\mathcal{C}(D)$ (see Section 5.6.3). Say $h = 2^l h_o$, where h_o is odd. If $f = \langle a, b, c\rangle \in \mathcal{C}(D)$, let $F = f^{h_o}$. Then either $F = 1_D$, or one of

$F, F^2, F^4, \ldots, F^{2^{l-1}}$ has order 2 in the group. A reduced member of a class of order 2 is ambiguous (this is the definition), so knowing h and f, it is a simple matter to construct an ambiguous form. If the ambiguous form constructed corresponds to 1_D or is $(2, 2, (1+n)/2)$ (in the case $n \equiv 1 \pmod 4$), then the factorization corresponding to our ambiguous form is trivial. Otherwise it is nontrivial.

So if the above scheme does not work with one choice of f in $\mathcal{C}(D)$, then presumably we could try again with another f. If we had a small set of generators of the class group, we could try anew with each generator and so factor n. (In fact, in this case, we would have enough ambiguous forms to find the complete prime factorization of n, by refining different factorizations through gcd's.) If we did not have available a small set of generators, we might instead take random choices of f.

The principal hurdle in applying the scheme to factor n is not coming up with an appropriate f in $\mathcal{C}(D)$, but in coming up with the class number h. We can actually get by with less. All we need in the above idea is the order of f in the class group.

Now, forgetting this for a moment, and actually going for the full order h of the class group, one might think that since we actually have a formula for the order of this group, given by (5.3), we are home free. However, this formula involves an infinite sum, and it is not clear how many terms we have to take to get a good enough approximation to make the formula useful.

Note that the infinite sum $L(1, \chi_D)$ that is in the class number formula (5.3) can be written, too, as an infinite product:

$$L(1, \chi_D) = \prod_p \left(1 - \frac{\chi_D(p)}{p}\right)^{-1},$$

where the product is over all primes. It is shown in [Shanks 1971], [Schoof 1982] that if the ERH is assumed (see Conjecture 1.4.2), and if

$$\tilde{L} = \prod_{p \le n^{1/5}} \left(1 - \frac{\chi_D(p)}{p}\right)^{-1}, \quad \tilde{h} = (w/\pi)\sqrt{|D|}\tilde{L},$$

then there is a computable number c such that $|h - \tilde{h}| < cn^{2/5}\ln^2 n$. If we go to the trouble to compute $\tilde{L}$ to some accuracy, we then have for our trouble an estimate $\tilde{h}$ to the class number h that is within $cn^{2/5}\ln^2 n$ of the truth. Then the Shanks baby-steps, giant-steps method discussed in Section 7.5 and Section 5.3 can then be used to find a multiple of the order of any given $f \in \mathcal{C}(D)$ that lies in the interval $(\tilde{h} - cn^{2/5}\ln^2 n, \tilde{h} + cn^{2/5}\ln^2 n)$ in time $O(n^{1/5}\ln n)$. Since the computation of $\tilde{L}$ can be accomplished in $O(n^{1/5})$ steps, we can then achieve a factorization of n, given an appropriate f, in $O(n^{1/5}\ln n)$ operations with integers the size of n.

If one is willing to assume the ERH, which seems a fair enough gamble in a factoring algorithm (if the method fails to factor your number, you have for your effort a disproof of the ERH, presumably something of far greater interest than the factorization you were attempting), one might ask what other information the ERH might give, other than the predictable convergence of the infinite product for $L(1, \chi_D)$. In fact, it can help in a second way. Assuming the ERH, there is a computable number c' such that the classes of the primitive reduced forms (a, b, c) of discriminant D, with $a \le c' \ln^2 |D|)$, generate the full class group $\mathcal{C}(D)$ (see [Schoof 1982]). Thus, there need be no uncertainty on the choice of f in the above scenario. Namely, just make all choices for f with a representative (a, b, c) with $a \le c' \ln^2 |D|$.

Assembling these ingredients, we have, then, a deterministic factoring algorithm with a complexity of $O\left(n^{1/5} \ln^3 n\right)$ operations with integers the size of n. The proof of correctness for this algorithm depends on the so-far unproved ERH.

Shanks goes further, and shows that on assumption of the ERH, one can actually compute the class number h, and the group structure for $\mathcal{C}(D)$, and in time $O\left(|D|^{1/5+\epsilon}\right)$.

Recently, [Srinivasan 1995] showed that there is a probabilistic algorithm to approximate L that is expected to give enough precision to approximate h again with an error of $O\left(|D|^{2/5+\epsilon}\right)$, after which the Shanks baby-steps, giant-steps method may take over. The Srinivasan probabilistic method gets the approximation in expected time $O\left(|D|^{1/5+\epsilon}\right)$, and so becomes a probabilistic factoring algorithm with expected running time $O\left(n^{1/5+\epsilon}\right)$. This algorithm is completely rigorous, depending on no unproved hypotheses. Her method also computes the class number and group structure in the expected time $O\left(|D|^{1/5+\epsilon}\right)$. However, unlike with factoring, which may be easily checked for correctness, there is no simple way to see whether Srinivasan's computation of the class number is correct, though it almost certainly is. As we shall see in the next chapter, there are faster, completely rigorous, probabilistic factoring algorithms. The Srinivasan method, though, stands as the fastest known completely rigorous probabilistic method for computing the class number $\mathcal{C}(D)$. ([Hafner and McCurley 1989] have a subexponential probabilistic method, but its analysis depends on the ERH.)

5.7 Exercises

5.1. Starting with Lenstra's Algorithm 4.2.11, develop a deterministic factoring method that takes at most $n^{1/3+o(1)}$ operations to factor n.

5.2. Show that for $p = 257$, the rho iteration $x = x^2 - 1 \bmod p$ has only three possible cycle lengths, namely 12, 7, 2. Show in general that for an arbitrary Pollard iteration $x = x^2 + a \bmod q$ for q prime, the expected (in some heuristic sense) number of cycle lengths is $O(\ln q)$. See Exercise 5.22 for an extension on the notion of cycle lengths.

5.3. Let G be a cyclic group of order n with generator g, and element t. Say our goal is to solve for the discrete logarithm l of t, that is, an integer l with

$g^l = t$. Assume that we somehow discover an instance $g^b = t^a$. Show that the desired logarithm is then given by

$$l = ((bu + kn)/d) \bmod n,$$

for some integer $k \in [0, d-1]$, where $d = \gcd(a, n)$ and u is a solution to the extended-Euclid relation $au + nv = d$.

This exercise shows that finding a logarithm for a nontrivial power of t is, if d is not too large, essentially equivalent to the original DL problem.

5.4. Suppose G is a finite cyclic group, you know the group order n, and you know the prime factorization of n. Show how the Shanks baby-steps, giant-steps method of Section 5.3 can be used to solve discrete logs in G in $O\left(\sqrt{p}\ln n\right)$ operations, where p is the largest prime factor of n. Give a similar bound for the space required.

5.5. As we have seen in the chapter, the basic Shanks baby-steps, giant-steps procedure can be summarized thus: Make respective lists for baby steps and giant steps, sort one list, then find a match by sequentially searching through the other list. As we know, solving $g^l = t$ (where g is a generator of the cyclic group of order n and t is an element) can be effected in this way in $O(n^{1/2}\ln n)$ operations (comparisons). But there is a so-called hash-table construction that heuristically alters this complexity (albeit slightly) and in practice works quite efficiently. A summary of such a method runs as follows:

(1) Construct the baby-step list, but in hash-table form.

(2) On each successive giant step look up (rapidly) the corresponding hash-table entry, seeking a match.

The present exercise is to work through—by machine—the following example of an actual DL solution. This example, unlike the fundamental Algorithm 5.3.1, uses some tricks that exploit the way machines tend to function, effectively reducing complexity in this way. For the prime $p = 2^{31} - 1$ and an explicitly posed DL problem, say to solve

$$g^l \equiv t \pmod{p},$$

we proceed as follows. Reminiscent of Algorithm 5.3.1 set $b = \lceil\sqrt{p}\rceil$, but in addition choose a special parameter $\beta = 2^{12}$ to create a baby-steps "hash table" whose r-th row, for $r \in [0, \beta-1]$, consists of all those residues $g^j \bmod p$, for $j \in [0, b-1]$, that have $r = (g^j \bmod p) \bmod \beta$. That is, the row of the hash table into which a power $g^j \bmod p$ is inserted depends *only* on that modular power's low $\lg\beta$ bits. Thus, in about $\sqrt{p}$ multiplies (successively, by g) we construct a hash table of β rows. As a check on the programming effort, for a specific choice $g = 7$ the $(r = 1271)$-th row should appear as

$$((704148727, 507),\ (219280631, 3371),\ (896259319, 4844)\ldots),$$

meaning, for example,

$$7^{507} \bmod p = 704148727 = (\ldots 010011110111)_2,$$

$$7^{3371} \bmod p = 219280631 = (\ldots 010011110111)_2,$$

and so on. After the baby-steps hash table is constructed, you can run through giant-step terms tg^{-ib} for $i \in [0, b-1]$ and, by inspecting only the low 12 bits of each of these terms, index directly into the table to discover a collision. For the example $t = 31$, this leads immediately to the DL solution

$$7^{723739097} \equiv 31 \pmod{2^{31} - 1}.$$

This exercise is a good start for working out out a general DL solver, which takes arbitrary input of p, g, l, t, then selects optimal parameters such as β. Incidentally, hash-table approaches such as this one have the interesting feature that the storage is essentially that of *one* list, not two lists. Moreover, if the hash-table indexing is thought of as one fundamental operation, the algorithm has operation complexity $O(p^{1/2})$; i.e., the $\ln p$ factor is removed. Note also one other convenience, which is that the hash table, once constructed, can be reused for another DL calculation (as long as g remains fixed).

5.6. [E. Teske] Let g be a generator of the finite cyclic group G, and let $h \in G$. Suppose $\#G = 2^m \cdot n$ with $m \geq 0$ and n odd. Consider the following walk:

$$h_0 = g * h, \qquad h_{k+1} = {h_k}^2.$$

The terms h_k are computed until $h_k = h_j$ for some $j < k$, or $h_k = 1$. Let us investigate whether this is a good walk for computing discrete logarithms.

(1) Let (α_k) and (β_k) be the sequences of exponents for g and h, respectively. That is, $h_k = g^{\alpha_k} * h^{\beta_k}$ for each k. Determine closed formulae for α_k and β_k.

(2) Determine all possible group elements h for which it can happen that $h_k = 1$ for some k. Determine the largest possible value of k for which this can happen.

(3) Determine the period λ of the sequence (h_k) under the assumption that $\#G$ is prime.

(4) Would you recommend this walk to use for discrete logarithm computation? If yes, why? If no, why not?

5.7. Here are tasks that allow practical testing of any implementation of the $p-1$ method, Algorithm 5.4.1.

(1) Use the basic algorithm with search bound $B = 1000$ to achieve the factorization

$$n = 67030883744037259 = 179424673 \cdot 373587883.$$

(2) Explain *why*, in view of the factorization of 373587882, your value of B worked.

(3) Again in view of the factorization of 373587882, write a second-stage version of the algorithm, this time finding the factor with $B = 100$ but second-stage bound $B' = 1000$. This program should be faster than the first instance, of course.

(4) Find a nontrivial factor of $M_{67} = 2^{67} - 1$ using $B = 100$, $B' = 2000$.

5.8. Here we describe an interesting way to effect a second stage, and end up asking an also interesting computational question. We have seen that a second stage makes sense if a hidden prime factor p of n has the form $p = zq+1$ where z is B-smooth and $q \in (B, B']$ is a single outlying prime. One novel approach ([Montgomery 1992a], [Crandall 1996a]) to a second-stage implementation is this: After a stage-one calculation of $b = a^{M(B)} \bmod n$ as described in the text, one can as a second stage accumulate some product (here, g, h run over some fixed range, or respective sets) like this one:

$$c = \prod_{g \neq h} \left(b^{g^K} - b^{h^K} \right) \bmod n$$

and take $\gcd(n, c)$, hoping for a nontrivial factor. The theoretical task here is to explain *why* this method works to uncover that outlying prime q, indicating a rough probability (based on q, K, and the range of g, h) of uncovering a factor because of a lucky instance $g^K \equiv h^K \pmod{q}$.

An interesting computational question arising from this "g^K" method is, how does one compute rapidly the chain

$$b^{1^K}, b^{2^K}, b^{3^K}, \ldots, b^{A^K},$$

where each term is, as usual, obtained modulo n? Find an algorithm that in fact generates the indicated "hyperpower" chain, for fixed K, in only $O(A)$ operations in $\mathbf{Z}_N$.

5.9. Show that equivalence of quadratic forms is an equivalence relation.

5.10. If two quadratic forms $ax^2 + bxy + cy^2$ and $a'x^2 + b'xy + c'y^2$ have the same range, must the coefficients (a', b', c') be related to the coefficients (a, b, c) as in (5.1) where $\alpha, \beta, \gamma, \delta$ are integers and $\alpha\delta - \beta\gamma = \pm 1$?

5.11. Show that equivalent quadratic forms have the same discriminant.

5.12. Show that the quadratic form that is the output of Algorithm 5.6.2 is equivalent to the quadratic form that is the input.

5.13. Show that if (a, b, c) is a reduced quadratic form of discriminant $D < 0$, then $a \le \sqrt{|D|/3}$.

5.14. Show that for input (A, B, C), the operation complexity of Algorithm 5.6.2 is $O(1 + \ln(\min\{A, C\}))$, with operations involving integers no larger than $4AC$.

5.15. Show that a positive integer n is a sum of two squares if and only if there is no prime $p \equiv 3 \pmod 4$ that divides n to an odd exponent. Using the fact that the sum of the reciprocals of the primes that are congruent to $3 \pmod 4$ diverges (Theorem 1.1.5), prove that the set of natural numbers that are representable as a sum of two squares has asymptotic density 0. (See Exercises 1.10, 1.88, and 3.17.)

5.16. Show that if p is a prime and $p \equiv 1 \pmod 4$, then there is a probabilistic algorithm to write p as a sum of two squares that is expected to succeed in polynomial time. In the case that $p \equiv 5 \pmod 8$, show how the algorithm can be made deterministic. Using the deterministic polynomial-time method in [Schoof 1985] for taking the square root of -1 modulo p, show how in the general case the algorithm can be made deterministic, and still run in polynomial time.

5.17. Suppose that (a, b, c), (a', b', c') are equivalent quadratic forms, n is a positive integer, $ax^2 + bxy + cy^2 = n$, and under the equivalence, x, y gets taken to x', y'. Let $u = 2ax + by$, $u' = 2a'x' + b'y'$. Show that $uy' \equiv u'y \pmod{2n}$.

5.18. Show that if (a, b, c) is a quadratic form, then for each integer $b' \equiv b \pmod{2a}$, there is an integer c' such that (a, b, c) is equivalent to (a, b', c').

5.19. Suppose $\langle a, b, c\rangle \in \mathcal{C}(D)$. Prove that $\langle a, b, c\rangle$ is the identity 1_D in $\mathcal{C}(D)$ if and only if (a, b, c) represents 1. Conclude that $\langle a, b, c\rangle * \langle c, b, a\rangle = 1_D$.

5.20. Study, and implement the McKee $O(n^{1/4+\epsilon})$ factoring algorithm as described in [McKee 1999]. The method is probabilistic, and is a kind of optimization of the celebrated Fermat method.

5.21. On the basis of the Dirichlet class number formula (5.3), derive the following formulae for π:

$$\pi = 2\prod_{p>2}\left(1 + \frac{(-1)^{(p-1)/2}}{p}\right)^{-1} = 4\prod_{p>2}\left(1 - \frac{(-1)^{(p-1)/2}}{p}\right)^{-1}.$$

From the mere fact that these formulae are well-defined, prove that there exist infinitely many primes of each of the forms $p = 4k + 1$ and $p = 4k + 3$. (Compare with Exercise 1.7.) As a computational matter, about how many primes would you need to attain a reliable value for π to a given number of decimal places?

5.8 Research problems

5.22. Here is a problem that arose in discussions with J. Pollard and for which the present authors know of no solution. Obtain the Pollard-rho complexity estimate $O(\sqrt{q})$, not in any of the traditional ways, but as a statistical expectation taken over the cycle lengths. That is, one should be able to show for prime modulus p and a random choice of Pollard-rho parameter that

$$\sum_m mP_m \approx \sqrt{p},$$

where m runs through the $O(\ln p)$ (really, very few!) cycle lengths (see Exercise 5.2) and P_m is the probability that a cycle (in which the iterates eventually find themselves) has length m. One interesting facet of this problem is that if one succeeds with some working heuristic for the expectation, one should also be able to extract the *variance* in the Pollard-rho run time. The reader might consult [Bach 1991] for more thoughts on the analysis of the rho method.

5.23. If a Pollard-rho iteration be taken not as $x = (x^2 + a) \bmod N$ but as

$$x = \left(x^{2K} + a\right) \bmod N,$$

it is an established heuristic that the expected number of iterations to uncover a hidden prime factor p of N is reduced from $c\sqrt{p}$ to

$$\frac{c\sqrt{p}}{\sqrt{\gcd(p-1, 2K) - 1}}.$$

For research involving this complexity reduction, it may be helpful first to work through this heuristic and explore some possible implementations based on the gcd reduction [Brent and Pollard 1981], [Montgomery 1987], [Crandall 1999d]. Note that when we *know* something about K the speedup is tangible, as in the application of Pollard-rho methods to Fermat or Mersenne numbers. (If K is small, it may be counterproductive to use an iteration $x = x^{2K} + a$, even if we know that $p \equiv 1 \pmod{2K}$, since the cost per iteration may not be outweighed by the gain of a shorter cycle.) However, it is when we do *not* know anything about K that really tough complexity issues arise.

So an interesting open issue is this: Given M machines each doing Pollard rho, and no special foreknowledge of K, what is the optimal way to assign respective values $\{K_m \ : \ m \in [1, \ldots, M]\}$ to said machines? Perhaps the answer is just $K_m = 1$ for the m-th machine, or maybe the K_m values should be just small distinct primes. It is also unclear how the K values should be alter—if they do at all—as one moves from an "independent machines" paradigm into a "parallel" paradigm, the latter discussed in Exercise 5.24. An intuitive glimpse of what is intended here goes like so: The McIntosh–Tardif factor of F_{18}, namely

$$81274690703860512587777 = 1 + 2^{23} \cdot 29 \cdot 293 \cdot 1259 \cdot 905678539$$

(which was found via ECM) could have been found via Pollard rho, especially if some "lucky" machine were iterating according to

$$x = x^{2^{23} \cdot 29} \bmod F_{18}.$$

In any complexity analysis, make sure to take into account the problem that the number of operations *per iteration* grows as $O(\ln K_m)$, the operation complexity of a powering ladder.

5.24. Analyze a particular idea for parallelization of the Pollard rho factoring method (not the parallelization method for discrete logarithms as discussed in the text) along the following lines. Say the j-th of M machines computes a Pollard sequence, from iteration $x = x^2 + a \bmod N$, with *common* parameter a but machine-dependent initial $x_1^{(j)}$ seed, as

$$\left\{x_i^{(j)} : i = 1, 2, \ldots, n\right\},$$

so we have such a whole length-n sequence for each $j \in [1, M]$. Argue that if we can calculate the product

$$Q = \prod_{i=1}^{n} \prod_{j=1}^{M} \prod_{k=1}^{M} \left(x_{2i}^{(j)} - x_i^{(k)}\right)$$

modulo the N to be factored, then the full product has about n^2M^2 algebraic factors, implying, in turn, about $p^{1/2}/M$ *parallel* iterations for discovering a hidden factor p. So the question comes down to this: Can one parallelize the indicated product, using some sort of fast polynomial evaluation scheme? The answer is yes, subject to some heuristic controversies, with details in [Crandall 1999d], where it is argued that with M machines one should be able to find a hidden factor p in

$$O\left(\sqrt{p}\,\frac{\ln^2 M}{M}\right)$$

parallel operations.

5.25. Recall that the Pollard-rho approach to DL solving has the feature that very little memory is required. What is more, variants of the basic rho approach are pleasantly varied. The present exercise is to work through a very simple such variant, with a view to solving the specific DL relation

$$g^l \equiv t \pmod{p},$$

where t and primitive root g are given as usual. First define a pseudorandom function on residues $z \bmod p$, for example,

$$f(z) = 2 + 3\theta(z - p/2),$$

that is, $f(z) = 2$ for $z < p/2$, and $f(z) = 5$ otherwise. Now define a sequence $x_1 = t, x_2, x_3, \ldots$ with

$$x_{n+1} = g^{f(x_n)} x_n t$$

for $n \geq 1$. The beautiful thing is that we can use two sequences $(w_n = x_{2n})$, (x_n) just as in Algorithm 5.2.1, with one sequence forging ahead of the other via twofold acceleration. We perform, then, these iterations and hope for a collision

$$x_{2n} \equiv x_n \pmod{p},$$

the point being that such a collision signals a relation

$$t^a \equiv g^b \pmod{p},$$

and we can use the result of Exercise 5.3 to infer the desired DL solution. In this way, using the explicit form for the pseudorandom f given above, solve by machine for the logarithm in such test cases as

$$11^{495011427} \equiv 3 \pmod{2^{31}-1},$$
$$17^{1629} \equiv 3 \pmod{2^{17}-1}.$$

An interesting research question is this: Just how varied are the Pollard-rho possibilities? We have now seen more than one way of creating Pollard sequences as mixtures of powers of x and g, but one can even consider fractional powers. For example, if a root chain can be established in Pollard-rho fashion

$$\sqrt{g^{e_n}\ldots\sqrt{g^{e_2}\sqrt{g^{e_1}t}}} \equiv \sqrt{g^{e_{2n}}\ldots\sqrt{g^{e_2}\sqrt{g^{e_1}t}}} \pmod{p},$$

where the powers e_n are random (except always chosen so that a square root along the chain can indeed be taken), then each side of the collision can be formally squared often enough to get a mixed relation in g, t as before. Though square-rooting is not inexpensive, this approach would be of interest if statistically short cycles for the root chains could somehow be generated.

5.26. In connection with the Pollard $p-1$ method, show that if n is composite and not a power, and if you are in possession of an integer $m < n^2$ such that $p-1|m$ for some prime $p|n$, then you can use this number m in a probabilistic algorithm to get a nontrivial factorization of n. Argue that the algorithm is expected to succeed in polynomial time (the number of arithmetic steps with integers the size of n is bounded by a power of $\ln n$).

5.27. Here we investigate the "circle group," defined for odd prime p as the set

$$C_p = \{(x, y) : x, y \in [0, p-1]; x^2 + y^2 \equiv 1 \pmod{p}\},$$

together with an operation "$\oplus$" defined by

$$(x, y) \oplus (x', y') = (xx' - yy', xy' + yx') \bmod p.$$

Show that the order of the circle group is

$$\#C_p = p - \left(\frac{-1}{p}\right).$$

Prove the corollary that this order is always divisible by 4. Explain how the $\oplus$ operation is equivalent to complex multiplication (for Gaussian integers) and discuss any algebraic connection between the circle group and the field $\mathbf{F}_{p^2}$.

Next, describe a factoring algorithm—which could be called a "$p \pm 1$" method—based on the circle group. One would start with an initial point $P_0 = (x_0.y_0)$, and evaluate multiples $[n]P_0$ in much the same style as we do in ECM. How does one even find an initial point? (In this connection see Exercise 5.15.) How efficient is your method, as compared to the standard $p-1$ method? In assessing efficiency, observe that a point may be doubled in only two field multiplies. How many multiplies does it take to add two arbitrary points?

Then, analyze whether a "hyperspherical" group factoring method makes sense. The group would be

$$H_p = \left\{(x, y, z, w) \; : \; x, y, z, w \in [0, p-1]; \; x^2 + y^2 + w^2 + z^2 \equiv 1 \pmod{p}\right\},$$

and the group operation would be quaternion hypercomplex multiplication. Show that the order of the group is

$$\#H_p = p^3 - p.$$

In judging the efficacy of such a factoring method, one should address at least the following questions. How, in this case, do we find an initial point (x_0, y_0, w_0, z_0) in the group? How many field operations are required for point doubling, and for arbitrary point addition?

Explore any algebraic connections of the circle and hyperspherical groups (and perhaps further relatives of these) with groups of matrices $\pmod{p}$. For example, all $n \times n$ matrices having determinant 1 modulo p form a group that can for better or worse be used to forge some kind of factoring algorithm. These relations are well known, including yet more relations with so-called cyclotomic factoring. But an interesting line of research is based on this question: How do we design *efficient* factoring algorithms, if any, using these group/matrix ideas? We already know that complex multiplication, for example, can be done in three multiplies instead of four, and large-matrix multiplication can be endowed with its own special speedups, such as Strassen recursion [Crandall 1994b] and number-theoretical transform acceleration [Yagle 1995]; see Exercise 9.84.

5.28. Investigate the possibility of modifying the polynomial evaluation method of Pollard and Strassen for application to the factorization of Fermat numbers $F_n = 2^{2^n} + 1$. Since we may restrict factor searches to primes of the form $p = k2^{n+2} + 1$, consider the following approach. Form a product

$$P = \prod_i \left(k_i 2^{n+2} + 1\right)$$

(all modulo F_n), where the $\{k_i\}$ constitute some set of cleverly chosen integers, with a view to eventual taking of $\gcd(F_n, P)$. The Pollard–Strassen notion of evaluating products of consecutive integers is to be altered: Now we wish to form the product over a special multiplier set. So investigate possible means for efficient creation of P. There is the interesting consideration that we should be able somehow to presieve the $\{k_i\}$, or even to alter the exponents $n+2$ in some i-dependent manner. Does it make sense to describe the multiplier set $\{k_i\}$ as a union of disjoint arithmetic progressions (as would result from a presieving operation)? One practical matter that would be valuable to settle is this: Does a Pollard–Strassen variant of this type have any hope of exceeding the performance of direct, conventional sieving (in which one simply checks $2^{2^n} \pmod p$ for various $p = k2^{n+2}+1$)? The problem is not without merit, since beyond F_{20} or thereabouts, direct sieving has been the only recourse to date for discovering factors of the mighty F_n.

Chapter 6

SUBEXPONENTIAL FACTORING ALGORITHMS

The methods of this chapter include two of the three basic workhorses of modern factoring, the quadratic sieve (QS) and the number field sieve (NFS). (The third workhorse, the elliptic curve method (ECM), is described in Chapter 7.) The quadratic sieve and number field sieve are direct descendants of the continued fraction factoring method of Brillhart and Morrison, which was the first subexponential factoring algorithm on the scene. The continued fraction factoring method, which was introduced in the early 1970s, allowed complete factorizations of numbers of around 50 digits, when previously, about 20 digits had been the limit. The quadratic sieve and the number field sieve, each with its strengths and domain of excellence, have pushed our capability for complete factorization from 50 digits to now over 150 digits for the size of numbers to be routinely factored. By contrast, the elliptic curve method has allowed the discovery of prime factors up to 50 digits and beyond, with fortunately weak dependence on the size of number to be factored. We include in this chapter a small discussion of rigorous factorization methods that in their own way also represent the state of the art. We also discuss briefly some subexponential discrete logarithm algorithms for the multiplicative groups of finite fields.

6.1 The quadratic sieve factorization method

Though first introduced in [Pomerance 1982], the quadratic sieve (QS) method owes much to prior factorization methods, including the continued-fraction method of [Morrison and Brillhart 1975]. See [Pomerance 1996b] for some of the history of the QS method and also the number field sieve.

6.1.1 Basic QS

Let n be an odd number with exactly k distinct prime factors. Then there are exactly 2^k square roots of 1 modulo n. This is easy in the case $k = 1$, and it follows in the general case from the Chinese remainder theorem; see Section 2.1.3. Two of these 2^k square roots of 1 are the old familiar ± 1. All of the others are interesting in that they can be used to split n. Indeed, if $a^2 \equiv 1 \pmod{n}$ and $a \not\equiv \pm 1 \pmod{n}$, then $\gcd(a-1, n)$ must be a nontrivial factor of n. To see this, note that $n|(a-1)(a+1)$, but n does not divide either factor, so part of n must divide $a - 1$ and part must divide $a + 1$.

For example, take the case $a = 11$ and $n = 15$. We have $a^2 \equiv 1 \pmod{n}$, and $\gcd(a-1, n) = 5$, a nontrivial factor of 15.

Consider the following three simple tasks: Find a factor of an even number, factor nontrivial powers, compute gcd's. The first task needs no comment! The second can be accomplished by extracting $\lfloor n^{1/k} \rfloor$ and seeing whether its k-th power is n, the root extraction being done via Newton's method and for k up to $\lg n$. The third simple task is easily done via Algorithm 2.1.2. Thus, we can "reduce" the factorization problem to finding nontrivial square roots of 1 for odd composites that are not powers. We write "reduce" in quotes since it is not much of a reduction—the two tasks are essentially computationally equivalent. Indeed, if we can factor n, an odd composite that is not a power, it is easy to play with this factorization and with gcd's to get a factorization $n = AB$ where A, B are greater than 1 and coprime; see the Exercises. Then let a be the solution to the Chinese remainder theorem problem posed thus:

$$a \equiv 1 \pmod{A}, \ , \ a \equiv -1 \pmod{B}.$$

We have thus created a nontrivial square root of 1 modulo n.

So we now set out on the task of finding a nontrivial square root of 1 modulo n, where n is an odd composite that is not a power. This task, in turn, is equivalent to finding a solution to $x^2 \equiv y^2 \pmod{n}$, where xy is coprime to n and $x \not\equiv \pm y \pmod{n}$. For then, $xy^{-1} \pmod{n}$ is a nontrivial square root of 1. However, as we have seen, any solution to $x^2 \equiv y^2 \pmod{n}$ with $x \not\equiv \pm y \pmod{n}$ can be used to split n.

The basic idea of the QS algorithm is to find congruences of the form $x_i^2 \equiv a_i \pmod{n}$, where $\prod a_i$ is a square, say y^2. If $x = \prod x_i$, then $x^2 \equiv y^2 \pmod{n}$. The extra requirement that $x \not\equiv \pm y \pmod{n}$ is basically ignored. If this condition works out, we are happy and can factor n. If it does not work out, we try the method again. We shall see that we actually can obtain many pairs of congruent squares, and assuming some kind of statistical independence, half of them or more should lead to a nontrivial factorization of n. It should be noted, though, right from the start, that QS is not a random algorithm. When we talk of statistical independence we do so heuristically. The numbers we are trying to factor don't seem to mind our lack of rigor, they get factored anyway.

Let us try this out on $n = 1649$, which is composite and not a power. Beginning as with Fermat's method, we take for the x_i's the numbers just above $\sqrt{n}$ (see Section 5.1.1):

$$41^2 = 1681 \equiv 32 \pmod{1649},$$
$$42^2 = 1764 \equiv 115 \pmod{1649},$$
$$43^2 = 1849 \equiv 200 \pmod{1649}.$$

With the Fermat method we would continue this computation until we reach 57^2, but with our new idea of combining congruences, we can stop with the above three calculations. Indeed, $32 \cdot 200 = 6400 = 80^2$, so we have

$$(41 \cdot 43)^2 \equiv 80^2 \pmod{1649}.$$

Note that $41 \cdot 43 = 1763 \equiv 114 \pmod{1649}$ and that $114 \not\equiv \pm 80 \pmod{1649}$, so we are in business. Indeed, $\gcd(114 - 80, 1649) = 17$, and we discover that $1649 = 17 \cdot 97$.

Can this idea be tooled up for big numbers? Say we look at the numbers $x^2 \bmod n$ for x running through integers starting at $\lceil \sqrt{n} \rceil$. We wish to find a nonempty subset of them with product a square. An obvious problem comes to mind: How does one search for such a subset?

Let us make some reductions in the problem to begin to address the issue of searching. First, note that if some $x^2 \bmod n$ has a large prime factor to the first power, then if we are to involve this particular residue in our subset with square product, there will have to be another $x'^2 \bmod n$ that has the same large prime factor. For example, in our limited experience above with 1649, the second residue is 115 which has the relatively large prime factor 23 (large compared with the prime factors of the other two residues), and indeed we threw this congruence away and did not use it in our product. So, what if we do this systematically and throw away any $x^2 \bmod n$ that has a prime factor exceeding B, say? That is, suppose we keep only the B-smooth numbers, (see Definition 1.4.7)? A relevant question is the following:

> How many positive B-smooth numbers are necessary before we are sure that the product of a nonempty subset of them is a square?

A moment's reflection leads one to realize that this question is really in the arena of linear algebra! Let us associate an "exponent vector" to a B-smooth number $m = \prod p_i^{e_i}$, where $p_1, p_2, \ldots, p_{\pi(B)}$ are the primes up to B and each exponent $e_i \ge 0$. The exponent vector is

$$\vec{v}(m) = (e_1, e_2, \ldots, e_{\pi(B)}).$$

If $m_1, m_2, \ldots, m_k$ are all B-smooth, then $\prod_{i=1}^k m_i$ is a square if and only if $\sum_{i=1}^k \vec{v}(m_i)$ has all even coordinates.

This last thought suggests we reduce the exponent vectors modulo 2 and think of them in the vector space $\mathbf{F}_2^{\pi(B)}$. The field of scalars of this vector space is $\mathbf{F}_2$ which has only the two elements 0, 1. Thus a linear combination of different vectors in this vector space is precisely the same thing as a subset sum; the subset corresponds to those vectors in the linear combination that have the coefficient 1. So the search for a nonempty subset of integers with product being a square is reduced to a search for a linear dependency in a set of vectors.

There are two great advantages of this point of view. First, we immediately have the theorem from linear algebra that a set of vectors is linearly dependent if there are more of them than the dimension of the vector space. So we have an answer: The creation of a product as a square requires at most $\pi(B) + 1$ positive B-smooth numbers. Second, the subject of linear algebra also comes equipped with efficient algorithms such as matrix reduction. So the issue of finding a linear dependency in a set of vectors comes down to row-reduction of the matrix formed with these vectors.

So we seem to have solved the "obvious problem" stated above for ramping up the 1649 example to larger numbers. We have a way of systematically handling our residues $x^2 \bmod n$, a theorem to tell us when we have enough of them, and an algorithm to find a subset of them with product being a square.

We have not, however, specified how the smoothness bound B is to be chosen, and actually, the above discussion really does not suggest that this scheme will be any faster than the method of Fermat.

If we choose B small, we have the advantage that we do not need many B-smooth residues to find a subset product that is a square. But if B is too small, the property of being B-smooth is so special that we may not find *any* B-smooth numbers. So we need to balance the two forces operating on the smoothness bound B: The bound should be small enough that we do not need too many B-smooth numbers to be successful, yet B should be large enough that the B-smooth numbers are arriving with sufficient frequency.

To try to solve this problem, we should compute what the frequency of B-smooth numbers will be as a function of B and n. Perhaps we can try to use (1.44), and assume that the "probability" that $x^2 \bmod n$ is B-smooth is about u^{-u}, where $u = \ln n / \ln B$.

There are two thoughts about this approach. First, (1.44) applies only to a total population of all numbers up to a certain bound, not a special subset. Are we so sure that members of our subset are just as likely to be smooth as is a typical number? Second, what exactly is the size of the numbers in our subset? In the above paragraph we just used the bound n when we formed the number u.

We shall overlook the first of these difficulties, since we are designing a heuristic factorization method. If the method works, our "conjecture" that our special numbers are just like typical numbers, as far as smoothness goes, gains some validity. The second of the difficulties, after a little thought, actually can be resolved in our favor. That is, we are wrong about the size of the residues $x^2 \bmod n$, they are actually *smaller* than n, much smaller.

Recall that we have suggested starting with $x = \lceil \sqrt{n} \rceil$ and running up from that point. But until we get to $\lceil \sqrt{2n} \rceil$, the residue $x^2 \bmod n$ is given by the simple formula $x^2 - n$. And if $\sqrt{n} < x < \sqrt{n} + n^{\epsilon}$, where $\epsilon > 0$ is small, then $x^2 - n$ is of order of magnitude $n^{1/2+\epsilon}$. Thus, we should revise our heuristic estimate on the likelihood of x leading to a B-smooth number to u^{-u} with u now about $\frac{1}{2} \ln n / \ln B$.

There is one further consideration before we try to use the u^{-u} estimate to pick out an optimal B and estimate the number of x's needed. That is, how long do we need to spend with a particular number x to see whether $x^2 - n$ is B-smooth? One might first think that the answer is about $\pi(B)$, since trial division with the primes up to B is certainly an obvious way to see whether a number is B-smooth. But in fact, there is a much better way to do this, a way that makes a big difference. We can use the sieving methods of Section 3.2.5 and Section 3.2.6 so that the average number of arithmetic operations spent per value of x is only about $\ln \ln B$, a very small bound indeed. These sieving methods require us to sieve by primes and powers of primes where the power is as high as could possibly divide one of the values $x^2 - n$. The primes p on which these powers are based are those for which $x^2 - n \equiv 0 \pmod{p}$ is solvable, namely the prime $p = 2$ and the odd primes $p \le B$ for which the Legendre symbol $\left(\frac{n}{p}\right) = 1$. And for each such odd prime p and each relevant power of p, there are two residue classes to sieve over. Let K be the number of primes up to B that over which we sieve. Then, heuristically, K is about $\frac{1}{2}\pi(B)$. We will be assured of a linear dependency among our exponent vectors once we have assembled $K + 1$ of them.

If the probability of a value of x leading to a B-smooth is u^{-u}, then the expected number of values of x to get one success is u^u, and the expected number of values to get $K + 1$ successes is $u^u(K + 1)$. We multiply this expectation by $\ln\ln B$, the amount of work on average to deal with each value of x. So let us assume that this all works out, and take the expression

$$T(B) = u^u(K + 1)\ln\ln B, \text{ where } u = \frac{\ln n}{2\ln B}.$$

We now attempt to find B as a function of n so as to minimize $T(B)$. Since $K \approx \frac{1}{2}\pi(B)$ is of order of magnitude $B/\ln B$ (see Theorem 1.1.4), we have that $\ln T(B) \sim S(B)$, where $S(B) = u\ln u + \ln B$. Putting in what u is we have that the derivative is given by

$$\frac{dS}{dB} = \frac{-\ln n}{2B\ln^2 B}(\ln\ln n - \ln\ln B - \ln 2 + 1) + \frac{1}{B}.$$

Setting this equal to zero, we find that $\ln B$ is somewhere between a constant times $\sqrt{\ln n}$ and a constant times $\sqrt{\ln n \ln\ln n}$, so that $\ln\ln B \sim \frac{1}{2}\ln\ln n$. Thus we find that the critical B and other entities behave as

$$\ln B \sim \frac{1}{2}\sqrt{\ln n \ln\ln n}, \quad u \sim \sqrt{\ln n/\ln\ln n}, \quad S(B) \sim \sqrt{\ln n \ln\ln n}.$$

We conclude that an optimal choice of the smoothness bound B is about $\exp\left(\frac{1}{2}\sqrt{\ln n \ln\ln n}\right)$, and that the running time with this choice of B is about B^2, that is, the running time for the above scheme to factor n should be about $\exp\left(\sqrt{\ln n \ln\ln n}\right)$.

We shall abbreviate this last function of n as follows:

$$L(n) = e^{\sqrt{\ln n \ln\ln n}}. \tag{6.1}$$

The above argument ignores the complexity of the linear algebra step, but it can be shown that this, too, is about B^2; see Section 6.1.3. Assuming the validity of all the heuristic leaps made, we have described a deterministic algorithm for factoring an odd composite n that is not a power. The running time is $L(n)^{1+o(1)}$. This function of n is subexponential; that is, it is of the form $n^{o(1)}$, and as such, it is a smaller-growing function of n than any of the complexity estimates for the factoring algorithms described in Chapter 5.

6.1.2 Basic QS: A summary

We have described the basic QS algorithm in the above discussion. We now give a summary description.

Algorithm 6.1.1 (Basic quadratic sieve). We are given an odd composite number n that is not a power. This algorithm attempts to give a nontrivial factorization of n.

1. [Initialization]
   $B = \lceil L(n)^{1/2} \rceil$; // Or tune B to taste.
   Set $p_1 = 2$ and $a_1 = 1$;
   Find the odd primes $p \le B$ for which $\left(\frac{n}{p}\right) = 1$, and label them $p_2, \dots, p_K$;
   for($2 \le i \le K$) find roots $\pm a_i$ with $a_i^2 \equiv n \pmod{p_i}$;
   // Find such roots via Algorithm 2.3.8 or 2.3.9.
2. [Sieving]
   Sieve the sequence $(x^2 - n)$, $x = \lceil \sqrt{n} \rceil, \lceil \sqrt{n} \rceil + 1, \dots$ for B-smooth values, until $K + 1$ such pairs $(x, x^2 - n)$ are collected in a set S;
   // See Sections 3.2.5, 3.2.6, and remarks (2), (3), (4).
3. [Linear algebra]
   for($(x, x^2 - n) \in S$) {
   Establish prime factorization $x^2 - n = \prod_{i=1}^{K} p_i^{e_i}$;
   $\vec{v}(x^2 - n) = (e_1, e_2, \dots, e_K)$; // Exponent vector.
   }
   Form the $(K+1) \times K$ matrix with rows being the various vectors $\vec{v}(x^2 - n)$ reduced $\bmod 2$;
   Use algorithms of linear algebra to find a nontrivial subset of the rows of the matrix that sum to the 0-vector $\pmod 2$, say $\vec{v}(x_1) + \vec{v}(x_2) + \cdots + \vec{v}(x_k) = \vec{0}$;
4. [Factorization]
   $x = x_1 x_2 \cdots x_k \bmod n$;
   $y = \sqrt{(x_1^2 - n)(x_2^2 - n) \dots (x_k^2 - n)} \bmod n$;
   // Infer this root directly from the known prime factorization of the perfect square $(x_1^2 - n)(x_2^2 - n) \dots (x_k^2 - n)$, see remark (6).
   $d = \gcd(x - y, n)$;
   return d;

There are several points that should be made about this algorithm:

(1) In practice, people generally use a somewhat smaller value of B than that given by the formula in step [Initialization]. Any value of B of order of magnitude $L(n)^{1/2}$ will lead to the same overall complexity, and there are various practical issues that mitigate towards a smaller value, such as the size of the matrix that one deals with in step [Linear algebra], and the size of the moduli one sieves with in comparison to cache size on the machine used in step [Sieving]. The optimal B-value is more of an art than a science, and is perhaps best left to experimentation.

(2) To do the sieving, one must know which residue classes to sieve for each p_i found in step [Initialization]. (For simplicity, we shall ignore the problem of sieving with higher powers of these primes. Such sieving is easy to do—one can use Algorithm 2.3.11, for example—but might also be ignored in practice, since it does not contribute much to the finding of B-smooth numbers.) For the odd primes p_i in step [Initialization], we have solved the congruence $x^2 \equiv n \pmod{p_i}$. This is solvable, since the p_i's have been selected in step [Initialization] precisely to have this property. Either Algorithm 2.3.8 or Algorithm 2.3.9 may be used to solve the congruence. Of course, for each solution, we also have the negative of this residue class as a second solution, so we sieve two residue classes for each p_i with p_i odd. (Though we could sieve with $p_1 = 2$ as indicated in the pseudocode, we do not have to sieve at all with 2 and other small primes; see the remarks in Section 3.2.5.)

(3) An important point is that the arithmetic involved in the actual sieving can be done through additions of approximate logarithms of the primes being sieved, as discussed in Section 3.2.5. In particular, one should set up a zero-initialized array of some convenient count of b bytes, corresponding to the first b of the x values. Then one adds a $\lg p_i$ increment (rounded to the nearest integer) starting at offsets x_i, x_i', the least integers $\geq \lceil \sqrt{n} \rceil$ that are congruent (mod p_i) to $a_i, -a_i$, respectively, and at every spacing p_i from there forward in the array. If necessary (i.e., not enough smooth numbers have been found) a new array is zeroed with its first element corresponding to $\lceil \sqrt{n} \rceil + b$, and continue in the same fashion. The threshold set for reporting a location with a B-smooth value is set as $\lfloor \lg |x^2 - n| \rfloor$, minus some considerable fudge, such as 20, to make up for the errors in the approximate logarithms, and other errors that might accrue from not sieving with small primes or higher powers. Any value reported must be tested by trial division to see if it is indeed B-smooth. This factorization plays a role in step [Linear algebra]. (To get an implementation working properly, it helps to test the logarithmic array entries against actual, hard factorizations.)

(4) Instead of starting at $\lceil \sqrt{n} \rceil$ and running up through the integers, consider instead the possibility of x running through a sequence of integers *centered* at $\sqrt{n}$. There is an advantage and a disadvantage to this thought. The advantage is that the values of the polynomial $x^2 - n$ are now somewhat smaller on average, and so presumably they are more likely to be B-smooth. The disadvantage is that some values are now negative, and the sign is an important consideration when forming squares. Squares not only have all their prime factors appearing with even exponents, they are also positive. This disadvantage can be handled very simply. We enlarge the exponent vectors by one coordinate, letting the new coordinate, say the zeroth one, be 1 if the integer is negative and 0 if it is positive. So, just like all of the other coordinates, we wish to get an even number of 1's. This has the effect of raising the dimension of our vector space

from K to $K + 1$. Thus the disadvantage of using negatives is that our vectors are 1 bit longer, and we need one more vector to be assured of a linear dependency. This disadvantage is minor; it is small compared to the advantage of smaller numbers in the sieve. We therefore go ahead and allow negative polynomial values.

(5) We have been ignoring the problem that there is no guarantee that the number d produced in step [Factorization] is a nontrivial divisor of n. Assuming some kind of randomness (which is certainly not the case, but may be a reasonable heuristic assumption), the "probability" that d is a nontrivial divisor is $1/2$ or larger; see Exercise 6.2. If we find a few more dependencies among our exponent vectors, and again assuming statistical independence, we can raise the odds for success. For example, say we sieve in step [Sieving] until $K + 11$ polynomial values are found that are B-smooth. Assuming that the dimension of our space is now $K+1$ (because we allow negative values of the polynomial; see above), there will be at least 10 independent linear dependencies. The odds that none will work to give a nontrivial factorization of n is smaller than 1 in 1000. And if these odds for failure are still too high for your liking, you can collect a few more B-smooth numbers for good measure.

(6) In step [Factorizaton] we have to take the square root of perhaps a very large square, namely $Y^2 = (x_1^2 - n)(x_2^2 - n) \cdots (x_k^2 - n)$. However, we are interested only in $y = Y \bmod n$. We can exploit the fact that we actually know the prime factorization of Y^2, and so we know the prime factorization of Y. We can thus compute y by using Algorithm 2.1.5 to find the residue of each prime power in Y modulo n, and then multiply these together, again reducing modulo n. We shall find that in the number field sieve, the square root problem cannot be solved so easily.

In the next few sections we shall discuss some of the principal enhancements to the basic quadratic sieve algorithm.

6.1.3 Fast matrix methods

With $B = \exp\left(\frac{1}{2}\sqrt{\ln n \ln\ln n}\right)$, we have seen that the time to complete the sieving stage of QS is (heuristically) $B^{2+o(1)}$. After this stage, one has about B vectors of length about B, with entries in the finite field $\mathbf{F}_2$ of two elements, and one wishes to find a nonempty subset with sum being the zero vector. To achieve the overall complexity of $B^{2+o(1)}$ for QS, we shall need a linear algebra subroutine that can find the nonempty subset within this time bound.

We first note that forming a matrix with our vectors and using Gaussian elimination to find subsets with sum being the zero vector has a time bound of $O\left(B^3\right)$ (assuming that the matrix is $B \times B$). Nevertheless, in practice, Gaussian elimination is a fine method to use for smaller factorizations. There are several reasons why the high-complexity estimate is not a problem in practice.

(1) Since the matrix arithmetic is over $\mathbf{F}_2$, it naturally lends itself to computer implementation. With w being the machine word length (typically 8 or 16 bits on older machines, 32 or 64 or even more bits on newer ones), we can deal with blocks of w coordinates in a row at a time, where one step is just a logical operation requiring very few clock cycles.

(2) The initial matrix is quite sparse, so at the start, before "fill in" occurs, there are few operations to perform, thus somewhat reducing the worst case time bound.

(3) If the number we are factoring is not too large, we can load the algorithm towards the sieving stage and away from the matrix stage. That is, we can choose a bound B that is somewhat too small, thus causing the sieving stage to run longer, but easing difficulties in the matrix stage. Space difficulties with higher values of B form another practical reason to choose B smaller than an otherwise optimal choice.

Concerning point (2), ways have been found to use Gaussian elimination in an "intelligent" way so as to preserve sparseness as long as possible, see [Odlyzko 1985] and [Pomerance and Smith 1992]. These methods are sometimes referred to as "structured-Gauss" methods.

As the numbers we try to factor get larger, the matrix stage of QS (and especially of the number field sieve; see Section 6.2) looms larger. The unfavorable complexity bound of Gaussian elimination ruins our overall complexity estimates, which assume that the matrix stage is not a bottleneck. In addition, the awkwardness of dealing with huge matrices seems to require large and expensive computers, computers for which it is not easy to get large blocks of time.

There have been suggested at least three alternative sparse-matrix methods intended to replace Gaussian elimination, two of which having already been well-studied in numerical analysis. These two, the conjugate gradient method and the Lanczos method, have been adapted to matrices with entries in a finite field. A third option is the coordinate recurrence method of [Wiedemann 1986]. This method is based on the Berlekamp–Massey algorithm for discovering the smallest linear recurrence relation in a sequence of finite field elements.

Each of these methods can be accomplished with a sparse encoding of the matrix, namely an encoding that lists merely the locations of the nonzero entries. Thus, if the matrix has N nonzero entries, the space required is $O(N \ln B)$. Since our factorization matrices have at most $O(\ln n)$ nonzero entries per row, the space requirement for the matrix stage of the algorithm, using a sparse encoding, is $O\left(B \ln^2 n\right)$.

Both the Wiedemann and Lanczos methods can be made rigorous. The running time for these methods is $O(BN)$, where N is the number of nonzero entries in the matrix. Thus, the time bound for the matrix stage of factorization algorithms such as QS is $B^{2+o(1)}$, equaling the time bound for sieving.

For a discussion of the conjugate gradient method and the Lanczos method, see [Odlyzko 1985]. For a study of the Lanczos method in a theoretical setting see [Teitelbaum 1998]. For some practical improvements to the Lanczos method see [Montgomery 1995].

6.1.4 Large prime variations

As discussed above and in Section 3.2.5, sieving is a very cheap operation. Unlike trial division, which takes time proportional to the number of trial divisors, that is, one "unit" of time per prime used as a trial, sieving takes less and less time per prime sieved as the prime modulus grows. In fact the time spent per sieve location, on average, for each prime modulus p is proportional to $1/p$. However, there are hidden costs for increasing the list of primes p with which we sieve. One is that it is unlikely we can fit the entire sieve array into memory on a computer, so we segment it. If a prime p exceeds the length of this part of the sieve, we have to spend a unit of time per segment to see whether this prime will "hit" something or not. Thus, once the prime exceeds this threshold, the $1/p$ "philosophy" of the sieve is left behind, and we spend essentially the same time for each of these larger primes: Sieving begins to resemble trial division. Another hidden cost is perhaps not so hidden at all. When we turn to the linear-algebra stage of the algorithm, the matrix will be that much bigger if more primes are used. Suppose we are using 10^6 primes, a number that is not inconceivable for the sieving stage. The matrix, if encoded as a binary (0,1) matrix, would have 10^{12} bits. Indeed, this would be a large object on which to carry out linear algebra! In fact, some of the linear algebra routines that will be used, see Section 6.1.3, involve a sparse encoding of the matrix, namely, a listing of where the 1's appear, since almost all of the entries are 0's. Nevertheless, space for the matrix is a worrisome concern, and it puts a limit on the size of the smoothness bound we take.

The analysis in Section 6.1.1 indicates a third reason for not taking the smoothness bound too large; namely, it would increase the number of reports necessary to find a linear dependency. Somehow, though, this reason is specious. If there is already a dependency around with a subset of our data, having more data should not destroy this, but just make it a bit harder to find, perhaps. So we should not take an overshooting of the smoothness bound as a serious handicap if we can handle the two difficulties mentioned in the above paragraph.

In its simplest form, the large-prime variation allows us a cheap way to somewhat increase our smoothness bound, by giving us for free many numbers that are almost B-smooth, but fail because they have one larger prime factor. This larger prime could be taken in the interval $(B, B^2]$. It should be noted from the very start that allowing for numbers that are B-smooth except for having one prime factor in the interval $(B, B^2]$ is *not* the same as taking B^2-smooth numbers. With B about $L(n)^{1/2}$, as suggested in Section 6.1.1, a typical B^2-smooth number near $n^{1/2+\epsilon}$ in fact has many prime factors in the interval $(B, B^2]$, not just one.

Be that as it may, the large-prime variation does give us something that we did not have before. By allowing sieve reports of numbers that are close to the threshold for B-smoothness, but not quite there, we can discover numbers that have one slightly larger prime. In fact, if a number has all the primes up to B removed from its prime factorization, and the resulting number is smaller than B^2, but larger than 1, then the resulting number must be a prime. It is this idea that is at work in the large-prime variation. Our sieve is not perfect, since we are using approximate logarithms and perhaps not sieving with small primes (see Section 3.2.5), but the added grayness does not matter much in the mass of numbers being considered. Some numbers with a large prime factor that might have been reported are possibly passed over, and some numbers are reported that should not have been, but neither problem is of great consequence.

So if we can obtain these numbers with a large prime factor for free, how then can we process them in the linear algebra stage of the algorithm? In fact, we should not view the numbers with a large prime as having longer exponent vectors, since this could cause our matrix to be too large. There is a very cheap way to process these large prime reports. Simply sort them on the value of the large prime factor. If any large prime appears just once in the sorted list, then this number cannot possibly be used to make a square for us, so it is discarded. Say we have k reports with the same large prime: $x_i^2 - n = y_i P$, for $i = 1, 2, \ldots, k$. Then

$$(x_1 x_i)^2 \equiv y_1 y_i P^2 \pmod{n}, \text{ for } i = 2, \ldots, k.$$

So when $k \geq 2$ we can use the exponent vectors for the $k-1$ numbers $y_1 y_i$, since the contribution of P^2 to the exponent vector, once it is reduced mod 2, is 0. That is, duplicate large primes lead to exponent vectors on the primes up to B. Since it is very fast to sort a list, the creation of these new exponent vectors is like a gift from heaven.

There is one penalty to using these new exponent vectors, though it has not proved to be a big one. The exponent vector for a $y_1 y_i$ as above is usually not as sparse as an exponent vector for a fully smooth report. Thus, the matrix techniques that take advantage of sparseness are somewhat hobbled. Again, this penalty is not severe, and every important implementation of the QS method uses the large-prime variation.

One might wonder how likely it is to have a pair of large primes matching. That is, when we sort our list, could it be that there are very few matches, and that almost everything is discarded because it appears just once? The birthday paradox from probability theory suggests that matches will not be uncommon, once one has plenty of large prime reports. In fact the experience that factorers have is that the importance of the large prime reports is nil near the beginning of the run, because there are very few matches, but as the data set gets larger, the effect of the birthday paradox begins, and the matches for the large primes blossom and become a significant source of rows for the final matrix.

It is noticed in practice, and this is supported too by theory, that the larger the large prime, the less likely for it to be matched up. Thus, most practitioners eschew the larger range for large primes, perhaps keeping only those in the interval $(B, 20B]$ or $(B, 100B]$.

Various people have suggested over the years that if one large prime is good, perhaps two large primes are better. This idea has been developed in [Lenstra and Manasse 1994], and they do, in fact, find better performance for larger factorizations if they use two large primes. The landmark factorization of the RSA129 challenge number mentioned in Section 1.1.2 was factored using this double large-prime variation.

There are various complications for the double large-prime variation that are not present in the single large-prime variation discussed above. If an integer in the interval $(1, B^2]$ has all prime factors exceeding B, then it must be prime: This is the fundamental observation used in the single large-prime variation. What if an integer in $(B^2, B^3]$ has no prime factor $\leq B$? Then either it is a prime, or it is the product of two primes each exceeding B. In essence, the double large prime variation allows for reports where the unfactored portion is as large as B^3. If this unfactored portion m exceeds B^2, a cheap pseudoprimality test is applied, say checking whether $2^{m-1} \equiv 1 \pmod{m}$; see Section 3.3.1. If m satisfies the congruence, it is *discarded*, since then it is likely to be prime, and also too large to be matched with another large prime. If m is proved composite by the congruence, it is then factored, say by the Pollard rho method; see Section 5.2.1. This will then allow reports that are B-smooth, except for two prime factors larger than B (and not much larger).

As one can see, this already requires much more work than the single large-prime variation. But there is more to come. One must search the reported numbers with a single large prime or two large primes for cycles, that is, subsets whose product is B-smooth, except for larger primes that all appear to even exponents. For example, say we have the reports $y_1P_1, y_2P_2, y_3P_1P_2$, where y_1, y_2, y_3 are B-smooth and P_1, P_2 are primes exceeding B (so we are describing here a cycle consisting of two single large prime reports and one double large prime report). The product of these three reports is $y_1y_2y_3P_1^2P_2^2$, whose exponent vector modulo 2 is the same as that for the B-smooth number $y_1y_2y_3$. Of course, there can be more complicated cycles than this, some even involving only double large-prime factorizations (though that kind will be infrequent). It is not as simple as before, to search through our data set for these cycles. For one, the data set is much larger than before and there is a possibility of being swamped with data. These problems are discussed in [Lenstra and Manasse 1994]. They find that with larger numbers they gain a more than twofold speed-up using the double large-prime variation. However, they also admit that they use a value of B that is perhaps smaller than others would choose. It would be interesting to see an experiment that allows for variations of all parameters involved to see which combination is the best for numbers of various sizes.

And what, then, of three large primes? One can appreciate that the added difficulties with two large primes increase still further. It may be worth it, but it seems likely that instead, using a larger B would be more profitable.

6.1.5 Multiple polynomials

In the basic QS method we let x run over integers near $\sqrt{n}$, searching for values $x^2 - n$ that are B-smooth. The reason we take x near $\sqrt{n}$ is to minimize the size of $x^2 - n$, since smaller numbers are more likely to be smooth than larger numbers. But for x near to $\sqrt{n}$, we have $x^2 - n \approx 2\left(x - \sqrt{n}\right)\sqrt{n}$, and so as x marches away from $\sqrt{n}$, so, too, do the numbers $x^2 - n$, and at a steady and rapid rate. There is thus built into the basic QS method a certain diminishing return as one runs the algorithm, with perhaps a healthy yield rate for smooth reports at the beginning of the sieve, but this rate declining perceptibly as one continues to sieve.

The multiple polynomial variation of the QS method allows one to get around this problem by using a family of polynomials rather than just the one polynomial $x^2 - n$. Different versions of using multiple polynomials have been suggested independently by Davis, Holdridge, and Montgomery; see [Pomerance 1985]. The Montgomery method is slightly better and is the way we currently use the QS algorithm. Basically, what Montgomery does is replace the variable x with a wisely chosen linear function in x.

Suppose a, b, c are integers with $b^2 - ac = n$. Consider the quadratic polynomial $f(x) = ax^2 + 2bx + c$. Then

$$af(x) = a^2x^2 + 2abx + ac = (ax + b)^2 - n, \tag{6.2}$$

so that

$$(ax + b)^2 \equiv af(x) \pmod{n}.$$

If we have a value of a that is a square times a B-smooth number and a value of x for which $f(x)$ is B-smooth, then the exponent vector for $af(x)$, once it is reduced modulo 2, gives us a row for our matrix. Moreover, the possible odd primes p that can divide $f(x)$ (and do not divide n) are those with $\left(\frac{p}{n}\right) = 1$, namely the same primes that we are using in the basic QS algorithm. (It is somewhat important to have the set of primes occurring not depend on the polynomial used, since otherwise, we will have more columns for our matrix, and thus need more rows to generate a dependency.)

We are requiring that the triple a, b, c satisfy $b^2 - ac = n$ and that a be a B-smooth number times a square. However, the reason we are using the polynomial $f(x)$ is that its values might be small, and so more likely to be smooth. What conditions should we put on a, b, c to have small values for $f(x) = ax^2 + 2bx + c$? Well, this depends on how long an interval we sieve on for the given polynomial. Let us decide beforehand that we will only sieve the polynomial for arguments x running in an interval of length $2M$. Also, by (6.2), we can agree to take the coefficient b so that it satisfies $|b| \leq \frac{1}{2}a$ (assuming a is positive). That is, we are ensuring our interval of length $2M$

for x to be precisely the interval $[-M, M]$. Note that the largest value of $f(x)$ on this interval is at the endpoints, where the value is about $(a^2M^2 - n)/a$, and the least value is at $x = 0$, being there about $-n/a$. Let us set the absolute values of these two expressions approximately equal to each other, giving the approximate equation $a^2M^2 \approx 2n$, so that $a \approx \sqrt{2n}/M$.

If a satisfies this approximate equality, then the absolute value of $f(x)$ on the interval $[-M, M]$ is bounded by $(M/\sqrt{2})\sqrt{n}$. This should be compared with the original polynomial $x^2 - n$ used in the basic QS method. On the interval $[\sqrt{n} - M, \sqrt{n} + M]$, the values are bounded by approximately $2M\sqrt{n}$. So we have saved a factor $2\sqrt{2}$ in size. But we have saved much more than that. In the basic QS method the values continue to grow, we cannot stop at a preset value M. But when we use a family of polynomials, we can continually change. Roughly, using the analysis of Section 6.1.1, we can choose $M = B = L(n)^{1/2}$ when we use multiple polynomials, but must choose $M = B^2 = L(n)$ when we use only one polynomial. So the numbers that "would be smooth" using multiple polynomials are smaller on average by a factor B. A heuristic analysis shows that using multiple polynomials speeds up the quadratic sieve method by roughly a factor $\frac{1}{2}\sqrt{\ln n \ln\ln n}$. When n is about 100 digits, this gives a savings of about a factor 17; that is, QS with multiple polynomials runs about 17 times as fast as the basic QS method. (This "thought experiment" has not been numerically verified, though there can be no doubt that using multiple polynomials is considerably faster in practice.)

However, there is one last requirement for the leading coefficient a: We need to find values of b, c to go along with it. If we can solve $b^2 \equiv n \pmod{a}$ for b, then we can ensure that $|b| \le a/2$, and we can let $c = (b^2 - n)/a$. Note that the methods of Section 2.3.2 will allow us to solve the congruence provided that we choose a such that a is odd, we know the prime factorization of a, and for each prime $p|a$, we have $\left(\frac{n}{p}\right) = 1$. One effective way to do this is to take various primes $p \approx (2n)^{1/4}/M^{1/2}$, with $\left(\frac{n}{p}\right) = 1$, and choose $a = p^2$. Then such values of a meet all the criteria we have set for them:

(1) We have a equal to a square times a B-smooth number.

(2) We have $a \approx \sqrt{2n}/M$.

(3) We can efficiently solve $b^2 \equiv n \pmod{a}$ for b.

The congruence $b^2 \equiv n \pmod{a}$ has two solutions, if we take $a = p^2$ as above. However, the two solutions lead to equivalent polynomials, so we use only one of the solutions, say the one with $0 < b < \frac{1}{2}a$.

6.1.6 Self initialization

In Section 6.1.5 we learned that it is good to change polynomials frequently. The question is, how frequently? One constraint, already implicitly discussed, is that the length, $2M$, of the interval on which we sieve a polynomial should be at least B, the bound for the moduli with which we sieve. If this is the only constraint, then a reasonable choice might then be to take M with $2M = B$.

For numbers in the range of 50 to 150 digits, typical choices for B are in the range 10^4 to 10^7, approximately. It turns out that sieving is so fast an operation, that if we changed polynomials every time we sieved B numbers, the overhead in making the change would be so time-consuming that overall efficiency would suffer. This overhead is principally to solve the *initialization problem*. That is, given a, b, c as in Section 6.1.5, for each odd prime $p \le B$ with $\left(\frac{n}{p}\right) = 1$, we have to solve the congruence

$$ax^2 + 2bx + c \equiv 0 \pmod{p}$$

for the two roots $r(p) \bmod p$ and $s(p) \bmod p$ (we assume here that p does not divide an). Thus, we have

$$r(p) = (-b + t(p))a^{-1} \bmod p, \;\; s(p) = (-b - t(p))a^{-1} \bmod p, \tag{6.3}$$

where

$$t(p)^2 \equiv n \pmod{p}.$$

For each polynomial, we can use the exact same residue $t(p)$ each time when we come to finding $r(p), s(p)$. So the principal work in using (6.3) is in computing $a^{-1} \bmod p$ for each p (say by Algorithm 2.1.4) and the two mod p multiplications. If there are many primes p for which this needs to be done, it is enough work that we do not want to do it too frequently.

The idea of self initialization is to amortize the work in (6.3) over several polynomials with the same value of a. For each value of a, we choose b such that $b^2 \equiv n \pmod{a}$ and $0 < b < a/2$, see Section 6.1.5. For each such b we can write down a polynomial $ax^2 + 2bx + c$ to use in QS, by letting $c = (b^2 - n)/a$. The number of choices for b for a given value of a is 2^{k-1}, where a has k distinct prime factors (assuming that a is odd, and for each prime $p|a$ we have $\left(\frac{n}{p}\right) = 1$). So, choosing a as the square of a prime, as suggested in Section 6.1.5, gives exactly 1 choice for b. Suppose instead we choose a as the product of 10 different primes p. Then there are $512 = 2^9$ choices for b corresponding to the given a, and so the $a^{-1} \pmod{p}$ computations need only be done once and then used for all 512 of the polynomials. Moreover, if none of the 10 primes used in a exceeds B, then it is not necessary to have them squared in a, their elimination is already built into the matrix step anyway.

There can be more savings with self initialization if one is willing to do some additional precomputation and store some files. For example, if one computes and stores the list of all $2t(p)a^{-1} \bmod p$ for all the primes p with which we sieve, then the computation to get $r(p), s(p)$ in (6.3) can be done with a single multiplication rather than 2. Namely, multiply $-b + t(p)$ by the stored value $a^{-1} \bmod p$ and reduce mod p. This gives $r(p)$. Subtracting the stored value $2t(p)a^{-1} \bmod p$ and adding p if necessary, we get $s(p)$.

It is even possible to eliminate the one multiplication remaining, by traversing the different solutions b using a Gray code; see Exercise 6.7. In fact, the Chinese remainder theorem, see Section 2.1.3, gives the different solutions b in the form $B_1 \pm B_2 \pm \cdots \pm B_k$. (If $a = p_1 p_2 \cdots p_k$, then B_i satisfies

$B_i^2 \equiv n \pmod{p_i}$ and $B_i \equiv 0 \pmod{a/p_i}$.) If we traverse the 2^{k-1} numbers $B_1 \pm B_2 \pm \ldots \pm B_k$ using a Gray code and precompute the lists $2B_ia^{-1} \bmod p$ for all p with which we sieve, then we can move from the sieving coordinates for one polynomial to the next doing merely some low-precision adds and subtracts for each p. One can get by with storing only the most frequently used files $2B_ia^{-1} \bmod a$ if space is at a premium. For example, storing this file only for $i = k$, which is in action every second step in the Gray code, we have initialization being very cheap half the time, and done with a single modular multiplication for each p (and a few adds and subtracts) the other half of the time.

The idea for self initialization was briefly sketched in [Pomerance et al. 1988] and more fully described in [Alford and Pomerance 1995] and [Peralta 1993]. In [Contini 1997] it is shown through some experiments that self initialization gives about a twofold speedup over standard implementations of QS using multiple polynomials.

6.1.7 Zhang's special quadratic sieve

What makes the quadratic sieve fast is that we have a polynomial progression of small quadratic residues. That they are quadratic residues renders them useful for obtaining congruent squares that can split n. That they form a polynomial progression (that is, consecutive values of a polynomial) makes it easy to discover smooth values, namely, via a sieve. And of course, that they are small makes them more likely to be smooth than random residues modulo n. One possible way to improve this method is to find a polynomial progression of even smaller quadratic residues. Recently, M. Zhang has found such a way, but only for special values of n, [Zhang 1998]. We call his method the special quadratic sieve, or SQS.

Suppose the number n we are trying to factor (which is odd, composite, and not a power) can be represented as

$$n = m^3 + a_2m^2 + a_1m + a_0, \tag{6.4}$$

where m, a_2, a_1, a_0 are integers, $m \approx n^{1/3}$. Actually, every number n can be represented in this way; just choose $m = \lfloor n^{1/3} \rfloor$, let $a_1 = a_2 = 0$, and let $a_0 = n - m^3$. We shall see below, though, that the representation (6.4) will be useful only when the a_i's are all small in absolute value, and so we are considering only special values of n.

Let b_0, b_1, b_2 be integer variables, and let

$$x = b_2m^2 + b_1m + b_0,$$

where m is as in (6.4). Since

$$\begin{aligned} m^3 &\equiv -a_2m^2 - a_1m - a_0 \pmod{n}, \\ m^4 &\equiv (a_2^2 - a_1)m^2 + (a_1a_2 - a_0)m + a_0a_2 \pmod{n}, \end{aligned}$$

we have

$$x^2 \equiv c_2m^2 + c_1m + c_0 \pmod{n}, \tag{6.5}$$

where

$$\begin{aligned} c_2 &= (a_2^2 - a_1)b_2^2 - 2a_2b_1b_2 + b_1^2 + 2b_0b_2, \\ c_1 &= (a_1a_2 - a_0)b_2^2 - 2a_1b_1b_2 + 2b_0b_1, \\ c_0 &= a_0a_2b_2^2 - 2a_0b_1b_2 + b_0^2. \end{aligned}$$

Since b_0, b_1, b_2 are free variables, perhaps they can be chosen so that they are small integers and that $c_2 = 0$. Indeed, they can. Let

$$b_2 = 2, \quad b_1 = 2b, \quad b_0 = a_1 - a_2^2 + 2a_2b - b^2,$$

where b is an arbitrary integer. With these choices of b_0, b_1, b_2 we have

$$x(b)^2 \equiv y(b) \pmod{n}, \tag{6.6}$$

where

$$\begin{aligned} x(b) &= 2m^2 + 2bm + a_1 - a_2^2 + 2a_2b - b^2, \\ y(b) &= \left(4a_1a_2 - 4a_0 - \left(4a_1 + 4a_2^2\right)b + 8a_2b^2 - 4b^3\right)m \\ &\qquad + 4a_0a_2 - 8a_0b + \left(a_1 - a_2^2 + 2a_2b - b^2\right)^2. \end{aligned}$$

The proposal is to let b run through small numbers, use a sieve to search for smooth values of $y(b)$, and then use a matrix of exponent vectors to find a subset of the congruences (6.6) to construct two congruent squares mod n that then may be tried for factoring n. If a_0, a_1, a_2, and b are all $O(n^\epsilon)$, where $0 \le \epsilon < 1/3$, and $m = O\left(n^{1/3}\right)$, then $y(b) = O(n^{1/3+3\epsilon})$. The complexity analysis of Section 6.1.1 gives a heuristic running time of

$$L(n)^{\sqrt{2/3+6\epsilon+o(1)}},$$

where $L(n)$ is defined in (6.1). If ϵ is small enough, this estimate beats the heuristic complexity of QS.

It may also be profitable to generalize (6.4) to

$$an = m^3 + a_2m^2 + a_1m + a_0.$$

The number a does not appear in the expressions for $x(b), y(b)$, but it does affect the size of the number m, which is now about $(an)^{1/3}$.

For example, consider the number $2^{601} - 1$. We have the two prime factors 3607 and 64863527, but the resulting number n_0 when these primes are divided into $2^{601} - 1$ is a composite of 170 decimal digits for which we know no factor. We have

$$2^2 \cdot 3607 \cdot 64863527 n_0 = 2^{603} - 2^2 = \left(2^{201}\right)^3 - 4,$$

so that we may take $a_0 = -4$, $a_1 = a_2 = 0$, $m = 2^{201}$. These assignments give the congruence (6.6) with

$$x(b) = 2m^2 + 2bm - b^2, \quad y(b) = (16 - 4b^3)m + 32b + b^4, \quad m = 2^{201}.$$

As the number b grows in absolute value, $y(b)$ is dominated by the term $-4b^3m$. It is not unreasonable to expect that b will grow as large as 2^{40}, in which case the size of $|y(b)|$ will be near 2^{323}. This does not compare favorably with the quadratic sieve with multiple polynomials, where the size of the numbers we sieve for smooths would be about $2^{20}\sqrt{n} \approx 2^{301}$. (This assumes a sieving interval of about 2^{20} per polynomial.)

However, we can also use multiple polynomials with the special quadratic sieve. For example, for the above number n_0, take $b_0 = -2u^2$, $b_1 = 2uv$, $b_2 = v^2$. This then implies that we may take

$$x(u,v) = v^2m^2 + 2uvm - 2u^2, \quad y(u,v) = (4v^4 - 8u^3v)m + 16uv^3 + 4u^4,$$

and let u, v range over small, coprime integers. (It is important to take u, v coprime, since otherwise, we shall get redundant relations.) If u, v are allowed to range over numbers with absolute value up to 2^{20}, we get about the same number of pairs as choices for b above, but the size of $|y(u,v)|$ is now about 2^{283}, a savings over the ordinary quadratic sieve. (There is a small additional savings, since we may actually consider the pair $\frac{n-1}{2}x(u,v), \frac{1}{4}y(u,v)$.)

It is perhaps not clear why the introduction of u, v may be considered as "multiple polynomials." The idea is that we may fix one of these letters, and sieve over the other. Each choice of the first letter gives a new polynomial in the second letter.

The assumption in the above analysis of a sieve of length 2^{40} is probably on the small side for a number the size of n_0. A larger sieve length will make SQS look poorer in comparison with ordinary QS.

It is not clear whether the special quadratic sieve, as described above, will be a useful factoring algorithm (as of this writing, it has not actually been tried out in significant settings). If the number n is not too large, the growth of the coefficient of m in $y(b)$ or $y(u,v)$ will dominate and make the comparison with the ordinary quadratic sieve poor. If the number n is somewhat larger, so that the special quadratic sieve starts to look better, as in the above example, there is actually another algorithm that may come into play and again majorize the special quadratic sieve. This is the number field sieve, something we shall discuss in the next section.

6.2 Number field sieve

We have encountered some of the inventive ideas of J. Pollard in Chapter 5. In 1988 (see [Lenstra and Lenstra 1993]) Pollard suggested a factoring method that was very well suited for numbers, such as Fermat numbers, that are close to a high power. Before long, this method had been generalized so that it could be used for general composites. Today, the number field sieve (NFS) stands as the asymptotically fastest heuristic factoring algorithm we know for "worst-case" composite numbers.

6.2.1 Basic NFS: Strategy

The quadratic sieve factorization method is fast because it produces small quadratic residues modulo the number we are trying to factor, and because we can use a sieve to quickly recognize which of these quadratic residues are smooth. The QS method would be faster still if the quadratic residues it produces could be arranged to be smaller, since then they would be more likely to be smooth, and so we would not have to sift through as many of them. An interesting thought in this regard is that it is not necessary that they be quadratic residues, only small! We have a technique through linear algebra of multiplying subsets of smooth numbers so as to obtain squares. In the quadratic sieve, we had only to worry about one side of the congruence, since the other side was already a square. In the number field sieve we use the linear algebra method on both sides of the key congruence.

However, our congruences will not start with two integers being congruent mod n. Rather, they will start with pairs $\theta, \phi(\theta)$, where θ lies in a particular algebraic number ring, and ϕ is a homomorphism from the ring to $\mathbf{Z}_n$. (These concepts will be described concretely, in a moment.) Suppose we have k such pairs $\theta_1, \phi(\theta_1), \ldots, \theta_k, \phi(\theta_k)$, such that the product $\theta_1 \cdots \theta_k$ is a square in the number ring, say γ^2, and there is an integer square, say v^2, such that $\phi(\theta_1) \cdots \phi(\theta_k) \equiv v^2 \pmod{n}$. Then if $\phi(\gamma) \equiv u \pmod{n}$ for an integer u, we have

$$u^2 \equiv \phi(\gamma)^2 \equiv \phi(\gamma^2) \equiv \phi(\theta_1 \cdots \theta_k) \equiv \phi(\theta_1) \cdots \phi(\theta_k) \equiv v^2 \pmod{n}.$$

That is, stripping away all of the interior expressions, we have the congruence $u^2 \equiv v^2 \pmod{n}$, and so could try to factor n via $\gcd(u - v, n)$.

The above ideas constitute the strategy of NFS. We now discuss the basic setup that introduces the number ring and the homomorphism ϕ. Suppose we are trying to factor the number n, which is odd, composite, and not a power. Let

$$f(x) = x^d + c_{d-1}x^{d-1} + \cdots + c_0$$

be an irreducible polynomial in $\mathbf{Z}[x]$, and let α be a complex number that is a root of f. We do not need to numerically approximate α; we just use the symbol "α" to stand for one of the roots of f. Our number ring will be $\mathbf{Z}[\alpha]$. This is computationally thought of as the set of ordered d-tuples $(a_0, a_1, \ldots, a_{d-1})$ of integers, where we "picture" such a d-tuple as the element $a_0 + a_1\alpha + \cdots a_{d-1}\alpha^{d-1}$. We add two such expressions coordinatewise, and we multiply via the normal polynomial product, but then reduce to a d-tuple via the identity $f(\alpha) = 0$. Another, equivalent way of thinking of the number ring $\mathbf{Z}[\alpha]$ is to realize it as $\mathbf{Z}[x]/(f(x))$, that is, involving polynomial arithmetic modulo $f(x)$.

The connection to the number n we are factoring comes via an integer m with the property that

$$f(m) \equiv 0 \pmod{n}.$$

We do need to know what the integer m is. We remark that there is a very simple method of coming up with an acceptable choice of $f(x)$ and

m. Choose the degree d for our polynomial. (We will later give a heuristic argument on how to choose d so as to minimize the running time to factor n. Experimentally, for numbers of around 130 digits, the choice $d = 5$ is acceptable.) Let $m = \lfloor n^{1/d} \rfloor$, and write n in base m, so that

$$n = m^d + c_{d-1}m^{d-1} + \cdots + c_0,$$

where each $c_j \in [0, m-1]$. (From Exercise 6.8 we have that if $1.5(d/\ln 2)^d < n$, then $n < 2m^d$, so the m^d-coefficient is indeed 1, as in the above display.) So the polynomial $f(x)$ falls right out of the base-m expansion of n: We have $f(x) = x^d + c_{d-1}x^{d-1} + \cdots + c_0$. This polynomial is self-evidently monic. But it may not be irreducible. Actually, this is an excellent situation in which to find ourselves, since if we have the nontrivial factorization $f(x) = g(x)h(x)$ in $\mathbf{Z}[x]$, then the integer factorization $n = g(m)h(m)$ is also nontrivial; see [Brillhart et al. 1981] and Exercise 6.9. Since polynomial factorization is relatively easy, see [Lenstra et al. 1982], [Cohen 2000, p. 139], one should factor f into irreducibles in $\mathbf{Z}[x]$. If the factorization is nontrivial, one has a nontrivial factorization of n. If f is irreducible, we may continue with NFS.

The homomorphism ϕ from $\mathbf{Z}[\alpha]$ to $\mathbf{Z}_n$ is defined by $\phi(\alpha)$ being the residue class $m \pmod{n}$. That is, ϕ first sends $a_0 + a_1\alpha + \cdots + a_{d-1}\alpha^{d-1}$ to the integer $a_0 + a_1 m + \cdots + a_{d-1}m^{d-1}$, and then reduces this integer mod n. It will be interesting to think of ϕ in this "two step" way, since we will also be dealing with the integer $a_0 + a_1 m + \cdots + a_{d-1}m^{d-1}$ before it is reduced.

The elements θ in the ring $\mathbf{Z}[\alpha]$ that we will consider will all be of the form $a - b\alpha$, where $a, b \in \mathbf{Z}$, with $\gcd(a, b) = 1$. Thus, we are looking for a set $\mathcal{S}$ of coprime integer pairs (a, b) such that

$$\prod_{(a,b)\in\mathcal{S}} (a - b\alpha) = \gamma^2, \text{ for some } \gamma \in \mathbf{Z}[\alpha],$$

$$\prod_{(a,b)\in\mathcal{S}} (a - bm) = v^2, \text{ for some } v \in \mathbf{Z}.$$

Then, if u is an integer such that $\phi(\gamma) \equiv u \pmod{n}$, then, as above, $u^2 \equiv v^2 \pmod{n}$, and we may try to factor n via $\gcd(u - v, n)$. (The pairs (a, b) in $\mathcal{S}$ are assumed to be coprime so as to avoid trivial redundancies.)

6.2.2 Basic NFS: Exponent vectors

How, then, are we supposed to find the set $\mathcal{S}$ of pairs (a, b)? The method resembles what we do in the quadratic sieve. There we have a single variable that runs over an interval. We use a sieve to detect smooth values of the given polynomial, and associate exponent vectors to these smooth values, using linear algebra to find a subset of them with product being a square. With NFS, we have two variables a, b. As with the special quadratic sieve (see Section 6.1.7), we can fix the first variable, and sieve over the other, then change to the next value of the first variable, sieve on the other, and so on.

But sieve what? To begin to answer this question, let us begin with a simpler question. Let us ignore the problem of having the product of the $a - b\alpha$ being a square in $\mathbf{Z}[\alpha]$ and instead focus just on the second property that $\mathcal{S}$ is supposed to have, namely, the product of the $a - bm$ is a square in $\mathbf{Z}$. Here, m is a fixed integer that we compute at the start. Say we let a, b run over pairs of integers with $0 < |a|, b \le M$, where M is some large bound (large enough so that there will be enough pairs a, b for us to be successful). Then we have just the degree-1 homogeneous polynomial $G(a, b) = a - bm$, which we sieve for smooth values, say B-smooth. We toss out any pair (a, b) found with $\gcd(a, b) > 1$. Once we have found more than $\pi(B) + 1$ such pairs, linear algebra modulo 2 can be used on the exponent vectors corresponding to the smooth values of $G(a, b)$ to find a subset of them whose product is a square.

This is all fine, but we are ignoring the hardest part of the problem: to *simultaneously* have our set of pairs (a, b) have the additional property that the product of $a - b\alpha$ is a square in $\mathbf{Z}[\alpha]$.

Let the roots of $f(x)$ in the complex numbers be $\alpha_1, \ldots, \alpha_d$, where $\alpha = \alpha_1$. The norm of an element $\beta = s_0 + s_1\alpha + \cdots + s_{d-1}\alpha^{d-1}$ in the algebraic number field $\mathbf{Q}[\alpha]$ (where the coefficients $s_0, s_1, \ldots, s_{d-1}$ are arbitrary rational numbers) is simply the product of the complex numbers $s_0 + s_1\alpha_j + \cdots + s_{d-1}\alpha_j^{d-1}$ for $j = 1, 2, \ldots, d$. This complex number, denoted by $N(\beta)$, is actually a rational number, since it is a symmetric expression in the roots $\alpha_1, \ldots, \alpha_d$, and the elementary symmetric polynomials in these roots are $\pm c_j$ for $j = 0, 1, \ldots, d - 1$, which are integers. In particular, if the rationals s_j are all actually integers, then $N(\beta)$ is an integer, too. (We shall later refer to what is called the trace of β. This is the sum of the conjugates $s_0 + s_1\alpha_j + \cdots + s_{d-1}\alpha_j^{d-1}$ for $j = 1, 2, \ldots, d$.)

The norm function N is also fairly easily seen to be multiplicative, that is, $N(\beta\beta') = N(\beta)N(\beta')$. An important corollary goes: If $\beta = \gamma^2$ for some $\gamma \in \mathbf{Z}[\alpha]$, then $N(\beta)$ is an integer square, namely the square of the integer $N(\gamma)$.

Thus, a *necessary* condition for the product of $a - b\alpha$ for (a, b) in $\mathcal{S}$ to be a square in $\mathbf{Z}[\alpha]$ is for the corresponding product of the integers $N(a - b\alpha)$ to be a square in $\mathbf{Z}$. Let us leave aside momentarily the question of whether this condition is also *sufficient* and let us see how we might arrange for the product of $N(a - b\alpha)$ to be a square.

We first note that

$$\begin{aligned}
N(a - b\alpha) &= (a - b\alpha_1) \cdots (a - b\alpha_d) \\
&= b^d(a/b - \alpha_1) \cdots (a/b - \alpha_d) \\
&= b^d f(a/b),
\end{aligned}$$

since $f(x) = (x - \alpha_1) \cdots (x - \alpha_d)$. Let $F(x, y)$ be the homogeneous form of f, namely,

$$F(x, y) = x^d + c_{d-1}x^{d-1}y + \cdots + c_0 y^d = y^d f(x/y).$$

Then $N(a - b\alpha) = F(a, b)$. That is, $N(a - b\alpha)$ may be viewed quite explicitly as a polynomial in the two variables a, b.

Thus, we can arrange for the product of $N(a - b\alpha)$ for $(a, b) \in \mathcal{S}$ to be a square by letting a, b run so that $|a|, |b| \le M$, using a sieve to detect B-smooth values of $F(a, b)$, form the corresponding exponent vectors, and use matrix methods to find the subset $\mathcal{S}$. And if we want $\mathcal{S}$ also to have the first property that the product of the $a - bm$ is also a square in $\mathbf{Z}$, then we alter the procedure to sieve for smooth values of $F(a, b)G(a, b)$, this product, too, being a polynomial in the variables a, b. For the smooth values we create exponent vectors with *two fields* of coordinates. The first field corresponds to the prime factorization of $F(a, b)$, and the second to the prime factorization of $G(a, b)$. These longer exponent vectors are then collected into a matrix, and again we can do linear algebra modulo 2. Before, we needed just $\pi(B) + 2$ vectors to ensure success. Now we need $2\pi(B) + 3$ vectors to ensure success, since each vector will have $2\pi(B)+2$ coordinates: the first half for the prime factorization of $F(a, b)$, and the second half for the prime factorization of $G(a, b)$. So we need only to collect twice as many vectors, and then we can accomplish both tasks simultaneously.

We return now to the question of sufficiency. That is, if $N(\beta)$ is a square in $\mathbf{Z}$ and $\beta \in \mathbf{Z}[\alpha]$, must it be true that β is a square in $\mathbf{Z}[\alpha]$? The answer is a resounding no. It is perhaps instructive to look at a simple example. Consider the case $f(x) = x^2 + 1$, and let us denote a root by the symbol "i" (as one might have guessed). Then $N(a+bi) = a^2+b^2$. If a^2+b^2 is a square in $\mathbf{Z}$, then $a + bi$ need not be a square in $\mathbf{Z}[i]$. For example, if a is a positive, nonsquare integer, then it is also a nonsquare in $\mathbf{Z}[i]$, yet $N(a) = a^2$ is a square in $\mathbf{Z}$.

Actually, the ring $\mathbf{Z}[i]$, known as the ring of Gaussian integers, is a well-understood ring with many beautiful properties in complete analogy to the ring $\mathbf{Z}$. The Gaussian integers are a unique factorization domain, as $\mathbf{Z}$ is. Each prime in $\mathbf{Z}[i]$ "lies over" an ordinary prime p in $\mathbf{Z}$. If the prime p is 1 (mod 4), it can be written in the form $a^2 + b^2$, and then $a + bi$ and $a - bi$ are the two different primes of $\mathbf{Z}[i]$ that lie over p. (Each prime has 4 "associates" corresponding to multiplying by the 4 units: $1, -1, i, -i$. Associated primes are considered the same prime, since the principal ideals they generate are exactly the same.) If the ordinary prime p is 3 (mod 4), then it remains prime in $\mathbf{Z}[i]$. And the prime 2 has the single prime $1 + i$ (and its associates) lying over it. For more on the arithmetic of the Gaussian integers, see [Niven et al. 1991].

So we can see, for example, that $5i$ is definitely not a square in $\mathbf{Z}[i]$, since it has the prime factorization $(2+i)(1+2i)$, and $2+i$ and $1+2i$ are different primes. (In contrast, $2i$ *is* a square, it is $(1+i)^2$.) However, $N(5i) = 25$, and of course, 25 is recognized as a square in $\mathbf{Z}$. The problem is that the norm function smashes together the two different primes $1+2i$ and $2+i$. We would like then to have some way to distinguish the different primes.

If our ring $\mathbf{Z}[\alpha]$ in the number field sieve were actually a unique factorization domain, our challenge would be much simpler: Just form exponent vectors based on the prime factorization of the various elements

$a - b\alpha$. There is a problem with units, and if we were to take this route, we would also want to find a system of "fundamental units" and have coordinates in our exponent vectors for each of these. (In the case of $\mathbf{Z}[i]$ the fundamental unit is rather trivial, it is just i, and we can take for distinguished primes in each associate class the one that is in the first quadrant but not on the imaginary axis.)

However, we shall see that the number field sieve can work just fine even if the ring $\mathbf{Z}[\alpha]$ is far from being a unique factorization domain, and even if we have no idea about the units.

For each prime p, let $R(p)$ denote the set of integers $r \in [0, p-1]$ with $f(r) \equiv 0 \pmod{p}$. For example, if $f(x) = x^2+1$, then $R(2) = \{1\}$, $R(3) = \{\ \}$, and $R(5) = \{2, 3\}$. Then if a, b are coprime integers,

$$F(a, b) \equiv 0 \pmod{p} \text{ if and only if } a \equiv br \pmod{p} \text{ for some } r \in R(p).$$

Thus, if we discover that $p|F(a, b)$, we also have a second piece of information, namely a number $r \in R(p)$ with $a \equiv br \pmod{p}$. (Actually, the sets $R(p)$ are used in the sieve that we use to factor the numbers $F(a, b)$. We may fix the number b and consider $F(a, b)$ as a polynomial in the variable a. Then when sieving by the prime p, we sieve the residue classes $a \equiv br \pmod{p}$ for multiples of p.) We keep track of this additional information in our exponent vectors. The field of coordinates of our exponent vectors that correspond to the factorization of $F(a, b)$ will have entries for each pair p, r, where p is a prime $\le B$, and $r \in R(p)$.

Let us again consider the polynomial $f(x) = x^2 + 1$. If $B = 5$, then exponent vectors for B-smooth members of $\mathbf{Z}[i]$ (that is, members of $\mathbf{Z}[i]$ whose norms are B-smooth integers) will have three coordinates, corresponding to the three pairs: (2,1), (5,2), and (5,3). Then

$$\begin{aligned} F(3, 1) &= 10 \text{ has the exponent vector } (1, 0, 1), \\ F(2, 1) &= 5 \text{ has the exponent vector } (0, 1, 0), \\ F(1, 1) &= 2 \text{ has the exponent vector } (1, 0, 0), \\ F(2, -1) &= 5 \text{ has the exponent vector } (0, 0, 1). \end{aligned}$$

Although $F(3, 1)F(2, 1)F(1, 1) = 100$ is a square, the exponent vectors allow us to see that $(3 + i)(2 + i)(1 + i)$ is *not* a square: The sum of the three vectors modulo 2 is $(0, 1, 1)$, which is not the zero vector. But now consider $(3 + i)(2 - i)(1 + i) = 8 + 6i$. The sum of the three corresponding exponent vectors modulo 2 is $(0, 0, 0)$, and indeed, $8 + 6i$ is a square in $\mathbf{Z}[i]$.

This method is not foolproof. For example, though i has the zero vector as its exponent vector in the above scheme, it is not a square. If this were the only problem, namely the issue of units, we could fairly directly find a solution. However, this is not the only problem.

Let $\mathcal{I}$ denote the ring of algebraic integers in the algebraic number field $\mathbf{Q}[\alpha]$. That is, $\mathcal{I}$ is the set of elements of $\mathbf{Q}[\alpha]$ that are the root of some monic polynomial in $\mathbf{Z}[x]$. The set $\mathcal{I}$ is closed under multiplication and addition.

That is, it is a ring; see [Marcus 1977]. In the case of $f(x) = x^2 + 1$, the algebraic integers in $\mathbf{Q}[i]$ constitute exactly the ring $\mathbf{Z}[i]$. The ring $\mathbf{Z}[\alpha]$ will always be a subset of $\mathcal{I}$, but in general, it will be a proper subset. For example, consider the case where $f(x) = x^2 - 5$. The ring of all algebraic integers in $\mathbf{Q}\left[\sqrt{5}\right]$ is $\mathbf{Z}\left[(1+\sqrt{5})/2\right]$, which properly contains $\mathbf{Z}\left[\sqrt{5}\right]$.

We now summarize the situation regarding the exponent vectors for the numbers $a - b\alpha$. We say that $a - b\alpha$ is B-smooth if its norm $N(a-b\alpha) = F(a,b)$ is B-smooth. For a, b coprime and $a - b\alpha$ being B-smooth, we associate to it an exponent vector $\vec{v}(a - b\alpha)$ that has entries $v_{p,r}(a - b\alpha)$ for each pair (p, r), where p is a prime number not exceeding B with $r \in R(p)$. (Later we shall use the notation $\vec{v}(a - b\alpha)$ for a longer vector that contains within it what is being considered here.) If $a \not\equiv br \pmod{p}$, then we define $v_{p,r}(a - b\alpha) = 0$. Otherwise $a \equiv br \pmod{p}$ and $v_{p,r}(a - b\alpha)$ is defined to be the exponent on p in the prime factorization of $F(a, b)$. We have the following important result.

Lemma 6.2.1. *If $\mathcal{S}$ is a set of coprime integer pairs a, b such that each $a - b\alpha$ is B-smooth, and if $\prod_{(a,b)\in\mathcal{S}}(a - b\alpha)$ is the square of an element in $\mathcal{I}$, the ring of algebraic integers in $\mathbf{Q}[\alpha]$, then*

$$\sum_{(a,b)\in\mathcal{S}} \vec{v}(a - b\alpha) \equiv \vec{0} \pmod{2}. \tag{6.7}$$

Proof. We begin with a brief discussion of what the numbers $v_{p,r}(a - b\alpha)$ represent. It is well known in algebraic number theory that the ring $\mathcal{I}$ is a Dedekind domain; see [Marcus 1977]. In particular, nonzero ideals of $\mathcal{I}$ may be uniquely factored into prime ideals. We also use the concept of norm of an ideal: If J is a nonzero ideal of $\mathcal{I}$, then $N(J)$ is the number of elements in the (finite) quotient ring $\mathcal{I}/J$. (The norm of the zero ideal is defined to be zero.) The norm function is multiplicative on ideals, that is, $N(J_1J_2) = N(J_1)N(J_2)$ for any ideals J_1, J_2 in $\mathcal{I}$. The connection with the norm of an element of $\mathcal{I}$ and the norm of the principal ideal it generates is beautiful: If $\beta \in \mathcal{I}$, then $N((\beta)) = |N(\beta)|$.

If p is a prime number and $r \in R(p)$, let $P_1, \dots, P_k$ be the prime ideals of $\mathcal{I}$ that divide the ideal $(p, \alpha - r)$. (This ideal is not the unit ideal, since $N(\alpha - r) = f(r)$, an integer divisible by p.) There are positive integers $e_1, \dots, e_k$ such that $N(P_j) = p^{e_j}$ for $j = 1, \dots, k$. The usual situation is that $k = 1$, $e_1 = 1$, and that $(p, \alpha - r) = P_1$. In fact, this scenario occurs whenever p does not divide the index of $\mathbf{Z}[\alpha]$ in $\mathcal{I}$; see [Marcus 1977]. However, we will deal with the general case.

Note that if $r' \in R(p)$ and $r' \neq r$, then the prime ideals that divide $(p, \alpha - r)$ are different from the prime ideals that divide $(p, \alpha - r')$, that is, the ideals $(p, \alpha - r')$ and $(p, \alpha - r)$ are coprime. This observation follows, since the integer $r - r'$ is coprime to the prime p. In addition, if a, b are integers, then $a - b\alpha \in (p, \alpha - r)$ if and only if $a \equiv br \pmod{p}$. To see this, write $a - b\alpha = a - br - b(\alpha - r)$, so that $a - b\alpha \in (p, \alpha - r)$ if and only if $a - br \in (p, \alpha - r)$, if and only if $a \equiv br \pmod{p}$. We need one further property: If a, b are coprime integers, $a \equiv br \pmod{p}$, and if P is a prime ideal of $\mathcal{I}$ that divides both (p) and $(a - b\alpha)$, then P divides $(p, \alpha - r)$; that is, P is one of the P_j. To see this, note that the hypotheses that a, b are coprime and $a \equiv br \pmod{p}$ imply $b \not\equiv 0 \pmod{p}$, so there is an integer c with $cb \equiv 1 \pmod{p}$. Then, since $a - b\alpha = a - br - b(\alpha - r) \in P$ and $a - br \equiv 0 \pmod{p}$, we have $b(\alpha - r) \in P$, so that $cb(\alpha - r) \in P$, and $\alpha - r \in P$. Thus, P divides $(p, \alpha - r)$, as claimed.

Suppose a, b are coprime integers and that $P_1^{a_1} \cdots P_k^{a_k}$ appears in the prime ideal factorization of $(a - b\alpha)$. As we have seen, if any of these exponents a_j are positive, it is necessary and sufficient that $a \equiv br \pmod{p}$, in which case all of the exponents a_j are positive and no other prime ideal divisor of (p) divides $(a - b\alpha)$. Thus the "p part" of the norm of $a - b\alpha$ is exactly the norm of $P_1^{a_1} \cdots P_k^{a_k}$; that is,

$$p^{v_{p,r}(a-b\alpha)} = N(P_1^{a_1} \cdots P_k^{a_k}) = p^{e_1 a_1 + \cdots + e_k a_k}.$$

Let $v_P(a - b\alpha)$ denote the exponent on the prime ideal P in the prime ideal factorization of $(a - b\alpha)$. Then from the above,

$$v_{p,r}(a - b\alpha) = \sum_{j=1}^{k} e_j v_{P_j}(a - b\alpha).$$

Now, if $\prod_{(a,b)\in\mathcal{S}}(a - b\alpha)$ is a square in $\mathcal{I}$, then the principal ideal it generates is a square of an ideal. Thus, for every prime ideal P in $\mathcal{I}$ we have that $\sum_{(a,b)\in\mathcal{S}} v_P(a - b\alpha)$ is even. We apply this principle to the prime ideals P_j dividing $(p, \alpha - r)$. We have

$$\sum_{(a,b)\in\mathcal{S}} v_{p,r}(a - b\alpha) = \sum_{j=1}^{k} e_j \sum_{(a,b)\in\mathcal{S}} v_{P_j}(a - b\alpha).$$

As each inner sum on the right side of this equation is an even integer, the integer on the left side of the equation must also be even. □

6.2.3 Basic NFS: Complexity

We have not yet given a full description of NFS, but it is perhaps worthwhile to envision why the strategy outlined so far leads to a fast factorization method, and to get an idea of the order of magnitude of the parameters to be chosen.

In both QS and NFS we are presented with a stream of numbers on which we may use a sieve to detect smooth values. When we have enough smooth values, we can use linear algebra on exponent vectors corresponding to the smooth values to find a nonempty subset of these vectors whose sum in the zero vector mod 2. Let us model the general problem as follows. We have a random sequence of positive integers bounded by X. How far does one expect to go in this sequence before a nontrivial subsequence has product being a square? The heuristic analysis in Section 6.1.1 gives an answer: It is at most $L(X)^{\sqrt{2}+o(1)}$, where the smoothness bound to achieve this is $L(X)^{1/\sqrt{2}}$. (We use here the notation of (6.1).) This heuristic upper bound can actually be rigorously proved as a two-sided estimate via the following theorem.

Theorem 6.2.2 (Pomerance 1996a). *Suppose $m_1, m_2, \ldots$ is a sequence of integers in $[1, X]$, each chosen independently and with uniform distribution. Let N be the least integer such that a nonempty subsequence from $m_1, m_2, \ldots, m_N$ has product being a square. Then the expected value for N is $L(X)^{\sqrt{2}+o(1)}$. The same expectation holds if we also insist that each m_j used in the product be B-smooth, with $B = L(X)^{1/\sqrt{2}}$.*

Thus, in some sense, smooth numbers are forced upon us, and are not merely an artifact. Interestingly, there is an identical theorem for the random variable N', being the least integer such that $m_1, m_2, \ldots, m_{N'}$ are "multiplicatively dependent", which means that there are integers $a_1, a_2, \ldots, a_{N'}$, not all zero, such that $\prod m_j^{a_j} = 1$. (Equivalently, the numbers $\ln m_1, \ln m_2, \ldots, \ln m_{N'}$ are linearly dependent over $\mathbf{Q}$.)

In the QS analysis, the bound X is $n^{1/2+o(1)}$, and this is where we get the complexity $L(n)^{1+o(1)}$ for QS. This complexity estimate is not a theorem, since the numbers we are looking at to form squares are not random—we just assume they are random for convenience in the analysis.

This approach, then, seems like a relatively painless way to do a complexity analysis. Just find the bound X for the numbers that we are trying to combine to make squares. The lower X is, the lower the complexity of the algorithm. In NFS the integers that we deal with are the values of the polynomial $F(x,y)G(x,y)$, where $F(x,y) = x^d + c_{d-1}x^{d-1}y + \cdots + c_0y^d$ and $G(x,y) = x - my$. We will ignore the fact that integers of the form $F(a,b)G(a,b)$ are already factored into the product of two numbers, and so may be more likely to be smooth than random numbers of the same magnitude, since this property has little effect on the asymptotic complexity.

Let us assume that the integer m in NFS is bounded by $n^{1/d}$, the coefficients c_j of the polynomial $f(x)$ are also bounded by $n^{1/d}$, and that we investigate values of a, b with $|a|, |b| \le M$. Then a bound for the numbers $|F(a,b)G(a,b)|$ is $2(d+1)n^{2/d}M^{d+1}$. If we call this number X, then from Theorem 6.2.2, we might expect to have to look at $L(X)^{\sqrt{2}+o(1)}$ pairs a, b to find enough to be used to complete the algorithm. Thus, M should satisfy the constraint $M^2 = L(X)^{\sqrt{2}+o(1)}$. Putting this into the equation $X = 2(d+1)n^{2/d}M^{d+1}$ and taking the logarithm of both sides, we have

$$\ln X \sim \ln(2(d+1)) + \frac{2}{d}\ln n + (d+1)\sqrt{\frac{1}{2}\ln X \ln\ln X}. \tag{6.8}$$

It is clear that the first term on the right is negligible compared to the third term. Suppose first that d is fixed; that is, we are going to analyze the complexity of NFS when we fix the degree of the polynomial $f(x)$, and assume that $n \to \infty$. Then the last term on the right of (6.8) is small compared to the left side, so (6.8) simplifies to

$$\ln X \sim \frac{2}{d} \ln n.$$

Hence the running time with d fixed is

$$L(X)^{\sqrt{2}+o(1)} = L(n)^{\sqrt{4/d}+o(1)},$$

which suggests that NFS will not do better than QS until we take $d = 5$ or larger.

Now let us assume that $d \to \infty$ as $n \to \infty$. Then we may replace the coefficient $d+1$ in the last term of (6.8) with d, getting

$$\ln X \sim \frac{2}{d} \ln n + d\sqrt{\frac{1}{2} \ln X \ln\ln X}.$$

Let us somewhat imprecisely change the "$\sim$" to "$=$" and try to choose d so as to minimize X. (An optimal X_0 will have the property that $\ln X_0 \sim \ln X$.) Taking the derivative with respect to the "variable" d, we have

$$\frac{X'}{X} = \frac{-2}{d^2} \ln n + \sqrt{\frac{1}{2} \ln X \ln\ln X} + \frac{dX'(1 + \ln\ln X)}{4X\sqrt{\frac{1}{2} \ln X \ln\ln X}}.$$

Setting $X' = 0$, we get

$$d = (2 \ln n)^{1/2}((1/2) \ln X \ln\ln X)^{-1/4},$$

so that

$$\ln X = 2(2 \ln n)^{1/2}((1/2) \ln X \ln\ln X)^{1/4}.$$

Then

$$(\ln X)^{3/4} = 2(2 \ln n)^{1/2}((1/2) \ln\ln X)^{1/4},$$

so that $\frac{3}{4} \ln\ln X \sim \frac{1}{2} \ln\ln n$. Substituting, we get

$$(\ln X)^{3/4} \sim 2(2 \ln n)^{1/2}((1/3) \ln\ln n)^{1/4},$$

or

$$\ln X \sim \frac{4}{3^{1/3}} (\ln n)^{2/3} (\ln\ln n)^{1/3}.$$

So the running time for NFS is

$$L(X)^{\sqrt{2}+o(1)} = \exp\left(\left((64/9)^{1/3} + o(1)\right) (\ln n)^{1/3} (\ln\ln n)^{2/3}\right).$$

The values of d that achieve this heuristic asymptotic complexity satisfy

$$d \sim \left(\frac{3 \ln n}{\ln\ln n}\right)^{1/3}.$$

One can see that "at infinity," NFS is far superior (heuristically) than QS. The low complexity estimate should motivate us to forge on and solve the remaining technical problems in connection with the algorithm.

If we could come up with a polynomial with smaller coefficients, the complexity estimate would be smaller. In particular, if the polynomial $f(x)$ has coefficients that are bounded by $n^{\epsilon/d}$, then the above analysis gives the complexity $L(n)^{\sqrt{(2+2\epsilon)/d+o(1)}}$ for fixed d; and for $d \to \infty$ as $n \to \infty$, it is $\exp\left(\left((32(1+\epsilon)/9)^{1/3}+o(1)\right)(\ln n)^{1/3}(\ln\ln n)^{2/3}\right)$. The case $\epsilon = o(1)$ is the "special" number field sieve; see Section 6.2.7.

6.2.4 Basic NFS: Obstructions

After this interlude into complexity theory, we return to the strategy of NFS. We are looking for some easily checkable condition for the product of $(a - b\alpha)$ for $(a, b) \in \mathcal{S}$ to be a square in $\mathbf{Z}[\alpha]$. Lemma 6.2.1 goes a long way to meet this condition, but there are several "obstructions" that remain. Suppose that (6.7) holds. Let $\beta = \prod_{(a,b)\in\mathcal{S}}(a - b\alpha)$.

(1) If the ring $\mathbf{Z}[\alpha]$ is equal to $\mathcal{I}$ (the ring of all algebraic integers in $\mathbf{Q}(\alpha)$), then we at least have the ideal (β) in $\mathcal{I}$ being the square of some ideal J. But it may not be that $\mathbf{Z}[\alpha] = \mathcal{I}$. So it may not be that (β) in $\mathcal{I}$ is the square of an ideal in $\mathcal{I}$.

(2) Even if $(\beta) = J^2$ for some ideal J in $\mathcal{I}$, it may not be that J is a principal ideal.

(3) Even if $(\beta) = (\gamma)^2$ for some $\gamma \in \mathcal{I}$, it may not be that $\beta = \gamma^2$.

(4) Even if $\beta = \gamma^2$ for some $\gamma \in \mathcal{I}$, it may not be that $\gamma \in \mathbf{Z}[\alpha]$.

Though these four obstructions appear forbidding, we shall see that two simple devices can be used to overcome all four. We begin with the last of the four. The following lemma is of interest here.

Lemma 6.2.3. *Let $f(x)$ be a monic irreducible polynomial in $\mathbf{Z}[x]$, with root α in the complex numbers. Let $\mathcal{I}$ be the ring of algebraic integers in $\mathbf{Q}(\alpha)$, and let $\beta \in \mathcal{I}$. Then $f'(\alpha)\beta \in \mathbf{Z}[\alpha]$.*

Proof. Our proof follows an argument in [Weiss 1963, Sections 3–7]. Let $\beta_0, \beta_1, \ldots, \beta_{d-1}$ be the coefficients of the polynomial $f(x)/(x - \alpha)$. That is, $f(x)/(x-\alpha) = \sum_{j=0}^{d-1} \beta_j x^j$. From Proposition 3-7-12 in [Weiss 1963], a result attributed to Euler, we have $\beta_0/f'(\alpha), \ldots, \beta_{d-1}/f'(\alpha)$ a basis for $\mathbf{Q}(\alpha)$ over $\mathbf{Q}$, each $\beta_j \in \mathbf{Z}[\alpha]$, and the trace of $\alpha^k\beta_j/f'(\alpha)$ is 1 if $j = k$, and 0 otherwise. (See Section 6.2.2 for the definition of trace. From this definition it is easy to see that the trace operation is $\mathbf{Q}$-linear, it takes values in $\mathbf{Q}$, and on elements of $\mathcal{I}$ it takes values in $\mathbf{Z}$.) Let $\beta \in \mathcal{I}$. There are rationals $s_0, \ldots, s_{d-1}$ such that $\beta = \sum_{j=0}^{d-1} s_j\beta_j/f'(\alpha)$. Then the trace of $\beta\alpha^k$ is s_k for $k = 0, \ldots, d-1$. So each $s_k \in \mathbf{Z}$. Thus, $f'(\alpha)\beta = \sum_{j=0}^{d-1} s_j\beta_j$ is in $\mathbf{Z}[\alpha]$. □

We use Lemma 6.2.3 as follows. Instead of holding out for a set $\mathcal{S}$ of coprime integers with $\prod_{(a,b)\in\mathcal{S}}(a-b\alpha)$ being a square in $\mathbf{Z}[\alpha]$, as we originally desired, we settle instead for the product being a square in $\mathcal{I}$, say γ^2. Then by Lemma 6.2.3, $f'(\alpha)\gamma \in \mathbf{Z}[\alpha]$, so that $f'(\alpha)^2\prod_{(a,b)\in\mathcal{S}}(a-b\alpha)$ is a square in $\mathbf{Z}[\alpha]$.

The first three obstructions are all quite different, but they have a common theme, namely well-studied groups. Obstruction (1) is concerned with the group $\mathcal{I}/\mathbf{Z}[\alpha]$. Obstruction (2) is concerned with the class group of $\mathcal{I}$. And obstruction (3) is concerned with the unit group of $\mathcal{I}$. A befuddled reader may well consult a text on algebraic number theory for full discussions of these groups, but as we shall see below, a very simple device will let us overcome these first three obstructions. Further, to understand how to implement the number field sieve, one needs only to understand this simple device. This hypothetical befuddled reader might well skip ahead a few paragraphs!

For obstruction (1), though the prime ideal factorization (into prime ideals in $\mathcal{I}$) of $\left(\prod_{(a,b)\in\mathcal{S}}(a-b\alpha)\right)$ may not have all even exponents, the prime ideals with odd exponents all lie over prime numbers that divide the index of $\mathbf{Z}[\alpha]$ in $\mathcal{I}$, so that the number of these exceptional prime ideals is bounded by the (base-2) logarithm of this index.

Obstruction (2) is more properly described as the ideal class group modulo the subgroup of squares of ideal classes. This is a 2-group whose rank is the 2-rank of the ideal class group, which is bounded by the (base-2) logarithm of the order of the class group, that is, the logarithm of the class number.

Obstruction (3) is again more properly described as the group of units modulo the subgroup of squares of units. This again is a 2-group, and its rank is $\le d$, the degree of $f(x)$. (We use here the famous Dirichlet unit theorem.)

The detailed analysis of these obstructions can be found in [Buhler et al. 1993]. We shall be content with the conclusion that though all are different, obstructions (1), (2), and (3) are all "small." There is a brute force way around these three obstructions, but there is also a beautiful and simple circumvention. The circumvention idea is due to Adleman and runs as follows. For a moment, suppose you somehow could not tell positive numbers from negative numbers, but you could discern prime factorizations. Thus both 4 and -4 would look like squares to you, since in their prime factorizations we have 2 raised to an even power, and no other primes are involved. However, -4 is not a square. Without using that it is negative, we can still tell that -4 is not a square by noting that it is not a square modulo 7. We can detect this via the Legendre symbol $\left(\frac{-4}{7}\right) = -1$. More generally, if q is an odd prime and if $\left(\frac{m}{q}\right) = -1$, then m is not a square. Adleman's idea is to use the converse statement, even though it is not a theorem! The trick is to think probabilistically. Suppose for a given integer m, we choose k distinct odd primes q at random in the range $q < |m|$. And suppose for each of the k test primes q we have $\left(\frac{m}{q}\right) = 1$. If m is not a square, then the probability of this event occurring is (heuristically) about 2^{-k}. So, if the event *does* occur and k

is large (say, $k > \lg |m|$), then it is reasonable to suppose that m actually is a square.

We wish to use this idea with the algebraic integers $a - b\alpha$, and the following result allows us to do so via ordinary Legendre symbols.

Lemma 6.2.4. *Let $f(x)$ be a monic, irreducible polynomial in $\mathbf{Z}[x]$ and let α be a root of f in the complex numbers. Suppose q is an odd prime number and s is an integer with $f(s) \equiv 0 \pmod{q}$ and $f'(s) \not\equiv 0 \pmod{q}$. Let $\mathcal{S}$ be a set of coprime integer pairs (a, b) such that q does not divide any $a - bs$ for $(a, b) \in \mathcal{S}$ and $f'(\alpha)^2 \prod_{(a,b)\in\mathcal{S}}(a - b\alpha)$ is a square in $\mathbf{Z}[\alpha]$. Then*

$$\prod_{(a,b)\in\mathcal{S}} \left(\frac{a - bs}{q}\right) = 1. \tag{6.9}$$

Proof. Consider the homomorphism ϕ_q from $\mathbf{Z}[\alpha]$ to $\mathbf{Z}_q$ where $\phi_q(\alpha)$ is the residue class $s \pmod{q}$. We have $f'(\alpha)^2 \prod_{(a,b)\in\mathcal{S}}(a - b\alpha) = \gamma^2$ for some $\gamma \in \mathbf{Z}[\alpha]$. By the hypothesis, $\phi_q(\gamma^2) \equiv f'(s)^2 \prod_{(a,b)\in\mathcal{S}}(a - bs) \not\equiv 0 \pmod{q}$. Then $\left(\frac{\phi_q(\gamma^2)}{q}\right) = \left(\frac{\phi_q(\gamma)^2}{q}\right) = 1$ and $\left(\frac{f'(s)^2}{q}\right) = 1$, so that

$$\left(\frac{\prod_{(a,b)\in\mathcal{S}}(a - bs)}{q}\right) = 1,$$

which implies that (6.9) holds. □

So again we have a necessary condition for squareness, while we are still searching for a sufficient condition. But we are nearly there. As we have seen, one might heuristically argue that if k is sufficiently large and if $q_1, \ldots, q_k$ are odd primes that divide no $N(a - b\alpha)$ for $(a, b) \in \mathcal{S}$ and if we have $s_j \in R(q_j)$ for $j = 1, \ldots, k$, where $f'(s_j) \not\equiv 0 \pmod{q_j}$, then

$$\sum_{(a,b)\in\mathcal{S}} \vec{v}(a - b\alpha) \equiv \vec{0} \pmod{2}$$

and

$$\prod_{(a,b)\in\mathcal{S}} \left(\frac{a - bs_j}{q_j}\right) = 1 \text{ for } j = 1, \ldots, k$$

imply that

$$\prod_{(a,b)\in\mathcal{S}} (a - b\alpha) = \gamma^2 \text{ for some } \gamma \in \mathcal{I}.$$

And how large is sufficiently large? Again, since the dimensions of obstructions (1), (2), (3) are all small, k need not be very large at all. We shall choose the polynomial $f(x)$ so that the degree d satisfies $d^{2d^2} < n$ (where n is the number we are factoring), and the coefficients of c_j of f all satisfy $|c_j| < n^{1/d}$. Under these conditions, it can be shown that the sum of the dimensions of the first three obstructions is less than $\lg n$; see [Buhler et al. 1993], Theorem 6.7. It is conjectured that it is sufficient to choose $k = \lfloor 3 \lg n \rfloor$ (with the k primes q_j chosen as the least possible). Probably a somewhat smaller value of k would also suffice, but this aspect is not a time bottleneck for the algorithm.

We use the pairs q_j, s_j to augment our exponent vectors with k additional entries. If $\left(\frac{a-bs_j}{q_j}\right) = 1$, the entry corresponding to q_j, s_j in the exponent vector for $a - b\alpha$ is 0. If the Legendre symbol is -1, the entry is 1. (This allows the translation from the multiplicative group $\{1, -1\}$ of order 2 to the additive group $\mathbf{Z}_2$ of order 2.) These augmented exponent vectors turn out now to be not only necessary, but also sufficient (in practice) for constructing squares.

6.2.5 Basic NFS: Square roots

Suppose we have overcome all the obstructions of the last section, and we now have a set $\mathcal{S}$ of coprime integer pairs such that $f'(\alpha)^2 \prod_{(a,b)\in\mathcal{S}}(a - b\alpha) = \gamma^2$ for $\gamma \in \mathbf{Z}[\alpha]$, and $\prod_{(a,b)\in\mathcal{S}}(a - bm) = v^2$ for $v \in \mathbf{Z}$. We then are nearly done, for if u is an integer with $\phi(\gamma) \equiv u \pmod{n}$, then $u^2 \equiv (f'(m)v)^2 \pmod{n}$, and we may attempt to factor n via $\gcd(u - f'(m)v, n)$.

However, a problem remains. The methods of the above sections allow us to find the set $\mathcal{S}$ with the above properties, but they do not say how we might go about finding the square roots γ and v. That is, we have squares, one in $\mathbf{Z}[\alpha]$, the other in $\mathbf{Z}$, and we wish to find their square roots.

The problem for v is simple, and can be done in the same way as in QS. From the exponent vectors, we can deduce easily the prime factorization of v^2, and from this, we can deduce even more easily the prime factorization of v. We actually do not need to know the integer v; rather, we need to know only its residue modulo n. For each prime power divisor of v, compute its residue mod n by a fast modular powering algorithm, say Algorithm 2.1.5. Then multiply these residues together in $\mathbf{Z}_n$, finally getting $v \pmod{n}$.

The more difficult, and more interesting, problem is the computation of γ. If γ is expressed as $a_0 + a_1\alpha + \cdots + a_{d-1}\alpha^{d-1}$, then an integer u that works is $a_0 + a_1 m + \cdots + a_{d-1}m^{d-1}$. Since again we are interested only in the residue $u \pmod{n}$, it means that we are interested only in the residues $a_j \pmod{n}$. This is good, since the integers $a_0, \ldots, a_{d-1}$ might well be very large, with perhaps about as many digits as the square root of the number of steps for the rest of the algorithm! One would not want to do much arithmetic with such huge numbers. Even if one computed only the algebraic integer γ^2, and did not worry about finding the square root γ, one would have to use the fast multiplication methods of Chapter 9 in order to keep the computation within the time bound of Section 6.2.3. And this does not even begin to touch how one would take the square root.

If we are in the special case where $\mathbf{Z}[\alpha] = \mathcal{I}$ and this ring is a unique factorization domain, we can use a method similar to the one sketched above for computing $v \pmod{n}$. But in the general case, our ring may be far from being a UFD.

One method, suggested in [Buhler et al. 1993], begins by finding a prime p such that $f(x)$ is irreducible modulo p. Then we solve for $\gamma \pmod p$ (that is, for the coefficients of γ modulo p). We do this as a computation in the finite field $\mathbf{Z}_p[x]/(f(x))$; see Section 2.2.2. The square root computation can follow along the lines of Algorithm 2.3.8; see Exercise 2.14. So this is a start, since we can actually find the residues $a_0 \pmod p, \ldots, a_{d-1} \pmod p$ fairly easily. Why not do this for other primes p, and then glue using the Chinese remainder theorem? There is a seemingly trivial problem with this overall approach. For each prime p for which we do this, there are two square roots, and we don't know how to choose the signs in the gluing. We could try every possibility, but if we use k primes, only 2 of the 2^k possibilities work. We may choose one of the solutions for one of the primes p, and then get it down to 2^{k-1} choices for the other primes, but this is small comfort if k is large.

There are at least two possible ways to overcome this problem of choosing the right signs. The method suggested in [Buhler et al. 1993] is not to use Chinese remaindering with different primes, but rather to use Hensel lifting to get solutions modulo higher and higher powers of the same fixed prime p; see Algorithm 2.3.11. When the power of p exceeds a bound for the coefficients a_j, it means we have found them. This is simpler than using the polynomial factorization methods of [Lenstra 1983], but at the top of the Hensel game when we have our largest prime powers, we are doing arithmetic with huge integers, and to keep the complexity bound under control we must use fast subroutines as in Chapter 9.

Another strategy, suggested in [Couveignes 1993], allows Chinese remaindering, but it works only for the case d odd. In this case, the norm of -1 is -1, so that we can set off right from the beginning and insist that we are looking for the choice for γ with positive norm. Since the prime factorization of $N(\gamma)$ is known from the exponent vectors, we may compute $N(\gamma) \pmod p$, where p is as above, a prime modulo which $f(x)$ is irreducible. When we compute γ_p that satisfies $\gamma_p^2 \equiv \gamma^2 \pmod p$, we choose γ_p or $-\gamma_p$ according to which has norm congruent to $N(\gamma) \pmod p$. This, then, allows a correct choice of signs for each prime p used. This idea does not seem to generalize to even degrees d.

As it turns out there is a heuristic approach for finding square roots that seems to work very well in practice, making this step of the algorithm not of great consequence for the overall running time. The method uses some of the ideas above, as well as some others. For details, see [Montgomery 1994], [Nguyen 1998].

6.2.6 Basic NFS: Summary algorithm

We now sum up the preceding sections by giving a reasonably concise description of the NFS. Due to the relative intricacy of the algorithm, we have chosen to use a fair amount of English description in the following display.

Algorithm 6.2.5 (Number field sieve). We are given an odd composite number n that is not a power. This algorithm attempts to find a nontrivial factorization of n.

1. [Setup]

   $d = \lfloor (3\ln n/\ln\ln n)^{1/3} \rfloor$; // This d has $d^{2d^2} < n$.

   $B = \lfloor \exp((8/9)^{1/3}(\ln n)^{1/3}(\ln\ln n)^{2/3}) \rfloor$;

   // Note that d, B can optionally be tuned to taste.

   $m = \lfloor n^{1/d} \rfloor$;

   Write n in base m: $n = m^d + c_{d-1}m^{d-1} + \cdots + c_0$;

   $f(x) = x^d + c_{d-1}x^{d-1} + \cdots + c_0$; // Establish the polynomial f.

   Attempt to factor $f(x)$ into irreducible polynomials in $\mathbf{Z}[x]$ using the factoring algorithm of [Lenstra et al. 1982] or a variant such as [Cohen 2000, p. 139];

   If $f(x)$ has the nontrivial factorization $g(x)h(x)$, return the (also nontrivial) factorization $n = g(m)h(m)$;

   $F(x,y) = x^d + c_{d-1}x^{d-1}y + \cdots + c_0y^d$; // Establish polynomial F.

   $G(x,y) = x - my$;

   for(prime $p \le B$) compute the set

   $R(p) = \{r \in [0, p-1] : f(r) \equiv 0 \pmod{p}\}$;

   $k = \lfloor 3\lg n \rfloor$;

   Compute the first k primes $q_1, \dots, q_k > B$ such that $R(q_j)$ contains some element s_j with $f'(s_j) \not\equiv 0 \pmod{q_j}$, storing the k pairs (q_j, s_j);

   $B' = \sum_{p \le B} \#R(p)$;

   $V = 1 + \pi(B) + B' + k$;

   $M = B$;

2. [The sieve]

   Use a sieve to find a set $\mathcal{S}'$ of coprime integer pairs (a,b) with $0 < |a|, b \le M$, and $F(a,b)G(a,b)$ being B-smooth, until $\#\mathcal{S}' > V$, or failing this, increase M and try again, or goto [Setup] and increase B;

3. [The matrix]

   // We shall build a $V \times \#\mathcal{S}'$ binary matrix, one row per (a,b) pair.

   // We shall compute $\vec{v}(a - b\alpha)$, the binary exponent vector for $a - b\alpha$ having V bits (coordinates) as follows:

   Set the first bit of $\vec{v}$ to 1 if $G(a,b) < 0$, else set this bit to 0;

   // The next $\pi(B)$ bits depend on the primes $p \le B$: Define p^γ as the power of p in the prime factorization of $|G(a,b)|$.

   Set the bit for p to 1 if γ is odd, else set this bit to 0;

   // The next B' bits are to correspond to the pairs p, r where p is a prime not exceeding B and $r \in R(p)$. We use the notation $v_{p,r}(a - b\alpha)$ defined prior to Lemma 6.2.1.

   Set the bit for p, r to 1 if $v_{p,r}(a - b\alpha)$ is odd, else set it to 0;

   // Next, the last k bits correspond to the pairs q_j, s_j.

   Set the bit for q_j, s_j to 1 if $\left(\frac{a - bs_j}{q_j}\right)$ is -1, else set it to 0;

Install the exponent vector $\vec{v}(a - b\alpha)$ as the next row of the matrix;
}

4. [Linear algebra]
   By some method of linear algebra (see Section 6.1.3), find a nonempty subset $\mathcal{S}$ of $\mathcal{S}'$ such that $\sum_{(a,b)\in\mathcal{S}} \vec{v}(a - b\alpha)$ is the 0-vector $\pmod 2$;
5. [Square roots]
   Use the known prime factorization of the integer square $\prod_{(a,b)\in\mathcal{S}}(a - bm)$ to find a residue $v \bmod n$ with $\prod_{(a,b)\in\mathcal{S}}(a - bm) \equiv v^2 \pmod n$;
   By some method, such as those of Section 6.2.5, find a square root γ in $\mathbf{Z}[\alpha]$ of $f'(\alpha)^2 \prod_{(a,b)\in\mathcal{S}}(a - b\alpha)$, and, via simple replacement $\alpha \to m$, compute $u = \phi(\gamma) \pmod n$;
6. [Factorization]
   return $\gcd(u - f'(m)v, n)$;

If the divisor of n that is reported in Algorithm 6.2.5 is trivial, one has the option of finding more linear dependencies in the matrix and trying again. If we run out of linear dependencies, one again has the option to sieve further to find more rows for the matrix, and so have more linear dependencies.

6.2.7 NFS: Further considerations

As with the basic quadratic sieve, there are many "bells and whistles" that may be added to the number field sieve to make it an even better factorization method. In this section we shall briefly discuss some of these improvements.

Free relations

Suppose p is a prime in the "factor base," that is, $p \le B$. Our exponent vectors have a coordinate corresponding to p as a possible prime factor of $a - bm$, and $\#R(p)$ further coordinates corresponding to integers $r \in R(p)$. (Recall that $R(p)$ is the set of residues $r \pmod p$ with $f(r) \equiv 0 \pmod p$.) On average, $\#R(p)$ is 1, but it can be as low as 0 (in the case that $f(x)$ has no roots $\pmod p$), or it can be as high as d, the degree of $f(x)$ (in the case that $f(x)$ splits into d distinct linear factors $\pmod p$)). In this latter case, we have that the product of the prime ideals $(p, \alpha - r)$ in the full ring of algebraic integers in $\mathbf{Q}[\alpha]$ is (p).

Suppose p is a prime with $p \le B$, and $R(p)$ has d members. Let us throw into our matrix an extra row vector $\vec{v}(p)$, which has 1's in the coordinates corresponding to p and to each pair p, r where $r \in R(p)$. Also, in the final field of k coordinates corresponding to the quadratic characters modulo q_j for $j = 1, \dots, k$, put a 0 in place j of $\vec{v}(p)$ if $\left(\frac{p}{q_j}\right) = 1$ and put a 1 in place j if $\left(\frac{p}{q_j}\right) = -1$. Such a vector $\vec{v}(p)$ is called a free relation, since it is found in the precomputations, and not in the sieving stage. Now, when we find a subset of rows that sum to the zero vector mod 2, we have that the subset corresponds to a set $\mathcal{S}$ of coprime pairs a, b and a set $\mathcal{F}$ of free relations. Let w be the product of the primes p corresponding to the free relations in $\mathcal{F}$.

Then it should be that

$$wf'(\alpha)^2 \prod_{(a,b)\in\mathcal{S}} (a - b\alpha) = \gamma^2, \text{ for some } \gamma \in \mathbf{Z}[\alpha],$$

$$wf'(m)^2 \prod_{(a,b)\in\mathcal{S}} (a - bm) = v^2, \text{ for some } v \in \mathbf{Z}.$$

Then if $\phi(\gamma) = u$, we have $u^2 \equiv v^2 \pmod{n}$, as before.

The advantage of free relations is that the more of them there are, the fewer relations need be uncovered in the time-consuming sieve stage. Also, the vectors $\vec{v}(p)$ are sparser than a typical exponent vector $\vec{v}(a,b)$, so including free relations allows the matrix stage to run faster.

So, how many free relations do we expect to find? A free relation corresponds to a prime p that splits completely in the algebraic number field $\mathbf{Q}(\alpha)$. Let g be the order of the splitting field of $f(x)$, that is, the Galois closure of $\mathbf{Q}(\alpha)$ in the complex numbers. It follows from the Chebotarev density theorem that the number of primes p up to a bound X that split completely in $\mathbf{Q}(\alpha)$ is asymptotically $\frac{1}{g}\pi(X)$, as $X \to \infty$. That is, on average, 1 out of every g prime numbers corresponds to a free relation. Assuming that our factor base bound B is large enough so that the asymptotics are beginning to take over (this is yet *another* heuristic, but reasonable, assumption), we thus should expect about $\frac{1}{g}\pi(B)$ free relations. Now, the order g of the splitting field could be as small as d, the degree of $f(x)$, or as high as $d!$. Obviously, the smaller g is, the more free relations we should expect. Unfortunately, the generic case is $g = d!$. That is, for most irreducible polynomials $f(x)$ in $\mathbf{Z}[x]$ of degree d, the order of the splitting field of $f(x)$ is $d!$. So, for example, if $d = 5$, we should expect only about $\frac{1}{120}\pi(B)$ free relations, if we choose our polynomial $f(x)$ according to the scheme in step [Setup] in Algorithm 6.2.5. Since our vectors have about $2\pi(B)$ coordinates, the free relations in this case would only reduce the sieving time by less than one-half of 1 per cent. But still, it is free, so to speak, and every little bit helps.

Free relations can help considerably more in the case of special polynomials $f(x)$ with small splitting fields. For example, in the factorization of the ninth Fermat number F_9, the polynomial $f(x) = x^5 + 8$ was used. The order of the splitting field here is 20, so free relations allowed the sieving time to be reduced by about 2.5%.

Partial relations

As in the quadratic sieve method, sieving in the number field sieve not only reveals those pairs a, b where both of the numbers $N(a - b\alpha) = F(a,b) = b^d f(a/b)$ and $a - bm$ are B-smooth, but also pairs a, b where one or both of these numbers are a B-smooth number times one somewhat larger prime. If we allow relations that have such large primes, at most one each for $N(a - b\alpha)$ and $a - bm$, we then have a data structure not unlike the quadratic sieve with the double large-prime variation; see Section 6.1.4. It has also been suggested that reports can be used with $N(a - b\alpha)$ having two large primes and $a - bm$

being B-smooth, and vice versa. And some even consider using reports where both numbers in question have up to two large prime factors. One wonders whether it would not be simpler and more efficient in this case just to increase the size of the bound B.

Nonmonic polynomials

It is specified in Algorithm 6.2.5 that the polynomial $f(x)$ chosen in step [Setup] be done so in a particular way, a way that renders f monic. The discussion in the above sections assumed that the polynomial $f(x)$ is indeed monic. In this case, where α is a root of $f(x)$, the ring $\mathbf{Z}[\alpha]$ is a subring of the ring of algebraic integers in $\mathbf{Q}(\alpha)$. In fact, we have more freedom in the choice of $f(x)$ than stated. It is necessary only that $f(x) \in \mathbf{Z}[x]$ be irreducible. It is not necessary that f be chosen in the particular way of step [Setup], nor is it necessary that f be monic. Primes that divide the leading coefficient of $f(x)$ have a somewhat suspect treatment in our exponent vectors. But we are used to this kind of thing, since also primes that divide the discriminant of $f(x)$ in the treatment of the monic case were suspect, and became part of the need for the quadratic characters in step [The matrix] of Algorithm 6.2.5 (discussed in Section 6.2.4). Suffice it to say that nonmonic polynomials do not introduce any significant new difficulties.

But why should we bother with nonmonic polynomials? As we saw in Section 6.2.3, the key to a faster algorithm is reducing the size of the numbers that over which we sieve in the hope of finding smooth ones. The size of these numbers in NFS depends directly on the size of the number m and the coefficients of the polynomial $f(x)$, for a given degree d. Choosing a monic polynomial we could arrange for m and these coefficients to be bounded by $n^{1/d}$. If we now allow nonmonic polynomials, we can choose m to be $\lceil n^{1/(d+1)} \rceil$. Writing n in base m, we have $n = c_d m^d + c_{d-1} m^{d-1} + \cdots + c_0$. This suggests that we use the polynomial $f(x) = c_d x^d + c_{d-1} x^{d-1} + \cdots + c_0$. The coefficients c_i are bounded by $n^{1/(d+1)}$, so both m and the coefficients are smaller by a factor of about $n^{1/(d^2+d)}$.

For numbers at infinity, this savings in the coefficient size is not very significant: The heuristic complexity of NFS stands roughly as before. (The asymptotic speedup is about a factor of $\ln^{1/6} n$.) However, we are still not factoring numbers at infinity, and for the numbers we are factoring, the savings is important.

Suppose $f(x) = c_d x^d + c_{d-1} x^{d-1} + \cdots + c_0$ is irreducible in $\mathbf{Z}[x]$ and that $\alpha \in \mathbf{C}$ is a root. Then $c_d \alpha$ is an algebraic integer. It is a root of $F(x) = x^d + c_{d-1} x^{d-1} + c_d c_{d-2} x^{d-2} + \cdots + c_d^{d-1} c_0$, which can be easily seen, since $F(c_d x) = c_d^{d-1} f(x)$. We conclude that if $\mathcal{S}$ is a set of coprime integer pairs a, b, if

$$\prod_{(a,b) \in \mathcal{S}} (a - b\alpha)$$

is a square in $\mathbf{Q}(\alpha)$, and if $\mathcal{S}$ has an *even number* of pairs, then

$$F'(c_d\alpha)^2 \prod_{(a,b)\in\mathcal{S}} (ac_d - bc_d\alpha)$$

is a square in $\mathbf{Z}[c_d\alpha]$, say γ^2. Finding the integral coefficients (modulo n) of γ with respect to the basis $1, c_d\alpha, \ldots, (c_d\alpha)^{d-1}$ then allows us as before to get two congruent squares modulo n, and so gives us a chance to factor n. (Note that if $F(x,y) = y^d f(x/y)$ is the homogenized form of $f(x)$, then $F(c_d x, c_d) = c_d F(c_d x)$, and so $F_x(c_d\alpha, c_d) = c_d F'(c_d\alpha)$. We thus may use $F_x(c_d\alpha, c_d)$ in place of $F'(c_d\alpha)$ in the above, if we wish.) So, using a nonmonic polynomial poses no great complications. To ensure that the cardinality of the set $\mathcal{S}$ is even, we can enlarge all of our exponent vectors by one additional coordinate, which is always set to be 1.

The above argument assumes that the coefficient c_d is coprime to n. However, it is a simple matter to check that c_d and n are coprime. And, since c_d is smaller than n in all the cases that would be considered, a nontrivial gcd would lead to a nontrivial splitting of n. For further details on how to use nonmonic polynomials, and also how to use homogeneous polynomials, [Buhler et al. 1993, Section 12].

There have been some exciting developments in polynomial selection, developments that were very important in the record 155-digit factorization of the famous RSA challenge number in late 1999. It turns out that a good polynomial makes so much difference that it is worthwhile to spend a considerable amount of resources searching through polynomial choices. For details on the latest strategies see [Murphy 1998, 1999].

Polynomial pairs

The description of NFS given in the sections above actually involves *two* polynomials, though we have emphasized only the single polynomial $f(x)$ for which we have an integer m with $f(m) \equiv 0 \pmod{n}$. It is more precisely the homogenized form of f that we considered, namely $F(x,y) = y^d f(x/y)$, where d is the degree of $f(x)$. The second polynomial is the rather trivial $g(x) = x - m$. Its homogenized form is $G(x,y) = yg(x/y) = x - my$. The numbers that we sieve looking for smooth values are the values of $F(x,y)G(x,y)$ in a box near the origin.

However, it is not necessary for the degree of $g(x)$ to be 1. Suppose we have two distinct, irreducible (not necessarily monic) polynomials $f(x), g(x) \in \mathbf{Z}[x]$, and an integer m with $f(m) \equiv g(m) \equiv 0 \pmod{n}$. Let α be a root of $f(x)$ in $\mathbf{C}$ and let β be a root of $g(x)$ in $\mathbf{C}$. Assuming that the leading coefficient c of $f(x)$ and C of $g(x)$ are coprime to n, we have homomorphisms $\phi : \mathbf{Z}[c\alpha] \to \mathbf{Z}_n$ and $\psi : \mathbf{Z}[C\beta] \to \mathbf{Z}_n$, where $\phi(c\alpha) \equiv cm \pmod{n}$ and $\psi(C\beta) \equiv Cm \pmod{n}$.

Suppose, too, that we have a set $\mathcal{S}$ consisting of an even number of coprime integer pairs a, b and elements $\gamma \in \mathbf{Z}[\alpha]$ and $\beta \in \mathbf{Z}[\beta]$ with

$$F_x(c\alpha, c)^2 \prod_{(a,b)\in\mathcal{S}} (ac - bc\alpha) = \gamma^2, \qquad G_x(C\beta, C)^2 \prod_{(a,b)\in\mathcal{S}} (aC - bC\beta) = \delta^2.$$

If $\mathcal{S}$ has $2k$ elements, and $\phi(\gamma) \equiv v \pmod{n}$, $\psi(\delta) \equiv w \pmod{n}$, then

$$\left(C^k G_x(Cm, C)v\right)^2 \equiv \left(c^k F_x(cm, c)w\right)^2 \pmod{n},$$

and so we may attempt to factor n via $\gcd(C^k G_x(Cm, C)u - c^k F_x(cm, c)v, n)$.

One may wonder why it is advantageous to use two polynomials of degree higher than 1. The answer is a bit subtle. Though the first-order desirable quality for the numbers that we sieve for smooth values is their size, there is a second-order quality that also has some significance. If a number near x is given to us as a product of two numbers near $x^{1/2}$, then it is more likely to be smooth than if it is a random number near x that is not necessarily such a product. If it is y-smoothness we are interested in and $u = \ln x / \ln y$, then this second-order effect may be quantified as about 2^u. That is, a number near x given as a product of two random numbers near $x^{1/2}$ is about 2^u times as likely to be y-smooth than is a random number near x. If we have two polynomials in the number field sieve with the same degree and with coefficients of the same magnitude, then their respective homogeneous forms have values that are of the same magnitude. It is the product of the two homogeneous forms that we are sieving for smooth values, so this 2^u philosophy seems to be relevant.

However, in the "ordinary" NFS as described in Algorithm 6.2.5, we are also looking for the product of two numbers to be smooth: One is the homogeneous form $F(a, b)$, and the other is the linear form $a - bm$. They do not have roughly equal magnitude. In fact, using the parameters suggested, $F(a, b)$ is about the $3/4$ power of the product, and $a - bm$ is about the $1/4$ power of the product. Such numbers also have an enhanced probability of being y-smooth, namely, $\left(4/3^{3/4}\right)^u$.

So, using two polynomials of the same degree $d \approx \frac{1}{2}(3 \ln n / \ln \ln n)^{1/3}$, and with coefficients bounded by about $n^{1/2d}$, we get an increased probability of smoothness over the choices in Algorithm 6.2.5 of about $\left(3^{3/4}/2\right)^u$. Now, u is about $2(3 \ln n / \ln \ln n)^{1/3}$, so that using the two polynomials of degree d saves a factor of about $(1.46)^{(\ln n / \ln \ln n)^{1/3}}$. While not altering the basic complexity, such a speedup represents significant savings.

The trouble, though, with using dual polynomials is finding them. Other than an exhaustive search, perhaps augmented with fast lattice techniques, no one has suggested a good way of finding such polynomials. For example, take the case of $d = 3$. We do not know any good method when given a large integer n of coming up with two distinct, irreducible, degree 3 polynomials $f(x), g(x)$, with coefficients bounded by $n^{1/6}$, say, and an integer m, perhaps very large, such that $f(m) \equiv g(m) \equiv 0 \pmod{n}$. A counting argument suggests that such polynomials should exist with coefficients even somewhat smaller, say bounded by about $n^{1/8}$.

Special number field sieve (SNFS)

Counting arguments show that for most numbers n, we cannot do very much better in finding polynomials than the simple-minded strategy of Algorithm 6.2.5. However, there are many numbers for which much better polynomials do exist, and if we can find such polynomials, then the complexity of NFS is significantly lowered. The special number field sieve (SNFS) refers to the cases of NFS where we are able to find extraordinarily good polynomials.

The SNFS has principally been used to factor many Cunningham numbers (these are numbers of the form $b^k \pm 1$ for $b = 2, 3, 5, 6, 7, 10, 11, 12$, see [Brillhart et al. 1988]). We have already mentioned the factorization of the ninth Fermat number, $F_9 = 2^{512} + 1$, by [Lenstra et al. 1993a]. They used the polynomial $f(x) = x^5 + 8$ and the integer $m = 2^{103}$, so that $f(m) = 8F_9 \equiv 0 \pmod{F_9}$. Even though we already knew the factor 2424833 of F_9 (found by A. E. Western in 1903), this was ignored. That is, the pretty nature of F_9 itself was used; the number $F_9/2424833$ is not so pretty!

What makes a polynomial extraordinary is that it has very small coefficients. If we have a number $n = b^k \pm 1$, we can create a polynomial as follows. Say we wish the degree of $f(x)$ to be 5. Write $k = 5l + r$, where r is the remainder when 5 is divided into k. Then $b^{5-r}n = b^{5(l+1)} \pm b^{5-r}$. Thus, we may use the polynomial $f(x) = x^5 \pm b^{5-r}$, and choose $m = b^{l+1}$. When k is large, the coefficients of $f(x)$ are very small in comparison to n.

A small advantage of a polynomial of the form $x^d + c$ is that the order of the Galois group is a divisor of $d\varphi(d)$, rather than having the generic value $d!$ for degree-d polynomials. Recall that the usefulness of free relations is proportional to the reciprocal of the order of the Galois group. Thus, free relations are more useful with special polynomials of the form $x^d + c$ than in the general case.

Sometimes a fair amount of ingenuity can go into the choosing of special polynomials. Take the case of $10^{193} - 1$, factored in 1996 by M. Elkenbracht-Huizing and P. Montgomery. They might have used the polynomial $x^5 - 100$ and $m = 10^{39}$, as suggested by the above discussion, or perhaps $10x^6 - 1$ and $m = 10^{32}$. However, the factorization still would have been a formidable. The number $10^{193} - 1$ was already partially factored. There is the obvious factor 9, but we also knew the factors

$$773,\ 39373,\ 561470969,\ 639701219449517,\ 4274417556076113498947,$$
$$26409540111952717487908689681403.$$

After dividing these known factors into $10^{193} - 1$, the resulting number n was still composite and had 108 digits. It would have been feasible to use either the quadratic sieve or the general NFS on n, but it seemed a shame not to use n's pretty ancestry. Namely, we know that 10 has a small multiplicative order modulo n. This leads us to the congruence $\left(10^{64}\right)^3 \equiv 10^{-1} \pmod{n}$, and to the congruence $\left(6 \cdot 10^{64}\right)^3 \equiv 6^3 \cdot 10^{-1} \equiv 108 \cdot 5^{-1} \pmod{n}$. Thus, for the polynomial $f(x) = 5x^3 - 108$ and $m = 6 \cdot 10^{64}$, we have $f(m) \equiv 0 \pmod{n}$. However, m is too large to profitably use the linear polynomial $x - m$. Instead, Elkenbracht-Huizing and Montgomery searched for a quadratic polynomial $g(x)$ with relatively small coefficients and with $g(m) \equiv 0 \pmod{n}$. This was done by considering the lattice of integer triples (A, B, C) with $Am^2 + Bm + C \equiv 0 \pmod{n}$. The task is to find a short vector in this lattice. Using techniques to find such short vectors, they came up with a choice for A, B, C all at most 36 digits long. They then used both $f(x)$ and $g(x) = Ax^2 + Bx + C$ to complete the factorization of n, finding that n is the product of two primes, the smaller being

$$44779828713128492805140830496526578289217495318108 7929.$$

Many polynomials

It is not hard to come up with many polynomials that may be used in NFS. For example, choose the degree d, let $m = \lceil n^{1/(d+1)} \rceil$, write n in base m, getting $n = c_d n^d + \cdots + c_0$, let $f(x) = c_d x^d + \cdots + c_0$, and let $f_j(x) = f(x) + jx - mj$ for various small integers j. Or one could look at the family $f_{j,k}(x) = f(x) + kx^2 - (mk - j)x - mj$ for various small integers k, j. Each of these polynomials evaluated at m is n.

One might use such a family to search for a particularly favorable polynomial, such as one where there is a tendency for many small primes to have multiple roots. Such a polynomial may have its homogeneous form being smooth more frequently than a polynomial where the small primes do not have this tendency.

But can all of the polynomials be used together? There is an obvious hindrance to doing this. Each time a new polynomial is introduced, the factor base must be extended to take into account the ways primes split for this polynomial. That is, each polynomial used must have its own field of coordinates in the exponent vectors, so that introducing more polynomials makes for longer vectors.

In [Coppersmith 1993] a way is found to (theoretically) get around this problem. He uses a large factor base for the linear form $a - bm$ and small factor bases for the various polynomials used. Specifically, if the primes up to B are used for the linear form, and k polynomials are used, then we use primes only up to B/k for each of these polynomials. Further, we consider only pairs a, b where both $a - bm$ is B-smooth and the homogeneous form of one of the polynomials is (B/k)-smooth. After B relations are collected, we (most likely) have more than enough to create congruent squares.

Coppersmith suggests first sieving over the linear form $a - bm$ for B-smooth numbers, and then individually checking at the homogeneous form of each polynomial used to see if the value at a, b is B/k-smooth. This check can be quickly done using the elliptic curve method (see Section 7.4). The elliptic curve method (ECM) used as a smoothness test is not as efficient in practice as sieving. However, if one wanted to use ECM in QS or NFS instead of sieving, the overall heuristic complexity would remain unchanged, the only difference coming in the $o(1)$ expression. In Coppersmith's variant of NFS he cannot efficiently use sieving to check his homogeneous polynomials for smoothness, since the pairs a, b that he checks for are irregularly spaced, being those where $a - bm$ has passed a smoothness test. (One might actually sieve over the letter j in the family $f_j(x)$ suggested above, but this will not be a long enough array to make the sieve economical.) Nevertheless, using ECM as a smoothness test allows one to use the same complexity estimates that one would have if one had sieved instead.

Assuming that about a total of B^2 pairs a, b are put into the linear form $a - bm$, at the end, a total of B^2k pairs of the linear form and the norm form of a polynomial are checked for simultaneous smoothness (the first being B-smooth, the second B/k-smooth). If the parameters are chosen so that at most B^2/k pairs a, b survive the first sieve, then the total time spent is not much more than B^2 total. This savings leads to a lower complexity in NFS. Coppersmith gives a heuristic argument that with an optimal choice of parameters the running time to factor n is $\exp\left((c + o(1))(\ln n)^{1/3}(\ln\ln n)^{2/3}\right)$, where

$$c = \frac{1}{3}\left(92 + 26\sqrt{13}\right)^{1/3} \approx 1.9019.$$

This compares with the value $c = (64/9)^{1/3} \approx 1.9230$ for the NFS as described in Algorithm 6.2.5. As mentioned previously, the smaller c in Coppersmith's method is offset by a "fatter" $o(1)$. This secondary factor likely makes the crossover point, after which Coppersmith's variant is superior, in the thousands of digits. Before we reach this point, NFS will probably have been replaced by far better methods. Nevertheless, Coppersmith's variant of NFS currently stands as the asymptotically fastest heuristic factoring method known.

There may yet be some practical advantage to using many polynomials. For a discussion, see [Elkenbracht-Huizing 1997].

6.3 Rigorous factoring

None of the factoring methods discussed so far in this chapter are rigorous. However, the subexponential ECM, discussed in the next chapter, comes close to being rigorous. Assuming a reasonable conjecture about the distribution in short intervals of smooth numbers, [Lenstra 1987] shows that ECM is expected to find the least prime factor p of the composite number n in $\exp((2 + o(1))\sqrt{\ln p \ln\ln p})$ arithmetic operations with integers the size of n, the "$o(1)$" term tending to 0 as $p \to \infty$. Thus, ECM requires only one heuristic "leap." In contrast, QS and NFS seem to require several heuristic leaps in their analyses.

It is of interest to see what is the fastest factoring algorithm that we can rigorously analyze. This is not necessarily of practical value, but seems to be required by the dignity of the subject!

The first issue one might address is whether a factoring algorithm is deterministic or probabilistic. Since randomness is such a powerful tool, we would expect to see lower complexity records for probabilistic factoring algorithms over deterministic ones, and indeed we do. The fastest deterministic factoring algorithm that has been rigorously analyzed is the Pollard–Strassen method. This uses fast polynomial evaluation techniques as discussed in Section 5.5, where the running time to factor n is seen to be $O\left(n^{1/4+o(1)}\right)$.

Assuming the ERH, see Conjecture 1.4.2, an algorithm of Shanks deterministically factors n in a running-time bound of $O(n^{1/5+o(1)})$. This method is described in Section 5.6.4.

That is it for rigorous, deterministic methods. What, then, of probabilistic methods? The first subexponential probabilistic factoring algorithm with a completely rigorous analysis was the "random-squares method" of [Dixon 1981]. His algorithm is to take random integers r in $[1, n]$, looking for those where $r^2 \bmod n$ is smooth. If enough are found, then congruent squares can be assembled, as in QS, and so a factorization of n may be attempted. The randomness of the numbers r that are used allows one to say rigorously how frequently the residues $r^2 \bmod n$ are smooth, and how likely the congruent squares assembled lead to a nontrivial factorization of n. Dixon showed that the expected running time for his algorithm to split n is bounded by $\exp\left((c + o(1))\sqrt{\ln n \ln\ln n}\right)$, where $c = \sqrt{8}$. Subsequent improvements by Pomerance and later by B. Vallée lowered c to $\sqrt{4/3}$.

The current lowest running-time bound for a rigorous probabilistic factoring algorithm is $\exp((1 + o(1))\sqrt{\ln n \ln\ln n})$. This is achieved by the "class-group-relations method" of [Lenstra and Pomerance 1992]. Previously, this time bound was achieved by A. Lenstra for a very similar algorithm, but the analysis required the use of the ERH. It is interesting that this time bound is exactly the same as that heuristically achieved by QS. Again the devil is in the "$o(1)$," making the class-group-relations method impractical in comparison.

It is interesting that both the improved versions of the random-squares method and the class-group-relations method use ECM as a subroutine to quickly recognize smooth numbers. One might well wonder how a not-yet-rigorously analyzed algorithm can be used as a subroutine in a rigorous algorithm. The answer is that one need not show that the subroutine always works, just that it works frequently enough to be of use. It can be shown rigorously that ECM recognizes most y-smooth numbers below x in $y^{o(1)} \ln x$ arithmetic operations with integers the size of x. There may be some exceptional numbers that are stubborn for ECM, but they are provably rare.

Concerning the issue of smoothness tests, a probabilistic algorithm announced in [Lenstra et al. 1993b] recognizes all y-smooth numbers n in $y^{o(1)} \ln n$ arithmetic operations. That is, it performs similarly as ECM, but unlike ECM, the complexity estimate is completely rigorous and there are provably no exceptional numbers.

6.4 Index-calculus method for discrete logarithms

In Chapter 5 we described some general algorithms for the computation of discrete logarithms that work in virtually any cyclic group for which we can represent group elements on a computer and perform the group operation. These exponential-time algorithms have the number of steps being about the square root of the group order. In certain specific groups we have more information that might be used profitably for DL computations. We have seen in this chapter the ubiquitous role of smooth numbers as an aid to factorization. In some groups sense can be made of saying that a group element is smooth, and when this is the case, it is often possible to perform DLs via a subexponential algorithm. The basic idea is embodied in the index-calculus method.

We first describe the index-calculus method for the multiplicative group of the finite field $\mathbf{F}_p$, where p is prime. Later we shall see how the method can be used for all finite fields.

The fact that subexponential methods exist for solving DLs in the multiplicative group of a finite field have led cryptographers to use other groups, the most popular being elliptic-curve groups; see Chapter 7.

6.4.1 Discrete logarithms in prime finite fields

Consider the multiplicative group $\mathbf{F}_p^*$, where p is a large prime. This group is cyclic, a generator being known as a primitive root (Definition 2.2.6). Suppose g is a primitive root and t is an element of the group. The DL problem for $\mathbf{F}_p^*$ is, given p, g, t to find an integer l with $g^l = t$. Actually, l is not well-defined by this equation, the integers l that work form a residue class modulo $p-1$. We write $l \equiv \log_g t \pmod{p-1}$.

What makes the index-calculus method work in $\mathbf{F}_p^*$ is that we do not have to think of g and t as abstract group elements, but rather as integers, and we may think of the equation $g^l = t$ as the congruence $g^l \equiv t \pmod{p}$. The index-calculus method consists of two principal stages. The first stage involves gathering "relations." These are congruences $g^r \equiv p_1^{r_1} \cdots p_k^{r_k} \pmod{p}$, where $p_1, \ldots, p_k$ are small prime numbers. Such a congruence gives rise to a congruence of discrete logarithms:

$$r \equiv r_1 \log_g p_1 + \cdots + r_k \log_g p_k \pmod{p-1}.$$

If there are enough of these relations, it may then be possible to use linear algebra to solve for the various $\log_g p_i$. After this precomputation, which is the heart of the method, the final discrete logarithm of t is relatively simple. If one has a relation of the form $g^R t \equiv p_1^{\tau_1} \cdots p_k^{\tau_k} \pmod{p}$, then we have that

$$\log_g t \equiv -R + \tau_1 \log_g p_1 + \cdots + \log_g p_k \pmod{p-1}.$$

Both kinds of relations are found via random choices for the numbers r, R. A choice for r gives rise to some residue $g^r \bmod p$, which may or may not factor completely over the small primes $p_1, \ldots, p_k$. Similarly, a choice for R gives rise to the residue $g^R t \bmod p$. By taking residues closest to 0 and allowing a factor -1 in a prime factorization, a small gain is realized. Note that we do not have to solve for the discrete logarithm of -1; it is already known as $(p-1)/2$. We summarize the index-calculus method for $\mathbf{F}_p^*$ in the following pseudocode.

Algorithm 6.4.1 (Index-calculus method for $\mathbf{F}_p^*$). We are given a prime p, a primitive root g, and a nonzero residue $t \pmod p$. This probabilistic algorithm attempts to find $\log_g t$.

1. [Set smoothness bound]
   Choose a smoothness bound B; // See text for reasonable B choices.
   Find the primes $p_1, \ldots, p_k$ in $[1, B]$;
2. [Search for general relations]
   Choose random integers r in $[1, p-2]$ until B cases are found with $g^r \bmod p$ being B-smooth;
   // It is slightly better to use the residue of $g^r \bmod p$ closest to 0.
3. [Linear algebra]
   By some method of linear algebra, use the relations found to solve for $\log_g p_1, \ldots, \log_g p_k$;
4. [Search for a special relation]
   Choose random integers R in $[1, p-2]$ and find the residue closest to 0 of $g^R t \pmod p$ until one is found with this residue being B-smooth;
   Use the special relation found together with the values of $\log_g p_1, \ldots, \log_g p_k$ found in step [Linear algebra] to find $\log_g t$;

This brief description raises several questions:

(1) How does one determine whether a number is B-smooth?

(2) How does one do linear algebra modulo the composite number $p - 1$?

(3) Are B relations an appropriate number so that there is a reasonable chance of success in step [Linear algebra]?

(4) What is a good choice for B?

(5) What is the complexity of this method, and is it really subexponential?

On question (1), there are several options including trial division, the Pollard rho method (Algorithm 5.2.1), and the elliptic curve method (Algorithm 7.4.2). Which method one employs affects the overall complexity, but with any of these methods, the index-calculus method is subexponential.

It is a bit tricky doing matrix algebra over $\mathbf{Z}_n$ with n composite. In step [Linear algebra] we are asked to do this with $n = p - 1$, which is composite for all primes $p > 3$. As with solving polynomial congruences, one idea is to reduce the problem to prime moduli. Matrix algebra over $\mathbf{Z}_q$ with q prime is just matrix algebra over a finite field, and the usual Gaussian methods work, as well as do various faster methods. As with polynomial congruences, one can also employ Hensel-type methods for matrix algebra modulo prime powers, and Chinese remainder methods for gluing powers of different primes. In addition, one does not have to work all that hard at the factorization. If some large factor of $p - 1$ is actually composite and difficult to factor further, one can proceed with the matrix algebra modulo this factor as if it were prime. If one is called to invert a nonzero residue, usually one will be successful, but if not, a factorization is found for free. So either one is successful in the matrix algebra, which is the primary goal, or one gets a factorization of the modulus, and so can restart the matrix algebra with the finer factors one has found.

Regarding question (3), it is likely that with somewhat more than $\pi(B)$ relations of the form $g^r \equiv p_1^{r_1} \cdots p_k^{r_k} \pmod{p}$, where $p_1, \ldots, p_k$ are all of the primes in $[1, B]$, that the various exponent vectors $(r_1, \ldots, r_k)$ found span the module $\mathbf{Z}_{p-1}^k$. So obtaining B of these vectors is a bit of overkill. In addition, it is not even necessary that the vectors span the complete module, but only that the vector corresponding to the relation found in step [Search for a special relation] be in the submodule generated by them. This idea, then, would make the separate solutions for $\log_g p_i$ in step [Linear algebra] unnecessary; namely, one would do the linear algebra only after the special relation is found.

The final two questions above can be answered together. Just as with the analysis of some of the factorization methods, we find that an asymptotically optimal choice for B is of the shape $L(p)^c$, where $L(p)$ is defined in (6.1). If a fast smoothness test is used, such as the elliptic curve method, we would choose $c = 1/\sqrt{2}$, and end up with a total complexity of $L(p)^{\sqrt{2}+o(1)}$. If a slow smoothness test is used, such as trial division, a smaller value of c should be chosen, namely $c = 1/2$, leading to a total complexity of $L(p)^{2+o(1)}$. If a smoothness test is used that is of intermediate complexity, one is led to an intermediate value of c and an intermediate total complexity.

At finite levels, the asymptotic analysis is only a rough guide, and good choices should be chosen by the implementer following some trial runs. For details on the index-calculus method for prime finite fields, see [Pomerance 1987b].

6.4.2 Discrete logarithms via smooth polynomials and smooth algebraic integers

What makes the index-calculus method successful, or even possible, for $\mathbf{F}_p$ is that we may think of $\mathbf{F}_p$ as $\mathbf{Z}_p$, and thus represent group elements with integers. It is *not* true that $\mathbf{F}_{p^d}$ is isomorphic to $\mathbf{Z}_{p^d}$ when $d > 1$, and so there is no convenient way to represent elements of nonprime finite fields with integers. As we saw in Section 2.2.2, we may view $\mathbf{F}_{p^d}$ as the quotient ring $\mathbf{Z}_p[x]/(f(x))$, where $f(x)$ is an irreducible polynomial in $\mathbf{Z}_p[x]$ of degree d. Thus, we may identify to each member of $\mathbf{F}_{p^d}^*$ a nonzero polynomial in $\mathbf{Z}_p[x]$ of degree less than d.

The polynomial ring $\mathbf{Z}_p[x]$ is like the ring of integers $\mathbf{Z}$ in many ways. Both are unique factorization domains, where the "primes" of $\mathbf{Z}_p[x]$ are the monic irreducible polynomials of positive degree. Both have only finitely many invertible elements (the residues $1, 2, \ldots, p-1$ modulo p in the former case, and the integers ± 1 in the latter case), and both rings have a concept of size. Indeed, though $\mathbf{Z}_p[x]$ is not an ordered ring, we nevertheless have a rudimentary concept of size via the degree of a polynomial. And so, we have a concept of "smoothness" for a polynomial: We say that a polynomial is b-smooth if each of its irreducible factors has degree at most b. We even have a theorem analogous to (1.44): The fraction of b-smooth polynomials in $\mathbf{Z}_p[x]$

of degree less than d is about u^{-u}, where $u = d/b$, for a wide range of the variables p, d, b.

Now obviously, this does not make too much sense when d is small. For example, when $d = 2$, everything is 1-smooth, and about $1/p$ of the polynomials are 0-smooth. However, when d is large the index-calculus method does work for discrete logarithms in $\mathbf{Z}_{p^d}^*$, giving a method that is subexponential; see [Lovorn Bender and Pomerance 1998].

What, then, of the cases when $d > 1$, but d is not large. There is an alternative representation of $\mathbf{F}_{p^d}$ that is useful in these cases. Suppose K is an algebraic number field of degree d over the field of rational numbers. Let O_K denote the ring of algebraic integers in K. If p is a prime number that is inert in K, that is, the ideal (p) in O_k is a prime ideal, then the quotient structure $O_K/(p)$ is isomorphic to $\mathbf{F}_{p^d}$. Thus we may think of members of the finite field as algebraic integers. And as we saw with the NFS factoring algorithm, it makes sense to talk of when an algebraic integer is smooth: Namely, it is y-smooth if all of the prime factors of its norm to the rationals are at most y.

Let us illustrate in the case $d = 2$ where p is a prime that is 3 (mod 4). We take $K = \mathbf{Q}[i]$, the field of Gaussian rationals, namely $\{a + bi : a, b \in \mathbf{Q}\}$. Then O_K is $\mathbf{Z}[i] = \{a + bi : a, b \in \mathbf{Z}\}$, the ring of Gaussian integers. We have that $\mathbf{Z}[i]/(p)$ is isomorphic to the finite field $\mathbf{F}_{p^2}$. So, the index-calculus method will still work, but now we are dealing with Gaussian integers $a + bi$ instead of ordinary integers.

In the case $d = 2$, the index-calculus method via a quadratic imaginary field can be made completely rigorous; see [Lovorn 1992]. The use of other fields are conjecturally acceptable, but the analysis of the index calculus method in these cases remains heuristic.

There are heuristic methods analogous to the NFS factoring algorithm to do discrete logs in any finite field $\mathbf{F}_{p^d}$, including the case $d = 1$. For a wide range of cases, the complexity is heuristically brought down to functions of the shape $\exp\left(c\left(\log p^d\right)^{1/3}\left(\log\log p^d\right)^{2/3}\right)$; see [Gordon 1993], [Schirokauer et al. 1996], and [Adleman 1994]. These methods may be thought of as grand generalizations of the index-calculus method, and what makes them work is a representation of group elements that allows the notion of smoothness. It is for this reason that cryptographers tend to eschew multiplicative groups of finite fields in favor of elliptic-curve groups. With elliptic-curve groups we have no convenient notion of smoothness, and the index-calculus method appears to be useless. For these groups, the best DL methods that universally work all take exponential time.

6.5 Exercises

6.1. You are given a composite number n that is not a power, and a nontrivial factorization $n = ab$. Describe an efficient algorithm for finding a nontrivial coprime factorization of n, that is, finding coprime integers A, B, both larger than 1, with $n = AB$.

6.2. Show that if n is odd, composite, and not a power, then at least half of the pairs x, y with $0 \le x, y < n$ and $x^2 \equiv y^2 \pmod{n}$ have $1 < \gcd(x - y, n) < n$.

6.3. Sometimes when one uses QS, the number n to be factored is replaced with kn for a small integer k. Though using a multiplier increases the magnitude of the residues being sieved for smoothness, there can be significant compensation. It can happen that k skews the set of sieving primes to favor smaller primes. Investigate the choice of a multiplier for using QS to factor

$$n = 18831998556192052 03.$$

In particular, compare the time for factoring this number n with the time for factoring $3n$. (That is, the number $3n$ is given to the algorithm which should eventually come up with a factorization $3n = ab$ where $3 < a < b$.) Next, investigate the choice of multiplier for using QS to factor

$$n = 215659417219997979398437139 63.$$

(If you are interested in actual program construction, see Exercise 6.13 for implementation issues.)

6.4. There are numerous factoring methods exploiting the idea of "small squares" as it is enunciated at the beginning of the chapter. While the QS and NFS are powerful manifestations of the idea, there are other, not so powerful, but interesting, methods that employ side factorizations of small residues, with eventual linear combination as in our QS discussion. One of the earlier methods of the class is the Brillhart–Morrison continued-fraction method (see [Cohen 2000] for a concise summary), which involves using the continued fraction expansion of $\sqrt{n}$ (or $\sqrt{kn}$ for a small integer k) for the generation of many congruences $Q \equiv x^2 \pmod{n}$ with $Q \ne x^2, |Q| = O(\sqrt{n})$. One attempts to factor the numbers Q to construct instances of $u^2 \equiv v^2 \pmod{n}$. An early triumph of this method was the 1974 demolition of F_7 by Brillhart and Morrison (see Table 1.3). In the size of the quadratic residues Q that are formed, the method is somewhat superior to QS. However, the sequence of numbers Q does not appear to be amenable to a sieve, so practitioners of the continued-fraction method have been forced to spend a fair amount of time per Q value, even though most of the Q are ultimately discarded for not being sufficiently smooth.

We shall not delve into the continued-fraction method further. Instead, we list here various tasks and questions intended to exemplify—through practice, algebra, and perhaps some entertainment!—the creation and use of "small squares" modulo a given n to be factored. We shall focus below on special numbers such as the Fermat numbers $n = F_k = 2^{2^k} + 1$ or Mersenne numbers $n = M_q = 2^q - 1$ because the manipulations are easier in many respects for such special forms; but, like the mighty NFS, the notions can for the most part be extended to more general composite n.

(1) Use the explicit congruences

$$\begin{aligned} 258883717^2 \bmod M_{29} &= -2 \cdot 3 \cdot 5 \cdot 29^2, \\ 301036180^2 \bmod M_{29} &= -3 \cdot 5 \cdot 11 \cdot 79, \\ 126641959^2 \bmod M_{29} &= 2 \cdot 3^2 \cdot 11 \cdot 79, \end{aligned}$$

to create an appropriate nontrivial congruence $u^2 \equiv v^2$ and thereby discover a factor of M_{29}.

(2) It turns out that $\sqrt{2}$ exists modulo each of the special numbers $n = F_k, k \geq 2$, and the numbers $n = M_q, q \geq 3$; and remarkably, one can give explicit such roots whether or not n is composite. To this end, show that

$$2^{3 \cdot 2^{k-2}} - 2^{2^{k-2}}, \qquad 2^{(q+1)/2}$$

are square roots of 2 in the respective Fermat, Mersenne cases. In addition, give an explicit, primitive fourth root of (-1) for the Fermat cases, and an explicit ((q mod 4)-dependent) fourth root of 2 in the Mersenne cases. Incidentally, these observations have actual application: One can now remove any power of 2 in a squared residue, because there is now a closed form for $\sqrt{2^k}$; likewise in the Fermat cases factors of (-1) in squared residues can be removed.

(3) Using ideas from the previous item, prove "by hand" the congruence

$$2(2^6 - 8)^2 \equiv (2^6 + 1)^2 \pmod{M_{11}},$$

and infer from this the factorization of M_{11}.

(4) It is a lucky fact that for a certain ω, a primitive fourth root of 2 modulo M_{43}, we have

$$\left(2704\omega^2 - 3\right)^2 \bmod M_{43} = 2^3 \cdot 3^4 \cdot 43^2 \cdot 2699^2.$$

Use this fact to discover a factor of M_{43}.

(5) For ω a primitive fourth root of -1 modulo $F_k, k \geq 2$, and with given integers a, b, c, d, set

$$x = a + b\omega + c\omega^2 + d\omega^3.$$

It is of interest that certain choices of a, b, c, d automatically give small squares—one might call them small "symbolic squares"—for any of the F_k indicated. Show that if we adopt a constraint

$$ad + bc = 0$$

then $x^2 \bmod F_k$ can be written as a polynomial in ω with degree less than 3. Thus for example

$$\left(-6 + 12\omega + 4\omega^2 + 8\omega^3\right)^2 \equiv 4(8\omega^2 - 52\omega - 43),$$

and furthermore, the coefficients in this congruence hold uniformly across all the Fermat numbers indicated (except that ω, of course, depends on the Fermat number). Using these ideas, provide a lower bound, for a given constant K, on how many "symbolic squares" can be found with

$$|x^2 \bmod F_k| < K\sqrt{F_k}.$$

Then provide a similar estimate for small squares modulo Mersenne numbers M_q.

(6) Pursuant to the previous item, investigate this kind of factoring for more general odd composites $N = \omega^4 + 1$ using the square of a fixed cubic form, e.g.

$$x = -16 + 8\omega + 2\omega^2 + \omega^3,$$

along the following lines. Argue that (-1) is always a square modulo N, and also that

$$x^2 \equiv 236 - 260\omega - \omega^2 \pmod N.$$

In this way discover a proper factor of

$$N = 16452725990417$$

by finding a certain square that is congruent, nontrivially, to x^2. Of course, the factorization of this particular N is easily done in other ways, but the example shows that certain forms $\omega^4 + 1$ are immediately susceptible to the present, small-squares formalism. Investigate, then, ways to juggle the coefficients of x in such a way that a host of other numbers $N = \omega^4 + 1$ become susceptible.

Related ideas on creating small squares, for factoring certain cubic forms, appear in [Zhang 1998].

6.5. Suppose you were in possession of a device such that if you give it a positive integer n and an integer a in $[1, n]$, you are told one solution to $x^2 \equiv a \pmod n$ if such a solution exists, or told that no solution exists if this is the case. If the congruence has several solutions, the device picks one of these by some method unknown to you. Assume that the device takes polynomial time to do its work; that is, the time it takes to present its answer is bounded by a constant times a fixed power of the logarithm of n. Show how, armed with such a device, one can factor via a probabilistic algorithm with expected running time being polynomial. Conversely, show that if you can factor in polynomial time, then you can build such a device.

6.6. Suppose you had a magic algorithm that given an N to be factored could routinely (and quickly, say in polynomial time per instance) find integers x satisfying

$$\sqrt{N} < x < N - \sqrt{N}, \quad x^2 \bmod N < N^\alpha,$$

for some fixed α. (Note that the continued-fraction method and the quadratic sieve do this essentially for $\alpha \approx 1/2$.) Assume, furthermore, that these "small square" congruences each require $O(\ln^\beta N)$ operations to discover. Give the (heuristic) complexity, then, for factoring via this magic algorithm.

6.7. A Gray code is a sequence of k-bit binary strings in such an order that when moving from one string to the next, one and only one bit flips to its opposite bit. Show that such a code—whether for the self-initialization QS option or any other application—can be generated with ease, using a function that employs exclusive-or "∧" and shift ">>" operators in the following elegant way:

$$g(n) = n \wedge (n >> 1).$$

This very simple generator is easily seen to yield, for example, a 3-bit Gray counter that runs:

$$(g(0), \ldots, g(7)) = (000, 001, 011, 010, 110, 111, 101, 100),$$

this counting chain clearly having exactly one bit flip on each iteration.

6.8. Show that if $n \ge 64$ and $m = \lfloor n^{1/3} \rfloor$, then $n < 2m^3$. More generally, show that if d is a positive integer, $n > 1.5(d/\ln 2)^d$, and $m = \lfloor n^{1/d} \rfloor$, then $n < 2m^d$.

6.9. The following result, which allows an integer factorization via a polynomial factorization, is shown in [Brillhart et al. 1981].

Theorem. *Let n be a positive integer, let m be an integer with $m \ge 2$, write n in base m as $n = f(m)$ where $f(x) = c_d x^d + c_{d-1} x^{d-1} + \cdots + c_0$, so that the c_i's are nonnegative integers less than m. Suppose $f(x)$ is reducible in $\mathbf{Z}[x]$, with $f(x) = g(x)h(x)$ where neither $g(x)$ nor $h(x)$ is a constant polynomial with value ± 1. Then $n = g(m)h(m)$ is a nontrivial factorization of n.*

This exercise is to prove this theorem using the following outline:

(1) Show that $f(m-1) > 0$.

(2) Prove the inequality

$$\left|\frac{f(z)}{z^{d-1}}\right| \ge \mathrm{Re}(c_d z) + c_{d-1} + \mathrm{Re}\left(\frac{c_{d-2}}{z}\right) - \sum_{j=3}^{d} \frac{c_{d-j}}{|z|^{j-1}}$$

and use it to show that $f(z) \ne 0$ for $\mathrm{Re}\, z \ge m - \frac{1}{2}$. (Use that each c_j satisfies $0 \le c_j \le m-1$ and that $c_d \ge 1$.)

(3) Assume that the factorization $n = g(m)h(m)$ is trivial, say $g(m) = \pm 1$. Let $G(x) = g(m + x - 1/2)$. Using that all the roots of $G(x)$ have negative real parts and that $G(x)$ is in $\mathbf{Q}[x]$, show that all the coefficients of $G(x)$ have the same sign, and so $\left|G(-\frac{1}{2})\right| < \left|G(\frac{1}{2})\right|$.

(4) Complete the proof using $|G(\frac{1}{2})| = 1$ and $|G(-\frac{1}{2})| > 0$.

6.10. Use the method of Exercise 6.9 to factor $n = 187$ using the base $m = 10$. Do the same with $n = 4189, m = 29$.

6.11. Generalize the $x(u,v), y(u,v)$ construction in Section 6.1.7 to arbitrary numbers n satisfying (6.4).

6.12. Give a heuristic argument for the complexity bound

$$\exp\left((c+o(1))(\ln n)^{1/3}(\ln\ln n)^{2/3}\right)$$

operations, with $c = (32/9)^{1/3}$, for the special number field sieve (SNFS).

6.13. Here we sketch some practical QS examples that can serve as guidance for the creation of truly powerful QS implementations. In particular, the reader who chooses to implement QS can use the following examples for program checking. Incidentally, each one of the examples below—except the last—can be effected on a typical symbolic processor possessed of multiprecision operations. So the exercise shows that numbers in the 30-digit region and beyond can be handled even without fast, compiled implementations.

(1) In Algorithm 6.1.1 let us take the very small example $n = 10807$ and, because this n is well below typical ranges of applicability of practical QS, let us force at the start of the algorithm the smoothness limit $B = 200$. Then you should find $k = 21$ appropriate primes, You then get a 21×21 binary matrix, and can Gaussian-reduce said matrix. Incidentally, packages exist for such matrix algebra, e.g., in the *Mathematica* language a matrix m can be reduced for such purpose with the single statement

```
r = NullSpace[Transpose[m], Modulus->2];
```

(although, as pointed out to us by D. Lichtblau one may optimize the overall operation by intervention at a lower level, using bit operations rather than (mod 2) reduction, say). With such a command, there is a row of the reduced matrix r that has just three 1's, and this leads to the relation:

$$3^4 \cdot 11^4 \cdot 13^4 \equiv 106^2 \cdot 128^2 \cdot 158^2 \pmod{n},$$

and thus a factorization of n.

(2) Now for a somewhat larger composite, namely $n = 7001 \cdot 70001$, try using the B assignment of Algorithm 6.1.1 *as is*, in which case you should have $B = 2305$, $k = 164$. The resulting 164×164 matrix is not too unwieldy in this day and age, so you should be able to factor n using the same approach as in the previous item.

(3) Now try to factor the Mersenne number $n = 2^{67} - 1$ but using smoothness bound $B = 80000$, leading to $k = 3962$. Not only will this example start testing your QS implementation in earnest, it will demonstrate how 12-digit factors can be extracted with QS in a matter of seconds or minutes (depending on the efficiency of the sieve and the matrix package). This is still somewhat slower than sheer sieving or say Pollard-rho methods, but of course, QS can be pressed *much* further, with its favorable asymptotic behavior.

(4) Try factoring the repunit

$$n = \frac{10^{29} - 1}{9} = 11111111111111111111111111111$$

using a forced parameter $B = 40000$, for which matrices will be about 2000×2000 in size.

(5) If you have not already for the above, implement Algorithm 6.1.1 in fast, compiled fashion to attempt factorization of, say, 100-digit composites.

6.14. In the spirit of Exercise 6.13, we here work through the following explicit examples of the NFS Algorithm 6.2.5. Again the point is to give the reader some guidance and means for algorithm debugging. We shall find that a particular obstruction—the square-rooting in the number field—begs to be handled in different ways, depending on the scale of the problem.

(1) Start with the simple choice $n = 10403$ and discover that the polynomial f is *reducible*, hence the very step [Setup] yields a factorization, with no sieving required.

(2) Use Algorithm 6.2.5 with initialization parameters *as is* in the pseudocode listing, to factor $n = F_5 = 2^{32} + 1$. (Of course, the SNFS likes this composite, but the exercise here is to get the general NFS working!) From the initialization we thus have $d = 2$, $B = 265$, $m = 65536$, $k = 96$, and thus matrix dimension $V = 204$. The matrix manipulations then accrue exactly as in Exercise 6.13, and you will obtain a suitable set $\mathcal{S}$ of (a, b) pairs. Now, for the small composite n in question (and the correspondingly small parameters) you can, in step [Square roots], just multiply out the product $\prod_{(a,b)\in\mathcal{S}}(a - b\alpha)$ to generate a Gaussian integer, because the assignment $\alpha = i$ is acceptable. Note how one is lucky for such ($d = 2$) examples, in that square-rooting in the number field is a numerical triviality. In fact, the square root of a Gaussian integer $c + di$ can be obtained by solving simple simultaneous relations. So for such small degree as $d = 2$, the penultimate step [Square roots] of Algorithm 6.2.5 is about as simple as can be.

(3) As a kind of "second gear" with respect mainly to the square-root obstacle, try next the same composite $n = F_5$ but force parameters $d = 4$, $B = 600$, which choices will result in successful NFS. Now, at the step [Square roots], you can again just multiply out the product of terms $(a - b\alpha)$ where now $\alpha = \sqrt{i}$, and you can then take the square root of the resulting element

$$s_0 + s_1\alpha + s_2\alpha^2 + s_3\alpha^3$$

in the number field. There are easy ways to do this numerically, for example a simple version of the deconvolution of Exercise 6.17 will work, or you can just use the Vandermonde scheme discussed later in the present exercise.

(4) Next, choose $n = 76409$ and this time force parameters as: $d = 2$, $B = 96$, to get a polynomial $f(x) = x^2 + 233x$. Then, near the end of the algorithm,

you can again multiply out the $(a - b\alpha)$ terms, then use simple arithmetic to take the number-field root and thereby complete the factorization.

(5) Just as in the last item, factor the repunit $n = 11111111111$ by initializing parameters thus: $d = 2$, $B = 620$.

(6) Next, for $n = F_6 = 2^{64} + 1$, force $d = 4$, $B = 2000$, and this time force even the parameter $k = 80$ for convenience. Use any of the indicated methods to take a square root in the number field with $\alpha = \sqrt{i}$.

(7) Now we can try a "third gear" in the sense of the square-root obstruction. Factor the repunit $n = (10^{17} - 1)/9 = 11111111111111111$ but by forcing parameters $d = 3$, $B = 2221$. This time, the square root needs be taken in a number field with a cube root of 1. It is at this juncture that we may as well discuss the Vandermonde matrix method for rooting. Let us form γ^2, that is the form $f'(\alpha)^2 \prod_{(a,b)\in\mathcal{S}}(a - b\alpha)$, simply by multiplying all relevant terms together modulo $f(\alpha)$. (Such a procedure would always work in principle, yet for large enough n the coefficients of the result γ^2 become unwieldy.) The Vandermonde matrix approach then runs like so. Write the entity to be square-rooted as

$$\gamma^2 = s_0 + s_1\alpha + \cdots + s_{d-1}\alpha^{d-1}.$$

Then, use the (sufficiently precise) d roots of f, call them $\alpha_1, \ldots, \alpha_d$, to construct the matrix of ascending powers of roots

$$H = \begin{pmatrix} 1 & \alpha_1 & {\alpha_1}^2 & \cdots & {\alpha_1}^{d-1} \\ 1 & \alpha_2 & {\alpha_2}^2 & \cdots & {\alpha_2}^{d-1} \\ \vdots & \vdots & \vdots & \ddots & \vdots \\ 1 & \alpha_d & \alpha_d^2 & \cdots & {\alpha_d}^{d-1} \end{pmatrix}.$$

Then take a sufficiently high-precision square roots of *real numbers*, that is, calculate the vector

$$\beta = \sqrt{Hs^T},$$

where $s = (s_0, \ldots, s_{d-1})$ is the vector of coefficients of γ^2, and the square root of the matrix-vector product is simply taken componentwise. Now the idea is to calculate matrix-vector products:

$$H^{-1} \begin{pmatrix} \pm\beta_0 \\ \pm\beta_1 \\ \vdots \\ \pm\beta_{d-1} \end{pmatrix},$$

where the $\pm$ ambiguities are tried one at a time, until the vector resulting from this multiplication by H^{-1} has all integer components. Such a vector will be a square root in the number field. To aid in any implementations, we give here an explicit, small example of this rooting method. Let us take the polynomial $f(x) = x^3 + 5x + 6$ and square-root the entity

$\gamma^2 = 117 - 366x + 46x^2$ modulo $f(x)$ (we are using preknowledge that the entity here really is a square). We construct the Vandermode matrix using zeros of f, namely $(\alpha_1, \alpha_2, \alpha_3) = \left(-1, \left(1 - i\sqrt{23}\right)/2, \left(1 + i\sqrt{23}\right)/2\right)$, as a numerical entity whose first row is $(1, -1, 1)$ with complex entries in the other rows. There needs to be enough precision, which for this present example is say 12 decimal digits. Then we take a (componentwise) square root and try the eight possible $(\pm)$ combinations

$$\gamma = H^{-1} \begin{pmatrix} \pm r_1 \\ \pm r_2 \\ \pm r_3 \end{pmatrix}, \quad \begin{pmatrix} r_1 \\ r_2 \\ r_3 \end{pmatrix} = \sqrt{H \begin{pmatrix} 177 \\ -366 \\ 46 \end{pmatrix}}.$$

Sure enough, one of these eight combinations is the vector

$$\gamma = \begin{pmatrix} 15 \\ -9 \\ -1 \end{pmatrix}$$

indicating that

$$\left(15 - 9x - x^2\right)^2 \bmod f(x) = 117 - 366x + 46x^2$$

as desired.

(8) Just as with Exercise 6.13, we can only go so far with symbolic processors and must move to fast, compiled programs to handle large composites. Still, numbers in the region of 30 digits can indeed be handled interpretively. Take the repunit $n = (10^{29} - 1)/9$, force $d = 4$, $B = 30000$, and this time force also $k = 100$, to see a successful factorization that is doable without fast programs. In this case, you can use any of the above methods for handling degree-4 number fields, still with brute-force multiplying-out for the γ^2 entity (although for the given parameters one already needs perhaps 3000-digit precision, and the advanced means discussed in the text and in Exercise 6.17 start to look tantalizing for the square-rooting stage).

The explicit tasks above should go a long way toward the polishing of a serious NFS implementation. However, there is more that can be done even for these relatively minuscule composites. For example, the free relations and other optimizations of Section 6.2.7 can help even for the above tasks, and should certainly be invoked for large composites.

6.15. Here we solve an explicit and simple DL problem to give an illustration of the index-calculus method (Algorithm 6.4.1). Take the prime $p = 2^{13} - 1$, primitive root $g = 17$, and say we want to solve $g^l \equiv 5 \pmod{p}$. Note the following congruences, which can be obtained rapidly by machine:

$$\begin{aligned} g^{3513} &\equiv 2^3 \cdot 3 \cdot 5^2 \pmod{p}, \\ g^{993} &\equiv 2^4 \cdot 3 \cdot 5^2 \pmod{p}, \\ g^{1311} &\equiv 2^2 \cdot 3 \cdot 5 \pmod{p}. \end{aligned}$$

(In principle, one can do this by setting a smoothness limit on prime factors of the residue, then just testing random powers of g.) Now solve the indicated DL problem by finding via linear algebra three integers a, b, c such that

$$g^{3513a+993b+1311c} \equiv 5 \pmod{p}.$$

6.6 Research problems

6.16. Investigate the following idea for forging a subexponential factoring algorithm. Observe first the amusing algebraic identity [Crandall 1996a]

$$F(x) = \left((x^2 - 85)^2 - 4176\right)^2 - 2880^2$$

$$= (x-13)(x-11)(x-7)(x-1)(x+1)(x+7)(x+11)(x+13),$$

so that F actually has 8 simple, algebraic factors in $\mathbf{Z}[x]$. Another of this type is

$$G(x) = ((x^2 - 377)^2 - 73504)^2 - 50400^2$$

$$= (x-27)(x-23)(x-15)(x-5)(x+5)(x+15)(x+23)(x+27),$$

and there certainly exist others. It appears on the face of it that for a number $N = pq$ to be factored (with primes $p \approx q$, say) one could simply take $\gcd(F(x) \bmod N, N)$ for random $x \pmod{N}$, so that N should be factored in about $\sqrt{N}/(2 \cdot 8)$ evaluations of F. (The extra 2 is because we can get by chance either p or q as a factor.) Since F is calculated via 3 squarings modulo N, and we expect 1 multiply to accumulate a new F product, we should have an operational gain of $8/4 = 2$ over naive product accumulation. The gain is even more when we acknowledge the relative simplicity of a modular squaring operation vs. a modular multiply. But what if we discovered an appropriate set $\{a_j\}$ of fixed integers, and defined

$$H(x) = (\cdots((((x^2 - a_1)^2 - a_2)^2 - a_3)^2 - a_4)^2 - \cdots)^2 - a_k^2,$$

so that a total of k squarings (we assume a_k^2 prestored) would generate 2^k algebraic factors? Can this successive-squaring idea lead directly to subexponential (if not polynomial-time) complexity for factoring? Or are there blockades preventing such a wonderful achievement? Another question is, noting that the above two examples (F, G) have disjoint roots, i.e., $F(x)G(x)$ has 16 distinct factors, can one somehow use two identities at a time to improve the gain? Yet another observation is, since all roots of $F(x)G(x)$ are odd, x can simply be incremented/decremented to $x \pm 1$, yielding a whole new flock of factors. Is there some way to exploit this phenomenon for more gain?

Incidentally, there are other identities that require, for a desired product of terms, fewer operations than one might expect. For example, we have another general identity which reads:

$$\frac{(n+8)!}{n!} = \left(204 + 270n + 111n^2 + 18n^3 + n^4\right)^2 - 16(9+2n)^2,$$

allowing for a product of 8 consecutive integers to be effected in 5 multiplies (not counting multiplications by constants). Thus, even if the pure-squaring ladder at the beginning of this exercise fails to allow generalization, there are perhaps other ways to proceed.

Theoretical work on such issues does exist; for example, [Dilcher 1999] discourses on the difficulty of creating longer squaring ladders of the indicated kind. Recently, D. Symes has discovered a $(k = 4)$ identity, with coefficients

$$(a_1, a_2, a_3, a_4) \\ = (67405, 3525798096, 533470702551552000, 469208209191321600),$$

so that 16 algebraic factors may be found with only 4 squarings.

6.17. Are there yet-unknown ways to extract square roots in number fields, as required for successful NFS? We have discussed in Section 6.2.5 some state-of-the-art approaches, and seen in Exercise 6.14 that some elementary means exist. Here we enumerate some further ideas and directions.

(1) The method of Hensel lifting mentioned in Section 6.2.5 is a kind of p-adic Newton method. But are there other Newton variants? Note as in Exercise 9.14 that one can extract, in principle, square roots without inversion, at least in the real-number field. Moreover, there is such a thing as Newton solution of *simultaneous* nonlinear equations. But a collection of such equations is what one gets if one simply writes down the relations for a polynomial squared to be another polynomial (there is a mod f complication but that can possibly be built into the Newton–Jacobian matrix for the solver).

(2) In number fields depending on polynomials of the simple form $f(x) = x^d+1$, one can actually extract square roots via "negacyclic deconvolution" (see Section 9.5.3 for the relevant techniques in what follows). Let the entity for which we know there exists a square root be written

$$\gamma^2 = \sum_{j=0}^{d-1} z_j \alpha^j$$

where α is a d-th root of (-1) (i.e., a root of f). Now, in signal processing terminology, we are saying that for some length-d signal γ to be determined,

$$z = \gamma \times_- \gamma,$$

where $\times_-$ denotes negacyclic convolution, and z is the signal consisting of the z_j coefficients. But we know how to do negacyclic convolution via fast transform methods. Writing

$$\Gamma_k = \sum_{j=0}^{d-1} \gamma_j \alpha^j \alpha^{-2kj},$$

one can establish the weighted-convolution identity

$$z_n = \alpha^{-n} \frac{1}{d} \sum_{k=0}^{d-1} \Gamma_k^2 \alpha^{+2nk}.$$

The deconvolution idea, then, is simple: Given the signal z to be square-rooted, transform this last equation above to obtain the Γ_k^2, then assign one of 2^{d-1} distinct choices of sign for the respective $\pm\sqrt{\Gamma_k^2}$, $k \in [1, d-1]$, then solve for γ_j via another transform. This negacyclic deconvolution procedure will result in a correct square root γ of γ^2. The research question is this: Since we know that number fields based on $f(x) = x^d + 1$ are easily handled in many other ways, can this deconvolution approach be generalized? How about $f(x) = x^d + c$, or even much more general f? It is also an interesting question whether the transforms above need to be floating-point ones (which does, in fact, do the job at the expense of the high precision), or whether errorless, pure-integer number-theoretical transforms can be introduced.

(3) For any of these various ideas, a paramount issue is how to avoid the rapid growth of coefficient sizes. Therefore one needs to be aware that a square-root procedure, even if it is numerically sound, has to somehow keep coefficients under control. One general suggestion is to combine whatever square-rooting algorithm with a CRT; that is, work somehow modulo many small primes simultaneously. In this way, machine parallelism may be possible as well. As we intimated in text, ideas of Couveignes and Montgomery have brought the square-root obstacle down to a reasonably efficient phase in the best prevailing NFS implementations. Still, it would be good to have a simple, clear, and highly efficient scheme that generalizes not just to cases of parity on the degree d, but also manages somehow to control coefficients and still avoid CRT reconstruction.

Chapter 7

ELLIPTIC CURVE ARITHMETIC

The history of what are called elliptic curves goes back well more than a century. Originally developed for classical analysis, elliptic curves have found their way into abstract and computational number theory, and now sit squarely as a primary tool. Like the prime numbers themselves, elliptic curves have the wonderful aspects of elegance, complexity, and power. Elliptic curves are not only celebrated algebraic constructs; they also provide considerable leverage in regard to prime number and factorization studies. Elliptic curve applications even go beyond these domains; for example, they have an increasingly popular role in modern cryptography, as we discuss in Section 8.1.3.

In what follows, our primary focus will be on elliptic curves over fields $\mathbf{F}_p$, with $p > 3$ an odd prime. One is aware of a now vast research field—indeed even an industry—involving fields $\mathbf{F}_{p^k}$ where $k > 1$ or (more prevalent in current applications) fields $\mathbf{F}_{2^k}$. Because the theme of the present volume is prime numbers, we have chosen to limit discussion to the former fields of primary interest. For more information in regard to the alternative fields, the interested reader may consult references such as [Seroussi et al. 1999] and various journal papers referenced therein.

7.1 Elliptic curve fundamentals

Consider the general equation of a degree-3 polynomial in two variables, with coefficients in a field F, set equal to 0:

$$ax^3 + bx^2y + cxy^2 + dy^3 + ex^2 + fxy + gy^2 + hx + iy + j = 0. \qquad (7.1)$$

To ensure that the polynomial is really of degree 3, we assume that at least one of a, b, c, d is nonzero. We also assume that the polynomial is absolutely irreducible; that is, it is irreducible in $\overline{F}[x, y]$, where $\overline{F}$ is the algebraic closure of F. One might consider the pairs $(x, y) \in F \times F$ that satisfy (7.1); they are called the affine solutions to the equation. Or one might consider the projective solutions. For these we begin with triples $(x, y, z) \in F \times F \times F$ (with x, y, z not all zero) that satisfy

$$ax^3 + bx^2y + cxy^2 + dy^3 + ex^2z + fxyz + gy^2z + hxz^2 + iyz^2 + jz^3 = 0. \quad (7.2)$$

Note that (x, y, z) is a solution if and only if (tx, ty, tz) is also a solution, for $t \in F$, $t \neq 0$. Thus, in the projective case, it makes more sense to talk of

$[x, y, z]$ being a solution, the notation indicating that we consider as identical any two solutions $(x, y, z), (x', y', z')$ of (7.2) if and only if there is a nonzero $t \in F$ with $x' = tx, y' = ty, z' = tz$.

The projective solutions of (7.2) are almost exactly the same as the affine solutions of (7.1). In particular, a solution (x, y) of (7.1) may be identified with the solution $[x, y, 1]$ of (7.2), and any solution $[x, y, z]$ of (7.2) with $z \neq 0$ may be identified with the solution $(x/z, y/z)$ of (7.1). The solutions $[x, y, z]$ with $z = 0$ do not correspond to any affine solutions, and are called the "points at infinity" for the equation.

Equations (7.1) and (7.2) are cumbersome. It is profitable to consider a change in variables that sends solutions with coordinates in F to like solutions, and vice versa for the inverse transformation. For example, consider the Fermat equation for exponent 3, namely,

$$x^3 + y^3 = z^3.$$

Assume we are considering solutions in a field F with characteristic not equal to 2 or 3. Letting $X = 12z$, $Y = 36(x - y)$, $Z = x + y$, we have the equivalent equation

$$Y^2 Z = X^3 - 432Z^3.$$

The inverse change of variables is $x = \frac{1}{72}Y + \frac{1}{2}Z$, $y = -\frac{1}{72}Y + \frac{1}{2}Z$, $z = \frac{1}{12}X$.

The projective curve (7.2) is considered to be "nonsingular" (or "smooth") over the field F if even over the algebraic closure of F there is no point $[x, y, z]$ on the curve where all three partial derivatives vanish. In fact, if the characteristic of F is not equal to 2 or 3, any nonsingular projective equation (7.2) with at least one solution in $F \times F \times F$ (with not all of the coordinates zero) may be transformed by a change of variables to the standard form

$$y^2 z = x^3 + axz^2 + bz^3, \quad a, b \in F, \tag{7.3}$$

where the one given solution of the original equation is sent to $[0, 1, 0]$. Further, it is clear that a curve given by (7.3) has just this one point at infinity, $[0, 1, 0]$. The affine form is

$$y^2 = x^3 + ax + b. \tag{7.4}$$

Such a form for a cubic curve is called a Weierstrass form. It is sometimes convenient to replace x with $(x + \text{constant})$, and so get another Weierstrass form:

$$y^2 = x^3 + Cx^2 + Ax + B, \quad A, B, C \in F. \tag{7.5}$$

If we have a curve in the form (7.4) and the characteristic of F is not 2 or 3, then the curve is nonsingular if and only if $4a^3 + 27b^2$ is not 0; see Exercise 7.3. If the curve is in the form (7.5), the condition that the curve be nonsingular is more complicated: It is that $4A^3 + 27B^2 - 18ABC - A^2C^2 + 4BC^3 \neq 0$.

Whether we are dealing with the affine form (7.4) or (7.5), we use the notation O to denote the one point at infinity $[0, 1, 0]$ that occurs for the projective form of the curve.

We now make the fundamental definition for this chapter.

Definition 7.1.1. A nonsingular cubic curve (7.2) with coefficients in a field F and with at least one point with coordinates in F (that are not all zero) is said to be an elliptic curve over F. If the characteristic of F is not 2 or 3, then the equations (7.4) and (7.5) also define elliptic curves over F, provided that $4a^3 + 27b^2 \neq 0$ in the case of equation (7.4) and $4A^3 + 27B^2 - 18ABC - A^2C^2 + 4BC^3 \neq 0$ in the case of equation (7.5). In these two cases, we denote by $E(F)$ the set of points with coordinates in F that satisfy the equation together with the point at infinity, denoted by O. So, in the case of (7.4),

$$E(F) = \left\{(x, y) \in F \times F : y^2 = x^3 + ax + b\right\} \cup \{O\},$$

and similarly for a curve defined by equation (7.5).

Note that we are concentrating on fields of characteristic not equal to 2 or 3. For fields such as $\mathbf{F}_{2^m}$ the modified equation (7.11) of Exercise 7.1 must be used (see, for example, [Koblitz 1994] for a clear exposition of this).

We use the form (7.5) because it is sometimes computationally useful in, for example, cryptography and factoring studies. Since the form (7.4) corresponds to the special case of (7.5) with $C = 0$, it should be sufficient to give any formulae for the form (7.5), allowing the reader to immediately convert to a formula for the form (7.4) in case the quadratic term in x is missing. However, it is important to note that equation (7.5) is overspecified because of an extra parameter. So in a word, the Weierstrass form (7.4) is completely general for curves over the fields in question, but sometimes our parameterization (7.5) is computationally convenient.

The following parameter classes will be of special practical importance:

(1) $C = 0$, giving immediately the Weierstrass form $y^2 = x^3 + Ax + B$. This parameterization is the standard form for much theoretical work on elliptic curves.

(2) $A = 1$, $B = 0$, so curves are based on $y^2 = x^3 + Cx^2 + x$. This parameterization has particular value in factorization implementations [Montgomery 1987], [Brent et al. 2000], and admits of arithmetic enhancements in practice.

(3) $C = 0$, $A = 0$, so the cubic is $y^2 = x^3 + B$. This form has value in finding particular curves of specified order (the number elements of the set E, as we shall see), and also allows practical arithmetic enhancements.

(4) $C = 0$, $B = 0$, so the cubic is $y^2 = x^3 + Ax$, with advantages as in (3).

The tremendous power of elliptic curves becomes available when we define a certain group operation, under which $E(F)$ becomes, in fact, an abelian group:

Definition 7.1.2. Let $E(F)$ be an elliptic curve defined by (7.5) over a field F of characteristic not equal to 2 or 3. Denoting two arbitrary curve points by $P_1 = (x_1, y_1), P_2 = (x_2, y_2)$ (not necessarily distinct), and denoting by O the point at infinity, define a commutative operation $+$ with inverse operation $-$ as follows:

(1) $-O = O$;

(2) $-P_1 = (x_1, -y_1)$;

(3) $O + P_1 = P_1$;

(4) if $P_2 = -P_1$, then $P_1 + P_2 = O$;

(5) if $P_2 \neq -P_1$, then $P_1 + P_2 = (x_3, y_3)$, with

$$x_3 = m^2 - C - x_1 - x_2,$$
$$-y_3 = m(x_3 - x_1) + y_1,$$

where the *slope* m is defined by

$$m = \begin{cases} \dfrac{y_2 - y_1}{x_2 - x_1}, & \text{if } x_2 \neq x_1 \\ \dfrac{3x_1^2 + 2Cx_1 + A}{2y_1}, & \text{if } x_2 = x_1. \end{cases}$$

The addition/subtraction operations thus defined have an interesting geometrical interpretation in the case that the underlying field F is the real number field. Namely, 3 points on the curve are collinear if and only if they sum to 0. This interpretation is generalized to allow for a double intersection at a point of tangency (unless it is an inflection point, in which case it is a triple intersection). Finally, the geometrical interpretation takes the view that vertical lines intersect the curve at the point at infinity. When the field is finite, say $F = \mathbf{F}_p$, the geometrical interpretation is not evident, as we realize $\mathbf{F}_p$ as the integers modulo p; in particular, the division operations for the slope m are inverses (mod p).

It is a beautiful outcome of the theory that the curve operations in Definition 7.1.2 define a group; furthermore, this group has special properties, depending on the underlying field. We collect such results in the following theorem:

Theorem 7.1.3 (Cassels). *An elliptic curve $E(F)$ together with the operations of Definition 7.1.2 is an abelian group. In the finite-field case the group $E(\mathbf{F}_{p^k})$ is either cyclic or isomorphic to a product of two cyclic groups:*

$$E \cong \mathbf{Z}_{d_1} \times \mathbf{Z}_{d_2},$$

with $d_1|d_2$ and $d_1|p^k - 1$.

That E is an abelian group is not hard to show, except that establishing associativity is somewhat tedious (see Exercise 7.7). The structure result for $E\left(\mathbf{F}_{p^k}\right)$ may be found in [Cassels 1966], [Silverman 1986], [Cohen 2000].

If the field F is finite, $E(F)$ is always a finite group, and the group order, $\#E(F)$, which is the number of points (x, y) on the affine curve plus 1 for the point at infinity, is a number that gives rise to fascinating and profound issues. Indeed, the question of order will arise in such domains as primality proving, factorization, and cryptography.

We define elliptic multiplication by integers in a natural manner: For point $P \in E$ and positive integer n, we denote the n-th multiple of the point by

$$[n]P = P + P + \cdots + P,$$

where exactly n copies of P appear on the right. We define $[0]P$ as the group identity O, the point at infinity. Further, we define $[-n]P$ to be $-[n]P$. From elementary group theory we know that when F is finite,

$$[\#E(F)]P = O,$$

a fact of paramount importance in practical applications of elliptic curves. This issue of curve order is addressed in more detail in Section 7.5. As regards any group, we may consider the order of an element. In an elliptic-curve group, the order of a point P is the least positive integer n with $[n]P = 0$, while if no such integer n exists, we say that P has infinite order. If $E(F)$ is finite, then every point in $E(F)$ has finite order dividing $\#E(F)$.

The fundamental relevance of elliptic curves for factorization will be the fact that, if one has a composite n to be factored, one can try to work on an elliptic curve over $\mathbf{Z}_n$, even though $\mathbf{Z}_n$ is not a field and treating it as such might be considered "illegal." When an illegal curve operation is encountered, it is exploited to find a factor of n. This idea of what we might call "pseudocurves" is the starting point of H. Lenstra's elliptic curve method (ECM) for factorization, whose details are discussed in Section 7.4. Before we get to this wonderful algorithm we first discuss "legal" elliptic curve arithmetic over a field.

7.2 Elliptic arithmetic

Armed with some elliptic curve fundamentals, we now proceed to develop practical algorithms for elliptic arithmetic. For simplicity we shall adopt a finite field $\mathbf{F}_p$ for prime $p > 3$, although generally speaking the algorithm structures remain the same for other fields. We begin with a simple method for finding explicit points (x, y) on a given curve, the idea being that we require the relevant cubic form in x to be a square modulo p:

Algorithm 7.2.1 (Finding a point on a given elliptic curve). For a prime $p > 3$ we assume an elliptic curve $E(\mathbf{F}_p)$ determined by cubic $y^2 = x^3 + ax + b$. This algorithm returns a point (x, y) on E.

1. [Loop]
   Choose random $x \in [0, p-1]$;
   $t = (x(x^2 + a) + b) \bmod p$; // Affine cubic form in x.
   if($\left(\frac{t}{p}\right) == -1$) goto [Loop]; // Via Algorithm 2.3.5.
   return $(x, \pm\sqrt{t} \bmod p)$; // Square root via Algorithm 2.3.8 or 2.3.9.

Either square root of the residue may be returned, since $(x,y) \in E(\mathbf{F}_p)$ implies $(x,-y) \in E(\mathbf{F}_p)$. Though the algorithm is probabilistic, the method can be expected to require just a few iterations of the do-loop. There is another important issue here: For certain problems where the y-coordinate is not needed, one can always check that some point $(x,?)$ exists—i.e., that x is a valid x-coordinate—simply by checking whether the Jacobi symbol $\left(\frac{t}{p}\right)$ is not -1.

These means of finding a point on a given curve are useful in primality proving and cryptography. But there is an interesting modified question: How can one find both a random curve *and* a point on said curve? This question is important in factorization. We defer this algorithm to Section 7.4, where "pseudocurves" with arithmetic modulo composite n are indicated.

But given a point P, or some collection of points, on a curve E, how do we add them pairwise, and most importantly, how do we calculate elliptic multiples $[n]P$? For these operations, there are several ways to proceed:

Option (1): Affine coordinates. Use the fundamental group operations of Definition 7.1.2 in a straightforward manner, this approach generally involving an inversion for a curve operation.

Option (2): Projective coordinates. Use the group operations, but for projective coordinates $[X,Y,Z]$ to avoid inversions. When $Z \neq 0$, $[X,Y,Z]$ corresponds to the affine point $(X/Z, Y/Z)$ on the curve. The point $[0,1,0]$ is O, the point at infinity.

Option (3): Modified projective coordinates. Use triples $\langle X,Y,Z\rangle$, where if $Z \neq 0$, this corresponds to the affine point $(X/Z^2, Y/Z^3)$ on the curve, plus the point $\langle 0,1,0\rangle$ corresponding to O, the point at infinity. This system also avoids inversions, and has a lower operation count than projective coordinates.

Option (4): X, Z coordinates, sometimes called Montgomery coordinates. Use coordinates $[X : Z]$, which are the same as the projective coordinates $[X,Y,Z]$, but with "Y" dropped. One can recover the x coordinate of the affine point when $Z \neq 0$ as $x = X/Z$. There are generally two possibilities for y, and this is left ambiguous. This option tends to work well in elliptic multiplication and when y-coordinates are not needed at any stage, as sometimes happens in certain factorization and cryptography work, or when the elliptic algebra must be carried out in higher domains where coordinates themselves can be polynomials.

Which of these algorithmic approaches is best depends on various side issues. For example, assuming an underlying field $\mathbf{F}_p$, if one has a fast inverse (mod p), one might elect option (1) above. On the other hand, if one has already implemented option (1) and wishes to reduce the expensive time for a (slow) inverse, one might move to (2) or (3) with, as we shall see, minor changes in the algorithm flow. If one wishes to build an implementation from scratch, option (4) may be indicated, especially in factorization of very large numbers with ECM, in which case inversion (mod n) for the composite n can be avoided altogether.

As for explicit elliptic-curve arithmetic, we shall start for completeness with option (1), though the operations for this option are easy to infer directly from Definition 7.1.2. An important note: The operations are given here and in subsequent algorithms for underlying field F, although further work with "pseudocurves" as in factorization of composite n involves using the ring $\mathbf{Z}_n$ with operations mod n instead of mod p, while extension to fields $\mathbf{F}_{p^k}$ involves straightforward polynomial or equivalent arithmetic, and so on.

Algorithm 7.2.2 (Elliptic addition: Affine coordinates). We assume an elliptic curve $E(F)$ (see note preceding this algorithm), given by the affine equation $Y^2 = X^3 + aX + b$, where $a, b \in F$ and the characteristic of the field F is not equal to 2 or 3. We represent points P as triples (x, y, z), where for an affine point, $z = 1$ and (x, y) lies on the affine curve, and for O, the point at infinity, $z = 0$ (the triples $(0, 1, 0), (0, -1, 0)$, both standing for the same point). This algorithm provides functions for point negation, doubling, addition, and subtraction.

1. [Elliptic negate function]
   $neg(P)$ return $(x, -y, z)$;
2. [Elliptic double function]
   $double(P)$ return $add(P, P)$;
3. [Elliptic add function]
```
add(P1, P2){
    if(z1 == 0) return P2;                    // Point P1 = O.
    if(z2 == 0) return P1;                    // Point P2 = O.
    if(x1 == x2) {
        if(y1 + y2 == 0) return (0, 1, 0);    // i.e., return O.
        m = (3x1^2 + a)(2y1)^-1;              // Inversion in the field F.
    } else {
        m = (y2 - y1)(x2 - x1)^-1;            // Inversion in the field F.
    }
    x3 = m^2 - x1 - x2;
    return (x3, m(x1 - x3) - y1, 1);
}
```
4. [Elliptic subtract function]
   $sub(P_1, P_2)$ return $add(P_1, neg(P_2))$;

In the case of option (2) using ordinary projective coordinates, consider the curve $Y^2Z = X^3 + aXZ^2 + bZ^3$ and points $P_i = [X_i, Y_i, Z_i]$ for $i = 1, 2$. Rule (5) of Definition 7.1.2, for $P_1 + P_2$ when $P_1 \neq \pm P_2$ and neither P_1, P_2 is O, becomes

$$P_3 = P_1 + P_2 = [X_3, Y_3, Z_3],$$

where

$$X_3 = \alpha\left(\gamma^2\zeta - \alpha^2\beta\right),$$

$$Y_3 = \frac{1}{2}\left(\gamma\left(3\alpha^2\beta - \gamma^2\zeta\right) - \alpha^3\delta\right),$$
$$Z_3 = \alpha^3\zeta,$$

and

$$\alpha = X_2Z_1 - X_1Z_2, \quad \beta = X_2Z_1 + X_1Z_2,$$
$$\gamma = Y_2Z_1 - Y_1Z_2, \quad \delta = Y_2Z_1 + Y_1Z_2, \quad \zeta = Z_1Z_2.$$

By holding on to the intermediate calculations of $\alpha^2, \alpha^3, \alpha^2\beta, \gamma^2\zeta$, the coordinates of $P_1 + P_2$ may be computed in 14 field multiplications and 8 field additions (multiplication by 1/2 can generally be accomplished by a shift or an add and a shift). In the case of doubling a point by rule (5), if $[2]P \neq O$, the projective equations for

$$[2]P = [2][X, Y, Z] = [X', Y', Z']$$

are

$$X' = \nu(\mu^2 - 2\lambda\nu),$$
$$Y' = \mu\left(3\lambda\nu - \mu^2\right) - 2Y_1^2\nu^2,$$
$$Z' = \nu^3,$$

where

$$\lambda = 2XY, \quad \mu = 3X^2 + aZ^2, \quad \nu = 2YZ.$$

So doubling can be accomplished in 13 field multiplications and 4 field additions. In both adding and doubling, no field inversions of variables are necessary.

When using projective coordinates and starting from a given affine point (u, v), one easily creates projective coordinates by tacking on a 1 at the end, namely, creating the projective point $[u, v, 1]$. If one wishes to recover an affine point from $[X, Y, Z]$ at the end of a long calculation, and if this is not the point at infinity, one computes Z^{-1} in the field, and has the affine point (XZ^{-1}, YZ^{-1}).

We shall see that option (3) also avoids field inversions. In comparison with option (2), the addition for option (3) is more expensive, but the doubling for option (3) is cheaper. Since in a typical elliptic multiplication $[n]P$ we would expect about twice as many doublings as additions, one can see that option (3) could well be preferable to option (2). Recalling the notation, we understand $\langle X, Y, Z\rangle$ to be the affine point $(X/Z^2, Y/Z^3)$ on $y^2 = x^3 + ax + b$ if $Z \neq 0$, and we understand $\langle 0, 1, 0\rangle$ to be the point at infinity. Again, if we start with an affine point (u, v) on the curve and wish to convert to modified projective coordinates, we just taçk on a 1 at the end, creating the point $\langle u, v, 1\rangle$. And if one has a modified projective point $\langle X, Y, Z\rangle$ that is not the point at infinity, and one wishes to find the affine point corresponding to it, one computes Z^{-1}, Z^{-2}, Z^{-3} and the affine point (XZ^{-2}, YZ^{-3}). The following algorithm performs the algebra for modified projective coordinates, option (3).

Algorithm 7.2.3 (Elliptic addition: Modified projective coordinates). We assume an elliptic curve $E(F)$ over a field F with characteristic $\neq 2, 3$ (but see the note preceding Algorithm 7.2.2), given by the affine equation $y^2 = x^3+ax+b$. For modified projective points of the general form $P = \langle X, Y, Z\rangle$, with $\langle 0, 1, 0\rangle, \langle 0, -1, 0\rangle$ both denoting the point at infinity $P = O$, this algorithm provides functions for point negation, doubling, addition, and subtraction.

1. [Elliptic negate function]
   $neg(P)$ return $\langle X, -Y, Z\rangle$;
2. [Elliptic double function]
   $double(P)$ {
   if($Y == 0$ or $Z == 0$) return $\langle 0, 1, 0\rangle$;
   $M = (3X^2 + aZ^4)$; $S = 4XY^2$;
   $X' = M^2 - 2S$; $Y' = M(S - X_2) - 8Y^4$; $Z' = 2YZ$;
   return $\langle X', Y', Z'\rangle$;
   }
3. [Elliptic add function]
   $add(P_1, P_2)$ {
   if($Z_1 == 0$) return P_2; // Point $P_1 = O$.
   if($Z_2 == 0$) return P_1; // Point $P_2 = O$.
   $U_1 = X_2 Z_1^2$; $U_2 = X_1 Z_2^2$;
   $S_1 = Y_2 Z_1^3$; $S_2 = Y_1 Z_2^3$;
   $W = U_1 - U_2$; $R = S_1 - S_2$;
   if($W == 0$) { // x-coordinates match.
   if($R == 0$) return $double(P_1)$;
   return $\langle 0, 1, 0\rangle$;
   }
   $T = U_1 + U_2$; $M = S_1 + S_2$;
   $X_3 = R^2 - TW^2$;
   $Y_3 = \frac{1}{2}((TW^2 - 2X_3)R - MW^3)$;
   $Z_3 = Z_1 Z_2 W$;
   return $\langle X_3, Y_3, Z_3\rangle$;
   }
4. [Elliptic subtract function]
   $sub(P_1, P_2)$ {
   return $add(P_1, neg(P_2))$;
   }

It should be stressed that in all of our elliptic addition algorithms, if arithmetic is in $\mathbf{Z}_n$, modular reductions are taken whenever intermediate numbers exceed the modulus. This option (3) algorithm (modified projective coordinates) obviously has more field multiplications than does option (1) (affine coordinates), but as we have said, the idea is to avoid inversions (see Exercise 7.9). It is to be understood that in implementing Algorithm 7.2.3 one should save some of the intermediate calculations for further use; not all of these are explicitly described in our algorithm display above. In particular, for the elliptic add function, the value W^2 used for X_3 is recalled in the calculation of W^3 needed for Y_3, as is the value of TW^2. If such care is taken, the function *double*() consumes 10 field multiplications. (However, for small a or the special case $a = -3$ in the field, this count of 10 can be reduced further; see Exercise 7.10.) The general addition function *add*(), on the other hand, requires 16 field multiplications, but there is an important modification of this estimate: When $Z_1 = 1$ only 11 multiplies are required. And this side condition is very common; in fact, it is forced to hold within certain classes of multiplication ladders. (In the case of ordinary projective coordinates discussed before Algorithm 7.2.3 assuming $Z_1 = 1$ reduces the 14 multiplies necessary for general addition also to 11.)

Having discussed options (1), (2), (3) for elliptic arithmetic, we are now at an appropriate juncture to discuss elliptic multiplication, the problem of evaluating $[n]P$ for integer n acting on points $P \in E$. One can, of course, use Algorithm 2.1.5 for this purpose. However, since doubling is so much cheaper than adding two unequal points, and since subtracting has the same cost as adding, the method of choice is a modified binary ladder, the so-called addition–subtraction ladder. For most numbers n the ratio of doublings to addition–subtraction operations is higher than for standard binary ladders as in Algorithm 2.1.5, and the overall number of calls to elliptic arithmetic is lower. Such a method is good whenever the group inverse (i.e., negation) is easy—for elliptic curves one just flips the sign of the y-coordinate. (Note that a yet different ladder approach to elliptic multiplication will be exhibited later, as Algorithm 7.2.7.)

Algorithm 7.2.4 (Elliptic multiplication: Addition–subtraction ladder). This algorithm assumes functions $double(), add(), sub()$ from either Algorithm 7.2.2 or 7.2.3, and performs the elliptic multiplication $[n]P$ for nonnegative integer n and point $P \in E$. We assume a B-bit binary representation of $m = 3n$ as a sequence of bits $(m_{B-1}, \ldots, m_0)$, and a corresponding B-bit representation (n_j) for n (which representation is zero-padded on the left to B bits), with $B = 0$ for $n = 0$ understood.

1. [Initialize]
```
    if(n == 0) return O;                    // Point at infinity.
    Q = P;
```
2. [Compare bits of $3n, n$]
```
    for(B − 2 ≥ j ≥ 1) {
        Q = double(Q);
        if((m_j, n_j) == (1,0)) Q = add(Q, P);
        if((m_j, n_j) == (0,1)) Q = sub(Q, P);
    }
    return Q;
```

The proof that this algorithm works is encountered later as Exercise 9.31. There is a fascinating open research area concerning the best way to construct a ladder. See Exercise 9.77 in this regard.

Before we discuss option (4) for elliptic arithmetic, we bring in an extraordinarily useful idea, one that has repercussions far beyond option (4).

Definition 7.2.5. If $E(F)$ is an elliptic curve over a field F, governed by the equation $y^2 = x^3 + Cx^2 + Ax + B$, and g is a nonzero element of F, then the quadratic twist of E by g is the elliptic curve over F governed by the equation $gy^2 = x^3 + Cx^2 + Ax + B$. By a change of variables $X = gx, Y = g^2y$, the Weierstrass form for this twist curve is $Y^2 = X^3 + gCX^2 + g^2AX + g^3B$.

We shall find that in some contexts it will be useful to leave the curve in the form $gy^2 = x^3 + Cx^2 + Ax + B$, and in other contexts, we shall wish to use the equivalent Weierstrass form.

An immediate observation is that if g, h are nonzero elements of the field F, then the quadratic twist of an elliptic curve by g gives a group isomorphic to the quadratic twist of the curve by gh^2. (Indeed, just let a new variable Y be hy. To see that the groups are isomorphic, a simple check of the formulae involved suffices.) Thus, if $\mathbf{F}_q$ is a finite field, there is really only one quadratic twist of an elliptic curve $E(\mathbf{F}_q)$ that is different from the curve itself. This follows, since if g is not a square in $\mathbf{F}_q$, then as h runs over the nonzero elements of $\mathbf{F}_q$, gh^2 runs over all of the nonsquares. This unique nontrivial quadratic twist of $E(\mathbf{F}_q)$ is sometimes denoted by $E'(\mathbf{F}_q)$, especially when we are not particularly interested in which nonsquare is involved in the twist.

Now for option (4), homogeneous coordinates with "Y" dropped. We shall discuss this for a twist curve $gy^2 = x^3 + Cx^2 + Ax + B$; see Definition 7.2.5. We first develop the idea using affine coordinates. Suppose P_1, P_2 are affine points on an elliptic curve $E(F)$ with $P_1 \neq \pm P_2$. One can write down via Definition 7.1.2 (generalized for the presence of "g") expressions for x_+, x_-, namely, the x-coordinates of $P_1 + P_2$ and $P_1 - P_2$, respectively. If these expressions are multiplied, one sees that the y-coordinates of P_1, P_2 appear only to even powers, and so may be replaced by x-expressions, using the defining curve $gy^2 = x^3 + Cx^2 + Ax + B$. Somewhat miraculously the resulting expression is subject to much cancellation, including the disappearance of the parameter g. The equations are stated in the following result of [Montgomery 1987, 1992a], though we generalize them here to a quadratic twist of any curve that is given by equation (7.5).

Theorem 7.2.6 (Generalized Montgomery identities). *Given an elliptic curve E determined by the cubic*

$$gy^2 = x^3 + Cx^2 + Ax + B,$$

and two points $P_1 = (x_1, y_1)$, $P_2 = (x_2, y_2)$, neither being O, denote by $x_\pm$ respectively the x-coordinates of $P_1 \pm P_2$. Then if $x_1 \neq x_2$, we have

$$x_+x_- = \frac{(x_1x_2 - A)^2 - 4B(x_1 + x_2 + C)}{(x_1 - x_2)^2},$$

whereas if $x_1 = x_2$ *and* $2P_1 \neq O$, *we have*

$$x_+ = \frac{(x_1^2 - A)^2 - 4B(2x_1 + C)}{4(x_1^3 + Cx_1^2 + Ax_1 + B)}.$$

Note that g is irrelevant in the theorem, in the sense that the algebra for combining x-coordinates is independent of g; in fact, one would only use g if a particular starting y-coordinate were involved, but of course the main thrust of Montgomery parameterization is to ignore y-coordinates. We remind ourselves that the case $C = 0$ reduces to the ordinary Weierstrass form given by (7.4). However, as Montgomery noted, the case $B = 0$ is especially pleasant: For example, we have the simple relation

$$x_+ x_- = \frac{(x_1 x_2 - A)^2}{(x_1 - x_2)^2}.$$

We shall see in what follows how this sort of relation leads to computationally efficient elliptic algebra.

The idea is to use an addition chain to arrive at $[n]P$, where whenever we are to add two unequal points P_1, P_2, we happen to know already what $P_1 - P_2$ is. This magic is accomplished via the Lucas chain already discussed in Section 3.5.3. In the current notation, we will have at intermediate steps a pair $[k]P, [k+1]P$, and from this we shall form either the pair $[2k]P, [2k+1]P$ or the pair $[2k+1]P, [2k+2]P$, depending on the bits of n. In either case, we perform one doubling and one addition. And for the addition, we already know the difference of the two points added, namely P itself.

To avoid inversions, we adopt the homogeneous coordinates of option (2), but we drop the "Y" coordinate. Since the coordinates are homogeneous, when we have the pair $[X : Z]$, it is only the ratio X/Z that is determined (when $Z \neq 0$). The point at infinity is recognized as the pair $[0 : 0]$. Suppose we have points P_1, P_2 in homogeneous coordinates on an elliptic curve given by equation (7.5), and P_1, P_2 are not O, $P_1 \neq P_2$. If

$$P_1 = [X_1, Y_1, Z_1], \ P_2 = [X_2, Y_2, Z_2],$$

$$P_1 + P_2 = [X_+, Y_+, Z_+], \ P_1 - P_2 = [X_-, Y_-, Z_-],$$

then on the basis of Theorem 7.2.6 it is straightforward to establish, in the case that $X_- \neq 0$, that we may take

$$X_+ = Z_- \left((X_1 X_2 - A Z_1 Z_2)^2 - 4B(X_1 Z_2 + X_2 Z_1 + C Z_1 Z_2) Z_1 Z_2 \right), \tag{7.6}$$
$$Z_+ = X_- (X_1 Z_2 - X_2 Z_1)^2.$$

These equations define the pair X_+, Z_+ as a function of the six quantities $X_1, Z_1, X_2, Z_2, X_-, Z_-$, with Y_1, Y_2 being completely irrelevant. We denote this function by

$$[X_+ : Z_+] = addh([X_1 : Z_1], [X_2 : Z_2], [X_- : Z_-]),$$

the "h" in the function name emphasizing the homogeneous nature of each $[X : Z]$ pair. The definition of *addh* can easily be extended to any case where $X_- Z_- \neq 0$. That is, it is possible to allow one of $[X_1 : Z_1]$, $[X_2 : Z_2]$ to be $[0 : 0]$. In particular, if $[X_1 : Z_1] = [0 : 0]$ and $[X_2 : Z_2]$ is not $[0 : 0]$, then we may define $addh([0 : 0], [X_2 : Z_2], [X_2 : Z_2])$ as $[X_2 : Z_2]$ (and so not use the above equations). We may proceed similarly if $[X_2 : Z_2] = [0 : 0]$ and $[X_1 : Z_1]$ is not $[0 : 0]$. In the case of $P_1 = P_2$, we have a doubling function

$$[X_+ : Z_+] = doubleh([X_1 : Z_1]),$$

where

$$\begin{aligned} X_+ &= \left(X_1^2 - AZ_1^2\right)^2 - 4B(2X_1 + CZ_1)Z_1^3, \\ Z_+ &= 4Z_1\left(X_1^3 + CX_1^2 Z_1 + AX_1 Z_1^2 + BZ_1^3\right). \end{aligned} \tag{7.7}$$

The function *doubleh* works in all cases, even $[X_1 : Z_1] = [0 : 0]$. Let us see, for example, how we might compute $[X : Z]$ for $[13]P$, with P a point on an elliptic curve. Say $[k]P = [X_k : Y_k]$. We have

$$[13]P = ([2]([2]P) + ([2]P + P)) + ([2]([2]P + P)),$$

which is computed as follows:

$$\begin{aligned} [X_2 : Z_2] &= doubleh([X_1 : Z_1]), \\ [X_3 : Z_3] &= addh([X_2 : Z_2], [X_1 : Z_1], [X_1 : Z_1]), \\ [X_4 : Z_4] &= doubleh([X_2 : Z_2]), \\ [X_6 : Z_6] &= doubleh([X_3 : Z_3]), \\ [X_7 : Z_7] &= addh([X_4 : Z_4], [X_3 : Z_3], [X_1 : Z_1]), \\ [X_{13} : Z_{13}] &= addh([X_7 : Z_7], [X_6 : Z_6], [X_1 : Z_1]). \end{aligned}$$

(For this to be accurate, we must assume that $X_1 \neq 0$.) In general, we may use the following algorithm, which essentially contains within it Algorithm 3.5.7 for computing a Lucas chain.

Algorithm 7.2.7 (Elliptic multiplication: Montgomery method). This algorithm assumes functions $addh()$ and $doubleh()$ as described above and attempts to perform the elliptic multiplication of nonnegative integer n by point $P = [X : \text{any} : Z]$, in $E(F)$, with $XZ \neq 0$, returning the $[X : Z]$ coordinates of $[n]P$. We assume a B-bit binary representation of $n > 0$ as a sequence of bits $(n_{B-1}, \ldots, n_0)$.

1. [Initialize]

```
    if(n == 0) return O;                    // Point at infinity.
    if(n == 1) return [X : Z];              // Return the original point P.
    if(n == 2) return doubleh([X : Z]);
```

```
2. [Begin Montgomery adding/doubling ladder]
      [U : V] = [X : Z];                                   // Copy coordinate.
      [T : W] = doubleh([X : Z]);
3. [Loop over bits of n, starting with next-to-highest]
      for(B − 2 ≥ j ≥ 0) {
          if(n_j == 1) {
              [U : V] = addh([T : W], [U : V], [X : Z]);
              [T : W] = doubleh([T : W]);
          } else {
              [T : W] = addh([U : V], [T : W], [X : Z]);
              [U : V] = doubleh([U : V]);
          }
      }
4. [Final calculation]
      if(n_0 == 1) return addh([U : V], [T : W], [X : Y]);
      return doubleh([U : V]);
```

Montgomery's rules when $B = 0$ make for an efficient algorithm, as can be seen from the simplification of the $addh()$ and $doubleh()$ function forms. In particular, the $addh()$ and $doubleh()$ functions can each be done in 9 multiplications. In the case $B = 0, A = 1$, an additional multiplication may be dropped from each. Other parameterizations do fairly well also; and even though the computations are nonoptimal for the general case of (7.4), they do work, so that a complete (except, without y coordinates) elliptic multiplication routine for all parameterizations discussed thus far can be effected with Algorithm 7.2.7.

We have noted that to get the affine x-coordinate of $[n]P$, one must compute XZ^{-1} in the field. When n is very large, the single inversion is, of course, not expensive in comparison. But such inversion can sometimes be avoided entirely. For example, if, as in factoring studies covered later, we wish to know whether $[n]P = [m]P$ in the elliptic-curve group, it is enough to check whether the cross product $X_n Z_m - X_m Z_n$ vanishes, and this is yet another inversion-free task. Similarly, there is a very convenient fact: If the point at infinity has been attained by some multiple $[n]P = O$, then the Z denominator will have vanished, and any further multiples $[mn]P$ will also have vanishing Z denominator. Because of this, one need not find the precise multiple when O is attained; the fact of $Z = 0$ propagates nicely through successive applications of the elliptic multiply functions.

We have observed that only x-coordinates of multiples $[n]P$ are processed in Algorithm 7.2.7, and that ignorance of y values is acceptable in certain implementations. It is not easy to add two arbitrary points with the homogeneous coordinate approach above, because of the suppression of y coordinates. But all is not lost: There is a useful result that tells very quickly whether the sum of two points can possibly be a given third point. That is, given *merely* the x-coordinates of two points P_1, P_2 the following algorithm can be used to determine the two x-coordinates for the pair $P_1 \pm P_2$, although which of the coordinates goes with the $+$ and which with $-$ will be unknown.

Algorithm 7.2.8 (Sum/difference without y-coordinates (Crandall)). For an elliptic curve E determined by the cubic

$$y^2 = x^3 + Cx^2 + Ax + B,$$

we are given the unequal x-coordinates x_1, x_2 of two respective points P_1, P_2. This algorithm returns a quadratic polynomial whose roots are (in unspecified order) the x-coordinates of $P_1 \pm P_2$.

1. [Form coefficients]
   $G = x_1 - x_2$;
   $\alpha = (x_1x_2 + A)(x_1 + x_2) + 2(Cx_1x_2 + B)$;
   $\beta = (x_1x_2 - A)^2 - 4B(x_1 + x_2 + C)$;
2. [Return quadratic polynomial]
   return $G^2X^2 - 2\alpha X + \beta$;
   // This polynomial vanishes for x_+, x_-, the x-coordinates of $P_1 \pm P_2$.

It turns out that the discriminant $4(\alpha^2 - \beta G^2)$ must always be square in the field, so that if one requires the explicit pair of x-coordinates for $P_1 \pm P_2$, one may calculate

$$\left(\alpha \pm \sqrt{\alpha^2 - \beta G^2}\right) G^{-2}$$

in the field, to obtain x_+, x_-, although again, which sign of the radical goes with which coordinate is unspecified (see Exercise 7.11). The algorithm thus offers a test of whether $P_3 = P_1 \pm P_2$ for a set of three given points with missing y-coordinates; this test has value in certain cryptographic applications, such as digital signature [Crandall 1996b]. Note that the missing case of the algorithm, $x_1 = x_2$ is immediate: One of $P_1 \pm P_2$ is O, the other has x-coordinate as in the last part of Theorem 7.2.6.

For more on elliptic arithmetic, including some interesting new ideas, see [Cohen et al. 1998]. The issue of efficient exponentiation ladders for elliptic arithmetic is discussed later, in Section 9.3.

7.3 The theorems of Hasse, Deuring, and Lenstra

A fascinating and difficult problem is that of finding the order of an elliptic curve group defined over a finite field, i.e., the number of points including O on an elliptic curve $E_{a,b}(F)$ for a finite field F. For field $\mathbf{F}_p$, with prime $p > 3$, we can immediately write out an exact expression for the order $\#E$ by observing, as we did in the simple Algorithm 7.2.1, that for (x, y) to be a point, the cubic form in x must be a square in the field. Using the Legendre symbol we can write

$$\#E\left(\mathbf{F}_p\right) = p + 1 + \sum_{x \in \mathbf{F}_p} \left(\frac{x^3 + ax + b}{p}\right) \tag{7.8}$$

as the required number of points $(x, y) \pmod p$ that solve the cubic $\pmod p$, with of course 1 added for the point at infinity. This equation may be generalized to fields $\mathbf{F}_{p^k}$ as follows:

$$\#E\left(\mathbf{F}_{p^k}\right) = p^k + 1 + \sum_{x \in \mathbf{F}_{p^k}} \chi(x^3 + ax + b),$$

where χ is the quadratic character for $\mathbf{F}_{p^k}$. (That is, $\chi(u) = 1, -1, 0$, respectively, depending on whether u is a nonzero square in the field, not a square, or 0.) A celebrated result of H. Hasse is the following:

Theorem 7.3.1 (Hasse). *The order $\#E$ of $E_{a,b}(\mathbf{F}_{p^k})$ satisfies*

$$\left|(\#E) - (p^k + 1)\right| \le 2\sqrt{p^k}.$$

This remarkable result strikes to the very heart of elliptic curve theory and applications thereof. Looking at the Hasse inequality for $\mathbf{F}_p$, we see that

$$p + 1 - 2\sqrt{p} < \#E < p + 1 + 2\sqrt{p}.$$

There is an attractive heuristic connection between this inequality and the alternative relation (7.8). Namely, think of the Legendre symbol $\left(\frac{x^3+ax+b}{p}\right)$ as a "random walk," i.e., a walk driven by coin flips of value ± 1 except for possible symbols $\left(\frac{0}{p}\right) = 0$. It is known from statistical theory that the expected absolute distance from the origin after summation of n such random ± 1 flips is proportional to $\sqrt{n}$. Certainly, the Hasse theorem gives the "right" order of magnitude for the excursions away from p for the possible orders of $\#E_{a,b}(\mathbf{F}_p)$. At a deeper heuristic level one must have caution, however: As mentioned in Section 1.4.2, the ratio of such a random walk's position to $\sqrt{n}$ can be expected to diverge something like $\ln\ln n$. The Hasse theorem says this cannot happen—the stated ratio is bounded by 2. Indeed, there are certain subtle features of Legendre-symbol statistics that reveal departure from randomness (see Exercise 2.38).

Less well known is a theorem due to [Deuring 1941], saying that for any integer $m \in (p+1-2\sqrt{p}, p+1+2\sqrt{p})$, there exists some pair (a, b) in the set

$$\{(a, b) \ : \ a, b \in \mathbf{F}_p;\ 4a^3 + 27b^2 \ne 0\}$$

such that $\#E_{a,b}(\mathbf{F}_p) = m$. What the Deuring theorem actually says is that the number of curves—up to isomorphism—of order m is the so-called Kronecker class number of $(p+1-m)^2 - 4m$. In [Lenstra 1987], these results of Hasseand Deuring are exploited to say something about the statistics of curve orders over a given field $\mathbf{F}_p$, as we shall now see.

In applications to factoring, primality testing, and cryptography, we are concerned with choosing a random elliptic curve and then asking for the likelihood of the curve order possessing a particular arithmetic property, such as being smooth, being easily factorable, or being prime. However, there are two possible ways of choosing a random curve. One is to just choose a, b at random and be done with it. But sometimes we also would like to have a random point on the curve. If one is working with a true elliptic curve over a finite field, points on it can easily be found via Algorithm 7.2.1. But if one is working over $\mathbf{Z}_n$ with n composite, the call to the square root in this algorithm is not likely to be useful. However, it is possible to completely bypass Algorithm 7.2.1 and find a random curve and a point on it by choosing the point before the curve is fully defined! Namely, choose a at random, then choose a point (x_0, y_0) at random, then choose b such that (x_0, y_0) is on the curve $y^2 = x^3 + ax + b$; that is, $b = y_0^2 - x_0^3 - ax_0$.

With these two approaches to finding a random curve, we can formalize the question of the likelihood of the curve order having a particular property. Suppose p is a prime larger than 3, and let $\mathcal{S}$ be a set of integers in the Hasse interval $(p+1-2\sqrt{p}, p+1+2\sqrt{p})$. For example, $\mathcal{S}$ might be the set of B-smooth numbers in the interval for some appropriate value of B (see Section 1.4.5), or $\mathcal{S}$ might be the set of prime numbers in the interval, or the set of doubles of primes. Let $N_1(\mathcal{S})$ be the number of pairs $(a, b) \in \mathbf{F}_p^2$ with $4a^3 + 27b^2 \neq 0$ and with $\#E_{a,b}(\mathbf{F}_p) \in \mathcal{S}$. Let $N_2(\mathcal{S})$ be the number of triples $(a, x_0, y_0) \in \mathbf{F}_p^3$ such that for $b = y_0^2 - x_0^3 - ax_0$, we have $4a^3 + 27b^2 \neq 0$ and $\#E_{a,b}(\mathbf{F}_p) \in \mathcal{S}$. What would we expect for the counts $N_1(\mathcal{S}), N_2(\mathcal{S})$? For the first count, there are p^2 choices for a, b to begin with, and each number $\#E_{a,b}(\mathbf{F}_p)$ falls in an interval of length $4\sqrt{p}$, so we might expect $N_1(\mathcal{S})$ to be about $\frac{1}{4}(\#\mathcal{S})p^{3/2}$. Similarly, we might expect $N_2(\mathcal{S})$ to be about $\frac{1}{4}(\#\mathcal{S})p^{5/2}$. That is, in each case we expect the probability that the curve order lands in the set $\mathcal{S}$ to be about the same as the probability that a random integer chosen from $(p+1-2\sqrt{p}, p+1+2\sqrt{p})$ lands in $\mathcal{S}$. The following theorem says that this is almost the case.

Theorem 7.3.2 (Lenstra). *There is a positive number c such that if $p > 3$ is prime and $\mathcal{S}$ is a set of integers in the interval $(p+1-2\sqrt{p}, p+1+2\sqrt{p})$ with at least 3 members, then*

$$N_1(\mathcal{S}) > c(\#\mathcal{S})p^{3/2}/\ln p, \quad N_2(\mathcal{S}) > c(\#\mathcal{S})p^{5/2}/\ln p.$$

This theorem is proved in [Lenstra 1987], where also upper bounds, of the same approximate order as the lower bounds, are given.

7.4 Elliptic curve method

A subexponential factorization method of great elegance and practical importance is the elliptic curve method (ECM) of H. Lenstra. The elegance will be self-evident. The practical importance lies in the fact that unlike QS or NFS, ECM complexity to factor a number n depends strongly on the size of the least prime factor of n, and only weakly on n itself. For this reason, many factors of truly gigantic numbers have been uncovered in recent years; many of these numbers lying well beyond the range of QS or NFS.

Later in this section we exhibit some explicit modern ECM successes that exemplify the considerable power of this method.

7.4.1 Basic ECM algorithm

The ECM algorithm uses many of the concepts of elliptic arithmetic developed in the preceding sections. However, we shall be applying this arithmetic to a construct $E_{a,b}(\mathbf{Z}_n)$, something that is not a true elliptic curve, when n is a composite number.

Definition 7.4.1. For elements a, b in the ring $\mathbf{Z}_n$, with $\gcd(n, 6) = 1$ and discriminant condition $\gcd(4a^3 + 27b^2, n) = 1$, an elliptic pseudocurve over the ring is a set

$$E_{a,b}(\mathbf{Z}_n) = \{(x, y) \in \mathbf{Z}_n \times \mathbf{Z}_n : y^2 = x^3 + ax + b\} \cup \{O\},$$

where O is the point at infinity. (Thus an elliptic curve over $\mathbf{F}_p = \mathbf{Z}_p$ from Definition 7.1.1 is also an elliptic pseudocurve.)

(Curves given in the form (7.5) are also considered as pseudocurves, with the appropriate discriminant condition holding.) We have seen in Section 7.1 that when n is prime, the point at infinity refers to the one extra projective point on the curve that does not correspond to an affine point. When n is composite, there are additional projective points not corresponding to affine points, yet in our definition of pseudocurve, we still allow only the one extra point, corresponding to the projective solution $[0, 1, 0]$. Because of this (intentional) shortchanging in our definition, the pseudocurve $E_{a,b}(\mathbf{Z}_n)$, together with the operations of Definition 7.1.2, does not form a group (when n is composite). In particular, there are pairs of points P, Q for which "$P + Q$" is undefined. This would be detected in the construction of the slope m in Definition 7.1.2; since $\mathbf{Z}_n$ is not a field when n is composite, one would be called upon to invert a nonzero member of $\mathbf{Z}_n$ that is not invertible. This group-law failure is the motive for the name "pseudocurve," yet, happily, there are powerful applications of the pseudocurve concept. In particular, Algorithm 2.1.4 (the extended Euclid algorithm), if called upon to find the inverse of a nonzero member of $\mathbf{Z}_n$ that is in fact noninvertible, will instead produce a nontrivial factor of n. It is Lenstra's ingenious idea that through this failure of finding an inverse, we shall be able to factor the composite number n.

We note in passing that the concept of elliptic multiplication on a pseudocurve depends on the addition chain used. For example, $[5]P$ may be perfectly well computable if one computes it via $P \to [2]P \to [4]P \to [5]P$, but the elliptic addition may break down if one tries to compute it via $P \to [2]P \to [3]P \to [5]P$. Nevertheless, if two different addition chains to arrive at $[k]P$ both succeed, they will give the same answer.

Algorithm 7.4.2 (Lenstra elliptic curve method (ECM)). Given a composite number n to be factored, $\gcd(n, 6) = 1$, and n not a proper power, this algorithm attempts to uncover a nontrivial factor of n. There is a tunable parameter B_1 called the "stage-one limit" in view of further algorithmic stages in the modern ECM to follow.

1. [Choose B_1 limit]
   $B_1 = 10000$; // Or whatever is a practical initial "stage-one limit" B_1.
2. [Find curve $E_{a,b}(\mathbf{Z}_n)$ and point $(x, y) \in E$]
   Choose random $x, y, a \in [0, n-1]$;
   $b = (y^2 - x^3 - ax) \bmod n$;
   $g = \gcd(4a^3 + 27b^2, n)$;
   if($g == n$) goto [Find curve ...];
   if($g > 1$) return g; // Factor is found.
   $E = E_{a,b}(\mathbf{Z}_n)$; $P = (x, y)$; // Elliptic pseudocurve and point on it.
3. [Prime-power multipliers]
   for($1 \le i \le \pi(B_1)$) { // Loop over primes p_i.
      Find largest integer a_i such that $p_i^{a_i} \le B_1$;
      for($1 \le j \le a_i$) { // j is just a counter.
         $P = [p_i]P$, halting the elliptic algebra if the computation of some d^{-1} for addition-slope denominator d signals a nontrivial $g = \gcd(n, d)$, in which case return g;
         // Factor is found.
      }
   }
4. [Failure]
   Possibly increment B_1; // See text.
   goto [Find curve ...];

What we hope with basic ECM is that even though the composite n allows only a pseudocurve, an illegal elliptic operation—specifically the inversion required for slope calculation from Definition 7.1.2—is a signal that for some prime $p|n$ we have

$$[k]P = O, \text{ where } k = \prod_{p_i^{a_i} \le B_1} p_i^{a_i},$$

with this relation holding on the legitimate elliptic curve $E_{a,b}(\mathbf{F}_p)$. Furthermore, we know from the Hasse Theorem 7.3.1 that the order $\#E_{a,b}(\mathbf{F}_p)$ is in the interval $(p+1-2\sqrt{p}, p+1+2\sqrt{p})$. Evidently, we can expect a factor if the multiplier k is divisible by $\#E(\mathbf{F}_p)$, which should, in fact, happen if *this order is B_1-smooth.* (This is not entirely precise, since for the order to be B_1-smooth it is required only that each of its prime factors be at most B_1, but in the above display, we have instead the stronger condition that each prime power divisor of the order is at most B_1. We could change the inequality defining a_i to $p_i^{a_i} \le n + 1 + 2\sqrt{n}$, but in practice the cost of doing so is too high for the meager benefit it may provide.) We shall thus think of the stage-one limit B_1 as a smoothness bound on actual curve orders in the group determined by the hidden prime factor p.

It is instructive to compare ECM with the Pollard $p-1$ method (Algorithm 5.4.1). In the $p-1$ method one has only the one group $\mathbf{Z}_p^*$ (with order $p-1$), and one is successful if this group order is B-smooth. With ECM one has a host of elliptic-curve groups to choose from randomly, each giving a fresh chance at success.

With these ideas, we may perform a heuristic complexity estimate for ECM. Suppose the number n to be factored is composite, coprime to 6, and not a proper power. Let p denote the least prime factor of n and let q denote another prime factor of n. Algorithm 7.4.2 will be successful in splitting n if we choose a, b, P in step [Find curve ...] and if for some value of k of the form

$$k = p_l^a \prod_{i<l} p_i^{a_i},$$

where $l \le \pi(B_1)$ and $a \le a_l$, we have

$$[k]P = O \text{ on } E_{a,b}(\mathbf{F}_p), \quad [k]P \ne O \text{ on } E_{a,b}(\mathbf{F}_q).$$

The likelihood of these two events occurring is dominated by the first, and so we shall ignore the second. As mentioned above, the first event will occur if $\#E_{a,b}(\mathbf{F}_p)$ is B_1-smooth. From Theorem 7.3.2, the probability $prob(B_1)$ of success is greater than

$$c\frac{\psi(p+1+2\sqrt{p}, B_1) - \psi(p+1-2\sqrt{p}, B_1)}{\sqrt{p}\ln p}.$$

Here the notation $\psi(x,y)$ is as in (1.42). Since it takes about B_1 arithmetic steps to perform the trial for one curve in step [Prime-power multipliers], we would like to choose B_1 so as to minimize the expression $B_1/prob(B_1)$. Assuming that $prob(B_1)$ is about the same as

$$c\,\frac{\psi(\frac{3}{2}p, B_1) - \psi(\frac{1}{2}p, B_1)}{p\ln p},$$

so that we can use the estimates discussed in Section 1.4.5, we have that this minimum occurs when

$$B_1 = \exp\left((\sqrt{2}/2 + o(1))\sqrt{\ln p \ln\ln p}\right),$$

and for this value of B_1, the complexity estimate $B_1/prob(B_1)$ is given by

$$\exp\left((\sqrt{2} + o(1))\sqrt{\ln p \ln\ln p}\right);$$

see Exercise 7.12. Of course, we do not know p to begin with, and so it would only be a divination to choose an appropriate value of B_1 to begin with in step [Choose B_1 limit]. Thus, the algorithm instructs us to start with a low B_1 value of 10000, and then possibly to raise this value in step [Failure]. In practice, what is done is that one value of B_1 is run sufficiently many times without success for one to become convinced that a higher value is called for, perhaps double the prior value, and this procedure is iterated. Of course, another option in step [Failure] is to abort and so give up on the factorization attempt completely. When the B_1 value is gradually increased in ECM, one then expects success when B_1 finally reaches the critical range displayed above, and that the time spent unsuccessfully with smaller B_1's is negligible in comparison.

So, in summary, the heuristic expected complexity of ECM to give a nontrivial factorization of n with least prime factor p is $L(p)^{\sqrt{2}+o(1)}$ arithmetic steps with integers the size of n, using the notation from (6.1). (Note that the error expression "$o(1)$" tends to 0 as p tends to infinity.) Thus, the larger the least prime factor of n, the more arithmetic steps are expected. The worst case occurs when n is the product of two roughly equal primes, in which case the expected number of steps can be expressed as $L(n)^{1+o(1)}$, which is exactly the same as the heuristic complexity of the quadratic sieve; see Section 6.1.1. However, due to the higher precision of a typical step in ECM, we generally prefer to use the QS method, or the NFS method, for worst-case numbers. If we are presented with a number n that is unknown to be in the worst case, it is usually recommended to try ECM first, and only after a fair amount of time is spent with this method should QS or NFS be initiated. But if the number n is so large that we know beforehand that QS or NFS would be out of the question, it leaves ECM as the only current option. Who knows, we may get lucky! Here, "luck" can play either of two roles: The number under consideration may indeed have a small enough prime factor to discover with ECM, or upon implementing ECM, we may hit upon a fortunate choice of parameters sooner than expected and find an impressive factor. In fact, one interesting feature of ECM is that the variance in the expected number of steps is large since we are counting on just one successful event to occur.

It is interesting that the heuristic complexity estimate for the ECM may be made completely rigorous except for the one assumption we made that integers in the Hasse interval are just as likely to be smooth as typical integers in the larger interval $(p/2, 3p/2)$; see [Lenstra 1987].

In the discussion following we describe some optimizations of ECM. These improvements do not materially affect the complexity estimate. but they do help considerably in practice.

7.4.2 Optimization of ECM

As with the Pollard $(p-1)$ method (Section 5.4), on which the ECM is based, there is a natural, second stage continuation. In view of the remarks following Algorithm 7.4.2, assume that the order $\#E_{a,b}(\mathbf{F}_p)$ is not B_1-smooth for whatever practical choice of B_1 has been made, so that the basic algorithm can be expected to fail to find a factor. But we might just happen to have

$$\#E(\mathbf{F}_p) = q \prod_{p_i^{a_i} \le B_1} p_i^{a_i},$$

where q is a prime exceeding B_1. When such a single outlying prime is part of the unknown factorization of the order, one need not have multiplied the current point by *every* prime in $(B_1, q]$. Instead, one can use the point

$$Q = \left[\prod_{p_i \le B_1} p_i^{a_i}\right] P,$$

which is the point actually "surviving" the stage-one ECM Algorithm 7.4.2, and check the points

$$[q_0]Q,\ [q_0 + \Delta_0]Q,\ [q_0 + \Delta_0 + \Delta_1]Q,\ [q_0 + \Delta_0 + \Delta_1 + \Delta_2]Q, \ldots,$$

where q_0 is the least prime exceeding B_1, and Δ_i are the differences between subsequent primes after q_0. The idea is that one can *store* some points

$$R_i = [\Delta_i]Q,$$

once and for all, then quickly process the primes beyond B_1 by successive elliptic additions of appropriate R_i. The primary gain to be realized here is that to multiply a point by a prime such as q requires $O(\ln q)$ elliptic operations, while addition of a precomputed R_i is, of course, one operation.

Beyond this "stage-two" optimization and variants thereupon, one may invoke other enhancements such as

(1) Special parameterization to easily obtain random curves.

(2) Choice of curves with order known to be divisible by 12 or 16 [Montgomery 1992a], [Brent et al. 2000].

(3) Enhancements of large-integer arithmetic and of the elliptic algebra itself, say by FFT.

(4) Fast algorithms applied to stage two, such as "FFT extension" which is actually a polynomial-evaluation scheme applied to sets of precomputed x-coordinates.

Rather than work through such enhancements with incremental algorithm exhibitions, we instead adopt a specific strategy: We shall discuss the above enhancements briefly, then exhibit a single, practical algorithm containing many of said enhancements.

On enhancement (1) above, a striking feature our eventual algorithm will enjoy is that one need not involve y-coordinates at all. In fact, the algorithm will use the Montgomery parameterization

$$gy^2 = x^3 + Cx^2 + x,$$

with elliptic multiplication carried out via Algorithm 7.2.7. Thus a point will have the general homogeneous form $P = [X, \text{any}, Z] = [X : Z]$ (see Section 7.2 for a discussion of the notation), and we need only track the residues $X, Z \pmod n$. As we mentioned subsequent to Algorithm 7.2.7, the appearance of the point-at-infinity O during calculation on a curve over $\mathbf{F}_p$, where $p|n$, is signified by the vanishing of denominator Z, and such vanishing propagates forever afterward during further evaluations of functions *addh*() and *doubleh*(). Thus, the parameterization in question allows us to continually check $\gcd(n, Z)$, and if this is ever greater than 1, it may well be the hidden factor p. In practice, we "accumulate" Z-coordinates, and take the gcd only rarely, for example after stage one, and as we shall see, one final time after a stage two.

On enhancement (2), it is an observation of Suyama that under Montgomery parameterization the group order $\#E$ is divisible by 4. But one can press further, to ensure that the order be divisible by $8, 12$, or even 16. Thus, in regard to enhancement (2) above, we can make good use of a convenient result [Brent et al. 2000]:

Theorem 7.4.3 (ECM curve construction). *Define an elliptic curve $E_\sigma(\mathbf{F}_p)$ to be governed by the cubic*

$$y^2 = x^3 + C(\sigma)x^2 + x,$$

where C depends on field parameter $\sigma \neq 0, 1, 5$ according to

$$\begin{aligned} u &= \sigma^2 - 5, \\ v &= 4\sigma, \\ C(\sigma) &= \frac{(v-u)^3(3u+v)}{4u^3v} - 2. \end{aligned}$$

Then the order of E_σ is divisible by 12, and moreover, either on E or a twist E' (see Definition 7.2.5) there exists a point whose x-coordinate is u^3v^{-3}.

Now we can ignite any new curve attempt by simply choosing a random σ. We use, then, Algorithm 7.2.7 with homogeneous x-coordinatization starting in the form $X/Z = u^3/v^3$, proceeding to ignore all y-coordinates throughout the factorization run. What is more, we do not even care whether an initial point is on E or its twist, again because y-coordinate ignorance is allowed.

On enhancements (3), there are ideas that can reduce stage-two computations. One trick that some researchers enjoy is to use a "birthday paradox" second stage, which amounts to using semirandom multiples for two sets of coordinates, and this can sometimes yield performance advantages [Brent et al. 2000]. But there are some ideas that apply in the scenario of simply checking all outlying primes q up to some "stage-two limit" $B_2 > B_1$, that is, without any special list-matching schemes. Here is a very practical method that reduces the computational effort asymptotically down to just two (or fewer) multiplies (mod n) for each outlying prime candidate. We have already argued above that if q_n, q_{n+1} are consecutive primes, one can add some stored multiple $[\Delta_n]Q$ to any current calculation $[q_n]Q$ to get the next point $[q_{n+1}]Q$, and that this involves just one elliptic operation per prime q_m. Though that may be impressive, we recall that an elliptic operation is a handful, say, of multiplies (mod n). We can bring the complexity down simply, yet dramatically, as follows. If we know, for some prime r, the multiple $[r]Q = [X_r : Z_r]$ and we have in hand a precomputed, stored set of difference multiples $[\Delta]Q = [X_\Delta : Z_\Delta]$, where Δ has run over some relatively small finite set $\{2, 4, 6, \ldots\}$; then a prime s near to but larger than r can be checked as the outlying prime, by noting that a "successful strike"

$$[s]Q = [r + \Delta]Q = O$$

can be tested by checking whether the cross product

$$X_r Z_\Delta - X_\Delta Z_r$$

has a nontrivial gcd with n. Thus, armed with enough multiples $[\Delta]Q$, and a few occasional points $[r]Q$, we can check outlying prime candidates with 3 multiplies (mod n) per candidate. Indeed, beyond the 2 multiplies for the cross product, we need to accumulate the product $\prod(X_r Z_\Delta - X_\Delta Z_r)$ in expectation of a final gcd of such a product with n. But one can reduce the work still further, by observing that

$$X_r Z_\Delta - X_\Delta Z_r = (X_r - X_\Delta)(Z_r + Z_\Delta) + X_\Delta Z_\Delta - X_r Z_r.$$

Thus, one can store precomputed values $X_\Delta, Z_\Delta, X_\Delta Z_\Delta$, and use isolated values of $X_r, Z_r, X_r Z_r$ for well-separated primes r, to bring the cost of stage two asymptotically down to 2 multiplies (mod n) per outlying prime candidate, one for the right-hand side of the identity above and one for accumulation.

As exemplified in [Brent et al. 2000], there are even more tricks for such reduction of stage-two ECM work. One of these is also pertinent to enhancement (3) above, and amounts to mixing into various identities the notion of transform-based multiplication (see Section 9.5.3). These methods are most relevant when n is sufficiently large, in other words, when n is in the region where transform-based multiply is superior to "grammar-school" multiply. In the aforementioned identity for cross products, one can actually store transforms (for example DFT's)

$$\hat{X}_r, \hat{Z}_r,$$

in which case the product $(X_r - X_\Delta)(Z_r + Z_\Delta)$ now takes only 1/3 of a (transform-based) multiply. This dramatic reduction is possible because the single product indicated is to be done in spectral space, and so is asymptotically free, the inverse transform alone accounting for the 1/3. Similar considerations apply to the accumulation of products; in this way one can get down to about 1 multiply per outlying prime candidate. Along the same lines, the very elliptic arithmetic itself admits of transform enhancement. Under the Montgomery parameterization in question, the relevant functions for curve arithmetic degenerate nicely and are given by equations (7.6) and (7.7); and again, transform-based multiplication can bring the 6 multiplies required for *addh*() down to 4 transform-based multiplies, with similar reduction possible for *doubleh*() (see remarks following Algorithm 7.4.4).

As for enhancements (4) above, Montgomery's polynomial-evaluation scheme (sometimes called an "FFT extension" because of the details of how one evaluates large polynomials via FFT) for stage two is basically to calculate two sets of points

$$S = \{[m_i]P : i = 1, \dots, d_1\}, \quad T = \{[n_j]P : j = 1, \dots, d_2\},$$

where P is the point surviving stage one of ECM, $d_1|d_2$, and the integers m_i, n_j are carefully chosen so that some combination $m_i \pm n_j$ hopefully divides the (single) outlying prime q. This happy circumstance is in turn detected by the fact of some x-coordinate of the S list matching with some x-coordinate of the T list, in the sense that the difference of said coordinates has a nontrivial gcd with n. We will see this matching problem in another guise—in preparation for Algorithm 7.5.1. Because Algorithm 7.5.1 may possibly involve too much machine memory, for sorting and so on, one may proceed to define a degree-d_1 polynomial

$$f(x) = \prod_{s \in S} (x - X(s)) \bmod n,$$

where the $X(\ )$ function returns the affine x-coordinate of a point. Then one may evaluate this polynomial at the d_2 points $x \in \{X(t) : t \in T\}$. Alternatively, one may take the polynomial gcd of this $f(x)$ and a $g(x) = \prod_t (x - X(t))$. In any case, one can seek matches between the S, T point sets in $O\left(d_2^{1+\epsilon}\right)$ ring operations, which is lucrative in view of the alternative of actually doing $d_1 d_2$ comparisons. Incidentally, Montgomery's idea is predated by an approach of [Montgomery and Silverman 1990] for extensions to the Pollard $(p-1)$ method.

When we invoke some such means of highly efficient stage-two calculations, a rule of thumb is that one should spend only a certain fraction (say 1/4 to 1/2, depending on many details) of one's total time in stage two. This rule has arisen within the culture of modern users of ECM, and the rule's validity can be traced to the machine-dependent complexities of the various per-stage operations. In practice, this all means that the stage-two limit should be roughly two orders of magnitude over the stage-one limit, or

$$B_2 \approx 100 B_1$$

This is a good practical rule, effectively reducing nicely the degrees of freedom associated with ECM in general. Now, the time to resolve one curve—with both stages in place—is a function only of B_1. What is more, there are various tabulations of what good B_1 values might be, in terms of "suspected" sizes of hidden factors of n [Silverman and Wagstaff 1993], [Zimmermann 2000].

We now exhibit a specific form of enhanced ECM, a form that has achieved certain factoring milestones and that currently enjoys wide use. While not every possible enhancement is presented here, we have endeavored to provide many of the aforementioned manipulations; certainly enough to forge a practical implementation. The following ECM variant incorporates various enhancements of Brent, Crandall, Montgomery, Woltman, and Zimmermann:

Algorithm 7.4.4 (Inversionless ECM). Given a composite number n to be factored, with $\gcd(n, 6) = 1$, this algorithm attempts to uncover a nontrivial factor of n. This algorithm is inversion-free, needing only large-integer multiply-mod (but see text following).

1. [Choose criteria]
   $B_1 = 10000$; // Stage-one limit (must be even).
   $B_2 = 100B_1$; // Stage-two limit (must be even).
   $D = 100$; // Total memory is about $3D$ size-n integers.
2. [Choose random curve E_σ]
   Choose random $\sigma \in [6, n-1]$; // Via Theorem 7.4.3.
   $u = (\sigma^2 - 5) \bmod n$;
   $v = 4\sigma \bmod n$;
   $C = ((v-u)^3(3u+v)/(4u^3v) - 2) \bmod n$;
   // Note: C determines curve $y^2 = x^3 + Cx^2 + x$,
   // yet, C can be kept in the form num/den.
   $Q = [u^3 \bmod n : v^3 \bmod n]$; // Initial point is represented $[X : Z]$.
3. [Perform stage one]
   for($1 \le i \le \pi(B_1)$) { // Loop over primes p_i.
      Find largest integer a such that $p_i^a \le B_1$;
      $Q = [p_i^a]Q$; // Via Algorithm 7.2.7, and perhaps use FFT enhancements (see text following).
   }
   $g = \gcd(Z(Q), n)$; // Point has form $Q = [X(Q) : Z(Q)]$.
   if($1 < g < n$) return g; // Return a nontrivial factor of n.
4. [Enter stage two] // Inversion-free stage two.
   $S_1 = doubleh(Q)$;
   $S_2 = doubleh(S_1)$;
   for($d \in [1, D]$) { // This loop computes $S_d = [2d]Q$.
      if($d > 2$) $S_d = addh(S_{d-1}, S_1, S_{d-2})$;
      $\beta_d = X(S_d)Z(S_d) \bmod n$; // Store the XZ products also.
   }
   $g = 1$;
   $B = B_1 - 1$; // B is odd.
   $T = [B - 2D]Q$; // Via Algorithm 7.2.7.
   $R = [B]Q$; // Via Algorithm 7.2.7.
   for($r = B$; $r < B_2$; $r = r + 2D$) {
      $\alpha = X(R)Z(R) \bmod n$;
      for(prime $q \in [r+2, r+2D]$) { //Loop over primes.
         $\delta = (q - r)/2$; // Distance to next prime.
         // Note the next step admits of transform enhancement.
         $g = g((X(R) - X(S_\delta))(Z(R) + Z(S_\delta)) - \alpha + \beta_\delta) \bmod n$;

```
        }
        (R, T) = (addh(R, S_D, T), R);
    }
    g = gcd(g, n);
    if(1 < g < n) return g;           // Return a nontrivial factor of n.
5. [Failure]
    goto [Choose random curve ...];   // Or increase B_1, B_2 limits, etc.
```

The particular stage-two implementation suggested here involves D difference multiples $[2d]Q$, and a stored XZ product for each such multiple, for a total of $3D$ stored integers of size n. The stage-two scheme as presented is asymptotically (for large n and large memory parameter D, say) two multiplications modulo n per outlying prime candidate, which can be brought down further if one is willing to perform large-integer inversions—of which the algorithm as presented is entirely devoid—during stage two. Also, it is perhaps wasteful to recompute the outlying primes over and over for each choice of elliptic curve. If space is available, these primes might all be precomputed via a sieve in step [Choose criteria]. Another enhancement we did not spell out in the algorithm is the notion that, when we check whether a cross product $XZ' - X'Z$ has nontrivial gcd with n, we are actually checking two-point combinations $P \pm P'$, since x-coordinates of plus or minus any point are the same. This means that if two primes are equidistant from a "pivot value" r, say q', r, q form an arithmetic progression, then checking one cross product actually resolves both primes.

To provide a practical ECM variant in the form of Algorithm 7.4.4, we had to stop somewhere, deciding what detailed and sophisticated optimizations to drop from the above presentation. Yet more optimizations beyond the algorithm have been effected by [Montgomery 1987, 1992a], [Zimmermann 2000], and [Woltman 2000] to considerable advantage. Various of Zimmermann's enhancements resulted in his discovery in 1998 of a 49-digit factor of $M_{2071} = 2^{2071} - 1$. Woltman has implemented (specifically for cases $n = 2^m \pm 1$) variants of the discrete weighted transform (DWT) Algorithms 9.5.16, 9.5.18, ideas for elliptic multiplication using Lucas-sequence addition chains as in Algorithm 3.5.7, and also the FFT-intervention technique of [Crandall and Fagin 1994], [Crandall 1999b], with which one carries out the elliptic algebra itself in spectral space. Along lines previously discussed, one can perform either of the relevant doubling or adding operations (respectively, $doubleh(), addh()$ in Algorithm 7.2.7) in the equivalent of 4 multiplies. In other words, by virtue of stored transforms, each of said operations requires only 12 FFTs, of which 3 such are equivalent to one integer multiply as in Algorithm 7.2.7, and thus we infer the 4-multiplies equivalence. A specific achievement along these lines is the discovery by C. Curry and G. Woltman, of a 53-digit factor of $M_{667} = 2^{667} - 1$. Because the data have considerable value for anyone who wishes to test an ECM algorithm, we give the explicit parameters as follows. Curry used the seed

$$\sigma = 8689346476060549,$$

and the stage limits

$$B_1 = 11000000, \; B_2 = 100B_1,$$

to obtain the factorization of $2^{677} - 1$ as

$$1943118631 \cdot 531132717139346021081 \cdot 978146583988637765536217 \cdot \\ 53625112691923843508117942311516428173021903300344567 \cdot P,$$

where the final factor P is a proven prime. This beautiful example of serious ECM effort—which as of this writing involves one of the largest ECM factors yet found—looms even more beautiful when one looks at the group order $\#E(\mathbf{F}_p)$ for the 53-digit p above (and for the given seed σ), which is

$$2^4 \cdot 3^9 \cdot 3079 \cdot 152077 \cdot 172259 \cdot 1067063 \cdot 3682177 \cdot 3815423 \cdot 8867563 \cdot 15880351.$$

Indeed, the largest prime factor here in $\#E$ is greater than B_1, and sure enough, as Curry and Woltman reported, the 53-digit factor of M_{677} was found in stage two. Note that even though those investigators used detailed enhancements and algorithms, one should be able to find this particular factor—using the hindsight embodied in the above parameters—to factor M_{667} with the explicit Algorithm 7.4.4. An even more recent success is the 54-digit factor of $n = b^4 - b^2 + 1$, where $b = 6^{43} - 1$, found in January 2000 by N. Lygeros and M. Mizony. Such a factorization can be given the same "tour" of group order and so on that we did above for the 53-digit discovery [Zimmermann 2000].

Other successes have accrued from the polynomial-evaluation method pioneered by Montgomery and touched upon previously. His method was used to discover a 47-digit factor of $5 \cdot 2^{256} + 1$, and for a time this stood as an ECM record of sorts. Although requiring considerable memory, the polynomial-evaluation approach can radically speed up stage two, as we have explained.

In case the reader wishes to embark on an ECM implementation—a practice that can be quite a satisfying one—we provide here some results consistent with the notation in Algorithm 7.4.4. The 33-decimal-digit Fermat factor listed in Section 1.3.2, namely

$$188981757975021318420037633 \mid F_{15},$$

was found in 1997 by Crandall and C. van Halewyn, with the following parameters: $B_1 = 10^7$ for stage-one limit, and the choice $B_2 = 50B_1$ for stage-two limit, with the lucky choice $\sigma = 253301772$ determining the successful elliptic curve E_σ. After the 33-digit prime factor p was uncovered, Brent resolved the group order of $E_\sigma(\mathbf{F}_p)$ as

$$\#E_\sigma(\mathbf{F}_p) = (2^5 \cdot 3 \cdot 1889 \cdot 5701 \cdot 9883 \cdot 11777 \cdot 5909317) \cdot 91704181,$$

where we have intentionally shown the "smooth" part of the order in parentheses, with outlying prime 91704181. It is clear that B_1 "could have been" taken to be about 6 million, while B_2 could have been about 100 million; but of course—in the words of C. Siegel—"one cannot guess the real difficulties of a problem before having solved it." The paper [Brent et al. 2000] indicates other test values for recent factors of other Fermat numbers. Such data are extremely useful for algorithm debugging. In fact, one can effect a very rapid program check by taking the explicit factorization of a known curve order, starting with a point P, and just multiplying in the handful of primes, expecting a successful factor to indicate that the program is good.

As we have discussed, ECM is especially suitable when the hidden prime factor is not too large, even if n itself is very large. In practice, factors discovered via ECM are fairly rare in the 30-decimal-digit region, yet more rare in the 40-digit region, and so far have a vanishing population at say 60 digits.

7.5 Counting points on elliptic curves

We have seen in Section 7.3 that the number of points on an elliptic curve defined over a prime finite field $\mathbf{F}_p$ is an integer in the interval $\left((\sqrt{p}-1)^2, (\sqrt{p}+1)^2\right)$. In this section we shall discuss how one may go about actually finding this integer.

7.5.1 Shanks–Mestre method

For small primes p, less than 1000, say, one can simply carry out the explicit sum (7.8) for $\#E_{a,b}(\mathbf{F}_p)$. But this involves, without any special enhancements (such as fast algorithms for computing successive polynomial evaluations), $O(p \ln p)$ field operations for the $O(p)$ instances of $(p-1)/2$-th powers. One can do asymptotically better by choosing a point P on E, and finding all multiples $[n]P$ for $n \in (p+1-2\sqrt{p}, p+1+2\sqrt{p})$, looking for an occurrence $[n]P = O$. (Note that this finds only a multiple of the order of P—it is the actual order if it occurs that the order of P has a unique multiple in the interval $(p+1-2\sqrt{p}, p+1+2\sqrt{p})$, an event that is not unlikely.) But this approach involves $O(\sqrt{p}\ln p)$ field operations (with a fairly large implied big-O constant due to the elliptic arithmetic), and for large p, say greater than 10^{10}, this becomes a cumbersome method. There are faster $O\left(\sqrt{p}\ln^k p\right)$ algorithms that do not involve explicit elliptic algebra (see Exercise 7.26), but these, too, are currently useless for primes of modern interest in the present context, say $p \approx 10^{50}$ and beyond, this rough threshold being driven in large part by practical cryptography. All is not lost, however, for there are sophisticated modern algorithms, and enhancements to same, that press the limit on point counting to more acceptable heights.

There is an elegant, often useful, $O(p^{1/4+\epsilon})$ algorithm for assessing curve order. We have already visited the basic idea in Algorithm 5.3.1, the *baby-*

steps, giant-steps method of Shanks (for discrete logarithms). In essence this algorithm exploits a marvelous answer to the following question: If we have two length-N lists of numbers, say $A = \{A_0, \dots, A_{N-1}\}$ and $B = \{B_0, \dots, B_{N-1}\}$, how many operations (comparisons) are required to determine whether $A \cap B$ is empty? And if nonempty, what is the precise intersection $A \cap B$? A naive method is simply to check A_1 against every B_i, then check A_2 against every B_i, and so on. This inefficient procedure gives, of course, an $O(N^2)$ complexity. Much better is the following procedure:

(1) Sort each list A, B, say into nondecreasing order;

(2) Track through the sorted lists, logging any comparisons.

As is well known, the sorting step (1) requires $O(N \ln N)$ operations (comparisons), while the tracking step (2) can be done in only $O(N)$ operations. Though the concepts are fairly transparent, we think it valuable to lay out an explicit and general list-intersection algorithm. In the following exposition the input sets A, B are multisets, that is, repetitions are allowed, yet the final output $A \cap B$ is a set devoid of repetitions. We shall assume a function *sort*() that returns a sorted version of a list, having the same elements, but arranged in nondecreasing order; for example, $sort(\{3,1,2,1\}) = \{1,1,2,3\}$.

Algorithm 7.5.1 (Finding the intersection of two lists). Given two finite lists of numbers $A = \{a_0, \dots, a_{m-1}\}$ and $B = \{b_0, \dots, b_{n-1}\}$, this algorithm returns the intersection set $A \cap B$, written in strictly increasing order. Note that duplicates are properly removed; for example, if $A = \{3,2,4,2\}, B = \{1,0,8,3,3,2\}$, then $A \cap B$ is returned as $\{2,3\}$.

1. [Initialize]

   $A = sort(A)$; // Sort into nondecreasing order.

   $B = sort(B)$;

   $i = j = 0$;

   $S = \{\ \}$; // Intersection set initialized empty.

2. [Tracking stage]

```
while((i < #A) and (j < #B)) {
    if(a_i <= b_j) {
        if(a_i == b_j) S = S ∪ {a_i};          // Append the match to S.
        i = i + 1;
        while((i < (#A) - 1) and (a_i == a_{i-1})) i = i + 1;
    } else {
        j = j + 1;
        while((j < (#B) - 1) and (b_j == b_{j-1})) j = j + 1;
    }
}
return S;                                       // Return intersection A ∩ B.
```

Note that we have laid out the algorithm for general cardinalities; it is not required that $\#A = \#B$. Because of the aforementioned complexity of sorting, the whole algorithm has complexity $O(Q \ln Q)$ operations, where $Q = \max\{\#A, \#B\}$. Incidentally, there are other compelling ways to effect a list intersection (see Exercise 7.13).

Now to Shanks's application of the list intersection notion to the problem of curve order. Imagine we can find a relation for a point $P \in E$, say

$$[p + 1 + u]P = \pm[v]P,$$

or, what amounts to the same thing because $-(x, y) = (x, -y)$ always, we find a match between the x-coordinates of $[p+1+u]P$ and vP. Such a match implies that

$$[p + 1 + u \mp v]P = O.$$

This would be a tantalizing match, because the multiplier here on the left must now be a multiple of the order of the point P, and might be the curve order itself. Define an integer $W = \lceil p^{1/4}\sqrt{2} \rceil$. We can represent integers k with $|k| < 2\sqrt{p}$ as $k = \beta + \gamma W$, where β ranges over $[0, W-1]$ and γ ranges over $[0, W]$. (We use the letters β, γ to remind us of Shanks's baby-steps and giant-steps, respectively.) Thus, we can form a list of x-coordinates of the points

$$\{[p + 1 + \beta]P : \beta \in [0, \dots, W - 1]\},$$

calling that list A (with $\#A = W$), and form a separate list of x-coordinates of the points

$$\{[\gamma W]P : \gamma \in [0, \dots, W]\},$$

calling this list B (with $\#B = W + 1$). When we find a match, we can test directly to see which multiple $[p + 1 + \beta \mp \gamma W]P$ (or both) is the point at infinity. We see that the generation of baby-step and giant-step points requires $O\left(p^{1/4}\right)$ elliptic operations, and the intersection algorithm has $O\left(p^{1/4} \ln p\right)$ steps, for a total complexity of $O\left(p^{1/4+\epsilon}\right)$.

Unfortunately, finding a vanishing point multiple is not the complete task; it can happen that more than one vanishing multiple is found (and this is why we have phrased Algorithm 7.5.1 to return *all* elements of an intersection). However, whenever the point chosen has order greater than $4\sqrt{p}$, the algorithm will find the unique multiple of the order in the target interval, and this will be the actual curve order. It occasionally may occur that the group has low exponent (that is, all points have low order), and the Shanks method will never find the true group order using just one point. There are two ways around this impasse. One is to iterate the Shanks method with subsequent choices of points, building up larger subgroups that are not necessarily cyclic. If the subgroup order has a unique multiple in the Hasse interval, this multiple is the curve order. The second idea is much simpler to implement and is based on the following result of J. Mestre; see [Cohen 2000], [Schoof 1995]:

Theorem 7.5.2 (Mestre). *For an elliptic curve $E(\mathbf{F}_p)$ and its twist $E'(\mathbf{F}_p)$ by a quadratic nonresidue mod p, we have*

$$\#E + \#E' = 2p + 2.$$

When $p > 457$, there exists a point of order greater than $4\sqrt{p}$ on at least one of the two elliptic curves E, E'. Furthermore, if $p > 229$, at least one of the two curves possesses a point P with the property that the only integer $m \in (p+1-2\sqrt{p}, p+1+2\sqrt{p})$ having $[m]P = O$ is the actual curve order.

Note that the relation $\#E + \#E' = 2p+2$ is an easy result (see Exercise 7.16) and that the real content of the theorem lies in the statement concerning a singleton m in the stated Hasse range of orders. It is a further easy argument to get that there is a positive constant c (which is independent of p and the elliptic curve) such that the number of points P satisfying the theorem exceeds $cp/\ln\ln p$—see Exercise 7.17—so that points satisfying the theorem are fairly common. The idea now is to use the Shanks method on E, and if this fails (because the point order has more than one multiple in the Hasse interval), to use it on E', and if this fails, to use it on E, and so on. According to the theorem, if we try this long enough, it should eventually work. This leads to an efficient point-counting algorithm for curves $E(\mathbf{F}_p)$ when p is up to, roughly speaking, 10^{30}. In the algorithm following, we denote by $x(P)$ the x-coordinate of a point P. In the convenient scenario where all x-coordinates are given by X/Z ratios, the fact of denominator $Z = 0$ signifies as usual the point at infinity:

Algorithm 7.5.3 (Shanks–Mestre assessment of curve order)**.**
Given an elliptic curve $E = E_{a,b}(\mathbf{F}_p)$, this algorithm returns the order $\#E$. For list $S = \{s_1, s_2, \ldots\}$ and entry $s \in S$, we assume an index function $ind(S, s)$ to return some index i such that $s_i = s$. Also, list-returning function $shanks()$ is defined at the end of the algorithm; this function modifies two global lists A, B of coordinates.

1. [Check magnitude of p]
   if($p \le 229$) return $p + 1 + \sum_x \left(\frac{x^3+ax+b}{p}\right)$; // Equation (7.8).
2. [Initialize Shanks search]
   Find a quadratic nonresidue $g \pmod{p}$;
   $W = \lceil p^{1/4}\sqrt{2}\rceil$; // Giant-step parameter.
   $(c, d) = (g^2a, g^3b)$; // Twist parameters.
3. [Mestre loop] // We shall find a P of Theorem 7.5.2.
   Choose random $x \in [0, p-1]$;
   $\sigma = \left(\frac{x^3+ax+b}{p}\right)$;
   if($\sigma == 0$) goto [Mestre loop];
   // Henceforth we have a definite curve signature $\sigma = \pm 1$.
   if($\sigma == 1$) $E = E_{a,b}$; // Set original curve.
   else {
   $E = E_{c,d}$;
   $x = gx$; // Set twist curve and valid x.
   }
   Define an initial point $P \in E$ to have $x(P) = x$;

```
    S = shanks(P, E);                     // Search for Shanks intersection.
    if(#S ≠ 1) goto [Mestre loop];        // Exactly one match is sought.
    Set s as the (unique) element of S;
    β = ind(A, s); γ = ind(B, s);          // Find indices of unique match.
    Choose sign in t = β ± γW such that [p + 1 + t]P == O on E;
    return p + 1 + σt;                    // Desired order of original curve E_{a,b}.
4. [Function shanks()]
    shanks(P, E) {                        // P is assumed on given curve E.
        A = {x([p + 1 + β]P) : β ∈ [0, W − 1]};          //Baby steps.
        B = {x([γW]P) : γ ∈ [0, W]};                     // Giant steps.
        return A ∩ B;                                    // Via Algorithm 7.5.1.
    }
```

Note that assignment of point P based on random x can be done either as $P = (x, y, 1)$, where y is a square root of the cubic form, or as $P = [x : 1]$ in case Montgomery parameterization—and thus, avoidance of y-coordinates—is desired. (In this latter parameterization, the algorithm should be modified slightly, to use notation consistent with Theorem 7.2.6.) Likewise, in the *shanks*() function, one may use Algorithm 7.2.7 (or more efficient, detailed application of the *addh*(), *doubleh*() functions) to get the desired point multiples in $[X : Z]$ form, then construct the A, B lists from numbers XZ^{-1}. One can even imagine rendering the entire procedure inversionless, by working out an analogue of baby-steps, giant-steps for lists of (x, z) *pairs*, seeking matches not of the form $x = x'$, rather of the form $xz' = zx'$.

The condition $p > 229$ for applicability of the Shanks–Mestre approach is not artificial: There is a scenario for $p = 229$ in which the existence of a singleton set s of matches is not guaranteed (see Exercise 7.18).

7.5.2 Schoof method

Having seen point-counting schemes of complexities ranging from $O\left(p^{1+\epsilon}\right)$ to $O\left(p^{1/2+\epsilon}\right)$ and $O\left(p^{1/4+\epsilon}\right)$, we next turn to an elegant point-counting algorithm due to Schoof, which algorithm has polynomial-time complexity $O\left(\ln^k p\right)$ for fixed k. The basic notion of Schoof is to resolve the order $\#E \pmod{l}$ for sufficiently many small primes l, so as to reconstruct the desired order using the CRT. Let us first look at the comparatively trivial case of $\#E \pmod 2$. Now, the order of a group is even if and only if there is an element of order 2. Since a point $P \neq O$ has $2P = O$ if and only if the calculated slope (from Definition 7.1.2) involves a vanishing y-coordinate, we know that points of order 2 are those of the form $P = (x, 0)$. Therefore, the curve order is even if and only if the governing cubic $x^3 + ax + b$ has roots in $\mathbf{F}_p$. This, in turn, can be checked via a polynomial gcd as in Algorithm 2.3.10.

To consider $\#E \pmod{l}$ for small primes $l > 2$, we introduce a few more tools for elliptic curves over finite fields. Suppose we have an elliptic curve $E(\mathbf{F}_p)$, but now we consider points on the curve where the coordinates are in the algebraic closure $\overline{\mathbf{F}}_p$ of $\mathbf{F}_p$. Raising to the p-th power is a field automorphism of $\overline{\mathbf{F}}_p$ that fixes elements of $\mathbf{F}_p$, so this automorphism, applied to the coordinates of a point $(x, y) \in E(\overline{\mathbf{F}}_p)$, takes this point to another point in $E(\overline{\mathbf{F}}_p)$. And since the rules for addition of points involve rational expressions of the $\mathbf{F}_p$-coefficients of the defining equation, this mapping is seen to be a group automorphism of $E(\overline{\mathbf{F}}_p)$. This is the celebrated Frobenius endomorphism Φ. So if $(x, y) \in E(\overline{\mathbf{F}}_p)$, then $\Phi(x, y) = (x^p, y^p)$ (and also $\Phi(O)$ is defined to be O), and this mapping is a group homomorphism of $E(\overline{\mathbf{F}}_p)$ to itself. One might well wonder what use it is to consider the algebraic closure of $\mathbf{F}_p$ when it is really the points defined over $\mathbf{F}_p$ itself that we are interested in. The connection comes from a beautiful theorem: If the order of the elliptic curve group $E(\mathbf{F}_p)$ is $p + 1 - t$, then

$$\Phi^2(P) - [t]\Phi(P) + [p]P = O$$

for every point $P \in E(\overline{\mathbf{F}}_p)$. That is, the Frobenius endomorphism satisfies a quadratic equation, and the trace (the sum of the roots of the polynomial $x^2 - tx + p$) is t, the number that will give us the order of $E(\mathbf{F}_p)$.

A second idea comes into play. For any positive integer n, consider those points P of $E(\overline{\mathbf{F}}_p)$ for which $[n]P = O$. This set is denoted by $E[n]$, and it consists of those points of order dividing n in the group, namely, the n-torsion points. Two easy facts about $E[n]$ are crucial: It is a subgroup of $E(\overline{\mathbf{F}}_p)$, and Φ maps $E[n]$ to itself. Thus, we have

$$\Phi^2(P) - [t \bmod n]\Phi(P) + [p \bmod n]P = O, \text{ for all } P \in E[n]. \qquad (7.9)$$

The brilliant idea of Schoof, see [Schoof 1985], [Schoof 1995], was to use this equation to compute the residue $t \bmod n$ by trial and error procedure until the correct value that satisfies (7.9) is found. To do this, the division polynomials are used. These polynomials both simulate elliptic multiplication and pick out n-torsion points.

Definition 7.5.4. To an elliptic curve $E_{a,b}(\mathbf{F}_p)$ we associate the division polynomials $\Psi_n(X, Y) \in \mathbf{F}_p[X, Y]/(Y^2 - X^3 - aX - b)$ defined as follows:

$$\begin{aligned}
&\Psi_{-1} = -1, \quad \Psi_0 = 0, \quad \Psi_1 = 1, \quad \Psi_2 = 2Y, \\
&\Psi_3 = 3X^4 + 6aX^2 + 12bX - a^2, \\
&\Psi_4 = 4Y\left(X^6 + 5aX^4 + 20bX^3 - 5a^2X^2 - 4abX - 8b^2 - a^3\right),
\end{aligned}$$

while all further cases are given by

$$\begin{aligned}
\Psi_{2n} &= \Psi_n\left(\Psi_{n+2}\Psi_{n-1}^2 - \Psi_{n-2}\Psi_{n+1}^2\right)/(2Y), \\
\Psi_{2n+1} &= \Psi_{n+2}\Psi_n^3 - \Psi_{n+1}^3\Psi_{n-1}.
\end{aligned}$$

Note that in division polynomial construction, any occurrence of powers of Y greater than the first power are to be reduced according to the relation $Y^2 = X^3 + aX + b$. Some computationally important properties of the division polynomials are collected here:

Theorem 7.5.5 (Properties of division polynomials). *The division polynomial $\Psi_n(X,Y)$ is, for n odd, a polynomial in X alone, while for n even it is Y times a polynomial in X alone. For n odd and not a multiple of p, we have $\deg(\Psi_n) = (n^2-1)/2$. For n even and not a multiple of p, we have that the degree of Ψ_n in the variable X is $(n^2-4)/2$. For a point $(x,y) \in E(\overline{\mathbf{F}}_p) \setminus E[2]$ we have $[n]P = O$ if and only if $\Psi_n(x) = 0$ (when n is odd) and $\Psi_n(x,y) = 0$ (when n is even). Further, if $(x,y) \in E(\overline{\mathbf{F}}_p) \setminus E[n]$, then*

$$[n](x,y) = \left(x - \frac{\Psi_{n-1}\Psi_{n+1}}{\Psi_n^2}, \frac{\Psi_{n+2}\Psi_{n-1}^2 - \Psi_{n-2}\Psi_{n+1}^2}{4y\Psi_n^3}\right).$$

Note that in the last statement, if $y = 0$, then n must be odd (since $y = 0$ signifies a point of order 2, and we are given that $(x,y) \notin E[n]$), so y^2 divides the numerator of the rational expression in the second coordinate. In this case, it is natural to take this expression as 0.

It is worth remarking that for odd prime $l \neq p$, there is a *unique* integer t in $[0, l-1]$ such that

$$\left(x^{p^2}, y^{p^2}\right) + [p \bmod l](x,y) = [t]\left(x^p, y^p\right) \text{ for all } (x,y) \in E[l] \setminus \{O\}. \quad (7.10)$$

Indeed, this follows directly from (7.9) and the consequence of Theorem 7.5.5 that $E(\overline{\mathbf{F}}_p)$ does indeed contain points of order l. If this unique integer t could be computed, we would then know that the order of $E(\mathbf{F}_p)$ is congruent to $p+1-t$ modulo l.

The computational significance of the relation is that using the division polynomials, it is feasible to test the various choices for t to see which one works. This is done as follows:

(1) Points are pairs of polynomials in $\mathbf{F}_p[X,Y]$.

(2) Since the points are on E, we may constantly reduce modulo $Y^2 - X^3 - aX - b$ so as to keep powers of Y no higher than the first power, and since the points we are considering are in $E[n]$, we may reduce also by the polynomial Ψ_n to keep the X powers in check as well. Finally, the coefficients are in $\mathbf{F}_p$, so that mod p reductions can be taken with the coefficients, whenever convenient. These three kinds of reductions may be taken in any order.

(3) High powers of X, Y are to be reduced by a powering ladder such as that provided in Algorithm 2.1.5, with appropriate polynomial mods taken along the way for continual degree reduction.

(4) The addition on the left side of (7.10) is to be simulated using the formulae in Definition 7.1.2.

On the face of it, explicit polynomial inversion—from the fundamental elliptic operation definition—would seem to be required. This could be accomplished via Algorithm 2.2.2, but it is not necessary to do so because of the following observation. We have seen in various elliptic addition algorithms previous that inversions can be avoided by adroit representations of coordinates. In actual practice, we have found it convenient to work either with the projective point representation of Algorithm 7.2.3 or a "rational" variant of same. We now describe the latter representation, as it is well suited for calculations involving division polynomials, especially in regard to the point-multiplication property in Theorem 7.5.5. We shall consider a point to be $P = (U/V, F/G)$, where U, V, F, G are all polynomials, presumably bivariate in X, Y. There is an alternative strategy, which is to use projective coordinates as mentioned in Exercise 7.29. In either strategy a simplification occurs, that in the Schoof algorithm we *always* obtain any point in a particular form; for example in the $P = (U/V, F/G)$ parameterization option used in the algorithm display below, one always has the form

$$P = (N(X)/D(X), YM(X)/C(X)),$$

because of the division polynomial algebra. One should think of these four polynomials, then, as reduced mod Ψ_n and mod p, in the sense of item (2) above. Another enhancement we have found efficient in practice is to invoke large polynomial multiply via our Algorithm 9.6.1 (or see alternatives as in Exercise 9.70), which is particularly advantageous because $\deg(\Psi_n)$ is so large, making ordinary polynomial arithmetic painful. Yet more efficiency obtains when we use our Algorithm 9.6.4 to achieve polynomial mod for these large-degree polynomials.

Algorithm 7.5.6 (Explicit Schoof algorithm for curve order). Let $p > 3$ be a prime. For curve $E_{a,b}(\mathbf{F}_p)$ this algorithm returns the value of $t \pmod l$, where l is a prime (much smaller than p) and the curve order is $\#E = p + 1 - t$. Exact curve order is thus obtained by effecting this algorithm for enough primes l such that $\prod l > 4\sqrt{p}$, and then using the Chinese remainder theorem to recover the exact value of t. We assume that for a contemplated ceiling $L \geq l$ on the possible l values used, we have precomputed the division polynomials $\Psi_{-1}, \ldots, \Psi_{L+1} \bmod p$, which can be made monic (via cancellation of the high coefficient modulo p) with a view to such as Algorithm 9.6.4.

1. [Check $l = 2$]
   if($l == 2$) {
      $g(X) = \gcd(X^p - X, X^3 + aX + b)$; // Polynomial gcd in $\mathbf{F}_p[X]$.
      if($g(X) == 1$) return 0; // $T \equiv 0 \pmod 2$, so order $\#E$ is even.
      return 1; // $\#E$ is odd.
   }
2. [Analyze relation (7.10)]
   $\overline{p} = p \bmod l$;
   $u(X) = X^p \bmod (\Psi_l, p)$;
   $v(X) = (X^3 + aX + b)^{(p-1)/2} \bmod (\Psi_l, p)$;
   // That is, $v(X) = Y^{p-1} \bmod (\Psi_l, p)$.
   $P_0 = (u(X), Yv(X))$; // $P_0 = (X^p, Y^p)$.
   $P_1 = (u(X)^p \bmod (\Psi_l, p), Yv(X)^{p+1} \bmod (\Psi_l, p))$;
   // $P_1 = (X^{p^2}, Y^{p^2})$.

```
Cast P_2 = [p̄](X,Y) in rational form (N(X)/D(X), YM(X)/C(X)), for
   example by using Theorem 7.5.5;
if(P_1 + P_2 == O) return 0;        // #E = p+1-t with t ≡ 0 (mod l).
P_3 = P_0;
for(1 ≤ k ≤ l/2) {
   if(X-coordinates of (P_1 + P_2) and P_3 match) {
      if(Y-coordinates also match) return k;      // Y-coordinate check.
      return l - k;
   }
   P_3 = P_3 + P_0;
}
```

In the addition tests above for matching of some coordinate between (P_1+P_2) and P_3, one is asking generally whether

$$(N_1/D_1, YM_1/C_1) + (N_2/D_2, YM_2/C_2) = (N_3/D_3, YM_3/C_3),$$

and such a relation is to be checked, of course, using the usual elliptic addition rules. The polynomial $P_1 + P_2$ on the left can be combined—using the elliptic rules of Algorithm 7.2.2, with the coordinates in that algorithm being now, of course, our polynomial ratios—into polynomial form $(N'/D', YM'/C')$, and this is compared with $(N_3/D_3, YM_3/C_3)$. For such comparison in turn one checks whether the cross products $(N_3D' - N'D_3)$ and $(M_3C' - M'C_3)$ both vanish mod (Ψ_l, p). As for the check on whether $P_1 + P_2 = O$, we are asking whether $M_1/C_1 = -M_2/C_2$, and this is also an easy cross product relation. The idea is that the entire implementation we are describing involves only polynomial multiplication and the mod (Ψ_l, p) reductions throughout. And as we have mentioned, both polynomial multiply and mod can be made quite efficient.

In case an attempt is made by the reader to implement Algorithm 7.5.6, we give here some small cases within the calculation, for purpose of, shall we say, "algorithm debugging." For $p = 101$ and the curve

$$Y^2 = X^3 + 3X + 4$$

over $\mathbf{F}_p$, the algorithm gives, for l selections $l = 2, 3, 5, 7$, the results $t \bmod 2 = 0$, $t \bmod 3 = 1$, $t \bmod 5 = 0$, $t \bmod 7 = 3$, from which we infer $\#E = 92$. (We might have skipped the prime $l = 5$, since the product of the other primes exceeds $4\sqrt{p}$.) Along the way we have, for example,

$$\Psi_3 = 98 + 16X + 6X^2 + X^4,$$

$$\left(X^{p^2}, Y^{p^2}\right) = \left(32 + 17X + 13X^2 + 92X^3,\ Y(74 + 96X + 14X^2 + 68X^3)\right),$$

$$[2](X,Y) = \left(\frac{12 + 53X + 89X^2}{16 + 12X + 4X^3},\ Y\frac{74 + 10X + 5X^2 + 64X^3}{27 + 91X + 96X^2 + 37X^3}\right),$$

$$(X^p, Y^p) = \left(70 + 61X + 83X^2 + 44X^3,\ Y(43 + 76X + 21X^2 + 25X^3)\right),$$

where it will be observed that every polynomial appearing in the point coordinates has been reduced mod (Ψ_3, p). (Note that $\overline{p}$ in step [Analyze ...] is 2, which is why we consider $[2](X, Y)$.) It turns out that the last point here is indeed the elliptic sum of the two points previous, consistent with the claim that $t \bmod 3 = 1$.

There is an important enhancement that we have intentionally left out for clarity. This is that prime powers work equally well. In other words, $l = q^a$ can be used directly in the algorithm (with the gcd for $l = 2$ ignored when $l = 4, 8, 16, \ldots$) to reduce the computation somewhat. All that is required is that the overall product of all prime-power values l used (but no more than one for each prime) exceed $4\sqrt{p}$.

We have been able to assess curve orders, via this basic Schoof scheme, for primes in the region $p \approx 10^{80}$, by using prime powers $l < 100$. It is sometimes said in the literature that there is little hope of using l much larger than 30, say, but with the aforementioned enhancements—in particular the large-polynomial multiply/mod algorithms covered in Chapter 9—the Schoof prime l can be pressed to 100 and perhaps beyond.

By not taking Algorithm 7.5.6 all the way to CRT saturation (that is, not handling quite enough small primes l to resolve the order), and by then employing a Shanks–Mestre approach to finish the calculation based on the new knowledge of the possible orders, one may, in turn, press this rough bound of 10^{80} further. However, it is a testimony to the power of the Schoof algorithm that, upon analysis of how far a "Shanks–Mestre boost" can take us, we see that only a few extra decimal digits—say 10 or 20 digits—can be added to the 80 digits we resolve using the Schoof algorithm alone. For such reasons, it usually makes more practical sense to enhance an existing Schoof implementation, rather than to piggyback a Shanks–Mestre atop it.

But can one carry out point counting for significantly larger primes? Indeed, the transformation of the Schoof algorithm into a "Schoof–Elkies–Atkin" (SEA) variant due to [Atkin 1986, 1988, 1992] and [Elkies 1991, 1997] with computational enhancements by [Morain 1995], [Couveignes and Morain 1994], [Couveignes et al. 1996], have achieved unprecedented point-counting performance. The essential improvement of Elkies was to observe that for some of the l (depending on a, b, p; in fact, for about half of possible l values), a certain polynomial f_l dividing Ψ_l but of degree only $(l-1)/2$ can be employed, and furthermore, that the Schoof relation of (7.10) can be simplified. The Elkies approach is to seek an eigenvalue λ with

$$(X^p, Y^p) = [\lambda](X, Y),$$

where all calculations are done mod (f_l, p), whence $\#E = p + 1 - t$ with

$$t \equiv \lambda + p/\lambda \pmod{l}.$$

Because the degrees of f_l are so small, this important discovery effectively pulls some powers of $\ln p$ off the complexity estimate, to yield $O(\ln^6 p)$ rather than the original Schoof complexity $O(\ln^8 p)$ [Schoof 1995]. (Note, however, that such estimates assume direct "grammar-school" multiplication of integers, and can be reduced yet further in the power of ln.) The SEA ideas certainly give impressive performance. Atkin, for example, used such enhancements to find in 1992, for the smallest prime having 200 decimal digits, namely

$$\begin{aligned} p = {} & 10000000000000000000000000000000000000000000000000\backslash \\ & 00000000000000000000000000000000000000000000000000\backslash \\ & 00000000000000000000000000000000000000000000000000\backslash \\ & 00000000000000000000000000000000000000000000000153, \end{aligned}$$

and the curve over $\mathbf{F}_p$ governed by the cubic

$$Y^2 = X^3 + 105X + 78153,$$

a point order

$$\begin{aligned} \#E = {} & 10000000000000000000000000000000000000000000000000\backslash \\ & 00000000000000000000000000000000000000000000000000\backslash \\ & 06789750288004224118080314365460277641928049641888\backslash \\ & 39991591392960032210630561760029050858613689631753. \end{aligned}$$

Amusingly, it is not too hard to agree that this choice of curve is "random" (even if the prime p is not): The $(a, b) = (105, 78153)$ parameters for this curve were derived from a postal address in France [Schoof 1995]. Subsequently, Morain was able to provide further computational enhancements, to find an explicit order for a curve over $\mathbf{F}_p$, with p a 500-decimal-digit prime [Morain 1995].

In this treatment we have, in regard to the powerful Schoof algorithm and its extensions, touched merely the tip of the proverbial iceberg. There is a great deal more to be said; a good modern reference for practical point-counting on elliptic curves is [Seroussi et al. 1999], and various implementations of the SEA continuations have been reported [Izu et al. 1998], [Scott 1999].

In his original paper [Schoof 1985] gave an application of the point-counting method to obtain square roots of an integer D modulo p in (not random, but deterministic) polynomial time, assuming that D is fixed. Though the commonly used random algorithms 2.3.8, 2.3.9 are much more practical, Schoof's point-counting approach for square roots establishes, at least for fixed D, a true deterministic polynomial-time complexity.

Incidentally, an amusing anecdote cannot be resisted here. As mentioned by [Elkies 1997], Schoof's magnificent point-counting algorithm was rejected in its initial paper form as being, in the referee's opinion, somehow unimportant. But with modified title, that title now ending with "... square roots mod p," the modified paper [Schoof 1985] was, as we appreciate, finally published.

7.5.3 Atkin–Morain method

We have addressed the question, given a curve $E = E_{a,b}(\mathbf{F}_p)$, what is $\#E$? A kind of converse question—which is of great importance in primality proving and cryptography is, can we find a suitable order $\#E$, and *then* specify a curve having that order? For example, one might want a prime order, or an order $2q$ for prime q, or an order divisible by a high power of 2. One might call this the study of "closed-form" curve orders, in the following sense: for certain representations $4p = u^2 + |D|v^2$, as we have encountered previously in Algorithm 2.3.13, one can write down immediately certain curve orders and also—usually with more effort—the a, b parameters of the governing cubic. These ideas emerged from the seminal work of A. O. L. Atkin in the latter 1980s and his later joint work with F. Morain.

In order to make sense of these ideas it is necessary to delve a bit into some additional theoretical considerations on elliptic curves. For a more thorough treatment, see [Atkin and Morain 1993b], [Cohen 2000], [Silverman 1986].

For an elliptic curve E defined over the complex numbers $\mathbf{C}$, one may consider the "endomorphisms" of E. These are group homomorphisms from the group E to itself that are given by rational functions. The set of such endomorphisms, denoted by $\mathrm{End}(E)$, naturally form a ring, where addition is derived from elliptic addition, and multiplication is composition. That is, if ϕ, σ are in $\mathrm{End}(E)$, then $\phi + \sigma$ is the endomorphism on E that sends a point P to $\phi(P) + \sigma(P)$, the latter "+" being elliptic addition; and $\phi \cdot \sigma$ is the endomorphism on E that sends P to $\phi(\sigma(P))$.

If n is an integer, the map $[n]$ that sends a point P on E to $[n]P$ is a member of $\mathrm{End}(E)$, since it is a group homomorphism and since Theorem 7.5.5 shows that $[n]P$ has coordinates that are rational functions of the coordinates of P. Thus the ring $\mathrm{End}(E)$ contains an isomorphic copy of the ring of integers $\mathbf{Z}$. It is often the case, in fact usually the case, that this is the whole story for $\mathrm{End}(E)$. However, sometimes there are endomorphisms of E that do not correspond to an integer. It turns out, though, that the ring $\mathrm{End}(E)$ is never too much larger than $\mathbf{Z}$: if it is not isomorphic to $\mathbf{Z}$, then it is isomorphic to an order in an imaginary quadratic number field. (An "order" is a subring of finite index of the ring of algebraic integers in the field.) In such a case it is said that E has complex multiplication, or is a CM curve.

Suppose E is an elliptic curve defined over the rationals, and when considered over the complex numbers has complex multiplication by an order in $\mathbf{Q}(\sqrt{D})$, where D is a negative integer. Suppose $p > 3$ is a prime that does not divide the discriminant of E. We then may consider E over $\mathbf{F}_p$ by reducing the coefficients of E modulo p. Suppose the prime p is a norm of an algebraic integer in $\mathbf{Q}(\sqrt{D})$. In this case it turns out that we can easily find the order of the elliptic-curve group $E(\mathbf{F}_p)$. The work in computing this order does not even require the coefficients of the curve E, one only needs the numbers D and p. And this work to compute the order is indeed simple; one uses the Cornacchia–Smith Algorithm 2.3.13. There is additional, somewhat harder, work to compute the coefficients of an equation defining E, but if one can see for some reason that the order will not be useful, this extra work can be short-circuited. This, in essence, is the idea of Atkin and Morain.

We now review some ideas connected with imaginary quadratic fields, and the dual theory of binary quadratic forms of negative discriminant. Some of these ideas were developed in Section 5.6. The (negative) discriminants D relevant to curve order assessment are defined thus:

Definition 7.5.7. A negative integer D is a fundamental discriminant if the odd part of D is square-free, and $|D| \equiv 3, 4, 7, 8, 11, 15 \pmod{16}$.

Briefly put, these are discriminants of imaginary quadratic fields. Now, associated with each fundamental discriminant is the class number $h(D)$. As we saw in Section 5.6.3, $h(D)$ is the order of the group $\mathcal{C}(D)$ of reduced binary quadratic forms of discriminant D. In Section 5.6.4 we mentioned how the baby-steps, giant-steps method of Shanks can be used to compute $h(D)$. The following algorithm serves to do this and to optionally generate the reduced forms, as well as to compute the Hilbert class polynomial corresponding to D. This is a polynomial of degree $h(D)$ with coefficients in $\mathbf{Z}$ such that the splitting field for the polynomial over $\mathbf{Q}(\sqrt{D})$ has Galois group isomorphic to the class group $\mathcal{C}(D)$. This splitting field is called the Hilbert class field for $\mathbf{Q}(\sqrt{D})$ and is the largest abelian unramified extension of $\mathbf{Q}(\sqrt{D})$. The Hilbert class field has the property that a prime number p splits completely in this field if and only if there are integers u, v with $4p = u^2 + |D|v^2$. In particular, since the Hilbert class field has degree $2h(D)$ over the rational field $\mathbf{Q}$, the proportion, among all primes, of primes p with $4p$ so representable is $1/2h(D)$, [Cox 1989].

We require a function (again, we bypass the beautiful and complicated foundations of the theory in favor of an immediate algorithm development)

$$\Delta(q) = q\left(1 + \sum_{n=1}^{\infty}(-1)^n\left(q^{n(3n-1)/2} + q^{n(3n+1)/2}\right)\right)^{24},$$

arising in the theory of invariants and modular forms [Cohen 2000], [Atkin and Morain 1993b]. (It is interesting that $\Delta(q)$ has the alternative and beautiful representation $q\prod_{n\geq 1}(1-q^n)^{24}$, but we shall not use this in what follows. The first given expression for $\Delta(q)$ is more amenable to calculation since the exponents grow quadratically.)

Algorithm 7.5.8 (Class number and Hilbert class polynomial). Given a (negative) fundamental discriminant D, this algorithm returns any desired combination of the class number $h(D)$, the Hilbert class polynomial $T \in \mathbf{Z}[X]$ (whose degree is $h(D)$), and the set of reduced forms (a, b, c) of discriminant D (whose cardinality is $h(D)$).

1. [Initialize]

```
    T = 1;
    b = D mod 2;
    r = ⌊√(|D|/3)⌋;
    h = 0;                                    // Zero class count.
    red = { };          // Empty set of primitive reduced forms.
```

2. [Outer loop on b]

```
    while(b ≤ r) {
```

```
        m = (b^2 - D)/4;
        for(1 ≤ a and a^2 ≤ m) {
            if(m mod a ≠ 0) continue;      // Continue 'for' loop to force a|m.
            c = m/a;
            if(b > a) continue;                        // Continue 'for' loop.
3. [Optional polynomial setup]
            τ = (-b + i√|D|)/(2a);   // Note precision (see text following).
            f = Δ(e^{4πiτ})/Δ(e^{2πiτ});                  // Note precision.
            j = (256f + 1)^3/f;                            // Note precision.
4. [Begin divisors test]
            if(b == a or c == a or b == 0) {
                T = T * (X - j);
                h = h + 1;                                  // Class count.
                red = red ∪ (a, b, c);                        // New form.
            } else {
                T = T * (X^2 - 2 Re(j)X + |j|^2);
                h = h + 2;                                  // Class count.
                red = red ∪ (a, ±b, c);                  // Two new forms.
            }
        }
    }
5. [Return values of interest]
    return (combination of) h, round(Re(T(x))), red;
```

This algorithm is straightforward in every respect except on the issue of floating-point precision. Note that the function Δ must be evaluated for complex q arguments. The theory shows that sufficient precision for the whole algorithm is essentially

$$\delta = \frac{\pi\sqrt{|D|}}{\ln 10} \sum \frac{1}{a}$$

decimal digits, where the sum is over all primitive reduced forms (a, b, c) of discriminant D [Atkin and Morain 1993b]. This means that a little more than δ digits (perhaps $\delta + 10$, as in [Cohen 2000]) should be used for the [Optional polynomial setup] phase, the ultimate idea being that the polynomial $T(x)$—consisting of possibly some linear factors and some quadratic factors—should have integer coefficients. Thus the final polynomial output in the form $round(\mathrm{Re}(T(x)))$ means that T is to be expanded, with the coefficients rounded so that $T \in \mathbf{Z}[X]$. Algorithm 7.5.8 can, of course, be used in a multiple-pass fashion: First calculate just the reduced forms, to estimate $\sum 1/a$ and thus the required precision, then start over and this time calculate the actual Hilbert class polynomial. In any event, the quantity $\sum 1/a$ is always $O\left(\ln^2 |D|\right)$.

For reader convenience, we give here some explicit polynomial examples from the algorithm, where T_D refers to the Hilbert class polynomial for discriminant D:

$$
\begin{aligned}
T_{-3} &= X, \\
T_{-4} &= X - 1728, \\
T_{-15} &= X^2 + 191025X - 121287375, \\
T_{-23} &= X^3 + 3491750X^2 - 5151296875X + 12771880859375.
\end{aligned}
$$

One notes that the polynomial degrees are consistent with the class numbers below. There are further interesting aspects of these polynomials. One is that the constant coefficient is always a cube. Also, the coefficients of T_D grow radically as one works through lists of discriminants. But one can use in the Atkin-Morain approach less unwieldy polynomials—the Weber variety—at the cost of some complications for special cases. These and many more optimizations are discussed in [Morain 1990], [Atkin and Morain 1993b].

In the Atkin–Morain order-finding scheme, it will be useful to think of discriminants ordered by their class numbers, this ordering being essentially one of increasing complexity. As simple runs of Algorithm 7.5.8 would show (without the polynomial option, say),

$$
\begin{aligned}
h(D) &= 1 \text{ for } D = -3, -4, -7, -8, -11, -19, -43, -67, -163; \\
h(D) &= 2 \text{ for } D = -15, -20, -24, -35, -40, -51, -52, -88, -91, -115, \\
&\qquad\qquad -123, -148, -187, -232, -235, -267, -403, -427; \\
h(D) &= 3 \text{ for } D = -23, -31, -59, \ldots .
\end{aligned}
$$

That the discriminant lists for $h(D) = 1, 2$ are in fact complete as given here is a profound result of the theory [Cox 1989]. We currently have complete lists for $h(D) \le 16$, see [Watkins 2000], and it is known, in principle at least, how to compute a complete list for any prescribed value of h. The effective determination of such lists is an extremely interesting computational problem.

To apply the Atkin–Morain method, we want to consider discriminants ordered, say, as above, i.e., lowest $h(D)$ first. We shall seek curve orders based on specific representations

$$4p = u^2 + |D|v^2,$$

whence, as we see in the following algorithm exhibition, the resulting possible curve orders will be simple functions of p, u, v. Note that for $D = -3, -4$ there are $6, 4$ possible orders, respectively, while for other D there are two possible orders. Such representations of $4p$ are to be attempted via Algorithm 2.3.13. If p is prime, the "probability" that $4p$ is so representable, given that $\left(\frac{D}{p}\right) = 1$, is $1/h(D)$, as mentioned above. In the following algorithm, either it is assumed that our discriminant list is finite, or we agree to let the algorithm run for some prescribed amount of time.

Algorithm 7.5.9 (CM method for generating curves and orders). We assume a list of fundamental discriminants $\{D_j < 0 : j = 1, 2, 3, \ldots\}$ ordered, say, by increasing class number $h(D)$, and within the same class number by increasing $|D|$. We are given a prime $p > 3$. The algorithm reports (optionally) possible curve orders or (also optionally) curve parameters for CM curves associated with the various D_j.

1. [Calculate nonresidue]
   Find a random quadratic nonresidue $g \pmod p$;
   if($p \equiv 1 \pmod 3$ and $g^{(p-1)/3} \equiv 1 \pmod p$) goto [Calculate nonresidue];
   // In case $D = -3$ is used, g must also be a noncube modulo p.
   $j = 0$;
2. [Discriminant loop]
   $j = j + 1$;
   $D = D_j$;
   if($\left(\frac{D}{p}\right) \neq 1$) goto [Discriminant loop];
3. [Seek a quadratic form for $4p$]
   Attempt to represent $4p = u^2 + |D|v^2$, via Algorithm 2.3.13, but if the attempt fails, goto [Discriminant loop];
4. [Option: Curve orders]
   if($D == -4$) report $\{p+1 \pm u,\ p+1 \pm 2v\}$; // 4 possible orders.
   if($D == -3$) report $\{p+1 \pm u,\ p+1 \pm (u \pm 3v)/2\}$; // 6 possible orders.
   if($D < -4$) report $\{p+1 \pm u\}$; // 2 possible orders.
5. [Option: Curve parameters]
   if($D == -4$) return $\{(a,b)\} = \{(-g^k \bmod p, 0) : k = 0, 1, 2, 3\}$;
   if($D == -3$) return $\{(a,b)\} = \{(0, -g^k \bmod p) : k = 0, 1, 2, 3, 4, 5\}$;
6. [Continuation for $D < -4$]
   Compute the Hilbert class polynomial $T \in \mathbf{Z}[X]$, via Algorithm 7.5.8;
   $S = T \bmod p$; // Reduce to polynomial $\in \mathbf{F}_p[X]$.
   Obtain a root $j \in \mathbf{F}_p$ of S, via Algorithm 2.3.10;
   $c = j(j - 1728)^{-1} \bmod p$;
   $r = -3c \bmod p$;
   $s = 2c \bmod p$;
7. [Return two curve-parameter pairs]
   return $\{(a,b)\} = \{(r,s), (rg^2 \bmod p, sg^3 \bmod p)\}$;

What the Atkin–Morain method prescribes is that for $D = -4, -3$ the governing cubics are given in terms of a quadratic nonresidue g, which is also a cubic nonresidue in the case $D = -3$, by

$$y^2 = x^3 - g^k x,\ k = 0, 1, 2, 3,$$
$$y^2 = x^3 - g^k,\ k = 0, 1, 2, 3, 4, 5,$$

respectively (i.e., there are respectively 4, 6 isomorphism classes of curves for these two D values); while for other discriminants D the relevant curve and its twist are

$$y^2 = x^3 - 3cg^{2k}x + 2cg^{3k}, \ k = 0, 1,$$

where c is given as in step [Continuation for $D < -4$]. The method, while providing much more generality than closed-form solutions such as Algorithm 7.5.10 below, is more difficult to implement, mainly because of the Hilbert class polynomial calculation.

Note the important feature that prior to the actual curve parameter calculations, we already know the possible curve orders involved. Thus in both primality proving and cryptography applications, we can analyze the possible orders before entering into the laborious (a, b) calculations, knowing that if a curve order is attractive for any reason, we can get those parameters at will. We take up this issue in Sections 7.6 and 8.1.

Let us work through an example of Algorithm 7.5.9 in action. Take the Mersenne prime

$$p = 2^{89} - 1,$$

for which we desire some possible curves $E_{a,b}(\mathbf{F}_p)$ and their orders. In the [Seek a quadratic form for $4p$] algorithm step above, we find via Algorithm 2.3.13 many representations for $4p$, just a few of which being

$$\begin{aligned} 4p &= 48215832688019^2 + 3 \cdot 7097266064519^2 \\ &= 37064361490164^2 + 163 \cdot 2600275098586^2 \\ &= 35649086634820^2 + 51 \cdot 4860853432438^2 \\ &= 27347149714756^2 + 187 \cdot 3039854240322^2 \\ &= 28743118396413^2 + 499 \cdot 1818251501825^2. \end{aligned}$$

For these exemplary representations the discriminants of interest are $D = -3, -163, -51, -187, -499$, respectively; and we repeat that there are plenty of other values of D one may use for this p. The relevant curve orders will generally be

$$p + 1 \pm u,$$

where u is the first number being squared in a given representation; yet there will be more possible orders for the $D = -3$ case. To illustrate the detailed algorithm workings, let us consider the case $D = -499$ above. Then in the [Option: curve parameters] step we obtain

$$\begin{aligned} T_{-499} = \ & 4671133182399954782798673154437441310949376 \\ & - 6063717825494266394722392560011051008x \\ & + 300510110807102620070672596992 0x^2 \\ & + x^3 . \end{aligned}$$

Note that, as must be, the constant term in this polynomial is a cube. Now this cubic can be reduced right away (mod p) to yield

$$\begin{aligned} S = T_{-499} \bmod p = {} & 489476008241378181249146744 \\ & + 356560280230433613294194825x \\ & + 166270576558338910192101 5x^2 \\ & + x^3, \end{aligned}$$

but we are illustrating the concept that one could in principle prestore the Hilbert class polynomials $T_{-D} \in \mathbf{Z}[X]$, reducing quickly to $S \in \mathbf{F}_p[X]$ whenever a new p is being analyzed. We are then to use Algorithm 2.3.10 to find a root j of $S = T \bmod p$. A root is found as

$$j = 431302127816045615339451868.$$

It is this value that ignites the curve parameter construction. We obtain

$$c = j/(j - 1728) \bmod p = 544175025087910210133176287,$$

and thus end up with two governing cubics (the required nonresidue g can be taken to be -1 for this p):

$$y^2 = x^3 + 224384983664339781949157472x \pm 469380030533130282816790463,$$

with respective curve orders

$$\#E = 2^{89} \pm 28743118396413.$$

Incidentally, which curve has which order is usually an easy computation: For given a, b parameters, find a point $P \in E$ and verify that $[\#E]P = O$, for one possibility for $\#E$ and not the other. In fact, if $p > 475$, Theorem 7.5.2 implies that either there is a point P on E with $[\#E']P \neq O$ (where E' is the twist of E) or there is a point Q on E' with $[\#E]Q \neq O$. Thus, randomly choosing points, first on one of the curves, then on the other, one should expect to soon be able to detect which order goes with which curve. In any case, many of the algorithms based on the Atkin–Morain approach can make use of points that simply have vanishing multiples, and it is not necessary to ascertain the full curve order.

We observe that the polynomial calculations for the deeper discriminants (i.e. possessed of higher class numbers) can be difficult. For example, there is the precision issue when using floating-point arithmetic in Algorithm 7.5.8. It is therefore worthwhile to contemplate means for establishing some *explicit* curve parameters for small $|D|$, in this way obviating the need for class polynomial calculations. To this end, we have compiled here a complete list of curve parameter sets for all D with $h(D) = 1, 2$:

D	r	s
-7	125	189
-8	125	98
-11	512	539
-19	512	513
-43	512000	512001
-67	85184000	85184001
-163	151931373056000	151931373056001
-15	$1225 - 2080\sqrt{5}$	5929
-20	$108250 + 29835\sqrt{5}$	174724
-24	$1757 - 494\sqrt{2}$	1058
-35	$-1126400 - 1589760\sqrt{5}$	2428447
-40	$54175 - 1020\sqrt{5}$	51894
-51	$75520 - 7936\sqrt{17}$	108241
-52	$1778750 + 5125\sqrt{13}$	1797228
-88	$181713125 - 44250\sqrt{2}$	181650546
-91	$74752 - 36352\sqrt{13}$	205821
-115	$269593600 - 89157120\sqrt{5}$	468954981
-123	$1025058304000 - 1248832000\sqrt{41}$	1033054730449
-148	$499833128054750 + 356500625\sqrt{37}$	499835296563372
-187	$91878880000 - 1074017568000\sqrt{17}$	4520166756633
-232	$1728371226151263375 - 11276414500\sqrt{29}$	1728371165425912854
-235	$7574816832000 - 190341944320\sqrt{5}$	8000434358469
-267	$3632253349307716000000 - 12320504793376000\sqrt{89}$	3632369580717474122449
-403	$16416107434811840000 - 4799513373120384000\sqrt{13}$	33720998998872514077
-427	$564510997315289728000 - 5784785611102784000\sqrt{61}$	609691617259594724421

Table 7.1 Explicit curve parameters of CM curves for class number 1 and 2

Algorithm 7.5.10 (Explicit CM curve parameters: Class numbers 1, 2). Given prime $p > 3$, this algorithm reports explicit CM curves $y^2 = x^3 + ax + b$ over $\mathbf{F}_p$, with orders as specified in the [Option: Curve orders] step of Algorithm 7.5.9. The search herein is exhaustive over all discriminants D of class numbers $h(D) = 1, 2$: the algorithm reports every set of CM curve parameters (a, b) for the allowed class numbers.

1. [Establish full discriminant list]

$$\Delta = \{-3, -4, -7, -8, -11, -19, -43, -67, -163,$$
$$-15, -20, -24, -35, -40, -51, -52, -88, -91, -115, -123,$$
$$-148, -187, -232, -235, -267, -403, -427\};$$

2. [Loop over representations]

for($D \in \Delta$) {

Attempt to represent $4p = u^2 + |D|v^2$, via Algorithm 2.3.13, but if the attempt fails, jump to next D;
Calculate a suitable nonresidue g of p as in step [Calculate nonresidue] of Algorithm 7.5.9;

3. [Handle $D = -3, -4$]
 if($D == -3$) return $\{(a, b)\} = \{(0, -g^k) : k = 0, \ldots, 5\}$;
 // Six curves $y^2 = x^3 - g^k$.
 if($D == -4$) return $\{(a, b)\} = \{(-g^k, 0) : k = 0, \ldots, 3\}$;
 // Four curves $y^2 = x^3 - g^k x$.
4. [Parameters for all other D with $h(D) = 1, 2$]
 Select a pair (r, s) from Table 7.1, using Algorithm 2.3.9 when square roots are required (mod p);
5. [Return curve parameters]
 report $\{(a, b)\} = \{(-3rs^3g^{2k}, 2rs^5g^{3k}) : k = 0, 1\}$;
 // The governing cubic will be $y^2 = x^3 - 3rs^3g^{2k}x + 2rs^5g^{3k}$.

}

There are several points of interest in connection with this algorithm. The specific parameterizations of Algorithm 7.5.10 can be calculated, of course, via the Hilbert class polynomials, as in Algorithm 7.5.8. However, having laid these parameters out explicitly means that one can proceed to establish CM curves very rapidly, with minimal programming overhead. It is not even necessary to verify that $4a^3 + 27b^2 \neq 0$, as is demanded for legitimate elliptic curves over $\mathbf{F}_p$. Yet another interesting feature is that the specific square roots exhibited in the algorithm always exist (mod p). What is more, the tabulated r, s parameters tend to enjoy interesting factorizations. In particular the s values tend to be highly smooth numbers (see Exercise 7.15 for more details on these various issues).

It is appropriate at this juncture to clarify by worked example how quickly Algorithm 7.5.10 will generate curves and orders. Taking the prime $p = \left(2^{31} + 1\right)/3$, we find by appeal to Algorithm 2.3.13 representations $4p = u^2 + |D|v^2$ for ten discriminants D of class number not exceeding two, namely, for $D = -3, -7, -8, -11, -67, -51, -91, -187, -403, -427$. The respective a, b parameters and curve orders work out, via Algorithm 7.5.10 as tabulated below.

D	E	$\#E$
-3	$y^2 = x^3 + 0x + 715827882$	715861972
	$y^2 = x^3 + 0x + 715827878$	715880649
	$y^2 = x^3 + 0x + 715827858$	715846561
	$y^2 = x^3 + 0x + 715827758$	715793796
	$y^2 = x^3 + 0x + 715827258$	715775119
	$y^2 = x^3 + 0x + 715824758$	715809207
-7	$y^2 = x^3 + 331585657x + 632369458$	715788584
	$y^2 = x^3 + 415534712x + 305115120$	715867184
-8	$y^2 = x^3 + 362880883x + 649193252$	715784194
	$y^2 = x^3 + 482087479x + 260605721$	715871574
-11	$y^2 = x^3 + 710498587x + 673622741$	715774393
	$y^2 = x^3 + 582595483x + 450980314$	715881375
-67	$y^2 = x^3 + 265592125x + 480243852$	715785809
	$y^2 = x^3 + 197352178x + 616767211$	715869959
-51	$y^2 = x^3 + 602207293x + 487817116$	715826683
	$y^2 = x^3 + 22796782x + 131769445$	715829085
-91	$y^2 = x^3 + 407640471x + 205746226$	715824963
	$y^2 = x^3 + 169421413x + 664302345$	715830805
-187	$y^2 = x^3 + 389987874x + 525671592$	715817117
	$y^2 = x^3 + 443934371x + 568611647$	715838651
-403	$y^2 = x^3 + 644736647x + 438316263$	715881357
	$y^2 = x^3 + 370202749x + 386613767$	715774411
-427	$y^2 = x^3 + 370428023x + 532016446$	715860684
	$y^2 = x^3 + 670765979x + 645890514$	715795084

For this particular run, the requisite quadratic nonresidue (and cubic nonresidue for the $D = -3$ case) was chosen as 5. Note that Algorithm 7.5.10 does not tell us which of the curve parameter pairs (a, b) goes with which order (from step [Option: Curve orders] of Algorithm 7.5.9). As mentioned above, this is not a serious problem: One finds a point P on one curve where a candidate order does not kill it, so we know that the candidate belongs to another curve. For the example listing above with $p = (2^{31} + 1)/3$, the orders shown were matched to the curves in just this way.

But one can, in principle, go a little further and specify theoretically which orders go with which curves, at least for discriminants D having $h(D) = 1$. There are explicit curves and orders in the literature [Rishi et al. 1984], [Padma and Ventkataraman 1996]. Many such results go back to the work of Stark, who connected the precise curve order $p+1-u$, when $4p = u^2 + |D|v^2$ and u is allowed to be positive or negative, with the Jacobi symbol $\left(\frac{u}{|D|}\right)$. Interesting refinements of this work are found in the modern treatment of [Morain 1998].

7.6 Elliptic curve primality proving (ECPP)

We have seen in Section 4.1 that a partial factorization of $n-1$ can lead to a primality proof for n. One might wonder whether elliptic-curve groups—given their variable group orders under the Hasse theorem 7.3.1—can be brought to bear for primality proofs. Indeed they can, as evidenced by a certain theorem, which is a kind of elliptic curve analogy to the Pocklington Theorem 4.1.3.

Before we exhibit the theorem, we recall Definition 7.4.1 of a pseudocurve $E(\mathbf{Z}_n)$. Recalling, too, the caveat about elliptic multiplication on a pseudocurve mentioned following the definition, we proceed with the following central result.

Theorem 7.6.1 (Goldwasser–Kilian ECPP theorem). *Let $n > 1$ be an integer coprime to 6, let $E(\mathbf{Z}_n)$ be a pseudocurve, and let s, m be positive integers with $s|m$. Assume that there exists a point $P \in E$ such that we can carry out the curve operations for $[m]P$ to find*

$$[m]P = O,$$

and for every prime q dividing s we can carry out the curve operations to obtain

$$[m/q]P \neq O.$$

Then for every prime p dividing n we have

$$\#E(\mathbf{F}_p) \equiv 0 \pmod{s}.$$

Moreover, if $s > \left(n^{1/4} + 1\right)^2$, then n is prime.

Proof. Let p be a prime factor of n. The calculations on the pseudocurve, when reduced modulo p, imply that s divides the order of P on $E(\mathbf{F}_p)$. This proves the first assertion. In addition, if $s > \left(n^{1/4} + 1\right)^2$, we may infer that $\#E(\mathbf{F}_p) > \left(n^{1/4} + 1\right)^2$. But the Hasse Theorem 7.3.1 implies that $\#E(\mathbf{F}_p) < \left(p^{1/2} + 1\right)^2$. We deduce that $p^{1/2} > n^{1/4}$, so that $p > n^{1/2}$. As n has all of its prime factors greater than its square root, n must be prime. □

7.6.1 Goldwasser–Kilian primality test

On the basis of Theorem 7.6.1, Goldwasser and Kilian demonstrated a primality testing algorithm with expected polynomial-time complexity for conjecturally all, and provably "most," prime numbers n. That is, a number n could be tested in an expected number of operations $O\left(\ln^k n\right)$ for an absolute constant k. Their idea is to find appropriate curves with orders that have large enough "probable prime" factors, and recurse on the notion that these factors should in turn be provably prime. In each recursive level but the last, Theorem 7.6.1 is used with s the probable prime factor of the curve order. This continues for smaller and smaller probable primes, until the number is so small it may be proved prime by trial division. This, in turn, justifies all previous steps, and establishes the primality of the starting number n.

Algorithm 7.6.2 (Goldwasser–Kilian primality test). Given a nonsquare integer $n > 2^{32}$ strongly suspected of being prime (in particular, $\gcd(n, 6) = 1$ and presumably n has already passed a probable prime test), this algorithm attempts to reduce the issue of primality of n to that of a smaller number q. The algorithm returns either the assertion "n is composite" or the assertion "If q is prime then n is prime," where q is an integer smaller than n.

1. [Choose a pseudocurve over $\mathbf{Z}_n$]
   Choose random $(a, b) \in [0, n-1]^2$ such that $\gcd(4a^3 + 27b^2, n) = 1$;
2. [Assess curve order]
   Via Algorithm 7.5.6 calculate the integer m that would be $\#E_{a,b}(\mathbf{Z}_n)$ if n is prime (however if the point-counting algorithm fails, return "n is composite");
   // If n is composite, Algorithm 7.5.6 could fail if each candidate for $t \pmod{l}$ is rejected or if the final curve order is not in the interval $(n + 1 - 2\sqrt{n}, n + 1 + 2\sqrt{n})$.
3. [Attempt to factor]
   Attempt to factor $m = kq$ where $k > 1$ and q is a probable prime exceeding $\left(n^{1/4} + 1\right)^2$, but if this cannot be done according to some time-limit criterion, goto [Choose a pseudocurve ...];
4. [Choose point on $E_{a,b}(\mathbf{Z}_n)$]
   Choose random $x \in [0, n-1]$ such that $Q = (x^3 + ax + b) \bmod n$ has $\left(\frac{Q}{n}\right) \neq -1$;
   Apply Algorithm 2.3.8 or 2.3.9 (with $a = Q$ and $p = n$) to find an integer y that would satisfy $y^2 \equiv Q \pmod{n}$ if n were prime;
   if($y^2 \bmod n \neq Q$) return "n is composite";
   $P = (x, y)$;
5. [Operate on point]
   Compute the multiple $U = [m/q]P$ (however if any illegal inversions occur, return "n is composite");
   if($U == O$) goto [Choose point ...];
   Compute $V = [q]U$ (however check the above rule on illegal inversions);
   if($V \neq O$) return "n is composite";
   return "If q is prime, then n is prime";

The correctness of Algorithm 7.6.2 follows directly from Theorem 7.6.1, with q playing the role of s in that theorem.

In practice one would iterate the algorithm, getting a chain of inferences, with the last number q so small it can be proved prime by trial division. If some intermediate q is composite, then one can retreat one level in the chain and apply the algorithm again. Iterating the Goldwasser–Kilian scheme not only provides a rigorous primality test but also generates a certificate of primality. This certificate can be thought of as the chain

$$(n = n_0, a_0, b_0, m_0, q_0, P_0),\ (q_0 = n_1, a_1, b_1, m_1, q_1, P_1), \ldots$$

consisting of consecutive n, a, b, m, q, P entities along the recursion. The primary feature of the certificate is that it can be published alongside, or otherwise associated with, the original n that is proven prime. This concise listing can then be used by anyone who wishes to verify that n is prime, using Theorem 7.6.1 at the various steps along the way. The reconstruction of the proof usually takes considerably less time than the initial run that finds the certificate. The certificate feature is nontrivial, since many primality proofs must be run again from scratch if they are to be checked.

It should be noted that the elliptic arithmetic in Algorithm 7.6.2 can be sped up using Montgomery coordinates $[X : Z]$ with "Y" dropped, as discussed in Section 7.2.

To aid in the reader's testing of any implementations, we now report a detailed example. Let us take the prime $p = 10^{20} + 39$. On the first pass of Algorithm 7.6.2, we use $n = p$ and obtain random parameters in step [Choose a pseudocurve ...] as

$$a = 69771859804340235254, \quad b = 10558409492409151218,$$

for which $4a^3 + 27b^2$ is coprime to n. The number that would be the order of $E_{a,b}(\mathbf{Z}_n)$ if n is indeed prime is found, via Algorithm 7.5.6 to be

$$m = \#E = 99999999985875882644 = 2^2 \cdot 59 \cdot 1182449 \cdot q,$$

where $2, 59, 1182449$ are known primes (falling below the threshold 2^{32} suggested in the algorithm description), and $q = 358348489871$ is a probable prime. Then, in step [Choose point ...] the random point obtained is

$$P = [X : Z] = [31689859357184528586 : 1],$$

where for practical simplicity we have adopted Montgomery parameterization, with a view to using Algorithm 7.2.7 for elliptic multiples. Accordingly, it was found that

$$U = [m/q]P = [69046631243878263311 : 1] \neq O,$$
$$V = [q]U = O.$$

Therefore, p is prime if q is. So now we assign $n = 358348489871$ and run again through Algorithm 7.6.2. In so doing the relevant values encountered are

$$a = 34328822753, \quad b = 187921935449,$$
$$m = \#E = 358349377736 = 2^3 \cdot 7 \cdot 7949 \cdot 805019,$$

where now all the factors fall under our 2^{32} threshold. For randomly chosen starting point

$$P = [X : Z] = [245203089935 : 1]$$

we obtain, with $q = 805019$,

$$U = [m/q]P = [260419245130 : 1] \neq O,$$
$$V = [q]P = O.$$

It follows that the original $p = 10^{20} + 39$ is prime. The relevant numbers are then collected as a primality certificate for this prime. It should be noted that for larger examples one should not expect to be lucky enough to get a good factorization of m on every attempt, though conjecturally the event should not be so very rare.

The study of the computational complexity of Algorithm 7.6.2 is interesting. Success hinges on the likelihood of finding a curve order that factors as in step [Attempt to factor]. Note that one is happy even if one finds an order $m = 2q$ where q is a prime. Thus, it can be shown via Theorem 7.3.2 that if

$$\pi\left(x + 1 + 2\sqrt{x}\right) - \pi\left(x + 1 - 2\sqrt{x}\right) > A\frac{\sqrt{x}}{\ln^c x}$$

for positive constants A, c, then the expected bit complexity of the algorithm is $O\left(\ln^{9+c} n\right)$; see [Goldwasser and Kilian 1986]. It is conjectured that the inequality holds with $A = c = 1$ and all sufficiently large values of x. In addition, using results in analytic number theory that say that such inequalities are *usually* true, it is possible to show that the Goldwasser–Kilian test (Algorithm 7.6.2) usually works, and does so in polynomial time. To remove this lacuna, one might note that sufficient information *is* known about primes in an interval of length $x^{3/4}$ near x. Using this, [Adleman and Huang 1992] were able to achieve a guaranteed expected polynomial time bound. In their scheme, a certificate chain is likewise generated, yet, remarkably, the initial primes in the chain actually increase in size, eventually to decay to acceptable levels. The decay is done via the Goldwasser–Kilian test as above, and the increase is designed so as to "gain randomness." The initial candidate n might be one for which the Goldwasser–Kilian test does not work (this would be evidenced by never having luck in factoring curve orders or just taking too long to factor), so the initial steps of "reducing" the primality of n to that of larger numbers is a way of replacing the given number n with a new number that is random enough so that the Goldwasser–Kilian test is expected to work for it. This "going up" is done via Jacobian varieties of hyperelliptic curves of genus 2.

7.6.2 Atkin–Morain primality test

The Goldwasser–Kilian Algorithm 7.6.2 is, in practice for large n under scrutiny, noticeably sluggish due to the point-counting step to assess $\#E$. Atkin found an elegant solution to this impasse, and together with Morain implemented a highly efficient elliptic curve primality proving (ECPP) scheme [Atkin and Morain 1993b]. The method is now in wide use. There are various ways to proceed in practice with this ECPP; we give just one here.

The idea once again is to find either "closed-form" curve orders, or at least be able to specify orders relatively quickly. One could conceivably use closed forms such as those of Algorithm 7.5.10, but one may well "run out of gas," not being able to find an order with the proper structure for Theorem 7.6.1. The Atkin–Morain approach is to find curves with complex multiplication, as in Algorithm 7.5.9. In this way, a crucial step (called [Assess curve order], in Algorithm 7.6.2) is a point of entry into the Atkin–Morain order/curve-finding Algorithm 7.5.9. A quick perusal will show the great similarity of Algorithm 7.6.3 below and Algorithm 7.6.2. The difference is that here one searches for appropriate curve orders *first*, and only then constructs the corresponding elliptic curve, both using Algorithm 7.5.9, while the Schoof algorithm 7.5.6 is dispensed with.

Algorithm 7.6.3 (Atkin–Morain primality test). Given a nonsquare integer $n > 2^{32}$ strongly suspected of being prime (in particular $\gcd(n, 6) = 1$ and presumably n has already passed a probable prime test), this algorithm attempts to reduce the issue of primality of n to that of a smaller number q. The algorithm returns either the assertion "n is composite" or the assertion "If q is prime, then n is prime," where q is an integer smaller than n. (Note similar structure of Algorithm 7.6.2.)

1. [Choose discriminant]
   Select a fundamental discriminant D by increasing value of $h(D)$ for which $\left(\frac{D}{n}\right) = 1$ and for which we are successful in finding a solution $u^2 + |D|v^2 = 4n$ via Algorithm 2.3.13, yielding possible curve orders m:
   $m \in \{n + 1 \pm u,\ n + 1 \pm 2v\}$, for $D = -4$,
   $m \in \{n + 1 \pm u,\ n + 1 \pm (u \pm 3v)/2\}$, for $D = -3$,
   $m \in \{n + 1 \pm u\}$, for $D < -4$;
2. [Factor orders]
   Find a possible order m that factors as $m = kq$, where $k > 1$ and q is a probable prime $> (n^{1/4} + 1)^2$ (however if this cannot be done according to some time-limit criterion, goto [Choose discriminant]);
3. [Obtain curve parameters]
   Using the parameter-generating option of Algorithm 7.5.9, establish the parameters a, b for an elliptic curve that would have order m if n is indeed prime;
4. [Choose point on $E_{a,b}(\mathbf{Z}_n)$]

Choose random $x \in [0, n-1]$ such that $Q = (x^3 + ax + b) \bmod n$ has $\left(\frac{Q}{n}\right) \neq -1$;
Apply Algorithm 2.3.8 or 2.3.9 (with $a = Q$ and $p = n$) to find an integer y that would satisfy $y^2 \equiv Q \pmod{n}$ if n were prime;
if($y^2 \bmod n \neq Q$) return "n is composite";
$P = (x, y)$;

5. [Operate on point]
Compute the multiple $U = [m/q]P$ (however if any illegal inversions occur, return "n is composite");
if($U == O$) goto [Choose point ...];
Compute $V = [q]U$ (however check the above rule on illegal inversions);
if($V \neq O$) return "n is composite";
return "If q is prime, then n is prime";

Note that if n is composite, then there is no guarantee that Algorithm 2.3.13 in step [Choose discriminant] will successfully find u, v, even if they exist. In this event, we continue with the next D, until we are eventually successful, or we lose patience and give up.

Let us work through an explicit example. Recall the Mersenne prime $p = 2^{89} - 1$ analyzed after Algorithm 7.5.9. We found a discriminant $D = -3$ for complex multiplication curves, for which D there turn out to be six possible curve orders. The recursive primality proving works, in this case, by taking $p + 1 + u$ as the order; in fact, this choice happens to work at every level like so:

$$
\begin{aligned}
p &= 2^{89} - 1, \\
D &= -3: \ u = 34753815440788, \ v = 20559283311750, \\
\#E &= p + 1 + u = 2^2 \cdot 3^2 \cdot 5^2 \cdot 7 \cdot 848173 \cdot p_2, \\
p_2 &= 115836285129447871, \\
D &= -3: \ u = 557417116, \ v = 225559526, \\
\#E &= p_2 + 1 + u = 2^2 \cdot 3 \cdot 7 \cdot 37 \cdot 65707 \cdot p_3,
\end{aligned}
$$

and we establish that $p_3 = 567220573$ is prime by trial division. What we have outlined is the essential "backbone" of a primality certificate for $p = 2^{89} - 1$. The full certificate requires, of course, the actual curve parameters (from step [Obtain curve parameters]) and relevant starting points (from step [Choose point ...]) in Algorithm 7.6.3.

Compared to the Goldwasser–Kilian approach, the complexity of the Atkin–Morain method is a cloudy issue. The added difficulty comes from the fact that the potential curve orders that one tries to factor have an unknown distribution. However, in practice, the method is excellent, and like the Goldwasser–Kilian method a complete and succinct certificate of primality is provided. Morain's implementation of variants of Algorithm 7.6.3 has achieved primality proofs for "random" primes of well over two thousand decimal digits, as we mentioned in Section 1.1.2.

7.7 Exercises

7.1. Find a bilinear transformation of the form

$$(x, y) \mapsto (\alpha x + \beta y, \gamma x + \delta y)$$

that renders the curve

$$y^2 + axy + by = x^3 + cx^2 + dx + e \tag{7.11}$$

into Weierstrass form (7.4). Indicate, then, where the fact of field characteristic not equal to 2 or 3 is required for the transformation to be legal.

7.2. Show that curve with governing cubic

$$Y^2 = X^3 + CX^2 + AX + B$$

has affine representation

$$y^2 = x^3 + (A - C^2/3)x + (B - AC/3 + 2C^3/27).$$

This shows that a Montgomery curve ($B = 0$) always has an affine equivalent. But the converse is false. Describe exactly under what conditions on parameters a, b in

$$y^2 = x^3 + ax + b$$

such an affine curve does possess a Montgomery equivalent with $B = 0$. Describe applications of this result, for example in cryptography or point-counting.

7.3. Show that the curve given by relation (7.4) is nonsingular over a field F with characteristic $\neq 2, 3$ if and only if $4a^3 + 27b^2 \neq 0$.

7.4. As in Exercise 7.3 the nonsingularity condition for affine curves is that the discriminant $4a^3 + 27b^2$ be nonzero in the field $\mathbf{F}_p$. Show that for the parameterization

$$Y^2 = X^3 + CX^2 + AX + B$$

the condition is different on a discriminant Δ, namely

$$\Delta = 4(A - C^2/3)^3 + 27(B - AC/3 + 2C^3/27)^2 \neq 0.$$

Then show that in the computationally useful Montgomery parameterization

$$Y^2 = X^3 + CX^2 + X$$

is nonsingular if and only if $C \neq 2$.

7.5. For an elliptic curve over $\mathbf{F}_p$, $p > 3$, with cubic

$$Y^2 = X^3 + CX^2 + AX + B$$

we define the j-invariant of E as

$$j(E) = 4096 \frac{(C^2 - 3A)^3}{\Delta},$$

where the discriminant Δ is given in Exercise 7.4. Carry out the following computational exercise. By choosing a conveniently small prime that allows hand computation or easy machine work (you might assess curve orders via the direct formula (7.8)), create a table of curve orders vs. j-invariants. Based on such empirical evidence, state an apparent connection between curve orders and j-invariant values. For an excellent overview of the beautiful theory of j-invariants and curve isomorphisms see [Seroussi et al. 1999] and numerous references therein, especially [Silverman 1986].

7.6. Here we investigate just a little of the beautiful classical theory of elliptic integrals and functions, with a view to the connections of same to the modern theory of elliptic curves. Good introductory references are [Namba 1984], [Silverman 1986], [Kaliski 1988]. One essential connection is the observation of Weierstrass that the elliptic integral

$$Z(x) = \int_x^\infty \frac{ds}{\sqrt{4s^3 - g_2 s - g_3}}$$

can be considered as a solution to an implicit relation

$$\wp_{g_2,g_3}(Z) = x,$$

where $\wp$ is the Weierstrass function. Derive, then, the differential equations

$$\wp(z_1 + z_2) = \frac{1}{4}\left(\frac{\wp'(z_1) - \wp'(z_2)}{\wp(z_1) - \wp(z_2)}\right)^2 - \wp(z_1) - \wp(z_2)$$

and that

$$\wp'(z)^2 = \wp^3(z) - g_2\wp(z) - g_3,$$

and indicate how the parameters g_2, g_3 need be related to the affine a, b curve parameters, to render the differential scheme equivalent to the affine scheme.

7.7. Prove the first statement of Theorem 7.1.3, that $E_{a,b}(F)$ together with the defined operations is an abelian group. A good symbolic processor for abstract algebra might come in handy, especially for the hardest part, which is proving associativity $(P_1 + P_2) + P_3 = P_1 + (P_2 + P_3)$.

7.8. Show that an abelian group of square-free order is cyclic. Deduce that if a curve order $\#E$ is square-free, then the elliptic-curve group is cyclic. This is an important issue for cryptographic applications [Kaliski 1991], [Morain 1992].

7.9. Compare the operation (multiplies only) counts in Algorithms 7.2.2, 7.2.3, with a view to the different efficiencies of doubling and (unequal point) addition. In this way, determine the threshold k at which an inverse must be faster than k multiplies for the first algorithm to be superior. In this connection see Exercise 7.25.

7.10. Show that if we conspire to have parameter $a = -3$ in the field, the operation count of the doubling operation of Algorithm 7.2.3 can be reduced yet further. Investigate the claim of [Solinas 1998] that "the proportion of elliptic curves modulo p that can be rescaled so that $a = p - 3$ is about $1/4$ if $p \equiv 1 \pmod 4$ and about $1/2$ if $p \equiv 3 \pmod 4$." Incidentally, the slight speedup for doubling may seem trivial but in practice will always be noticed, because doubling operations constitute a significant portion of a typical point-multiplying ladder.

7.11. Prove that the elliptic addition test, Algorithm 7.2.8, works. Establish first, for the coordinates $x_\pm$ of $P_1 \pm P_2$, respectively, algebraic relations for the sum and product $x_+ + x_-$ and $x_+ x_-$, using Definition 7.1.2 and Theorem 7.2.6. The resulting relations should be entirely devoid of y dependence. Now from these sum and product relations, infer the quadratic relation.

7.12. Work out the heuristic expected complexity bound for ECM as discussed following Algorithm 7.4.2.

7.13. Recall the method, relevant to the second stage of ECM, and touched upon in the text, for finding a match between two lists but without using Algorithm 7.5.1. The idea is first to form a polynomial

$$f(x) = \prod_{i=0}^{m-1} (x - A_i),$$

then evaluate this at the n values in B; i.e., evaluate for $x = B_j, j = 0, \ldots, n-1$. The point is, if a zero of f is found in this way, we have a match (some B_j equals A_i). Give the computational complexity of this polynomial method for finding $A \cap B$. How does one handle duplicate matches in this polynomial setting? Note the related material in Sections 5.5, 9.6.3.

7.14. By analyzing the trend of "record" ECM factorizations, estimate in what calendar year we shall be able to discover 70-digit factors via ECM. ([Zimmermann 2000] has projected the year 2010, for example.)

7.15. Verify claims made in reference to Algorithm 7.5.10, as follows. First, show how the tabulated parameters r, s were obtained. For this, one uses the fact of the class polynomial being at most quadratic, and notes also that a defining cubic $y^2 = x^3 + Rx/S + T/S$ can be cleared of denominator S by multiplying through by S^6. Second, use quadratic reciprocity to prove that every explicit square root in the tabulated parameters does, in fact, exist. For this, one presumes that a representation $4p = u^2 + |D|v^2$ has been found for p. Third, show that $4a^3 + 27b^2$ cannot vanish (mod p). This could be done case by case, but it is easier to go back to Algorithm 7.5.9 and see how the final a, b parameters actually arise. Finally, factor the s values of the tabulated data to verify that they tend to be highly smooth. How can this smoothness be explained?

7.16. Recall that for elliptic curve $E_{a,b}(\mathbf{F}_p)$ a twist curve E' of E is governed by a cubic

$$y^2 = x^3 + g^2ax + g^3b,$$

where $\left(\frac{g}{p}\right) = -1$. Show that the curve orders are related thus:

$$\#E + \#E' = 2p + 2.$$

7.17. Suppose the largest order of an element in a finite abelian group G is m. Show there is an absolute constant $c > 0$ (that is, c does not depend on m or G) such that the proportion of elements of G with order m is at least $c/\ln\ln(3m)$. (The presence of the factor 3 is only to ensure that the double log is positive.) This result is relevant to the comments following Theorem 7.5.2 and also to some results in Chapter 3.

7.18. Consider, for $p = 229$, the curves E, E' over $\mathbf{F}_p$ governed respectively by

$$y^2 = x^3 - 1,$$

$$y^2 = x^3 - 8,$$

the latter being a twist curve of the former. Show that $\#E = 252, \#E' = 208$ with respective group structures

$$E \cong \mathbf{Z}_{42} \times \mathbf{Z}_6,$$

$$E' \cong \mathbf{Z}_{52} \times \mathbf{Z}_4.$$

Argue thus that every point $P \in E$ has $[252]P = [210]P = O$, and similarly every point $P \in E'$ has $[208]P = [260]P = O$, and therefore that for any point on either curve there is no unique m in the Hasse interval with $[m]P = O$. See [Schoof 1995] for this and other special cases pertaining to the Mestre theorems.

7.19. Here we investigate the operation complexity of the Schoof Algorithm 7.5.6. Derive the bound $O\left(\ln^8 p\right)$ on operation complexity for Schoof's original method, assuming grammar-school polynomial multiplication (which in turn has complexity $O(de)$ field operations for degrees d, e of operands). Explain why the Schoof–Elkies–Atkin (SEA) method continuation reduces this to $O\left(\ln^6 p\right)$. (To deduce such reduction, one need only know the degree of an SEA polynomial, which is $O(l)$ rather than $O(l^2)$ for the prime l.) Describe what then happens to the complexity bound if one also invokes a fast multiplication method not only for integers but also for polynomial multiplication (see text following Algorithm 7.5.6), and perhaps also a Shanks–Mestre boost. Finally, what can be said about *bit* complexity to resolve curve order for a prime p having n bits?

7.20. Elliptic curve theory can be used to establish certain results on sums of cubes in rings. By way of the Hasse Theorem 7.3.1, prove that if $p > 7$ is prime, then every element of $\mathbf{F}_p$ is a sum of two cubes. By analyzing, then, prime powers, prove the following conjecture (which was motivated numerically and communicated by D. Copeland): Let d_N be the density of representables (as (cube+cube)) in the ring $\mathbf{Z}_N$. Then

if $63|N$ then $d_N = 25/63$, otherwise
if $7|N$ then $d_N = 5/7$, or
if $9|N$ then $d_N = 5/9$,
and in all other cases $d_N = 1$.

An extension is: Study sums of higher powers (see Exercise 9.80).

7.21. Here is an example of how symbolic exercise can tune one's understanding of the workings a specific, tough algorithm. It is sometimes possible actually to carry out what we might call a "symbolic Schoof algorithm," to obtain exact results on curve orders, in the following fashion. Consider an elliptic curve $E_{0,b}(\mathbf{F}_p)$ for $p > 3$, and so governed by the cubic

$$y^2 = x^3 + b.$$

We shall determine the order (mod 3) of any such curve, yet do this via symbolic manipulations alone; i.e., without the usual numerical calculations associated with Schoof implementations. Perform the following proofs, *without* the assistance of computing machinery (although a symbolic machine may be valuable in checking one's algebra):

(1) Argue that with respect to the division polynomial Ψ_3, we have

$$x^4 \equiv -4bx \pmod{\Psi_3}.$$

(2) Prove that for $k > 0$,

$$x^{3k} \equiv (-4b)^{k-1}x^3 \pmod{\Psi_3}.$$

This reduction ignites a chain of exact results for the Frobenius relation, as we shall see.

(3) Show that x^p can now be given the closed form

$$x^p \equiv (-4b)^{\lfloor p/3 \rfloor} x^{p \bmod 3} \pmod{\Psi_3},$$

where our usual mod notation is in force, so $p \bmod 3 = 1$ or 2.

(4) Show that x^{p^2} can also be written down exactly as

$$x^{p^2} \equiv (-4b)^{(p^2-1)/3} x \pmod{\Psi_3},$$

and argue that for $p \equiv 2 \pmod 3$ the congruence here boils down to $x^{p^2} \equiv x$, independent of b.

(5) By way of binomial series and the reduction relation from (2) above, establish the following general identity for positive integer d and $\gamma \not\equiv 0 \pmod{p}$:

$$(x^3+\gamma)^d \equiv \gamma^d \left(1 - \frac{x^3}{4b}\left((1-4b/\gamma)^d - 1\right)\right) \pmod{\Psi_3}.$$

(6) Starting with the notion that $y^p \equiv y(x^3+b)^{(p-1)/2}$, resolve the power y^p as

$$y^p \equiv y b^{(p-1)/2} q(x) \pmod{\Psi_3},$$

where $q(x) = 1$ or $(1 + x^3/(2b))$ as $p \equiv 1, 2 \pmod{3}$, respectively.

(7) Show that we always have, then,

$$y^{p^2} \equiv y \pmod{\Psi_3}.$$

Now, given the above preparation, argue from Theorem 7.5.5 that for $p \equiv 2 \pmod{3}$ we have, independent of b,

$$\#E \equiv p + 1 \equiv 0 \pmod{3}.$$

Finally, for $p \equiv 1 \pmod{3}$ argue, on the basis of the remaining possibilities for the Frobenius

$$(c_1 x, y) + [1](x, y) = t(c_2 x, y c_3)$$

for b-dependent parameters c_i, that the curve order (mod 3) depends on the quadratic character of b (mod p) in the following way:

$$\#E \equiv p + 1 + \left(\frac{b}{p}\right) \equiv 2 + \left(\frac{b}{p}\right) \pmod{3}.$$

An interesting research question is: How far can this "symbolic Schoof" algorithm be pushed (see Exercise 7.30)?

7.22. For the example prime $p = \left(2^{31}+1\right)/3$ and its curve orders displayed after Algorithm 7.5.10, which is the best order to use to effect an ECPP proof that p is prime?

7.23. Use some variant of ECPP to prove primality of every one of the ten consecutive primes claimed in Exercise 1.84.

7.24. Here we apply ECPP ideas to primality testing of Fermat numbers $F_m = 2^{2^m}+1$. By considering representations

$$4F_m = u^2 + 4v^2,$$

prove that if F_m is prime, then there are four curves $\pmod{F_m}$

$$y^2 = x^3 - 3^k x; \quad k = 0, 1, 2, 3,$$

having, in some ordering, the curve orders

$$2^{2^m} + 2^{m/2+1} + 1,$$

$$2^{2^m} - 2^{m/2+1} + 1,$$

$$2^{2^m} - 1,$$

$$2^{2^m} + 3.$$

Prove by computer that F_7 (or some even larger Fermat number) is composite, by exhibiting on one of the four curves a point P that is *not* annihilated by any of the four orders. One should perhaps use the Montgomery representation in Algorithm 7.2.7, so that initial points need have only their x-coordinates checked for validity (see explanation following Algorithm 7.2.1). Otherwise, the whole exercise is doomed because one usually cannot even perform square-rooting for composite F_m, to obtain y coordinates.

Of course, the celebrated Pepin primality test (Theorem 4.1.2) is much more efficient in the matter of weeding out composites, but the notion of CM curves is instructive here. In fact, when the above procedure is invoked for $F_4 = 65537$, one finds that indeed, every one of the four curves has an initial point that is annihilated by one of the four orders. Thus we might regard 65537 as a "probable" prime in the present sense. Just a little more work, along the lines of the Atkin ECPP algorithm, will complete a primality proof for this largest known Fermat prime.

7.8 Research problems

7.25. With a view to the complexity tradeoffs between Algorithms 7.2.2, 7.2.3, 7.2.7, analyze the complexity of field inversion. One looks longingly at expressions $x_3 = m^2 - x_1 - x_2$, $y_3 = m(x_1 - x_3) - y_1$, in the realization that if only inversion were "free," the affine approach would surely be superior. However, known inversion methods are quite expensive. One finds in practice that inversion times tend to be one or two orders of magnitude greater than multiply-mod times. [De Win et al. 1998] explain that it is very hard even to bring down the cost of inversion (modulo a typical cryptographic prime $p \approx 2^{200}$) to 20 multiplies. But there are open questions. What about primes of special form, or lookup tables? The lookup notion stems from the simple fact that if y can be found such that $xy \equiv z \pmod p$ for some z whose inverse is already known, then $x^{-1} \bmod p = yz^{-1} \bmod p$. In connection with the complexity issue see Algorithm 9.4.5 and Exercise 2.11.

Another research direction is to attempt implementation of the interesting Sorenson-class methods for k-ary (as opposed to binary) gcd's [Sorenson 1994], which methods admit of an extended form for modular inversion.

7.26. For an elliptic curve $E(\mathbf{F}_p)$, prime p with governing cubic

$$y^2 = x(x+1)(x+c)$$

(and $c \not\equiv 0, 1 \pmod p$), show by direct appeal to the order relation (7.8) that $\#E = p + 1 - T$, where

$$T = \sum_{n=0}^{Q} c^n \binom{Q}{n}^2,$$

with $Q = (p-1)/2$ and we interpret the sum to lie modulo p in $(-2\sqrt{p}, 2\sqrt{p})$. (One way to proceed is to write the Legendre symbol in relation (7.8) as a $(p-1)/2$-th power, then formally sum over x.) Then argue that

$$T \equiv F(1/2, 1/2, 1; c)|_Q \pmod p,$$

where F is the standard Gauss hypergeometric function and the notation signifies that we are to take the hypergeometric series $F(A, B, C; z)$ only through the z^Q term inclusive. Also derive the formal relation

$$T = (1-c)^{Q/2} P_Q\left(\frac{1 - c/2}{\sqrt{1-c}}\right),$$

where P_Q is the classical Legendre polynomial of order Q. Using known transformation properties of such special series, find some closed-form curve orders. For example, taking $p \equiv 1 \pmod 4$ and the known evaluation

$$P_Q(0) = \binom{Q}{Q/2}$$

one can derive that curve order is $\#E = p + 1 \pm 2a$, where the prime p is represented as $p = a^2 + b^2$. Actually, this kind of study connects with algebraic number theory; for example, the study of binomial coefficients (mod p) [Crandall et al. 1997] is useful in the present context.

Observe that the hypergeometric series can be evaluated in $O\left(\sqrt{p}\ln^2 p\right)$ field operations, by appeal to fast series evaluation methods [Borwein and Borwein 1987] (and see Algorithm 9.6.7). This means that, at least for elliptic curves of the type specified, we have yet another point-counting algorithm whose complexity lies essentially between naive residue counting and the Shanks–Mestre algorithm. There is yet one more possible avenue of exploration: The DAGM of Exercise 2.39 might actually apply to truncated hypergeometric series (mod p) in some sense, which we say because the *classical* AGM—for real arguments—is a rapid means of evaluating such as the hypergeometric form above [Borwein and Borwein 1987].

7.27. Along the lines of Exercise 7.26, show that for a prime $p \equiv 1 \pmod 8$, the elliptic curve E with governing cubic

$$y^2 = x^3 + \frac{3}{\sqrt{2}}x^2 + x$$

has order

$$\#E = p + 1 - \left(2^{(p-1)/4}\binom{\frac{p-1}{2}}{\frac{p-1}{4}} \bmod_{\pm} p\right),$$

where the $\bmod_{\pm}$ notation means that we take the signed residue nearest 0. Does this observation have any value for factoring of Fermat numbers? Here are some observations. We do know that any prime factor of a composite F_n is $\equiv 1 \pmod 8$, and that $3/\sqrt{2}$ can be written modulo any Fermat number $F_n > 5$ as $3(2^{3m/4} - 2^{m/4})^{-1}$, with $m = 2^n$; moreover, this algebra works modulo any prime factor of F_n. In this connection see [Atkin and Morain 1993a], who show how to construct advantageous curves when potential factors p are known to have certain congruence properties.

7.28. Implement the ECM variant of [Peralta and Okamoto 1996], in which composite numbers $n = pq^2$ with p prime, q odd, are attacked efficiently. Their result depends on an interesting probabilistic way to check whether $x_1 \equiv x_2 \pmod p$; namely, choose a random r and check whether the Jacobi symbol equality

$$\left(\frac{x_1 + r}{n}\right) = \left(\frac{x_2 + r}{n}\right)$$

holds, which check can be performed, remarkably, in ignorance of p.

7.29. Here is a fascinating line of research in connection with Schoof point counting, Algorithm 7.5.6. First, investigate the time and space (memory) tradeoffs for the algorithm, as one decides upon one of the following representation options: (a) the rational point representations $(N(x)/D(x), yM(x)/C(x))$ as we displayed; (b) a projective description $(X(x,y), Y(x,y), Z(x,y))$ along the lines of Algorithm 7.2.3; or (c) an affine representation. Note that these options have the *same* basic asymptotic complexity, but we are talking here about implementation advantages, e.g., the implied big-O constants.

Such analyses have led to actual packages, not only for the "vanilla Schoof" Algorithm 7.5.6, but the sophisticated SEA variants. Some such packages are highly efficient, able to resolve the curve order for a 200-bit value of p in a matter of minutes. For example, there is the implementation of [Scott 1999], which uses projective coordinates and the Shoup method (see Exercise 9.70) for polynomial multiplication and for the SEA extension uses precomputed polynomials.

But there is another tantalizing option: Employ Montgomery representation, as in Algorithm 7.2.7, for which the Schoof relation

$$\left(x^{p^2}, y^{p^2}\right) + [k](x,y) = [t](x^p, y^p)$$

can be analyzed in x-coordinates alone. One computes x^{p^2} (but no powers of y), uses division polynomials to find the x-coordinate of $[k](x,y)$ (and perhaps the $[t]$ multiple as well), and employs Algorithm 7.2.8 to find doubly-ambiguous values of t. This having been done, one has a "partial-CRT" scenario that is itself of research interest. In such a scenario, one knows not a specific $t \bmod l$ for each small prime l, but a *pair* of t values for each l. At first it may seem that we need twice as many small primes, but not really so.

If one has, say, n smaller primes $l_1, \ldots, l_n$ one can perform at most 2^n elliptic multiplies to see which genuine curve order annihilates a random point. One might say that for large n this is too much work, but one could just use the x-coordinate arithmetic only on some of the larger l. So the research problem is this: Given that x-coordinate (Montgomery) arithmetic is less expensive than full (x, y) versions, how does one best handle the ambiguous t values that result? Besides the 2^n continuation, is there a Shanks–Mestre continuation that starts from the partial-CRT decomposition? Note that in all of this analysis, one will sometimes get the advantage that $t = 0$, in which case there is no ambiguity of $(p + 1 \pm t) \bmod l$.

7.30. In Exercise 7.21 was outlined "symbolic" means for carrying out Schoof calculations for an elliptic curve order. Investigate whether the same manipulations can be effected, again (mod 3), for curves governed by

$$y^2 = x^3 + ax,$$

or for that matter, curves having *both* a, b nonzero—which cases you would expect to be difficult. Investigate whether any of these ideas can be effected for small primes $l > 3$.

7.31. Describe how one may use Algorithm 7.5.10 to create a relatively simple primality-proving program, in which one would search only for discriminant-D curves with $h(D) = 1, 2$. The advantage of such a scheme is obvious: The elliptic curve generation is virtually immediate for such discriminants. The primary disadvantage, of course, is that for large probable primes under scrutiny, a great deal of effort must go into factoring the severely limited set of curve orders (one might even contemplate an ECM factoring engine, to put extra weight on the factoring part of ECPP). Still, this could be a fine approach for primes of a few hundred binary bits or less. For one thing, neither floating-point class-polynomial calculations nor massive polynomial storage nor sophisticated root-finding routines would be required.

7.32. There is a way to simplify somewhat the elliptic curve computations for ECPP. Argue that Montgomery parameterization (as in Algorithm 7.2.7) can certainly be used for primality proofs of some candidate n in the ECPP Algorithms 7.6.2 or 7.5.9, *provided* that along with the conditions of nonvanishing for multiples $(X', Z') = [m/q](X, Z)$, we always check $\gcd(Z', n)$ for possible factors of n.

Describe, then, some enhancements to the ECPP algorithms that we enjoy when Montgomery parameterization is in force. For example, finding a point on a curve is simpler, because we only need a valid x-coordinate, and so on.

7.33. Here is a peculiar form of "rapid ECPP" that can—if one has sufficient luck—work to effect virtually instantaneous primality proofs. Recall, as in Corollary 4.1.4, that if a probable prime n has $n - 1 = FR$ where the factored part F exceeds $\sqrt{n}$ (or in various refinements exceeds an even lesser bound), then a primality proof can be effected quickly. Consider instead a scenario in which the same "FR" decomposition is obtained, but we are lucky to be able to write

$$R = \alpha F + \beta,$$

with a representation $4\alpha = \beta^2 + |D|\gamma^2$ existing for fundamental discriminant $-|D|$. Show that, under these conditions, if n is prime, there then exists a CM curve E for discriminant $-|D|$, with curve order given by the attractive relation

$$\#E = \alpha F^2.$$

Thus, we might be able to have F nearly as small as $n^{1/4}$, and still effect an ECPP result on n.

Next, show that a McIntosh–Wagstaff probable prime of the form $n = (2^q+1)/3$ always has a representation with discriminant $D = -8$, and give the corresponding curve order. Using these ideas, prove that $(2^{313}+1)/3$ is prime, taking account of the fact that the curve order in question is $\#E = (2/3)h^2$, where h is

$$3^2 \cdot 5 \cdot 7 \cdot 13^2 \cdot 53 \cdot 79 \cdot 157 \cdot 313 \cdot 1259 \cdot 1613 \cdot 2731 \cdot 3121 \cdot 8191 \cdot 21841 \cdot 121369 \cdot 22366891.$$

Then prove another interesting corollary: If

$$n = 2^{2r+2m} + 2^{r+m+1} + 2^{2r} + 1$$

is prime, then the curve E in question has

$$\#E = 2^{2r}(2^{2m} + 1).$$

In this manner, and by analyzing the known algebraic factors of $2^{2m}+1$ when m is odd, prove that

$$n = 2^{576} + 2^{289} + 2^2 + 1$$

is prime.

For more information on "rapid" primality proofs, see [Pomerance 1987a] and the discussion in [Williams 1998, p. 366] in regard to numbers of certain ternary form.

7.34. An interesting problem one may address after having found a factor via an ECM scheme such as Algorithm 7.4.4 is this: What is the actual group order that allowed the factor discovery?

One approach, which has been used by [Brent et al. 2000], is simply to "backtrack" on the stage limits until the precise largest- and second-largest primes are found, and so on until the group order is completely factored.

But another way is simply to obtain, via Algorithm 7.5.6, say, the actual order. To this end, work out the preparatory curve algebra as follows. First, show that if a curve is constructed according to Theorem 7.4.3, then the rational initial point $x/z = u^3/v^3$ satisfies

$$x^3 + Cx^2z + xz^2 = \left(\sigma^2 - 5\right)^3 \left(125 - 105\sigma^2 - 21\sigma^4 + \sigma^6\right)^2$$

in the ring. Then deduce that the order of the curve is either the order of

$$y^2 = x^3 + ax + b,$$

or the order of the twist, depending respectively on whether $(\frac{\sigma^3 - 5\sigma}{p}) = 1$ or -1, where affine parameters a, b are computed from

$$\gamma = \frac{(v-u)^3(3u+v)}{4u^3v} - 2,$$

$$a = 1 - \frac{1}{3}\gamma^2,$$

$$b = \frac{2}{27}\gamma^2 - \frac{1}{3}\gamma.$$

These machinations suggest a straightforward algorithm for finding the order of the curve that discovered a factor p. Namely, one uses the starting seed σ, calculates again if necessary the u, v field parameters, then applies the above formulae to get an affine curve parameter pair (a, b), which in turn can be used directly in the Schoof algorithm.

Here is an explicit example of the workings of this method. The McIntosh–Tardif factor

$$p = 81274690703860512587777$$

of F_{18} was found with seed parameter $\sigma = 16500076$. One finds with the above formulae that

$$a = 26882295688729303004012,$$

$$b = 10541033639146374421403,$$

and Algorithm 7.5.6 determines the curve order as

$$\begin{aligned}\#E &= 81274690703989163570820\\ &= 2^2 \cdot 3 \cdot 5 \cdot 23 \cdot 43 \cdot 67 \cdot 149 \cdot 2011 \cdot 2341 \cdot 3571 \cdot 8161.\end{aligned}$$

Indeed, looking at the two largest prime factors here, we see that the factor could have been found with respective stage limits as low as $B_1 = 4000$, $B_2 = 10000$. R. McIntosh and C. Tardif actually used 100000, 4000000, respectively, but as always with ECM, what we might call post-factoring hindsight is a low-cost commodity. Note also the explicit verification that the Brent parameterization method indeed yields a curve whose order is divisible by twelve, as expected.

If you are in possession of sufficiently high-precision software, here is another useful test of the above ideas. Take the known prime factor $p = 4485296422913$ of F_{21}, and for the specific Brent parameter $\sigma = 1536151048$, find the elliptic-curve group order $\pmod{p}$, and show that stage limits $B_1 = 60000$, $B_2 = 3000000$ (being the actual pair used originally in practice to drive this example of hindsight) suffice to discover the factor p.

Chapter 8

THE UBIQUITY OF PRIME NUMBERS

It is often remarked that prime numbers finally found a legitimate practical application in the domain of cryptography. The cryptographic relevance is not disputed, but there are many other applications of the majestic primes. Some applications are industrial—such as applications in numerical analysis, applied mathematics, and other applied sciences—while some are of the "conceptual feedback" variety, in which primes and their surrounding concepts are used in theoretical work outside of, say, pure number theory. In this lucrative research mode, primes are used within algorithms that might appear *a priori* independent of primes, and so on. It seems fair to regard the prime number concept as ubiquitous, since the primes appear in so very many disparate domains of thought.

8.1 Cryptography

On the face of it, the prime numbers apply to cryptography by virtue of the extreme difficulty of certain computations. Two such problems are factoring and the discrete logarithm problem. We shall discuss practical instances of these problems in the field of cryptography, and also discuss elliptic curve generalizations.

8.1.1 Diffie–Hellman key exchange

In a monumental paper [Diffie and Hellman 1976], those authors observed the following "one-way function" behavior of certain group operations. For a given integer $x \geq 0$ and an element g of $\mathbf{F}_p^*$, the computation of

$$h = g^x$$

in the field (so, involving continual (mod p) reductions) is generally of complexity $O(\ln x)$ field operations. On the other hand, solving this equation for x, assuming g, h, p given, is evidently very much harder. As x is an exponent, and since we are taking something like a logarithm in this latter problem, the extraction of the unknown x is known as the discrete logarithm (DL) problem. Though the forward (exponentiation) direction is of polynomial-time complexity, no general method is known for obtaining the DL with anything like that efficiency. Some DL algorithms are discussed in Chapter 5 and in [Schirokauer et al. 1996].

An immediate application of this "one-way" feature of exponentiation is a cryptographic algorithm so simple that we simply state it in English without formal exhibition. Say you want individuals to have their own passwords to allow entry onto a computer system or information channel. A universal prime p and primitive root g are chosen for the whole system of users. Now each individual user "thinks up" his or her secret password x, an integer, and computes $h = g^x \bmod p$, finally storing his or her h value on the system itself. Thus for the array of users, there is a stored array of h values on the system. Now when it is time to gain entry to the system, a user need only type the "password" x, and the system exponentiates this, comparing the result to that user's h. The scheme is all very simple, depending on the difficulty of looking at an h and inferring what was the password x for that h.

Not quite so obvious, but equally elegant, is the Diffie–Hellman key exchange scheme, which allows two individuals to create a common encryption key:

Algorithm 8.1.1 (Diffie–Hellman key exchange). Two individuals, Alice and Bob, agree on a prime p and a generator $g \in \mathbf{F}_p^*$. This algorithm allows Alice and Bob to establish a mutual key (mod p), with neither individual being able (under DL difficulty) to infer each other's secret key.

1. [Alice generates public key]
   Alice chooses random $a \in [2, p-2]$; // Alice's secret key.
   $x = g^a \bmod p$; // x is Alice's public key.
2. [Bob generates public key]
   Bob chooses random $b \in [2, p-2]$; // Bob's secret key.
   $y = g^b \bmod p$; // y is Bob's public key.
3. [Each individual creates the same mutual key]
   Bob computes $k = x^b \bmod p$;
   Alice computes $k = y^a \bmod p$; // The two k-values are identical.

This mutual key creation works, of course, because

$$(g^a)^b = (g^b)^a = g^{ab},$$

and all of this goes through with the usual reductions (mod p). There are several important features of this basic Diffie–Hellman key exchange notion. First, note that in principle Alice and Bob could have avoided random numbers; choosing instead a memorable phrase, slogan, whatever, and made those into respective secret values a, b. Second, note that the public keys $g^a, g^b \bmod p$ can be made public in the sense that—under DL difficulty—it is safe literally to publish such values to the world. Third, on the issue of what to do with the mutual key created in the algorithm, actual practical applications often involve the use of the mutual key to encrypt/decrypt long messages, say through the expedient of a standard block cipher such as DES [Schneier 1996]. Though it is easy to break the Diffie–Hellman scheme given a fast DL method, it is unclear whether the two problems are equivalent. That is, if an oracle could tell you g^{ab} on input of g^a and g^b, could you use this oracle to quickly solve for discrete logarithms?

8.1.2 RSA cryptosystem

Soon after the Diffie–Hellman ideas, the now prevalent RSA cryptosystem was invented by Rivest, Shamir, and Adleman [Rivest et al. 1978].

Algorithm 8.1.2 (RSA private/public key generation). In this algorithm we generate an individual's private and associated public keys for the RSA cryptosystem.

1. [Choose primes]
   Choose two distinct primes p, q under prevailing safety criteria (see text);
2. [Generate public key]
   $N = pq$;
   $\varphi = (p-1)(q-1)$; // Euler totient of N.
   Choose random integer $E \in [3, N-2]$ coprime to φ;
   Report public key as (N, E); // User publishes this key.
3. [Generate private key]
   $D = E^{-1} \bmod \varphi$;
   Report private key as D; // User keeps D secret.

The primary observation is that because of the difficulty of factoring $N = pq$, the public integer N does not give an easy prescription for the private primes p, q. Furthermore, it is known that if one knows integers D, E in $[1, n-1]$ with $DE \equiv 1 \pmod{\varphi}$, then one can factor N in (probabilistic) polynomial time [Long 1981] (cf. Exercise 5.26). In the above algorithm it is fashionable to choose approximately equal private primes p, q, but some cryptographers suggest further safety tests. In fact, one can locate in the literature a host of potential drawbacks for certain p, q choices. There is a brief but illuminating listing of possible security flaws that depend on the magnitudes and other number-theoretical properties of p, q in [Williams 1998, p. 391]. The reference [Bressoud and Wagon 2000, p. 249] also lists RSA pitfalls. See also Exercise 8.2 for a variety of RSA security issues.

Having adopted the notion that the public key is the hard-to-break (i.e., difficult to factor) composite integer $N = pq$, we can proceed with actual encryption of messages, as follows:

Algorithm 8.1.3 (RSA encryption/decryption). We assume that Alice possesses a private key D_{A} and public key $(N_{\mathrm{A}}, E_{\mathrm{A}})$ from Algorithm 8.1.2. Here we show how another individual (Bob) can encrypt a message x (thought of as an integer in $[0, N_{\mathrm{A}})$) to Alice, and how Alice can decrypt said message.

1. [Bob encrypts]
   $y = x^{E_{\mathrm{A}}} \bmod N_{\mathrm{A}}$; // Bob is using Alice's public key.
   Bob then sends y to Alice;
2. [Alice decrypts]

Alice receives encrypted message y;
$x = y^{D_A} \bmod N_A$; // Alice recovers the original x.

It is not hard to see that, as required for Algorithm 8.1.3 to work, we must have

$$x^{DE} \equiv x \pmod{N}.$$

This, in turn, follows from the fact that $DE = 1 + k\varphi$ by construction of D itself, so that $x^{DE} = x(x^{\varphi})^k \equiv x \cdot 1^k = x \pmod{N}$, when $\gcd(x, N) = 1$. In addition, it is easy to see that $x^{DE} \equiv x \pmod{N}$ continues to hold even when $\gcd(x, N) > 1$.

Now with the RSA scheme we envision a scenario in which a great number of individuals all have their respective public keys (N_i, E_i) literally published—as one might publish individual numbers in a telephone book. Any individual may thereby send an encrypted message to individual j by casually referring to the public (N_j, E_j) and doing a little arithmetic. But can the recipient j know from whom the message was encrypted and sent? It turns out, yes, to be quite possible, using a clever digital signature method:

Algorithm 8.1.4 (RSA signature: Simple version). We assume that Alice possesses a private key D_A and public key (N_A, E_A) from Algorithm 8.1.2. Here we show how another individual (Bob) having private key D_B and public key (N_B, E_B) can "sign" a message x (thought of as an integer in $[0, \min\{N_A, N_B\})$).

1. [Bob encrypts with signature]
   $s = x^{D_B} \bmod N_B$; // Bob creates signature from message.
   $y = s^{E_A} \bmod N_A$; // Bob is using here Alice's public key.
   Bob then sends y to Alice;
2. [Alice decrypts]
   Alice receives signed/encrypted message y;
   $s = y^{D_A} \bmod N_A$; // Alice uses her private key.
   $x = s^{E_B} \bmod N_B$; // Alice recovers message using Bob's public key.

Note that in the final stage, Alice uses Bob's public key, the idea being that—up to the usual questions of difficulty or breakability of the scheme—only Bob could have originated the message, because only he knows private key D_B. But there are weaknesses in this admittedly elegant signature scheme. One such is this: If a forger somehow prepares a "factored message" $x = x_1x_2$, and somehow induces Bob to send Alice the signatures y_1, y_2 corresponding to the component messages x_1, x_2, then the forger can later pose as Bob by sending Alice $y = y_1y_2$, which is the signature for the composite message x. In a sense, then, Algorithm 8.1.4 has too much symmetry. Such issues can be resolved nicely by invoking a "message digest," or hash function, at the signing stage [Schneier 1996], [Menezes et al. 1997]. Such standards as SHA-1 provide such a hash function H, where if x is plaintext, $H(x)$ is an integer (often much smaller, i.e., having many fewer bits, than x). In this way certain methods for breaking signatures—or false signing—would be suppressed. A signature scheme involving a hash function goes as follows:

Algorithm 8.1.5 (RSA encrypt-with-signature: More practical version). We assume that Bob possesses a private key D_B and public key (N_B, E_B) from Algorithm 8.1.2. Here we show how Alice can recover Bob's plaintext message x (thought of as an integer in some appropriate interval) and also verify Bob's signature. We assume the existence of message digest function H, such as from the SHA-1 standard.

1. [Bob encrypts with signature]
   $y = x^{E_A} \bmod N_A$; // Bob encrypts, using Alice's public key.
   $y_1 = H(x)$; // y_1 is the "hash" of plaintext x.
   $s = y_1^{D_B} \bmod N_B$; // Bob creates signature s.
   Bob sends (y, s) (i.e., combined message/signature) to Alice;
2. [Alice decrypts]
   Alice receives (y, s);
   $x = y^{D_A} \bmod N_A$; // Alice decrypts to recover plaintext x.
3. [Alice processes signature]
   $y_2 = s^{E_B} \bmod N_B$;
   if($y_2 == H(x)$) Alice accepts signature;
      else Alice rejects signature;

We note that there are practical variants of this algorithm that do not involve actual encryption; e.g., if plaintext security is not an issue while only authentication is, one can simply concatenate the plaintext and signature, as (x, s) for transmission to Alice. Note also there are alternative, yet practical signature schemes that depend instead on a so-called redundancy function, as laid out, for example, in [Menezes et al. 1997].

8.1.3 Elliptic curve cryptosystems (ECCs)

The mid-1980s saw the emergence of yet another fascinating cryptographic idea, that of using elliptic curves in cryptosystems [Miller 1987], [Koblitz 1987]. Basically, elliptic curve cryptography (ECC) involves a public curve $E_{a,b}(F)$ where F is a finite field. Prevailing choices are $F = \mathbf{F}_p$ for prime p, and $F = \mathbf{F}_{2^k}$ for suitable integers k. We shall focus primarily on the former fields $\mathbf{F}_p$, although much of what we describe works for finite fields in general. The central idea is that given points $P, Q \in E$ such that the relation

$$Q = [k]P$$

holds for some integer k, it should be hard in general to extract the elliptic discrete logarithm (EDL), namely a value for the integer multiplier k. There is by now a considerable literature on the EDL problem, of which just one example work is [Lim and Lee 1997], in which it is explained why the group order's character (prime or composite, and what kind of factorization) is important as a security matter.

The Diffie–Hellman key exchange protocol (see Algorithm 8.1.1) can be used in a cyclic subgroup of any group. The following algorithm is Diffie–Hellman for elliptic-curve groups.

Algorithm 8.1.6 (ECC key exchange). Two individuals, Alice and Bob, agree on a public elliptic curve E and a public point $P \in E$ whose point order is n. (In many scenarios, n is prime, or admits of a large prime factor.) This algorithm produces a mutual key.

1. [Alice generates public key]
   Alice chooses random $K_{\rm A} \in [2, n-2]$; // Alice's secret key.
   $Q = [K_{\rm A}]P$; // Point Q is Alice's public key.
2. [Bob generates public key]
   Bob chooses random $K_{\rm B} \in [2, n-2]$; // Bob's secret key.
   $R = [K_{\rm B}]P$; // Point R is Bob's public key.
3. [Each individual creates the unique mutual key]
   Bob computes point $K = [K_{\rm B}]Q$;
   Alice computes point $K = [K_{\rm A}]R$. // Results agree.

That the mutual key is unique follows directly from the group rules, as

$$[K_{\rm B}]([K_{\rm A}]P) = [K_{\rm B}K_{\rm A}]P = [K_{\rm A}K_{\rm B}]P = [K_{\rm A}]([K_{\rm B}]P).$$

Again the notion of the difficulty of Bob, say, discovering Alice's private key $K_{\rm A}$ is presumably the difficulty of EDL. That is, if EDL is easy, then the ECC key exchange is not secure; and, it is thought that the converse is true as well. Note that in ECC implementations, private keys are integers, usually roughly the size of p (but could be larger than p—recall that the group order $\#E$ can itself slightly exceed p), while public keys and the exchanged mutual key are *points*. Typically, some bits of a mutual key would be used in, say, a block cipher; for example, one might take the bits of the x-coordinate.

A primary result in regard to the EDL problem is the so-called "MOV theorem," which states essentially that the EDL problem over $\mathbf{F}_p$ is equivalent to the normal DL problem over $\mathbf{F}^*_{p^B}$, for some B [Menezes et al. 1993]. There is a practical test for the estimated level of security in an ECC system—call this level the MOV threshold—see [Solinas 1998]. In practice, the MOV threshold B is "about 10," but depends, of course, on the prevailing complexity estimate for the DL problem in finite fields. Note, however, that "supersingular" curves, having order $\#E = p+1$, are particularly susceptible, having EDL complexity known to be no worse than that of the DL problem in $\mathbf{F}_{p^k}$, some $k \le 6$ [Menezes et al. 1993]. Such curves can be ruled out *a priori* for the reason stated.

There is also the so-called Semaev–Smart–Satoh–Araki attack, when the order is $\#E = p$, based on p-adic arithmetic. (The 1998 announcement of [Smart 1999] caused a noticeable ripple in the cryptography field, although the theoretical knowledge is older than the announcement; see [Semaev 1998], [Satoh and Araki 1998].) Another possible security flaw in elliptic curve cryptography is yet unresolved, and has to do with CM curves. In [Müller and Paulus 1998] it is argued that perhaps certain CM curves are susceptible to index-calculus attacks.

Incidentally, the question of how one finds elliptic curves of prime order (and so having elements of prime order) is itself interesting. One approach is just to generate random curves and assess their orders via Algorithm 7.5.6. Another is to use Algorithm 7.5.9 or 7.5.10 to generate possible orders, and when a prime order is found, go ahead and specify a curve with that order. But there are clever variants of these basic approaches (see Exercise 8.23). It should be remarked that some cryptographers accept curves of order $\#E = fr$, where f may consist of small prime factors while r is a large prime. For such curves, one still prefers to find points of the prime order r, and this can be done very simply:

Algorithm 8.1.7 (Find a point of prime order). Given an elliptic curve $E_{a,b}(\mathbf{F}_p)$ of order $\#E = fr$, where r is prime, this algorithm endeavors to find a point $P \in E$ of order r.

1. [Find starting point]
   Choose a random point $P \in E$, via Algorithm 7.2.1;
2. [Check multiple]
   $Q = [f]P$;
   if($Q == O$) goto [Find starting point];
   return Q; // A point of prime order r.

The algorithm is admittedly almost trivial, but important in cryptography applications. One such application is elliptic signature. There is a standard elliptic-curve digital signature scheme that runs like so, with the prerequisite of a point of prime order evident right at the outset:

Algorithm 8.1.8 (Elliptic curve digital signature algorithm (ECDSA)). This algorithm provides functions for key generation, signing, and verification of messages. A message is generally denoted by M, an integer, and it is assumed that a suitable hash function h is in hand.

1. [Alice generates key]
   Alice chooses a curve E, whose order $\#E = fr$ with r a "large" prime;
   Alice finds point $P \in E$ of order r, via Algorithm 8.1.7;
   Alice chooses random $d \in [2, r-2]$;
   $Q = [d]P$;
   Alice publishes public key (E, P, r, Q); // Private key is d.
2. [Alice signs]
   Alice chooses random $k \in [2, r-2]$;
   $(x_1, y_1) = [k]P$;
   $R = x_1 \bmod r$; // Note that $R \neq 0$.
   $s = k^{-1}(h(M) + Rd) \bmod r$;
   if($s == 0$) goto [Alice signs];
   Alice's signature is the pair (R, s), transmitted with message M;
3. [Bob verifies]
   Bob obtains Alice's public key (E, P, r, Q);
   $w = s^{-1} \bmod r$;

$u_1 = h(M)w \bmod r$;
$u_2 = Rw \bmod r$;
$(x_0, y_0) = [u_1]P + [u_2]Q$;
$v = x_0 \bmod r$;
if($v == R$) Bob accepts signature;
    else Bob rejects signature;

This algorithm is modeled on an older DSA standard, and amounts to the natural elliptic-curve variant of DSA. Modern details and issues are discussed in [Johnson and Menezes 1999]. The hash value $h(M)$ is, technically speaking, supposed to be effected via another standard, the SHA-1 hash function [Jurišić and Menezes 1997]. Those authors also discuss the interesting issue of security. They conclude that a 1024-bit DSA system is about as secure as a 160-bit ECDSA system. If valid, such an observation shows once again that, on our current knowledge, the EDL problem is about as hard as a computational number-theoretical problem can be.

Incidentally, there is a different way to effect a signature scheme with elliptic curves, which is the El Gamal scheme. We do not write out the algorithm—it is less standard than the above ECDSA scheme (but no less interesting))—but the essentials lie in Algorithm 8.1.10. Also, the theoretical ideas are found in [Koblitz 1987].

We have mentioned, in connection with RSA encryption, the practical expedient of using the sophisticated methods (RSA, ECC) for a key exchange, then using the mutually understood key in a rapid block cipher, such as DES, say. But there is another fascinating way to proceed with a kind of "direct" ECC scheme, based on the notion of embedding plaintext as points on elliptic curves. In this fashion, all encryption/decryption proceeds with nothing but elliptic algebra at all phases.

Theorem 8.1.9 (Plaintext-embedding theorem). *For prime $p > 3$ let E denote an elliptic curve over $\mathbf{F}_p$, with governing cubic*

$$y^2 = x^3 + ax + b.$$

Let X be any integer in $[0, p-1]$. Then X is either an x-coordinate of some point on E, or on the twist curve E' whose governing cubic is $gy^2 = x^3 + ax + b$, for some g with $\left(\frac{g}{p}\right) = -1$. Furthermore, if $p \equiv 3 \pmod 4$, and we assign

$$s = X^3 + aX + b \bmod p,$$

$$Y = s^{(p+1)/4} \bmod p,$$

then (X, Y) is a point on either E, E', respectively, as

$$Y^2 \equiv s, \ -s \pmod p,$$

where in the latter case we take the governing cubic for E' to be $-y^2 = x^3 + ax + b$.

This theorem is readily proved via the same twist algebra that we encountered in Theorem 7.5.2 and Exercise 7.16, and leads to the following algorithm for direct-embedding encryption:

Algorithm 8.1.10 (Direct-embedding ECC encryption). This algorithm allows encryption/decryption using exclusively elliptic algebra, i.e., with no intermediary cipher, via the direct embedding of plaintext onto curves. We assume that Alice and Bob have agreed upon a public curve $E_{a,b}(\mathbf{F}_p)$ with its twist curve E', on which lie respectively public points P, P'. In addition, it is assumed that Bob has generated respective public keys $P_{\rm B} = [K_{\rm B}]P, P'_{\rm B} = [K_{\rm B}]P'$, as in Algorithm 8.1.6. We denote by X a parcel of plaintext (an integer in $[0, \ldots, p-1]$) that Alice wishes to encrypt for Bob.

1. [Alice embeds plaintext X]
   Alice determines the curve E or E' on which X is a valid x-coordinate (and, if y-coordinates are relevant, computes such number Y) via Theorem 8.1.9, taking the curve to be E if X is on both curves;
   // See Exercise 8.5.
   Depending respectively on which curve E, E' is in force, Alice sets respectively:
   $d = 0$ or 1; // Curve-selecting bit.
   $Q = P$ or P';
   $Q_{\rm B} = P_{\rm B}$ or $P'_{\rm B}$.
   Alice chooses random $r \in [2, p-2]$;
   $U = [r]Q_{\rm B} + (X, Y)$; // Elliptic add, to obfuscate plaintext.
   $C = [r]Q$; // The "clue" for undoing the obfuscation.
   Alice transmits a parcel (encrypted message, clue, bit) as (U, C, d);
2. [Bob decrypts to get plaintext X]
   Bob inspects d to determine on which curve elliptic algebra will proceed;
   $(X, Y) = U - [K_{\rm B}]C$; // Private key applied with elliptic subtract.
   Bob now recovers the plaintext as the x-coordinate X;

This method will be recognized as an El Gamal embedding scheme, where we have made some improvements over previous renditions [Koblitz 1987], [Kaliski 1988]. Note that the last part of Theorem 8.1.9 allows Algorithm 8.1.10 to proceed efficiently when the field characteristic has $p \equiv 3 \pmod 4$. In practical implementations of Algorithm 8.1.10, there are two further substantial improvements one may invoke. First, the y-coordinates are not needed if one uses Montgomery coordinates (Algorithm 7.2.7) throughout and carefully applies Algorithm 7.2.8 at the right junctures. Second, the "clue" point C of the algorithm effectively doubles the transmitted data size. This, too, can be avoided by carefully setting up a random number exchange protocol, so that the random number r itself is deterministically kept in synchrony by the two parties. (The authors are indebted to B. Garst for this observation, which in fact has led to a U. S. Patent [Crandall and Garst 2001].) See Exercise 8.3 for more detail on such enhancements. If properly done, one obtains a fairly efficient, elegant direct-embedding scheme with—asymptotically speaking—no data expansion.

8.1.4 Coin-flip protocol

In cryptography, a protocol is essentially an algorithm specifying—in a certain order—the steps that involved parties must take. We have seen key-exchange and related protocols already. Here we investigate an intriguing cultural application of number-theoretical protocols. How can one toss a coin, fairly, over the telephone? Or play poker among n individuals, playing "blind" on a network? We assume the worst: That no party trusts any other, yet a decision has to be reached, as one would so reach it via a coin toss, with one party calling heads or tails. It turns out that such a remote tossing is indeed possible, using properties of certain congruences.

Incidentally, the motivation for even having a coin-flip protocol is obvious, when one imagines a telephone conversation—say between two hostile parties involved in a lawsuit—in which some important result accrues on the basis of a coin flip, meaning a random bit whose statistics cannot be biased by either party. Having one party claim they just flipped a head, and therefore won the toss, is clearly not good enough. Everyone must be kept honest, and this can be done via adroit application of congruences involving primes or certain composites. Here is one way to proceed, where we have adapted some ideas from [Bressoud and Wagon 2000] on simple protocols:

Algorithm 8.1.11 (Coin-flip protocol). Alice and Bob wish to "flip a fair coin," using only a communication channel. They have agreed that if Bob guesses correctly, below, then Bob wins, otherwise Alice wins.

1. [Alice selects primes]
   Alice chooses two large primes $p < q$, forms the number $n = pq$, and chooses a random prime r such that $\left(\frac{n}{r}\right) = -1$;
2. [Alice sends Bob partial information]
   Alice sends Bob n and r;
3. [Bob chooses]
   Bob makes a choice between "the smaller prime factor of n is a quadratic residue mod r" and "the larger prime factor of n is a quadratic residue mod r" and sends this choice to Alice;
4. [Alice announces winner]
   Alice announces whether Bob is correct or not, and sends him the primes p, q so that Bob can see for himself that she is not cheating;

It is interesting to investigate the cryptographic integrity of this algorithm; see Exercise 8.7. Though we have cast the above algorithm in terms of winner and loser, it is clear that Alice and Bob could use the same method just to establish a random bit, say "0" if Alice wins and "1" if Bob wins. There are many variants to this kind of coin-flip protocol. For example, there is a protocol in [Schneier 1996] in which four square roots of a number $n = pq$

are generated by Alice and sent to Bob, with Bob having generated a random square modulo n. This scenario is not as simple as Algorithm 8.1.11, but it is replete with interesting issues; e.g., one can extend it to handle the peculiar Micali scenario in which Bob intentionally loses [Schroeder 1997]. There are also algorithms based on Blum integers and, generally, the fact of a product pq allowing multiple roots (see Exercise 8.6). These ideas can be extended in a natural way to a poker-playing protocol in which a number of players claim what poker hands they possess, and so on [Goldwasser and Micali 1982].

8.2 Random-number generation

The problem of generating random numbers goes back, of course, to the dawn (1940s, say) of the computer age. It has been said that to generate random numbers via machine arithmetic is to live, in the words of J. von Neumann, "in a state of sin." Though machines can ensure nearly random statistics in many senses, there is the problem that conventional machine computation is deterministic, so the very notion of randomness is suspect in the world of Turing machines and serial programs. If the reader wonders what kind of technology could do better in the matter of randomness (though still not "purely" random in the sense of probability theory), here is one exotic example: Aim a microwave receiving dish at the remote heavens, listening to the black-body "fossil" radiation from the early cosmos, and digitize that signal to create a random bitstream. We are not claiming the cosmos is truly "random," but one does expect that a signal from remote regions is as "unknowable" as can be.

In modern times, the question of true randomness has more import than ever, as cryptographic systems in particular often require numbers that are as random, or as seemingly random, as can be. A deterministic generator that generates what looks to an eavesdropper like random numbers can be used to build a simple cryptosystem. Create a random bitstream. To encrypt a message, take the logical exclusive-or of bits of the message with bits of the random bitstream. To decrypt, do the exclusive-or operation again, against the same random bitstream. This cryptosystem is unbreakable, unless certain weaknesses are present—such as, the message is longer than the random stream, or the same random stream is reused on other messages, or the eavesdropper has special knowledge of the generator, and so on. In spite of such practical pitfalls, the scheme illustrates a fundamental credo of cryptography: Somehow, use something an eavesdropper does not know.

It seems that just as often as a new random-number generator is developed, so, too, is some older scheme shown to be nonrandom enough to be, say, "insecure," or yield misleading results in Monte Carlo simulations. We shall give a brief tour of random number generation, with a view, as usual, to the involvement of prime numbers.

8.2.1 Modular methods

The veritable workhorse of the random number generation industry has been the linear-congruential generator. This method uses an integer iteration

$$x_{n+1} = (ax_n + b) \bmod m,$$

where a, b, m are integer constants with $m > 1$, which recursion is to be ignited by an initial "seed," say x_0. To this day there continue to appear research results on the efficacy of this and related generators. One variant is the multiplicative congruential generator, with recursion

$$x_{n+1} = (cx_n) \bmod m,$$

where in this case the seed x_0 is assumed coprime to m. In applications requiring a *random*() function that returns samples out of the real interval $[0, 1)$, the usual expedient is simply to use x_n/m.

Recurrences, like the two above, are eventually periodic. For random number generation it is desirable to use a recursion of some long period. It is easy to see that the linear-congruential generator has period at most m and the multiplicative congruential generator has period at most $m-1$. The linear case can—under certain constraints on the parameters—have the full period m for the sequence (x_n), while the multiplicative variety can have period $m-1$. Fundamental rules for the behavior of such generators are embodied in the following theorem:

Theorem 8.2.1 (Lehmer). *The linear-congruential generator determined by*

$$x_{n+1} = (ax_n + b) \bmod m$$

has period m *if and only if*

(1) $\gcd(b, m) = 1$,

(2) $p|a-1$ *whenever prime* $p|m$,

(3) $4|a-1$ *if* $4|m$.

Furthermore, the multiplicative congruential generator determined by

$$x_{n+1} = (cx_n) \bmod m$$

has period $m-1$ *if and only if*

(1) m *is prime,*

(2) c *is a primitive root of* m,

(3) $x_0 \not\equiv 0 \pmod{m}$.

Many computer systems still provide the linear scheme, even though there are certain flaws, as we shall discuss.

First we give an explicit, standard linear-congruential generator:

Algorithm 8.2.2 (32-bit random-number generator (Knuth, Lewis)). This algorithm provides seeding and random functions for a certain generator known to have fairly good statistical behavior. We take $M = 2^{32}$ as the generator modulus, and will speedily effect operations modulo M by logical "and" (&) with $M-1$. One first calls the $seed()$ procedure, then calls $random()$ successively to get random numbers.

1. [Procedure *seed*]
```
seed() {
    Choose starting seed x;                // x is an integer in [0, M − 1].
    return;
}
```
2. [Function *random*]
```
random() {
    x = (1664525x + 1013904223) & (M − 1);
    return x;                              // New random number.
}
```

Note that the "and" operation with $M-1$ is simply the taking of the low 32 bits of the number involved. Along similar lines, the popular generator

$$x_{n+1} = (16807x_n) \bmod M_{31},$$

where $M_{31} = 2^{31} - 1$ is a Mersenne prime, has enjoyed a certain success in passing many (but not all) experimental tests [Park and Miller 1988], [Press et al. 1996].

An interesting optimization of certain congruential generators has been forwarded by [Wu 1997]. The recursion is

$$x_{n+1} = \left((2^{30} - 2^{19})x_n\right) \bmod M_{61},$$

where the fact of M_{61} being a Mersenne prime allows some rapid arithmetic.

Algorithm 8.2.3 (Fast, 61-bit random generator). This algorithm provides seeding and random functions for the Wu generator, modulus $M = 2^{61} - 1$ and multiplier $c = 2^{30} - 2^{19}$. Though modular multiplications occur in principle, the explicit operations below are relegated to addition/subtraction, left/right shifts (<< / >>, respectively), and logical "and" (&) which acts as a certain mod operation.

1. [Procedure *seed*]
```
seed() {
    Choose starting seed x;                // x is an integer in [1, M − 1].
    return;
}
```
2. [Function *random*]
```
random() {
    x = (x >> 31) + ((x << 30)&M) − (x >> 42) − ((x << 19)&M);
    if(x < 0) x = x + M;
```

```
        return x;                                   // New random number.
    }
```

Thanks to the shifts and "and" operations, this algorithm involves no explicit multiplication or division. Furthermore, the generator fares well under some established statistical tests [Wu 1997]. Of course, this generator can be generalized, yet as with any machine generator, caution should be taken in choosing the parameters; for example, the parameters c, M should be chosen so that c is a primitive root for the prime M to achieve long period. We should also add an important caution: Very recent experiments and analyses have uncovered weaknesses in the generator of the type in Algorithm 8.2.3. Whereas this kind of generator evidently does well on spectral tests, there are certain bit-population statistics with respect to which such generators are unsatisfactory [L'Ecuyer and Simard 1999]. Even so, there are still good reasons to invoke such a generator, such as its very high speed, ease of implementation, and good performance on some, albeit not all, statistical tests.

Variants to these congruential generators abound. One interesting development concerns generators with extremely long periods. A result along such lines concerns random number generation via matrix–vector multiplication. If $\mathbf{T}$ is a $k \times k$ matrix, and $\vec{x}$ a k-component vector, we may consider the next vector in a generator's iteration to be $\vec{x} = \mathbf{T}\vec{x}$, say, with some rule for extracting bits or values from the current vector.

Theorem 8.2.4 (Golomb). *For prime p, denote by $\mathbf{M}_k(p)$ the group of nonsingular $k \times k$ matrices $\pmod p$, and let $\vec{x}$ be a nonzero vector in $\mathbf{Z}_p^k$. Then the iterated sequence*

$$\vec{x},\ \mathbf{T}\vec{x},\ \mathbf{T}^2\vec{x},\ \ldots$$

has period $p^k - 1$ if and only if the order of $\mathbf{T} \in \mathbf{M}_k(p)$ is $p^k - 1$.

This elegant theorem can be applied in the same fashion as we have constructed the previous iterative generators. However, as [Golomb 1982] and [Marsaglia 1991] point out, there are much more efficient ways to provide extreme periods. In this case it is appropriate for the key theorem to follow the algorithm description, because of the iterative nature of the generator.

Algorithm 8.2.5 (Long-period random generator).
This algorithm assumes input integers $b \ge 2$ and $r > s > 0$ and produces an iterative sequence of pseudorandom integers, each calculated from r previous values and a running carry bit c. We start with a (vector) seed/carry entity $\vec{v}$ with its first r components assumed in $[0, b-1]$, and last component $c = 0$ or 1.

1. [Procedure *seed*]

```
    seed() {
        Choose parameters b ≥ 2 and r > s > 0;
```

```
        Initialize a seed vector/carry: v⃗ = (v_1, ..., v_r, c);
        return;
    }
2. [Function random]
    random() {
        x = v_s − v_r − c;          // Get new x as function of previous values.
        if(x < 0) {
            x = x + b;
            c = 1;                  // A 'borrow' has occurred.
        } else c = 0;
        v⃗ = (x, v_1, ..., v_{r−1}, c);   // Shift the old v_r into oblivion.
        return x;                   // New random number.
    }
```

In practice, this algorithm can be impressive, to say the least. For example, using input parameters $b = 2^{64}, r = 30, s = 6$, so that we shall iterate

$$x_0 = v_6 - v_{30} - c,$$

with mod, carry, and shift understood from Algorithm 8.2.5, the period turns out to be

$$P \approx 10^{578},$$

which is one of myriad striking examples of the following theorem:

Theorem 8.2.6 (Marsaglia). *The random-number generator of Algorithm 8.2.5 has period*

$$P = \varphi(b^r - b^s + 1).$$

Thus, the period for our previous explicit example is really

$$\varphi\left(2^{64\cdot 30} - 2^{64\cdot 6} + 1\right) = 2^{64\cdot 30} - 2^{64\cdot 6} \approx 10^{578},$$

the argument of φ being prime. Note that a number produced by the generator can repeat without the subsequent number repeating; it is the vector $\vec{v}$ internal to the algorithm that is key to the length of the period. As there are on the order of b^r possible vectors $\vec{v}$, the Marsaglia theorem above makes some intuitive sense.

Another iterative generator is the discrete exponential generator determined by

$$x_{n+1} = g^{x_n} \pmod{N},$$

for given g, x_0, N. It has been studied by [Blum et al. 1986], [Lagarias 1990], [Friedlander et al. 2001], and some rigorous results pertaining to security are known. It is often of interest to generate a secure random bit with as little computation as possible. It had been known that if just one bit is chosen from each x_n, then this is in a sense secure, but at the cost of much computation to generate each bit. In [Patel and Sundaram 1998], it is shown that most of the bits of x_n can be kept, and the result is still cryptographically secure. There is thus much less computation per bit.

There are many other generators in current use, such as shift-register, chaotic, and cellular-automata (CA) generators. Some generators have been cryptographically "broken," notably the simpler congruential ones, even if the linear congruence is replaced with higher polynomial forms [Lagarias 1990]. One dilemma that besets researchers in this field is that the generators that may well be quite "secure," such as the discrete exponential variety that in turn depends on the DL problem for its security, are sluggish. Incidentally, there are various standard randomness tests, especially as regard random generation of binary bits, which can often be invoked to demolish—alternatively to bestow some measure of confidence upon—a given generator [Menezes et al. 1997].

On the issue of security, an interesting idea due to V. Miller is to use a linear-congruential generator, but with elliptic addition. Given an elliptic curve E over a finite field, one might choose integer a and point $B \in E$ and iterate

$$P_{n+1} = [a]P_n + B, \tag{8.1}$$

where the addition is elliptic addition and now the seed will be some initial point $P_0 \in E$. One might then use the x-coordinate of P_n as a random field element. This scheme is not as clearly breakable as is the ordinary linear congruential scheme. It is of interest that certain multipliers a, such as powers of two, would be relatively efficient because of the implied simplicity of the elliptic multiplication ladder. Then, too, one could perhaps use reduced operations inherent in Algorithm 7.2.8. In other words, use only x-coordinates and live with the ambiguity in $[a]P \pm B$, never actually adding points *per se*, but having to take square roots.

Incidentally, a different approach to the use of elliptic curves for random generators appears in [Gong et al. 1999], where the older ideas of shift registers and codewords are generalized to curves over $\mathbf{F}_{2^m}$ (see Exercise 8.25).

Along the same lines, let us discuss for a moment the problem of random bit generation. Surely, one can contemplate using some bit—such as the lowest bit—of a "good" random-number generator. But one wonders, for example, whether the calculation of Legendre symbols appropriate to point-finding on elliptic curves,

$$\left(\frac{x^3 + ax + b}{p}\right) = \pm 1,$$

with x running over consecutive integers in an interval and with the (rare) zero value thrown out, say, constitute a statistically acceptable random walk of ± 1 values. And one wonders further whether the input of x into a Legendre-symbol machine, but *from* a linear-congruential or other generator, provides extra randomness in any statistical sense.

Such attempts at random bit streams should be compared statistically to the simple exclusive-or bit generators. An example given in [Press et al. 1996] is based on the primitive polynomial (mod 2)

$$x^{18} + x^5 + x^2 + x + 1.$$

(A polynomial over a finite field F is *primitive* if it is irreducible and if a root is a cyclic generator for the multiplicative group of the finite field generated by the root.) If one has a "current" bit x_{-1}, and labels the previous 17 bits $x_{-2}, x_{-3}, \ldots, x_{-18}$, then the shifting logic appropriate to the given polynomial is to form a new bit x_0 according to the logic

$$\begin{aligned} x_0 &= x_{-18}, \\ x_{-5} &= x_{-5} \wedge x_0, \\ x_{-2} &= x_{-2} \wedge x_0, \\ x_{-1} &= x_{-1} \wedge x_0, \end{aligned}$$

where "$\wedge$" is the exclusive-or operator (equivalent to addition in the even-characteristic field). Then all of the indices are shifted so that the new x_{-1}—the new current bit—is the x_0 from the above operations. An explicit algorithm is the following:

Algorithm 8.2.7 (Simple and fast random-bit generator). This algorithm provides seeding and random functions for a random-bit generator based on the polynomial $x^{18} + x^5 + x^2 + x + 1$ over $\mathbf{F}_2$.

1. [Procedure *seed*]

```
seed() {
    h = 2^17;                          // 100000000000000000 binary.
    m = 2^0 + 2^1 + 2^4;               // Mask is 10011 binary.
    Choose starting integer seed x in [1, 2^18];
    return;
}
```

2. [Function *random* returning 0 or 1]

```
random() {
    if((x & h) ≠ 0) {      // The bitwise "and" of x, h is compared to 0.
        x = ((x ∧ m) << 1) | 1;  // "Exclusive-or" (∧) and "or" (|) taken.
        return 1;
    }
    x = x << 1;
    return 0;
}
```

The reference [Press et al. 1996] has a listing of other polynomials (mod 2) for selected degrees up through 100.

In any comprehensive study of random number generation, one witnesses the conceptual feedback involving prime numbers. Not only do many proposed random-number generators involve primes *per se*, but many of the algorithms—such as some of the ones appearing in this book—use recourse to suitable random numbers. But if one lifts the requirement of statistically testable randomness as it is usually invoked, there is quite another way to use random sequences. It is to these alternatives—falling under the rubric of quasi-Monte Carlo (qMC)—to which we next turn.

8.3 Quasi-Monte Carlo (qMC) methods

Who would have guessed, back in the times of Gauss, Euler, Legendre, say, that primes would attain some practical value in the financial-market analysis of the latter 20th century? We refer here not to cryptographic uses—which certainly do emerge whenever money is involved—but quasi-Monte Carlo science which, loosely speaking, is a specific form of Monte Carlo (i.e., statistically motivated) analysis. Monte Carlo calculations pervade the fields of applied science.

The essential idea behind Monte Carlo calculation is to sample some large continuous (or even discrete, if need be) space—in doing a multidimensional integral, say—with random samples. Then one hopes that the "average" result is close to the true result one would obtain with the uncountable samples theoretically at hand. It is intriguing that number theory—in particular prime-number study—can be brought to bear on the science of quasi-Monte Carlo (qMC). The techniques of qMC differ from traditional Monte Carlo in that one does not seek expressly random sequences of samples. Instead, one attempts to provide quasirandom sequences that do not, in fact, obey the strict statistical rules of randomness, but instead have certain uniformity features attendant on the problem at hand.

Although it is perhaps overly simplistic, a clear way to envision the difference between random and qMC is this: Random points when dropped can be expected to exhibit "clumps" and "gaps," whereas qMC points generally *avoid* each other to minimize clumping and tend to *occupy* previous gaps. For these reasons qMC points can be—depending on the spatial dimension and precise posing of the problem—superior for certain tasks such as numerical integration, min–max problems, and statistical estimation in general.

8.3.1 Discrepancy theory

Say that one wants to know the value of an integral over some D-dimensional domain R, namely

$$I = \int\int \cdots \int_R f(\vec{x})\, d^D\vec{x},$$

but there is no reasonable hope of a closed-form, analytic evaluation. One might proceed in Monte Carlo fashion, by dropping a total of N "random" vectors $\vec{x} = (x_1, \ldots, x_D)$ into the integration domain, then literally adding up the corresponding integrand values to get an average, and then multiplying by the measure of R to get an approximation, say I', for the exact integral I. On the general variance principles of statistics, we can expect the error to behave no better than

$$|I' - I| = O\left(\frac{1}{\sqrt{N}}\right),$$

where of course, the implied big-O constant depends on the dimension D, the integrand f, and the domain R. It is interesting that the power law $N^{-1/2}$, though, is independent of D. By contrast, a so-called "grid" method, in which we split the domain R into grid points, can be expected to behave no better than

$$|I' - I| = O\left(\frac{1}{N^{1/D}}\right),$$

which growth can be quite unsatisfactory, especially for large D. In fact, a grid scheme—with few exceptions—makes practical sense only for 1- or perhaps 2-dimensional numerical integration, unless there is some special consideration like well-behaved integrand, extra reasons to use a grid, and so on. It is easy to see why Monte Carlo methods using random point sets have been used for decades on numerical integration problems in $D \geq 3$ dimensions.

But there is a remarkable way to improve upon direct Monte Carlo, and in fact obtain errors such as

$$|I' - I| = O\left(\frac{\ln^D N}{N}\right),$$

or sometimes with $\ln^{D-1}$ powers appearing instead, depending on the implementation (we discuss this technicality in a moment). The idea is to use low-discrepancy sequences, a class of quasi-Monte Carlo (qMC) sequences (some authors *define* a low-discrepancy sequence as one for which the behavior of $|I' - I|$ is bounded as above; see Exercise 8.28). We stress again, an important observation is that qMC sequences are *not* random in the classical sense. In fact, the points belonging to qMC sequences tend to avoid each other (see Exercise 8.11).

We start our tour of qMC methods with a definition of discrepancy, where it is understood that vectors drawn out of regions R consist of real-valued components.

Definition 8.3.1. Let P be a set of at least N points in the (unit D-cube) region $R = [0,1]^D$. The discrepancy of P with respect to a family F of Lebesgue-measurable subregions of R is defined (neither D_N nor D_N^* is to be confused with dimension D) by

$$D_N(F;P) = \sup_{\phi \in F} \left| \frac{\chi(\phi;P)}{N} - \mu(\phi) \right|,$$

where $\chi(\phi;P)$ is the number of points of P lying in ϕ, and μ denotes Lebesgue measure. Furthermore, the extreme discrepancy of P is defined by

$$D_N(P) = D_N(G;P),$$

where G is the family of subregions of the form $\prod_{i=1}^{D}[u_i, v_i)$. In addition, the star discrepancy of P is defined by

$$D_N^*(P) = D_N(H; P),$$

where H is the family of subregions of the form $\prod_{i=1}^{D}[0, v_i)$. Finally, if $S \subset R$ is a countably infinite sequence $S = (\vec{x}_1, \vec{x}_2, \ldots)$, we define the various discrepancies $D_N(S)$ always in terms of the first N points of S.

The definition is somewhat notation-heavy, but a little thought reveals what is being sought, an assessment of "how fairly" a set P samples a region. One might have thought on the face of it that a simple equispaced grid of points would have optimal discrepancy, but in more than one dimension such intuition is misleading, as we shall see. One way to gain insight into the meaning of discrepancy is to contemplate the theorem: A countably infinite set S is equidistributed in $R = [0, 1]^D$ if and only if the star discrepancy (alternatively, the extreme discrepancy) vanishes as $N \to \infty$. It is also the case that the star and extreme discrepancies are not that different; in fact, it can be shown that for any P of the above definition we have

$$D_N^*(P) \le D_N(P) \le 2^D D_N^*(P).$$

Such results can be found in [Niederreiter 1992], [Tezuka 1995].

The importance of discrepancy—in particular the star discrepancy D^*—is immediately apparent on the basis of the following central result, which may be taken to be the centerpiece of qMC integration theory. We shall refer here to the Hardy–Krause bounded variation, which is an estimate $H(f)$ on the excursions of a function f. We shall not need the precise definition for H (see [Niederreiter 1992]), since the computational aspect of qMC depends mainly on the rest of the overall variation term:

Theorem 8.3.2 (Koksma–Hlawka). *If a function f has bounded variation $H(f)$ on $R = [0, 1]^D$, and S is as in Definition 8.3.1, then*

$$\left| \frac{1}{N} \sum_{\vec{x} \in S} f(\vec{x}) - \int_{\vec{r} \in R} f(\vec{r})\, d^D\vec{r} \right| \le H(f) D_N^*(S).$$

What is more, this inequality is optimal in the following sense: For any N-point $S \subset R$ and any $\epsilon > 0$, there exists a function f with $H(f) = 1$ such that the left-hand side of the inequality is bounded below by $D_N^*(S) - \epsilon$.

This beautiful result connects multidimensional integration errors directly to the star discrepancy D_N^*. The quest for accurate qMC sequences will now hinge on the concept of discrepancy. Incidentally, one of the many fascinating theoretical results beyond Theorem 8.3.2 is the assessment of Wozniakowski of "average" case error bounds on the unit cube. As discussed in [Wozniakowski 1991], the statistical ensemble average—in an appropriately rigorous sense—of the integration error is closely related to discrepancy, verifying once and for all that discrepancy is of profound practical importance. Moreover, there are some surprising new results that go some distance, as we shall see, to explain why actual qMC experiments are sometimes fare much better—provide far more accuracy—than the discrepancy bounds imply.

A qMC sequence S should generally be one of low D^*, and it is in the construction of such S that number theory becomes involved. The first thing we need to observe is that there is a subtle distinction between a point-set discrepancy and the discrepancy of a sequence. Take $D = 1$ dimension for example, in which case the point set

$$P = \left\{ \frac{1}{2N}, \frac{3}{2N}, \ldots, 1 - \frac{1}{2N} \right\}$$

has $D_N^*(P) = 1/(2N)$. On the other hand, there *exists no countably infinite sequence* S that enjoys the property $D_N^*(S) = O(1/N)$. In fact, it was shown by [Schmidt 1972] that if S is countably infinite, then for infinitely many N,

$$D_N^*(S) \ge c\frac{\ln N}{N},$$

where c is an absolute constant (i.e., independent of N and S). Actually, the constant can be taken to be $c = 3/50$ [Niederreiter 1992], but the main point is that the requirement of an infinite qMC sequence, from which a researcher may draw arbitrarily large numbers of contiguous samples, gives rise to special considerations of error. The point set P above with its discrepancy $1/(2N)$ is allowed because, of course, the members of the sequence themselves depend on N.

8.3.2 Specific qMC sequences

We are now prepared to construct some low-star-discrepancy sequences. A primary goal will be to define a practical low-discrepancy sequence for any given prime p, by counting in a certain clever fashion through base-p representations of integers. We shall start with a somewhat more general description for arbitrary base-B representations. For more than one dimension, a set of pairwise coprime bases will be used.

Definition 8.3.3. For an integer base $B \ge 2$, the van der Corput sequence for base B is the sequence

$$S_B = (\rho_B(n)), \quad n = 0, 1, 2, \ldots,$$

where ρ_B is the radical-inverse function, defined on nonnegative integers n, with presumed base-B representation $n = \sum_i n_i B^i$, by:

$$\rho_B(n) = \sum_i n_i B^{-i-1}$$

These sequences are easy to envision and likewise easy to generate in practice; in fact, their generation is easier than one might suspect. Say we desire the van der Corput sequence for base $B = 2$. Then we simply count from $n = 0$, in binary

$$n = 0, 1, 10, 11, 100, \ldots,$$

and form the reversals of the bits to obtain (also in binary)

$$S = (0.0, 0.10, 0.01, 0.11, 0.001, \ldots).$$

To put it symbolically, if we are counting and happen to be at integer index

$$n = n_k n_{k-1} \ldots n_1 n_0,$$

then the term $\rho_B(n) \in S$ is given by reversing the digits thus:

$$\rho_B(n) = 0.n_0 n_1 \ldots n_k.$$

It is known that every van der Corput sequence has

$$D_N^*(S_B) = O\left(\frac{\ln N}{N}\right),$$

where the implied big-O constant depends only on B. It turns out that $B = 3$ has the smallest such constant, but the main point affecting implementations is that the constant generally increases for larger bases B [Faure 1981].

For $D > 1$ dimensions, it is possible to generate qMC sequences based on the van der Corput forms, in the following manner:

Definition 8.3.4. Let $\bar{B} = \{B_1, B_2, \ldots, B_D\}$ be a set of pairwise-coprime bases, each $B_i > 1$. We define the Halton sequence for bases $\bar{B}$ by

$$S_{\bar{B}} = (\vec{x}_n), \quad n = 0, 1, 2, \ldots,$$

where

$$\vec{x}_n = (\rho_{B_1}(n), \ldots, \rho_{B_D}(n)).$$

In other words, a Halton sequence involves a specific base for each vector coordinate, and the respective bases are to be pairwise coprime. Thus for example, a qMC sequence of points in the $(D = 3)$-dimensional unit cube can be generated by choosing prime bases $\{B_1, B_2, B_3\} = \{2, 3, 5\}$ and counting $n = 0, 1, 2, \ldots$ in those bases simultaneously, to obtain

$$\begin{aligned}
\vec{x}_0 &= (0,\ 0,\ 0),\\
\vec{x}_1 &= (1/2,\ 1/3,\ 1/5),\\
\vec{x}_2 &= (1/4,\ 2/3,\ 2/5),\\
\vec{x}_3 &= (3/4,\ 1/9,\ 3/5),
\end{aligned}$$

and so on. The manner in which these points deposit themselves in the unit 3-cube is interesting. We can see once again the basic, qualitative aspect of successful qMC sequences: The points tend to drop into regions where "they have not yet been." Contrast this to direct Monte Carlo methods, whereby—due to unbiased randomness—points will not only sometimes "clump" together, but sometimes leave "gaps" as the points accumulate in the domain of interest.

The Halton sequences are just one family of qMC sequences, as we discuss in the next section. For the moment, we exhibit a typical theorem that reveals information about how discrepancy grows as a function of the dimension:

Theorem 8.3.5 (Halton discrepancy). *Denote by $S_{\bar{B}}$ a Halton sequence for bases $\bar{B}$. Then the star discrepancy of the sequence satisfies*

$$D_N^*(S_{\bar{B}}) < \frac{D}{N} + \frac{1}{N} \prod_{i=1}^{D} \left(\frac{B_i - 1}{2 \ln B_i} \ln N + \frac{B_i + 1}{2} \right).$$

A rather intricate proof can be found in [Niederreiter 1992]. We observe that the theorem provides an explicit upper bound for the implied big-O constant in

$$D_N^*(S_{\bar{B}}) = O\left(\frac{\ln^D N}{N} \right),$$

an error behavior foreshadowed in the introductory remarks of this section. What is more, we can see the (unfortunate) effect of larger bases supposedly contributing more to the discrepancy (we say supposedly because this is just an upper bound); indeed, this effect for larger bases is seen in practice. We note that there is a so-called N-point Hammersley point set, for which the leading component of $\vec{x}_n$ is $x_0 = n/N$, while the rest of $\vec{x}_n$ is a $(D-1)$-dimensional Halton vector. This set is now N-dependent, so that it cannot be turned into an infinite sequence. However, the Hammersley set's discrepancy takes the slightly superior form

$$D_N^* = O\left(\frac{\ln^{D-1} N}{N} \right),$$

showing how N-dependent sets can offer a slight complexity reduction.

8.3.3 Primes on Wall Street?

Testing a good qMC sequence, say estimating the volume of the unit D-ball, is an interesting exercise. The Halton qMC sequence gives good results for moderate dimensions, say for D up to about 10. One advantage of the Halton sequence is that it is easy to jump ahead, so as to have several or many computers simultaneously sampling from disjoint segments of the sequence. The following algorithm shows how one can jump in at the n-th term, and how to continue sequentially from there. To make the procedure especially efficient, the digits of the index in the various bases under consideration are constantly updated as we proceed from one index to the next.

Algorithm 8.3.6 (Fast qMC sequence generation). This algorithm generates D-dimensional Halton-sequence vectors. Let $p_1, \ldots, p_D$ denote the first D primes. For starting index n, a $seed()$ procedure creates $\vec{x}_n$ whose components are for clarity denoted by $\vec{x}_n[1], \ldots, \vec{x}_n[D]$. Then a $random()$ function may be used to generate subsequent vectors $\vec{x}_{n+1}, \vec{x}_{n+2}, \ldots$, where we assume an upper bound of N for all indices. For high efficiency, global digits $(d_{i,j})$ are initially seeded to represent the starting index n, then upon subsequent calls to a $random()$ function, are incremented in "odometer" fashion for subsequent indices exceeding n.

1. [Procedure $seed$]

```
seed(n) {                                   // n is the desired starting index.
    for(1 ≤ i ≤ D) {
        K_i = ⌈ln(N+1)/ln p_i⌉;              // A precision parameter.
        q_{i,0} = 1;
        k = n;
        x[i] = 0;                            // x⃗ is the vector x⃗_n.
        for(1 ≤ j ≤ K_i) {
            q_{i,j} = q_{i,j-1}/p_i;         // q_{i,j} = p_i^{-j}.
            d_{i,j} = k mod p_i;             // The d_{i,j} start as base-p_i digits of n.
            k = (k - d_{i,j})/p_i;
            x[i] = x[i] + d_{i,j} q_{i,j};
        }
    }
    return;                                  // x⃗_n now available as (x[1],...,x[D]).
}
```

2. [Function $random$]

```
random() {
    for(1 ≤ i ≤ D) {
        for(1 ≤ j ≤ K_i) {
            d_{i,j} = d_{i,j} + 1;           // Increment the "odometer."
            x[i] = x[i] + q_{i,j};
            if(d_{i,j} < p_i) break;         // Exit loop when all carries complete.
            d_{i,j} = 0;
            x[i] = x[i] - q_{i,j-1};
        }
    }
    return (x[1],...,x[D]);                  // The new x⃗.
}
```

It is plain upon inspection that this algorithm functions as an "odometer," with ratcheting of base-p_m digits consistent with Definition 8.3.4. Note the parameters K_i, where K_i is the maximum possible number of digits, in base p_i, for an integer index j. This K_i must be set in terms of some N that is at least the value of any j that would ever be reached. This caution, or an equivalent one, is necessary to limit the precision of the reverse-radix base expansions.

Algorithm 8.3.6 is usually used in floating-point mode, i.e., with stored floating-point inverse powers $q_{i,j}$ but integer digits $n_{i,j}$. However, there is nothing wrong in principle with an exact generator in which actual integer powers are kept for the $q_{i,j}$. In fact, the integer mode can be used for testing of the algorithm, in the following interesting way. Take, for example, $N = 1000$, so vectors $\vec{x}_0, \ldots, \vec{x}_{999}$ are allowed, and choose $D = 2$ dimensions so that the primes 2,3 are involved. Then call *seed*(701), which sets the variable x to be the vector

$$\vec{x}_{701} = (757/1024, 719/729).$$

Now, calling *random*() exactly 9 times produces

$$\vec{x}_{710} = (397/1024, 674/729),$$

and sure enough, we can test the integrity of the algorithm by going back and calling *seed*(710) to verify that starting over thus with seed value 701+9 gives precisely the $\vec{x}_{710}$ shown.

It is of interest that Algorithm 8.3.6 really is fast, at least in this sense: In practice, it tends to be faster even than calling a system's built-in random-number function. And this advantage has meaning even outside the numerical-integration paradigm. When one really wants an equidistributed, random number in $[0, 1)$, say, a system's random function should certainly be considered, especially if the natural tendency for random samples to clump and separate is supposed to remain intact. But for many statistical studies, one simply wants some kind if irregular "coverage" of $[0, 1)$, one might say a "fair" coverage that does not bias any particular subinterval, in which case such a fast qMC algorithm should be considered.

Now we may get a multidimensional integral by calling, in a very simple way, the procedures of Algorithm 8.3.6:

Algorithm 8.3.7 (qMC multidimensional integration). Given a dimension D, and integrable function $f : R \to R$, where $R = [0, 1]^D$, this algorithm estimates the multidimensional integral

$$I = \int_{\vec{x}\in R} f(\vec{x})\, d^D\vec{x},$$

via the generation of N_0 qMC vectors, starting with the n-th of a sequence $(\vec{x}_0, \vec{x}_1, \ldots, \vec{x}_n, \ldots, \vec{x}_{n+N_0-1}, \ldots)$. It is assumed that Algorithm 8.3.6 is initialized with an index bound $N \geq n + N_0$.

1. [Initialize via Algorithm 8.3.6]

    seed(n); // Start the qMC process, to set a global $\vec{x} = \vec{x}_n$.
    $I = 0$;

2. [Perform qMC integration]

    // Function *random*() updates a global qMC vector (Algorithm 8.3.6).
    for($0 \leq j < N_0$) $I = I + f(random())$;
    return I/N_0; // An estimate for the integral.

Let us give an example of the application of such an algorithm. To assess the volume of the unit D-ball, which is the ball of radius 1, we can take f in terms of the Heaviside function θ (which is 1 for positive arguments, 0 for negative arguments, and 1/2 at 0),

$$f(\vec{x}) = \theta(1/4 - (\vec{x} - \vec{y}) \cdot (\vec{x} - \vec{y})),$$

with $\vec{y} = (1/2, 1/2, \ldots, 1/2)$, so that f vanishes everywhere outside a ball of radius 1/2. (This is the largest ball that fits inside the cube R.) The estimate of the unit D-ball volume will thus be $2^D I$, where I is the output of Algorithm 8.3.7 for the given, sphere-defining function f.

As we have intimated before, it is a wondrous thing to see firsthand how much better a qMC algorithm of this type can do, when compared to a direct Monte Carlo trial. One beautiful aspect of the fundamental qMC concept is that parallelism is easy: In Algorithm 8.3.7, just start each of, say, M machines at a different starting seed, ideally in such a way that some contiguous sequence of NM total vectors is realized. This option is, of course, the point of having a seed function in the first place. Explicitly, to obtain a one-billion-point integration, each of 100 machines would use the above algorithm *as is* with $N = 10^7$, except that machine 0 would start with $n = 0$ (and hence start by calling $seed(0)$), the second machine would start $n = 1$, through machine 99, which would start with $n = 99$. The final integral would be the average of the 100 machine estimates.

Here is a typical numerical comparison: We shall calculate the number π with qMC methods, and compare with direct Monte Carlo. Noting that the exact volume of the unit D-ball is

$$V_D = \frac{\pi^{D/2}}{\Gamma(1 + D/2)},$$

let us denote by $V_D(N)$ the calculated volume after N vectors are generated, and denote by π_N the "experimental" value for π obtained by solving the volume formula for π in terms of V_D. We shall do two things at once: Display the typical convergence and convey a notion of the inherent parallelism. For primes $p = 2, 3, 5$, so that we are assessing the 3-ball volume, the result of Algorithm 8.3.7 is displayed in Table 8.1.

What is displayed in the left-hand column is the total number of points "dropped" into the unit D-cube, while the second column is the associated, *cumulative* approximation to π. We say cumulative because one may have run each interval of 10^6 counts on a separate machine, yet we display the right-hand column as the answer obtained by combining the machines up to that N value inclusive. For example, the result π_5 can be thought of either as the result after $5 \cdot 10^6$ points are generated, or equivalently, after 5 separate machines each do 10^6 points. In the latter instance, one would have called the $seed(n)$ procedure with 5 different initial seeds to start each respective machine's interval. How do these data compare with direct Monte Carlo? The rough answer is that one can expect the error in the last ($N = 10^7$) row of

$N/10^6$	π_N
1	3.14158
2	3.14154
3	3.14157
4	3.14157
5	3.14158
6	3.14158
7	3.14158
8	3.141590
9	3.14158
10	3.1415929

Table 8.1 Approximations to π via prime-based qMC (Halton) sequence, using primes $p = 2, 3, 5$, the volume of the unit 3-ball is assessed for various cumulative numbers of qMC points, $N = 10^6$ through $N = 10^7$. We have displayed decimal digits only through the first incorrect digit.

a similar Monte Carlo table to be in the third or so digit to the right of the decimal (because $\log_{10}\sqrt{N}$ is about 3.5 in this case). This superiority of qMC to direct methods—which is an advantage of several orders of magnitude—is typical for "millions" of points and moderate dimensions.

Now to the matter of Wall Street, meaning the phenomenon of computational finance. If the notion of very large dimensions D for integration has seemed fanciful, one need only cure that skepticism by observing the kind of calculation that has been attempted in connection with risk management theory and other aspects of computational finance. For example, 25-dimensional integrals relevant to financial computation, of the form

$$I = \int \cdots \int_{\vec{x} \in R} \cos|\vec{x}|\, e^{-\vec{x}\cdot\vec{x}}\, d^D\vec{x},$$

were analyzed by [Papageorgiu and Traub 1997], with the conclusion that, surprisingly enough, qMC methods (in their case, using the Faure sequences) would outperform direct Monte Carlo methods, in spite of the asymptotic estimate $O((\ln^D N)/N)$, which does not fare too well in practice against $O(1/\sqrt{N})$ when $D = 25$. In other treatments, for example [Paskov and Traub 1995], integrals with dimension as high as $D = 360$ are tested. As those authors astutely point out, their integrals (involving collateralized mortgage obligation, or CMO in the financial language) are good test cases because the integrand has a certain computational complexity and so—in their words—"it is crucial to sample the integrand as few times as possible." As intimated by [Boyle et al. 1995] and various other researchers, whether or not a qMC is superior to a direct Monte Carlo in some high dimension D depends very much on the actual calculation being performed. The general sentiment is that numerical analysts not from the financial world *per se* tend to use integrals that present the more difficult challenge for the qMC methods. That is, financial integrands are often "smoother" in practice.

Just as interesting as the qMC technique itself is the controversy that has simmered in the qMC literature. Some authors believe that the Halton sequence—the one on which we have focused as an example of prime-based qMC—is inferior to, say, the Sobol [Bratley and Fox 1988] or Faure [Niederreiter 1992] sequences. And as we have indicated above, this assessment tends to depend strongly on the domain of application. Yet there is some theoretical motivation for the inferiority claims; namely, it is a theorem [Faure 1982] that the star discrepancy of a Faure sequence satisfies

$$D_N^* \le \frac{1}{D!}\left(\frac{p-1}{2\ln p}\right)^D \frac{\ln^D N}{N},$$

where p is the least prime greater than or equal to D. Whereas a D-dimensional Halton sequence can be built from the first D primes, and this Faure bound involves the next prime, still the bound of Theorem 8.3.5 is considerably worse. What is likely is that both bounding theorems are not best-possible results. In any case, the prime numbers once again enter into discrepancy theory and its qMC applications.

As has been pointed out in the literature, there is the fact that qMC's error growth of $O\left((\ln^D N)/N\right)$ is, for sufficiently large D, and sufficiently small N, or practical combinations of D, N magnitudes, worse than direct Monte Carlo's $O\left(1/\sqrt{N}\right)$. Thus, some researchers do not recommend qMC methods unconditionally. One controversial problem is that in spite of various theorems such as Theorem 8.3.5 and the Faure bound above, we still do not know how the "real-world" constants in front of the big-O terms really behave. Some recent developments address this controversy. One such development is the discovery of "leaped" Halton sequences. In this technique, one can "break" the unfortunate correlation between coordinates for the D-dimensional Halton sequence. This is done in two possible ways. First, one adopts a permutation on the inverse-radix digits of integers, and second, if the base primes are denoted by $p_0, \ldots, p_{D-1}$, then one chooses yet another distinct prime p_D and uses only every p_D-th vector of the usual Halton sequence. This is claimed to improve the Halton sequence dramatically for high dimension, say $D = 40$ to 400 [Kocis and Whiten 1997]. It is of interest that these authors found a markedly good distinct prime p_D to be 409, a phenomenon having no explanation. Another development, due to [Crandall 1999a], involves the use of a reduced set of primes—even when D is large—and using the resulting lower-dimensional Halton sequence as a vector parameter for a D-dimensional space-filling curve. In view of the sharply base-dependent bound of Theorem 8.3.5, there is reason to believe that this technique of involving only small primes carries a distinct statistical advantage in higher dimensions.

While the notion of discrepancy is fairly old, there always seem to appear new ideas pertaining to the generation of qMC sets. One promising new approach involves the so-called (t, m, s)-nets [Owen 1995, 1997a, 1997b],

[Tezuka 1995], [Veach 1997]. These are point clouds that have "minimal fill" properties. For example, a set of $N = b^m$ points in s dimensions is called a (t, m, s)-net if every justified box of volume b^{t-m} has exactly b^t points. Yet another intriguing connection between primes and discrepancy appears in the literature (see [Joe 1999] and references therein). This notion of "number-theoretical rules" involves approximations of the form

$$\int_{[0,1]^D} f(\vec{x})\, d^D\vec{x} \;\approx\; \frac{1}{p}\sum_{j=0}^{p-1} f\left(\left\{\frac{j\vec{K}}{p}\right\}\right),$$

where here $\{\vec{y}\}$ denotes the vector composed of the fractional parts of $\vec{y}$, and $\vec{K}$ is some chosen constant vector having each component coprime to p. Actually, composite numbers can be used in place of p, but the analysis of what is called L_2 discrepancy, and the associated typical integration error, goes especially smoothly for p prime. We have mentioned these new approaches to underscore the notion that qMC is continually undergoing new development. And who knows when or where number theory or prime numbers in particular will appear in qMC theories of the future?

In closing this section, we mention a new result that may explain why qMC experiments sometimes do "so well." Take the result of [Sloan and Wozniakowski 1998], in which the authors remark that some errors (such as those in Traub's qMC for finance in $D = 360$ dimensions) appear to have $O(1/N)$ behavior, i.e., independent of dimension D. What the authors actually prove is that there exist classes of integrand functions for which suitable low-discrepancy sequences provide overall integration errors of order $O(1/N^\rho)$ for some real $\rho \in [1, 2]$.

8.4 Diophantine analysis

Herein we discuss Diophantine analysis, which loosely speaking is the practice of discovering integer solutions to various equations. We have mentioned elsewhere Fermat's last theorem (FLT), for which one seeks solutions to

$$x^p + y^p = z^p,$$

and how numerical attacks alone have raised the lower bound on p into the millions (Section 1.3.3, Exercise 9.68). This is a wonderful computational problem—speaking independently, of course, of the marvelous FLT proof by A. Wiles—but there are many other similar explorations. Many such adventures involve a healthy mix of theory and computation.

For instance, there is the Catalan equation for p, q prime and x, y positive integers,

$$x^p - y^q = 1,$$

of which the *only* known solution is the trivial yet attractive

$$3^2 - 2^3 = 1.$$

Observe that in seeking Diophantine solutions here we are simply addressing the problem of whether there exist higher instances of consecutive *powers*. An accessible treatment of the history of this Catalan problem is [Ribenboim 1994], and the paper [Mignotte 2001] documents more recent developments. Using the theory of linear forms of logarithms of algebraic numbers, R. Tijdeman showed in 1976 that the Catalan equation has at most finitely many solutions; in fact,

$$y^q < e^{e^{e^{e^{730}}}},$$

as discussed in [Guy 1994]. Thus, the complete resolution of the Catalan problem is reduced to a (huge!) computation. Shortly afterward Tijdeman's great theorem, M. Langevin showed that any solution must have the exponents $p, q < 10^{110}$. This bound on the exponents has continued to fall, the latest result, due to Mignotte (based on work of C. Bennett, J. Blass, A. Glass, D. Meronk, and R. Steiner), being that $\max\{p, q\} < 7.78 \times 10^{16}$. In 1997, Y. Roy and Mignotte, using difficult theorems on class numbers of certain fields, together with extensive computation, showed that $\min\{p, q\} > 10^5$. An exciting new result of P. Mihailescu has been of help in raising the Mignotte–Roy lower bound on p, q. Mihailescu shows that if p, q are Catalan exponents other than the pair 2, 3, then p is a Wieferich prime base q and vice versa; that is,

$$p^{q-1} \equiv 1 \pmod{q^2} \quad \text{and} \quad q^{p-1} \equiv 1 \pmod{p^2}.$$

Using this criterion, Mignotte has shown that $\min\{p, q\} > 10^7$. One looks forward to the day when the interval $(10^7, 7.78 \times 10^{16})$ can finally be exhausted.

Related both to Fermat's last theorem and the Catalan problem is the Diophantine equation

$$x^p + y^q = z^r, \tag{8.2}$$

where x, y, z are positive, coprime integers and exponents p, q, r are positive integers with $1/p+1/q+1/r \leq 1$. The Fermat–Catalan conjecture asserts that there are at most finitely many such powers x^p, y^q, z^r in (8.2). The following are the only known examples:

$$\begin{aligned}
1^p + 2^3 &= 3^2 \quad (p \geq 7)\\
2^5 + 7^2 &= 3^4\\
13^2 + 7^3 &= 2^9\\
2^7 + 17^3 &= 71^2\\
3^5 + 11^4 &= 122^2\\
33^8 + 1549034^2 &= 15613^3\\
1414^3 + 2213459^2 &= 65^7\\
9262^3 + 15312283^2 &= 113^7\\
17^7 + 76271^3 &= 21063928^2\\
43^8 + 96222^3 &= 30042907^2.
\end{aligned}$$

(The latter 5 examples were found by F. Beukers and D. Zagier.) There is a cash prize (the Beal Prize) for a proof that (8.2) has no solutions at all when $p, q, r \geq 3$; see [Mauldin 2000].) It is known [Darmon and Granville 1995] that for p, q, r *fixed* with $1/p + 1/q + 1/r \leq 1$, the equation (8.2) has at most finitely many coprime solutions x, y, z.

The Fermat–Catalan conjecture is a special case of the notorious ABC conjecture. Let $\gamma(n)$ denote the largest square-free divisor of n. The ABC conjecture asserts that for each $\epsilon > 0$ there are at most finitely many coprime, positive integer triples a, b, c with

$$a + b = c, \quad \gamma(abc) < c^{1-\epsilon}.$$

Though much work in Diophantine equations is extraordinarily deep, there are many satisfying exercises that use such concepts as quadratic reciprocity to limit Diophantine solutions. For example, one can prove that

$$y^2 = x^3 + k \tag{8.3}$$

has no integral solutions whatever if $k = (4n - 1)^3 - 4m^2$, $m \neq 0$, and no prime dividing m is congruent to 3 (mod 4) (see Exercise 8.12).

Aside from interesting analyses of specific equations, there is a profound general theory of Diophantine equations. The saga of this decades-long investigation is fascinating. A fundamental question, posed at the turn of the last century as Hilbert's "tenth problem," asks for a *general* algorithm that will determine the solutions to an arbitrary Diophantine equation. In the decades long attack on this problem, a central notion was that of a Diophantine set, which is a set S of positive integers such that some multivariate polynomial $P(X, Y_1, \ldots, Y_l)$ exists with coefficients in $\mathbf{Z}$ with the property that $x \in S$ if and only if $P(x, y_1, \ldots, y_l) = 0$ has a positive integer solution in the y_j. It is not hard to prove the theorem of H. Putnam from 1960, see [Ribenboim 1996, p. 189], that a set S of positive integers is Diophantine if and only if there is a multivariate polynomial Q with integer coefficients such that the set of its positive values at nonnegative integer arguments is exactly the set S.

Armed with this definition of a Diophantine set, formal mathematicians led by Putnam, Davis, Robinson, and Matijasevič established the striking result that *the set of prime numbers is Diophantine.* That is, they showed that there exists a polynomial P—with integer coefficients in some number of variables—such that as its variables range over the nonnegative integers, the set of positive values of P is *precisely* the set of primes.

One such polynomial given explicitly by Jones, Sato, Wada, and Wiens in 1976 (see [Ribenboim 1996]) is

$$\begin{aligned}
&(k+2)\Big(1 - (wz + h + j - q)^2 - ((gk + 2g + k + 1)(h + j) + h - z)^2 \\
&- (2n + p + q + z - e)^2 - \left(16(k+1)^3(k+2)(n+1)^2 + 1 - f^2\right)^2 \\
&- \left(e^3(e+2)(a+1)^2 + 1 - o^2\right)^2 - \left(a^2y^2 - y^2 + 1 - x^2\right)^2
\end{aligned}$$

$$
\begin{aligned}
&- \left(16r^2y^4(a^2-1)+1-u^2\right)^2 \\
&- \left(((a+u^4-u^2a)^2-1)(n+4dy)^2+1-(x+cu)^2\right)^2 \\
&- (n+l+v-y)^2 - \left(a^2l^2-l^2+1-m^2\right)^2 - (ai+k+1-l-i)^2 \\
&- (p+l(a-n-1)+b(2an+2a-n^2-2n-2)-m)^2 \\
&- (q+y(a-p-1)+s(2ap+2a-p^2-2p-2)-x)^2 \\
&- \left.(z+pl(a-p)+t(2ap-p^2-1)-pm)^2\right).
\end{aligned}
$$

This polynomial has degree 25, and it conveniently has 26 variables, so that the letters of the alphabet can each be used! An amusing consequence of such a prime producing polynomial is that any prime p can be presented with a proof of primality that uses only $O(1)$ *arithmetic* operations. Namely, supply the 26 values of the variables used in the above polynomial that gives the value p. However, the number of *bit* operations for this verification can be enormous.

Hilbert's "tenth problem" was eventually solved—with the answer being that there can be no algorithm as sought—with the final step being Matijasevič's proof that every listable set is Diophantine. But along the way, for more than a half century, the set of primes was at center stage in the drama [Matijasevič 1971], [Davis 1973].

Diophantine analysis, though amounting to the historical underpinning of all of number theory, is still today a fascinating, dynamic topic among mathematicians and recreationalists. One way to glimpse the generality of the field is to make use of network resources such as [Weisstein 2000].

8.5 Quantum computation

It seems appropriate to have in this applications chapter a brief discussion of what may become a dominant computational paradigm for the 21st century. We speak of quantum computation, which is to be thought of as a genuine replacement for computer processes as we have previously understood them. The first basic notion is a distinction between classical Turing machines (TMs) and quantum Turing machines (QTMs). The older TM model is the model of every prevailing computer of today, with the possible exception of very minuscule, tentative and experimental QTMs, in the form of small atomic experiments and so on. (Although one could argue that nature has been running a massive QTM for billions of years.) The primary feature of a TM is that it processes "serially," in following a recipe of instructions (a program) in a deterministic fashion. (There is such a notion as a probabilistic TM behaving statistically, but we wish to simplify this overview and will avoid that conceptual pathway.) On the other hand, a QTM would be a device in which a certain "parallelism" of nature would be used to effect computations with truly unprecedented efficiency. That parallelism is, of course, nature's way of behaving according to laws of quantum mechanics. These laws involve many counterintuitive concepts. As students of quantum theory know, the microscopic phenomena in question do not occur as in the macroscopic world. There is the particle–wave duality (is an electron a wave or a particle or both?), the notion of amplitudes, probability, interference—not just among waves but among actual parcels of matter—and so on. The next section is a very brief outline of quantum computation concepts, intended to convey some qualitative features of this brand new science.

8.5.1 Intuition on quantum Turing machines (QTMs)

Because QTMs are still overwhelmingly experimental, not having solved a single "useful" problem so far, we think it appropriate to sketch, mainly by analogy, what kind of behavior could be expected from a QTM. Think of holography, that science whereby a solid three-dimensional object is cast onto a planar "hologram." What nature does is actually to "evaluate" a 3-dimensional Fourier transform whose local power fluctuations determine what is actually developed on the hologram. Because light moves about one foot in a nanosecond (10^{-9} seconds), one can legitimately say that when a laser light beam strikes an object (say a chess piece) and the reflections are mixed with a reference beam to generate a hologram, "nature performed a huge FFT in a couple of nanoseconds." In a qualitative but striking sense, a known $O(N \ln N)$ algorithm (where N would be sufficiently many discrete spatial points to render a high-fidelity hologram, say) has turned into more like an $O(1)$ one. Though it is somewhat facetious to employ our big-O notation in this context, we wish only to make the point that there is parallelism in the light-wave-interference model that underlies holography. On the film plane of the hologram, the final light intensity depends on *every* point on the chess piece. This is the holographic, one could say "parallel," aspect. And QTM proposals are reminiscent of this effect.

We are not saying that a laboratory hologram setup is a QTM, for some ingredients are missing in that simplistic scenario. For one thing, modern QTM theory has two other important elements beyond the principle of quantum interference; namely, probabilistic behavior, and a theoretical foundation involving operators such as unitary matrices. For another thing, we would like any practical QTM to bear not just on optical experiments, but also on some of the very difficult tasks faced by standard TMs—tasks such as the factoring of large integers. As have been a great many new ideas, the QTM notion was pioneered in large measure by the eminent R. Feynman, who observed that quantum-mechanical model calculations tend, on a conventional TM, to suffer an exponential slowdown. Feynman even devised an explicit model of a QTM based on individual quantum registers [Feynman 1982, 1985]. The first formal definition was provided by [Deutsch 1982, 1985], to which current formal treatments more or less adhere. An excellent treatment—which sits conveniently between a lay perspective and a mathematical one—is [Williams and Clearwater 1998]. On the more technical side of the physics, and some of the relevant number-theoretical ideas, a good reference is [Ekert and Jozsa

1996]. For a very accessible lay treatment of quantum computation, see [Hey 1999], and for course-level material see [Preskill 1999].

Let us add a little more quantum flavor to the idea of laser light calculating an FFT, nature's way. There is in quantum theory an ideal system called the quantum oscillator. Given a potential function $V(x) = x^2$, the Schrödinger equation amounts to a prescription for how a wave packet $\psi(x, t)$, where t denotes time, moves under the potential's influence. The classical analogue is a simple mass-on-a-spring system, giving smooth oscillations of period τ, say. The quantum model also has oscillations, but they exhibit the following striking phenomenon: After one quarter of the classical period τ, an initial wave packet *evolves into its own Fourier transform.* This suggests that you could somehow load data into a QTM as an initial function $\psi(x, 0)$, and later read off $\psi(x, \tau/4)$ as an FFT. (Incidentally, this idea underlies the discussion around the Riemann-ζ representation (8.5).) What we are saying is that the laser hologram scenario has an analogue involving particles and dynamics. We note also that wave functions ψ are complex amplitudes, with $|\psi|^2$ being probability density, so this is how statistical features of quantum theory enter into the picture.

Moving now somewhat more toward the quantitative, and to prepare for the rest of this section, we presently lay down a few specific QTM concepts. It is important right at the outset, especially when number-theoretical algorithms are involved, to realize that an exponential number of quantities may be "polynomially stored" on a QTM. For example, here is how we can store in some fashion—in a so-called quantum register—*every* integer $a \in [0, q-1]$, in only $\lg q$ so-called qbits. At first this seems impossible, but recall our admission that the quantum world can be notoriously counterintuitive. A mental picture will help here. Let $q = 2^d$, so that we shall construct a quantum register having d qbits. Now imagine a line of d individual ammonia molecules, each molecule being NH_3 in chemical notation, thought of as a tetrahedron formed by the three hydrogens and a nitrogen apex. The N apex is to be thought of as "up" or "down," 1 or 0, i.e., either above or below the three H's. Thus, any d-bit binary number can be represented by a collective orientation of the molecules. But what about representing all possible binary strings of length d? This turns out to be easy, because of a remarkable quantum property: An ammonia molecule can be in *both* $1, 0$ states at the same time. One way to think of this is that lowest-energy states—called ground states—are symmetrical when the geometry is. A container of ammonia in its ground state has each molecule somehow "halfway present" at each $0, 1$ position. In theoretical notation we say that the ground state of one ammonia qbit (molecule, in this model) is given by:

$$\phi = \frac{1}{\sqrt{2}} \left(\,|\, 0 \,\rangle + \,|\, 1 \,\rangle \,\right),$$

where the "bra-ket" notation $|\ \rangle$ is standard (see the aforementioned quantum-theoretical references). The notation reminds us that a state belongs to an abstract Hilbert space, and only an inner product can bring this back to a measurable number. For example, given the ground state ϕ here, the probability that we *find* the molecule in state $\mid 0 \rangle$ is the squared inner product

$$\mid \langle\, 0 \mid \phi \,\rangle \mid^2 = \left| \frac{1}{\sqrt{2}} \langle\, 0 \mid 0 \,\rangle \right|^2 = \frac{1}{2},$$

i.e., 50 per cent chance that the nitrogen atom is measured to be "down." Now back to the whole quantum register of d qbits (molecules). If each molecule is in the ground state ϕ, then in some sense every single d-bit binary string is represented. In fact, we can describe the state of the entire register as [Shor 1999]

$$\psi = \frac{1}{2^{d/2}} \sum_{a=0}^{2^d-1} \mid a \,\rangle,$$

where now $|a\rangle$ denotes the composite state given by the molecular orientations corresponding to the binary bits of a; for example, for $d = 5$ the state $|10110\rangle$ is the state in which the nitrogens are oriented "up, down, up, up, down." This is not so magical as it sounds, when one realizes that now the probability of finding the entire register in a particular state $a \in [0, 2^d - 1]$ is just $1/2^d$. It is this sense in which every integer a is stored—the collection of all a values is a "superposition" in the register.

Given a state that involves every integer $a \in [0, q-1]$, we can imagine acting on the qbits with unitary operators. For example, we might alter the 0th and 7th qbits by acting on the two states with a matrix operator. An immediate physical analogy here would be the processing of two input light beams, each possibly polarized up or down, via some slit interference experiment (having polaroid filters within) in which two beams are output. Such a unitary transformation preserves overall probabilities by redistributing amplitudes between states.

Using appropriate banks of unitary operators, it turns out that if $q > n$, and x be a chosen residue (mod n), then one can also form the state

$$\psi' = \frac{1}{2^{d/2}} \sum_{a=0}^{2^d-1} \mid x^a \bmod n \,\rangle,$$

again as a superposition. The difference now is that if we ask for the probability that the entire register be found in state $\mid b \,\rangle$, that probability is zero unless b is an a-th power residue modulo n.

We end this very brief conceptual sketch by noting that the sovereign of all divide-and-conquer algorithms, namely the FFT, can be given a concise QTM form. It turns out that by employing unitary operators, all of them pairwise as above, in a specific order, one can create the state

$$\psi'' = \frac{1}{\sqrt{q}} \sum_{a=0}^{q-1} e^{2\pi i a c/q} \mid c \,\rangle,$$

and this allows for many interesting algorithms to go through on QTMs—at least in principle—with polynomial-time complexity. For the moment, we remark that addition, multiplication, division, modular powering and FFT can all be done in time $O(d^\alpha)$, where d is the number of qbits in each of (finitely many) registers and α is some appropriate power. The aforementioned references have all the details for these fundamental operations. Though nobody has carried out the actual QTM arithmetic—only a few atomic sites have been built so far in laboratories—the literature descriptions are clear: We expect nature to be able to perform massive parallelism on d-bit integers, in time only a power of d.

8.5.2 The Shor quantum algorithm for factoring

Just as we so briefly overviewed the QTM concept, we now also briefly discuss some of the new quantum algorithms that pertain to number-theoretical problems. It is an astute observation of [Shor 1994, 1999] that one may factor n by finding the exponent orders of random integers (mod n) via the following proposition.

Proposition 8.5.1. *Suppose the odd integer $n > 1$ has exactly k distinct prime factors. For a randomly chosen member y of $\mathbf{Z}_n^*$ with multiplicative order r, the probability that r is even and that $y^{r/2} \not\equiv -1 \pmod n$ is at least $1 - 1/2^{k-1}$.*

(See Exercise 8.18, for a slightly stronger result.) The implication of this proposition is that one can—at least in principle—factor n by finding "a few" integers y with corresponding (even) orders r. For having done that, we look at

$$\gcd(y^{r/2} - 1, n)$$

for a nontrivial factor of n, which should work with good chance, since $y^r - 1 = (y^{r/2}+1)(y^{r/2}-1) \equiv 0 \pmod n$; in fact this will work with probability at least $1 - 1/2^{k-1}$, and this expression is not less than $1/2$, provided that n is neither a prime nor a prime power.

So the Shor algorithm comes down to finding the orders of random residues modulo n. For a conventional TM, this is a stultifying problem—a manifestation of the discrete logarithm (DL) problem. But for a QTM, the natural parallelism renders this residue-order determination not so difficult. We paraphrase a form of Shor's algorithm, drawing from the treatments of [Williams and Clearwater 1998], [Shor 1999]. We stress that an appropriate machine has not been built, but if it were the following algorithm is expected to work. And, there is nothing preventing one trying the following on a conventional Turing machine; and then, of course, experiencing an exponential slowdown for which QTMs have been proposed as a remedy.

Algorithm 8.5.2 (Shor quantum algorithm for factoring). Given an odd integer n that is neither prime nor a power of a prime, this algorithm attempts to return a nontrivial factor of n via quantum computation.

1. [Initialize]
   Choose $q = 2^d$ with $n^2 \le q < 2n^2$;
   Fill a d-qbit quantum register with the state:
$$\psi_1 = \frac{1}{\sqrt{q}} \sum_{a=0}^{q-1} |\, a \,\rangle \; ;$$
2. [Choose a base]
   Choose random $x \in [2, n-2]$ but coprime to n;
3. [Create all powers]
   Using quantum powering on ψ_1, fill a second register with
$$\psi_2 = \frac{1}{\sqrt{q}} \sum_{a=0}^{q-1} |\, x^a \bmod n \,\rangle \; ;$$
4. [Perform a quantum FFT]
   Apply FFT to the second quantum register, to obtain
$$\psi_3 = \frac{1}{q} \sum_{a=0}^{q-1} \sum_{c=0}^{q-1} e^{2\pi i a c/q} \,|\, c \,\rangle \,|\, x^a \bmod n \,\rangle \; ;$$
5. [Detect periodicity in x^a]
   Measure the state ψ_3, and employ (classical TM) side calculations to infer the period r as the minimum power enjoying $x^r \equiv 1 \pmod{n}$;
6. [Resolution]
   if(r odd) goto [Choose a base];
   Use Proposition 8.5.1 to attempt to produce a nontrivial factor of n. On failure, goto [Choose a base];

We have been intentionally brief in the final steps of the algorithm. The details for these last stages are laid out splendidly in [Shor 1999]. The core idea underlying the [Detect periodicity ...] step is this: After the FFT step, the machine should be found in a final state $|\, c \,\rangle|\, x^k \bmod n \,\rangle$ with probability

$$P_{c,k} = \left| \frac{1}{q} \sum_{\substack{a=0 \\ x^a \equiv x^k \pmod{n}}}^{q-1} e^{2\pi i a c/q} \right|^2 = \left| \frac{1}{q} \sum_{b=0}^{\lfloor (q-k-1)/r \rfloor} e^{2\pi i (br+k)c/q} \right|^2 . \tag{8.4}$$

This expression, in turn, can be shown to exhibit "spikes" at certain r-dependent values of c. From these spikes—which we presume would all show up simultaneously upon measurement of the QTM machine's state—one can infer after a quick side calculation the period r. See Exercises 8.18, 8.19, 8.20, 8.31 for some more of the relevant details. As mentioned in the latter exercise, the discrete logarithm (DL) problem also admits of a QTM polynomial-time solution.

Incidentally, quantum computers are not the only computational engines that enjoy the status of being talked about but not yet having been built to any practical specification. Recently, A. Shamir described a "Twinkle" device to factor numbers [Shamir 1999]. The proposed device is a special-purpose optoelectronic processor that would implement either the QS method or the NFS method. Yet another road on which future computing machines could conceivably travel is the "DNA computing" route, the idea being to exploit the undeniable processing talent of the immensely complex living systems that have evolved for eons [Paun et al. 1998]. If one wants to know not so much the mathematical but the cultural issues tied up in futuristic computing, a typical lay collection of pieces concerning DNA, molecular, and quantum computing is the May-June 2000 issue of the MIT magazine *Technology Review*.

8.6 Curious, anecdotal, and interdisciplinary references to primes

Just as practical applications of prime numbers have emerged in the cryptographic, statistical, and other computational fields, there are likewise applications in such disparate domains as engineering, physics, chemistry, and biology. Even beyond that, there are amusing anecdotes that collectively signal a certain awareness of primes in a more general, we might say lay, context. Beyond the scientific connections, there are what may be called the "cultural" connections. Being cognizant of the feasibility of filling an entire separate volume with interdisciplinary examples, we elect to close this chapter with a very brief mention of some exemplary instances of the various connections.

One of the pioneers of the interdisciplinary aspect is M. Schroeder, whose writings over the last decade on many connections between engineering and number theory continue to fascinate [Schroeder 1997]. Contained in such work are interdisciplinary examples. To name just a few, fields $\mathbf{F}_q$ as they pertain to the technology of error-correcting codes, discrete Fourier transforms (DFTs) over fields relevant to acoustics, the use of the Möbius μ and other functions in science, and so on. To convey a hint of how far the interdisciplinary connections can reach, we hereby cite Schroeder's observation that certain astronomical experiments to verify aspects of Einstein's general relativity involved such weak signals that error-correcting codes (and hence finite fields) were invoked. This kind of argument shows how certain cultural or scientific achievements do depend, at some level, on prime numbers. A pleasingly recreational source for interdisciplinary prime-number investigations is [Caldwell 1999].

In biology, prime numbers appear in contexts such as the following one, from [Yoshimura 1997]. We quote the author directly in order to show how prime numbers can figure into a field or a culture, without much of the standard number-theoretical language, rather with certain intuitive inferences relied upon instead:

> Periodical cicadas (Magicicada spp.) are known for their strikingly synchronized emergence, strong site tenacity, and unusually long (17- and 13-yr) life cycles for insects. Several explanations have been proposed for the origin and maintenance of synchronization. However, no satisfactory explanations have been made for the origins of the prime-numbered life cycles. I present an evolutionary hypothesis of a forced developmental delay due to climate cooling during ice ages. Under this scenario, extremely low adult densities, caused by their extremely long juvenile stages, selected for synchronized emergence and site tenacity because of limited mating opportunities. The prime numbers (13 and 17) were selected for as life cycles because these cycles were least likely to coemerge, hybridize, and break down with other synchronized cycles.

It is interesting that the literature predating Yoshimura is fairly involved, with at least three different explanations of why prime-numbered life cycles such as 13 and 17 years would evolve. Any of the old and new theories should, of course, exploit the fact of minimal divisors for primes, and indeed the attempts to do this are evident in the literature (see, for example, the various review works referenced in [Yoshimura 1997]). To convey a notion of the kind of argument one might use for evolution of prime life cycles, imagine a predator with a life cycle of 2 years—an even number—synchronized, of course, to the solar-driven seasons, with periodicity of those 2 years in most every facet of life such as reproduction and death. Because this period does not divide a 13- or 17-year one, the predators will from time to time go relatively hungry. This is not the only type of argument—for some such arguments do not involve predation whatsoever, rather depend on the internal competition and fitness of the prime-cycle species itself—but the lack of divisibility is always present, as it should be, in any evolutionary argument. In a word, such lines of thought must explain among other things why a life cycle with a substantial number of divisors has led to extinction.

Another appearance of the noble primes—this time in connection with molecular biology—is in [Yan et al. 1991]. These authors infer that certain amino acid sequences in genetic matter exhibit patterns expected of (binary representations of) prime numbers. In one segment they say:

> Additively generated numbers can be primes or nonprimes. Multiplicatively generated numbers are nonprimes ("composites" in number theory terminology). Thus, prime numbers are more creative than nonprimes .... The creativeness and indivisibility of prime numbers leads one to infer that primes smaller than 64 are the number equivalents of amino acids; or that amino acids are such Euclid units of living molecules.

The authors go on to suggest Diophantine rules for their theory. The present authors do not intend to critique the interdisciplinary notion that composite numbers somehow contain less information (are less profound) than the primes. Rather, we simply point out that some thought has gone into this connection with genetic codes.

Let us next mention some involvements of prime numbers in the particular field of physics. We have already touched upon the connection of quantum computation and number-theoretical problems. Aside from that, there is the fascinating history of the Hilbert–Pólya conjecture, saying in essence that the behavior of the Riemann zeta function on the critical line $\mathrm{Re}(s) = 1/2$ depends somehow on a mysterious (complex) Hermitian operator, of which the critical zeros would be eigenvalues. Any results along these lines—even partial results—would have direct implications about prime numbers, as we saw in Chapter 1. The study of the distribution of eigenvalues of certain matrices has been a strong focus of theoretical physicists for decades. In the early 1970s, a chance conversation between F. Dyson, one of the foremost researchers on the physics side of random matrix work, and H. Montgomery, a number theorist investigating the influence of critical zeros of the zeta function on primes, led them to realize that some aspects of the distribution of eigenvalues of random matrices are very close to those of the critical zeros. As a result, it is widely conjectured that the mysterious operator that would give rise to the properties of ζ is of the Gaussian unitary ensemble (GUE) class. A relevant $n \times n$ matrix G in such a theory has $G_{aa} = x_{aa}\sqrt{2}$ and for $a > b$, $G_{ab} = x_{ab} + iy_{ab}$, together with the Hermitian condition $G_{ab} = G^*_{ba}$; where every x_{ab}, y_{ab} is a Gaussian random variable with unit variance, mean zero. The works of [Odlyzko 1987, 1992, 1994, 2000b] show that the statistics of consecutive critical zeros are in many ways equivalent—experimentally speaking—to the theoretical distribution of eigenvalues of a large such matrix G. In particular, let $\{z_n : n = 1, 2, \ldots\}$ be the collection of the (positive) imaginary parts of the critical zeros of ζ, in increasing order. It is known from the deeper theory of the ζ function that the quantity

$$\delta_n = \frac{z_{n+1} - z_n}{2\pi} \ln \frac{z_n}{2\pi}$$

has mean value 1. But computer plots of the histogram of δ values show a remarkable agreement for the same (theoretically known) statistic on eigenvalues of a GUE matrix. Such comparisons have been done on over 10^8 zeros neighboring z_N where $N \approx 10^{20}$ (though the work of [Odlyzko 2000b] involves 10^{10} zeros of even greater height). The situation is therefore compelling: There may well be an operator whose eigenvalues are precisely the Riemann critical zeros (scaled by the logarithmic factor). But the situation is not as clean as it may appear. For one thing, Odlyzko has plotted the Fourier transform

$$\sum_{N+1}^{N+40000} e^{ixz_n},$$

and it does not exhibit the decay (in x) expected of GUE eigenvalues. In fact, there are spikes reported at $x = p^k$, i.e., at prime-power frequencies. This is expected from a number-theoretical perspective. But from the physics perspective, one can say that the critical zeros exhibit "long-range correlation," and it has been observed that such behavior would accrue if the critical zeros were not random GUE eigenvalues *per se*, but eigenvalues of some unknown Hamiltonian appropriate to a chaotic-dynamical system. In this connection, a great deal of fascinating work—by M. Berry and others—under the rubric of "quantum chaology" has arisen [Berry 1987].

There are yet other connections between the Riemann ζ and concepts from physics. For example, in [Borwein et al. 2000] one finds mention of an amusing connection between the Riemann ζ and quantum oscillators. In particular, as observed by Crandall in 1991, there exists a quantum wave function $\psi(x,0)$—smooth, devoid of any zero crossings on the x axis—that after a finite time T of evolution under the Schrödinger equation becomes a "crinkly" wave function $\psi(x,T)$ with infinitely many zero crossings, and these zeros are precisely the zeros of $\zeta(1/2+ix)$ on the critical line. In fact, for the wave function at the special time T in question, the specific eigenfunction expansion evaluates as

$$\psi(x,T) = f\left(\frac{1}{2}+ix\right)\zeta\left(\frac{1}{2}+ix\right) = e^{-x^2/(2a^2)}\sum_{n=0}^{\infty} c_n(-1)^n H_{2n}(x/a), \quad (8.5)$$

for some positive real a and a certain sequence (c_n) of real coefficients depending on a, with H_m being the standard Hermite polynomial of order m. Here, $f(s)$ is an analytic function of s having no zeros. It is amusing that one may truncate the n-summation at some N, say, and numerically obtain—now as zeros of a degree-$2N$ polynomial—fairly accurate critical zeros. For example, for $N = 27$ (so polynomial degree is 54) an experimental result appears in [Borwein et al. 2000] in which the first seven critical zeros are obtained, the first of which being to 10 good decimals. In this way one can in principle approximate arbitrarily closely the Riemann critical zeros as the eigenvalues of a Hessenberg matrix (which in turn are zeros of a particular polynomial). A fascinating phenomenon occurs in regard to the Riemann hypothesis, in the following way. If one truncates the Hermite sum above, say at $n = N$, then one expects $2N$ complex zeros of the resulting, degree-$2N$ polynomial in x. But in practice, only some of these $2N$ zeros are real (i.e., such that $\frac{1}{2}+ix$ is on the Riemann critical line). For large N, and again experimentally, the rest of the polynomial's zeros are "expelled" a good distance away from the critical line. The Riemann hypothesis, if it is to be cast in language appropriate to the Hermite expansion, must somehow address this expulsion of nonreal polynomial zeros away from the real axis. Thus the Riemann hypothesis can be cast in terms of quantum dynamics in some fashion, and it is not out of the question that this kind of interdisciplinary approach could be fruitful.

An anecdote cannot be resisted here; this one concerns the field of engineering. Peculiar as it may seem today, the scientist and engineer van der Pol did, in the 1940s, exhibit tremendous courage in his "analog" manifestation of an interesting Fourier decomposition. An integral used by van der Pol was a special case ($\sigma = 1/2$) of the following relation, valid for

$s = \sigma + it$, $\sigma \in (0,1)$ [Borwein et al. 2000]:

$$\zeta(s) = s \int_{-\infty}^{\infty} e^{-\sigma\omega} \left(\lfloor e^{\omega} \rfloor - e^{\omega} \right) e^{-i\omega t}\, d\omega.$$

Van der Pol actually built and tested an electronic circuit to carry out the requisite transform in analog fashion for $\sigma = 1/2$, [van der Pol 1947]. In today's primarily digital world it yet remains an open question whether the van der Pol approach can be effectively used with, say, a fast Fourier transform to approximate this interesting integral. In an even more speculative tone, one notes that in principle, at least, there could exist an analog device—say an extremely sophisticated circuit—that sensed the prime numbers, or something about such numbers, in this fashion.

At this juncture of our brief interdisciplinary overview, a word of caution is in order. One should not be led into a false presumption that theoretical physicists always endeavor to legitimate the prevailing conjectural models of the prime numbers or of the Riemann ζ function. For example, in the study of [Shlesinger 1986], it is argued that *if* the critical behavior of ζ corresponds to a certain "fractal random walk" (technically, if the critical zeros determine a Levy flight in a precise, stochastic sense), then fundamental laws of probability are violated unless the Riemann hypothesis is *false*.

Also, one should not think that the appearance of primes in physics is relegated to studies of the Riemann ζ function. Indeed, [Vladimirov et al. 1994] authored an entire volume on the subject of p-adic field expansions in theoretical physics. They say:

> Elaboration of the formalism of mathematical physics over a p-adic number field is an interesting enterprise apart from possible applications, as it promotes deeper understanding of the formalism of standard mathematical physics. One can think there is the following principle. *Fundamental physical laws should admit of formulation invariant under a choice of a number field.*

(The italics are theirs.) This quotation echoes the cooperative theme of the present section. Within this interesting reference one can find further references to p-adic quantum gravity and p-adic Einstein-relativistic equations.

Physicists have from time to time even performed "prime number experiments." For example, [Wolf 1997] takes a signal, call it $x = (x_0, x_1, \ldots, x_{N-1})$, where a component x_j is the count of primes over some interval. Specifically,

$$x_j = \pi((j+1)M) - \pi(jM),$$

where M is some fixed interval length. Then is considered the DFT

$$X_k = \sum_{j=0}^{N-1} x_j e^{-2\pi ijk/N},$$

of which the zeroth Fourier component is

$$X_0 = \pi(MN).$$

The interesting thing is that this particular signal exhibits the spectrum (the behavior in the index k) of "$1/f$" noise—actually, we could call it "pink" noise. Specifically, Wolf claims that

$$|X_k|^2 \sim \frac{1}{k^\alpha} \tag{8.6}$$

with exponent $\alpha \sim 1.64\ldots$. This means that in the frequency domain (i.e., behavior in Fourier index k) the power law involves, evidently, a fractional power. Wolf suggests that perhaps this means that the prime numbers are in a "self-organized critical state," pointing out that all possible (even) gaps between primes conjecturally occur so that there is no natural "length" scale. Such properties are also inherent in well-known complex systems that are also known to exhibit $1/k^\alpha$ noise. Though the power law may be imperfect in some asymptotic sense, Wolf finds it to hold over a very wide range of M, N. For example, $M = 2^{16}, N = 2^{38}$ gives a compelling and straight line on a $(\ln |X_k|^2, \ln k)$ plot with slope ≈ -1.64. Whether or not there will be a coherent theory of this exponent law (after all, it could be an empirical accident that has no real meaning for very large primes), the attractive idea here is to connect the behavior of complex systems with that of the prime numbers (see Exercise 8.29).

As for cultural (nonscientific, if you will) connections, there exist many references to the importance of very small primes such as $2, 3, 5, 7$; such references ranging from the biblical to modern, satirical treatments. As just one of myriad examples of the latter type of writing, there is the piece by [Paulos 1995], in *Forbes* financial magazine, called "High 5 Jive," being about the number 5, humorously laying out misconceptions that can be traced to the fact of five fingers on one hand. The number 7 also receives a great deal of airplay, as it were. In a piece by [Stuart 1996] in, of all things, a medical journal, the "magic of seven" is touted; for example, "The seven ages of man, the seven seas, the seven deadly sins, the seven league boot, seventh heaven, the seven wonders of the world, the seven pillars of wisdom, Snow White and the seven dwarves, 7-Up ... ." The author goes on to describe how the Hippocratic healing tradition has for eons embraced the number 7 as important, e.g., in the number of days to bathe in certain waters to regain good health. It is of interest that the very small primes have, over thousands of years, provided fascination and mystique to all peoples, regardless of their mathematical persuasions. Of course, much the same thing could be said about certain small composites, like $6, 12$. However, it would be interesting to know once and for all whether fascination with primes *per se* has occurred over the millennia because the primes are so dense in low-lying regions, or because the general population has an intuitive understanding of the special stature of the primes, thus prompting the human imagination to seize upon such numbers.

And there are numerous references to prime numbers in music theory and musicology, sometimes involving somewhat larger primes. For example, from an article of [Warren 1995] we read:

> Sets of 12 pitches are generated from a sequence of five consecutive prime numbers, each of which is multiplied by each of the three largest numbers in the sequence. Twelve scales are created in this manner, using the prime sequences up to the set (37, 41, 43, 47, 53). These scales give rise to pleasing dissonances that are exploited in compositions assisted by computer programs as well as in live keyboard improvisations.

And here is the abstract of a paper concerning musical correlations between primes and Fibonacci numbers [Dudon 1987] (note that the mention below of Fibonacci numbers is not the standard one, but closely related to it):

> The Golden scale is a unique unequal temperament based on the Golden number. The equal temperaments most used, $5, 7, 12, 19, 31, 50$, etc., are crystallizations through the numbers of the Fibonacci series, of the same universal Golden scale, based on a geometry of intervals related in Golden proportion. The author provides the ratios and dimensions of its intervals and explains the specific intonation interest of such a cycle of Golden fifths, unfolding into microtonal coincidences with the first five significant prime numbers ratio intervals (3:5:7:11:13).

From these and other musicology references it appears that not just the very smallest primes, rather also some two-digit primes, play a role in music theory. Who can tell whether larger primes will one day appear in such investigations, especially given how forcefully the human–machine–algorithm interactions have emerged in modern times?

8.7 Exercises

8.1. Explain quantitatively what R. Brent meant when he said that to remember the digits of 65537, you recite the mnemonic

> *"Fermat prime, maybe the largest."*

Along the same lines, to which factor of which Fermat number does the following mnemonic of J. Pollard apply?

> *"I am now entirely persuaded to employ rho method, a handy trick, on gigantic composite numbers."*

8.2. Over the years many attacks on the RSA cryptosystem have been developed, some of these attacks being elementary but some involving deep number-theoretical notions. Analyze one or more RSA attacks as follows:

(1) Say that a security provider wishes to live easily, dishing out the *same* modulus $N = pq$ for each of U users. A trusted central authority, say,

establishes for each user $u \in [1, U]$ a unique private key D_u and public key (N, E_u). Argue carefully exactly why the entire system is insecure.

(2) Show that Alice could fool (an unsuspecting) Bob into signing a bogus (say harmful to Bob) message x, in the following sense. Referring to Algorithm 8.1.4, say that Alice chooses a random r and can get Bob to sign and send back the "random" message $x' = r^{E_{\mathrm{B}}} x \bmod N_{\mathrm{B}}$. Show that Alice can then readily compute an s such that $s^{E_{\mathrm{B}}} \bmod N_{\mathrm{B}} = x$, so that Alice would possess a signed version of the harmful x.

(3) Here we consider a small-private-exponent attack based on an analysis of [Wiener 1990]. Consider an RSA modulus $N = pq$ with $q < p < 2q$. Assume the usual condition $ED \bmod \varphi(N) = 1$, but we shall restrict the private exponent by $D < N^{1/4}/3$. Show first that

$$|N - \varphi(N)| < 3\sqrt{N}.$$

Show then the existence of an integer k such that

$$\left| \frac{E}{N} - \frac{k}{D} \right| < \frac{1}{2D^2}.$$

Argue now that the private key D can be obtained (since you know the public pair N, E) in polynomial effort (operation count bounded by a power of $\ln N$).

(4) So-called timing attacks have also been developed. If a machine calculates numbers such as x^D using a power ladder whose square and multiply operations take *different* but fixed times, one can glean information about the exponent D. Say that you demand of a cryptosystem the generation of many signatures $x_i^D \bmod N$ for i running through some set, and that you store the respective times T_i required for the signing system to give the i-th signature. Then do the same timing experiment but for each x_i^3, say. Describe how correlations between the sets $\{t_i\}$ and $\{T_i\}$ can be used to determine bits of the private exponent D.

We have given above just a smattering of RSA attack notions. There are also attacks based on lattice reduction [Coppersmith 1997] and interesting issues involving the (incomplete) relation between factoring and breaking RSA [Boneh and Venkatesan 1998]. There also exist surveys on this general topic [Boneh 1999]. We are grateful to D. Cao for providing some ideas for this exercise.

8.3. We have noted that both y-coordinates and the "clue" point are not fundamentally necessary in the transmission of embedded encryption from Algorithm 8.1.10. With a view to Algorithm 7.2.8 and the Miller generator, equation (8.1), work out an explicit, detailed algorithm for direct embedding but with neither y-coordinates nor data expansion (except that one will still need to transmit the sign bit d—an asymptotically negligible expansion). You might elect to use a few more "parity bits," for example in Algorithm 7.2.8 you may wish to specify one of two quadratic roots, and so on.

8.4. Describe how one may embed any plaintext integer $X \in \{0, \dots, p-1\}$ on a *single* given curve, by somehow counting up from X as necessary, until $X^3 + aX + b$ is a quadratic residue (mod p). One such scheme is described in [Koblitz 1987].

8.5. In Algorithm 8.1.10 when is it the case that X is the x-coordinate of a point on both curves E, E'?

8.6. Devise a coin-flip protocol based on the idea that if n is the product of two different odd primes, then quadratic residues modulo n have 4 square roots of the form $\pm a, \pm b$. Further computing these square roots, given the quadratic residue, is easy when one knows the prime factorization of n and, conversely, when one has the 4 square roots, the factorization of n is immediate. Note in this connection the Blum integers of Exercise 2.22, which integers are often used in coin-flip protocols. References are [Schneier 1996] and [Bressoud and Wagon 2000, p. 146].

8.7. Explore the possibility of cryptographic defects in Algorithm 8.1.11. For example, Bob could cheat if he could quickly factor n, so the fairness of the protocol, as with many others, should be predicated on the presumed difficulty in factoring the number n that Alice sends. Is there any way for Alice to cheat by somehow misleading Bob into preferring one of the primes over the other? If Bob knows or guesses that Alice is choosing the primes p, q, r at random in a certain range, is there some way for him to improve his chances? Is there any way for either party to lose on purpose?

8.8. It is stated after Algorithm 8.1.11 that a coin-flip protocol can be extended to group games such as poker. Choose a specific protocol (from the text algorithm or such references as in Exercise 8.6), and write out *explicitly* a design for "telephone poker," in which there is, over a party-line phone connection, a deal of say 5 cards per person, hands eventually claimed, and so on. It may be intuitively clear that if flipping a coin can be done, so can this poker game, but the exercise here is to be explicit in the design of a full-fledged poker game.

8.9. Prove that the verification step of Algorithm 8.1.8 works, and discuss both the probability of a false signature getting through and the difficulty of forging.

8.10. Design a random-number generator based on a one-way function. It turns out that *any* suitable one-way function can be used to this effect. One reference is [Håstad et al. 1999]; another is [Lagarias 1990].

8.11. Implement the Halton-sequence fast qMC Algorithm 8.3.6 for dimension $D = 2$, and plot graphically a cloud of some thousands of points in the unit square. Comment on the qualitative (visual) difference between your plot and a plot of simple random coordinates.

8.12. Prove the claim concerning equation (8.3) under the stated conditions on k. Start by analyzing the Diophantine equation (mod 4), concluding that $x \equiv 1 \pmod 4$, continuing on with further analysis (mod 4) until a Legendre symbol $\left(\frac{-4m^2}{p}\right)$ is encountered for $p \equiv 3 \pmod 4$. (See, for example, [Apostol 1976, Section 9.8].)

8.13. Show that the ABC conjecture is false with $\epsilon = 0$. In fact, show that there are infinitely many coprime triples a, b, c of positive integers with $a + b = c$ and $\gamma(abc) = o(c)$. (As before, $\gamma(n)$ is the largest square-free divisor of n.)

8.14. [Tijdeman] Show that the ABC conjecture implies the Fermat–Catalan conjecture.

8.15. [Silverman] Show that the ABC conjecture implies that there are infinitely many primes p that are not Wieferich primes.

8.16. Say $q_1 < q_2 < \dots$ is the sequence of powers. That is, $q_1 = 1$, $q_2 = 4$, $q_3 = 8$, $q_4 = 9$, and so on. It is not known if the gaps $q_{n+1} - q_n$ tend to infinity with n, but show that this is indeed the case if the ABC conjecture is assumed. In fact, show on the ABC conjecture that for each $\epsilon > 0$, we have $q_{n+1} - q_n > n^{1/12-\epsilon}$ for all sufficiently large values of n.

8.17. Show that there is a polynomial in two variables with integer coefficients whose values at positive integral arguments coincide with the set of positive composite numbers. Next, starting from the Lagrange theorem that every positive integer is a sum of 4 squares (see Exercise 9.42), exhibit a polynomial in 8 variables with integer coefficients such that its values at all integral arguments constitute the set of positive composites.

8.18. Suppose the integer n of Proposition 8.5.1 has the distinct prime factors $p_1, \dots, p_k$, where $2^{s_i} \| p_i - 1$ and $s_1 \le \dots \le s_k$. Show that the relevant probability is then

$$1 - 2^{-(s_1+\cdots+s_k)}\left(1 + \frac{2^{s_1 k} - 1}{2^k - 1}\right)$$

and that this expression is not less than $1 - 2^{1-k}$. (Compare with Exercise 3.15.)

8.19. Complete one of the details for Shor factoring, as follows. We gave as relation (8.4) the probability $P_{c,k}$ of finding our QTM in the composite state $\mid c \rangle \mid x^k \rangle$. Explain quantitatively how the probability (for a fixed k, with c the running variable) should show spikes corresponding to solutions d to the Diophantine approximation

$$\left|\frac{c}{q} - \frac{d}{r}\right| \le \frac{1}{2q}.$$

Explain, then, how one can find d/r in lowest terms from (measured) knowledge of appropriate c. Note that if $\gcd(d, r)$ happens to be 1, this procedure gives the exact period r for the algorithm, and we know that two random integers are coprime with probability $6/\pi^2$.

On the computational side, model (on a classical TM, of course) the spectral behavior of the QTM occurring at the end of Algorithm 8.5.2, using the following exemplary input. Take $n = 77$, so that the [Initialization] step sets $q = 8192$. Now choose (we are using hindsight here) $x = 3$, for which the period turns out to be $r = 30$ after the [Detect periodicity ...] step. Of course, the whole point of the QTM is to measure this period physically, and quickly! To continue along and model the QTM behavior, use a (classical) FFT to make a graphical plot of c versus the probability $P_{c,1}$ from formula (8.4). You should see very strong spikes at certain c values. One of these values is $c = 273$, for example. Now from the relation

$$\left| \frac{273}{8192} - \frac{d}{r} \right| \le \frac{1}{2q}$$

one can derive the result $r = 30$ (the literature explains continued-fraction methods for finding the relevant approximants d/r). Finally, extract a factor of n via $\gcd(x^{r/2} - 1, n)$. These machinations are intended show the flavor of the missing details in the presentation of Algorithm 8.5.2; but beyond that, these examples pave the way to a more complete QTM emulation (see Exercise 8.20). Note the instructive phenomenon that even this small-n factoring emulation-via-TM requires FFT lengths into the thousands; yet a true QTM might require only a dozen or so qbits.

8.20. It is a highly instructive exercise to cast Algorithm 8.5.2 into a detailed form that incorporates our brief overview and the various details from the literature (including the considerations of Exercise 8.19).

A second task that lives high on the pedagogical ladder is to emulate a QTM with a standard TM program implementation, in a standard language. Of course, this will not result in a polynomial-time factorer, but only because the TM does what a QTM could do, yet the former involves an exponential slowdown. For testing, you might start with input numbers along the lines of Exercise 8.19. Note that one still has unmentioned options. For example, one could emulate very deeply and actually model quantum interference, or one could just use classical arithmetic and FFTs to perform the algebraic steps of Algorithm 8.5.2.

8.8 Research problems

8.21. Prove or disprove the claim of physicist D. Broadhurst that the number

$$P = \frac{2^{903} 5^{682}}{514269} \int_0^\infty dx \, \frac{x^{906} \sin(x \ln 2)}{\sinh(\pi x/2)} \left(\frac{1}{\cosh(\pi x/5)} + 8 \sinh^2(\pi x/5) \right)$$

is not only an integer, but in fact a prime number. This kind of integral shows up in the theory of multiple zeta functions, which theory in turn has application in theoretical physics, in fact in quantum field theory (and we mean here physical fields, not the fields of algebra!).

Since the 1st printing of the present book, Broadhurst has used a publicly available primality-proof package to establish that P is indeed prime. One research extension, then, is to find—with proof—an even larger prime having this kind of trigonometric-integral representation.

8.22. Here we explore a connection between prime numbers and fractals. Consider the infinite-dimensional Pascal matrix P with entries

$$P_{i,j} = \binom{i+j}{i},$$

for both i and j running through 0, 1, 2, 3, ...; thus the classical Pascal triangle of binomial coefficients has its apex packed into the upper-left corner of P, like so:

$$P = \begin{pmatrix} 1 & 1 & 1 & 1 & \cdots \\ 1 & 2 & 3 & 4 & \cdots \\ 1 & 3 & 6 & 10 & \cdots \\ 1 & 4 & 10 & 20 & \cdots \\ \vdots & \vdots & \vdots & \vdots & \ddots \end{pmatrix}.$$

There are many interesting features of this P matrix (see [Higham 1996, p. 520]), but for this exercise we concentrate on its fractal structure modulo primes.

Define the matrix $Q_n = P \bmod n$, where the mod operation is taken elementwise. Now imagine a geometrical object created by coloring each zero element of Q_n black, and all nonzero elements white. Imagine further that this object is the full infinite-dimensional Q_n matrix, but compressed into a finite planar square, so that we get, if you will, a kind of "snowflake" with many holes of black within a fabric of white. Now, argue that for prime modulus p, so that the mod matrix is Q_p, the fractal dimension of the "snowflake" object is given by

$$\delta = \frac{\ln(p(p+1)/2)}{\ln p}.$$

Technically, this is a "box dimension," and for this and other dimension definitions one source is [Crandall 1994b] and references therein. (Hint: The basic method for getting δ is to count how many nonzero elements there are in an upper-left $p^k \times p^k$ submatrix of Q_p, and see how this scales with the submatrix size p^{2k}.) Thus for example, the Pascal triangle modulo 2 has dimension $\delta = (\ln 3)/(\ln 2)$ and the triangle modulo 3 has dimension $\delta = (\ln 6)/(\ln 3)$. The case $p = 2$ here gives the famous Sierpinski gasket, a well-studied object in the theory of fractals. It is sometimes said that such a "gasket" amounts to "more than a line but less than the plane." Clarify this vague statement in quantitative terms, by looking at the numerical magnitude of the dimension δ.

Extensions to this fractal-dimension exercise abound. For example, one finds that for prime p, in the upper-left $p \times p$ submatrix of Q_p, the number of nonzero elements is always a triangular number. (A triangular number is a number of the form $1 + 2 + \ldots + n = n(n+1)/2$.) Question is, for what *composite* n does the upper-left $n \times n$ submatrix have a triangular number of nonzero elements? And here is an evidently tough question: What is the fractal dimension if we consider the object in "gray-scale," that is, instead of white/black pixels that make up the gasket object, we calculate δ using proper weight of an element of Q_p not as binary but as its actual residue in $[0, p-1]$?

8.23. In the field of elliptic curve cryptography (ECC) it is important to be able to construct elliptic curves of *prime* order. Describe how to adapt the Schoof method, Algorithm 7.5.6, so that it "sieves" curve orders, looking for such a prime order. In other words, curve parameters a, b would be chosen randomly, say, and small primes L would be used to "knock out" a candidate curve as soon as $p+1-t$ is ascertained as composite. Assuming that the Schoof algorithm has running time $O\left(\ln^k p\right)$, estimate the complexity of this sieving scheme as applied to finding just *one* elliptic curve of prime order. Incidentally, it may not be efficient overall to use maximal prime powers $L = 2^a, 3^b$, etc. (even though as we explained these do work in the Schoof algorithm) for such a sieve. Explain why that is. Note that some of the complexity issues herein are foreshadowed in Exercise 7.29 and related exercises of that chapter.

If one did implement a "Schoof sieve" to find a curve of prime order, the following example would be useful in testing the software:

$$p = 2^{113} - 133, \quad a = -3, \quad b = 10018.$$

Now, for the following moduli (we give here some prime-power L values even though, as we said, that is not necessarily an efficient approach)

$$7, 11, 13, 17, 19, 23, 25, 27, 29, 31, 32, 37, 41, 43,$$

the curve order $\#E = p + 1 - t$ has values $t \bmod L$ as

$$2, 10, 3, 4, 6, 11, 14, 9, 26, 1, 1, 10, 8, 8,$$

leading to the *prime* curve order

$$\#E = 10384593717069655112027224311117371.$$

Note that the task of finding curves for which both the order $p+1-t$ and the twist order $p+1+t$ are prime is more difficult, not unlike the task of finding twin primes as opposed to primes. A research problem: Prove via the methods of analytic number theory that there is a positive constant c such that for most primes p there are at least $c\sqrt{p}/\ln^2 p$ integers t with $0 < t < 2\sqrt{p}$, such that $p + 1 \pm t$ are both prime.

8.24. Work out software that very stringently tests random-number generators. The basic idea is simple: Assume an input stream of integers, say,

But the implementation is hard: There are spectral tests, collision tests, general statistical tests, normality tests, and so on. The idea is that the software would give a "score" to the generated stream, and thereby select "good" random number generators. Of course, goodness itself could even be context-dependent. For example, a good random generator for numerical integration in computational physics might be a cryptographically bad generator, and so on. One thing to note during such a research program is the folklore that chaos-based generators are cryptographically risky. To this end, one might consider the measurement of fractal dimension and Lyapunov exponents of generated pseudorandom sequences as something to add to one's test arsenal.

8.25. Investigate elliptic-curve-based random generation. Possible research directions are indicated in the text after iteration (8.1), including the possibility of casting the Gong–Berson–Stinson generator scheme ([Gong et al. 1999]) into a form suitable for curves over odd-characteristic fields.

8.26. Investigate possibilities for random generators that have even longer periods than the Marsaglia example of the text. For example, [Brent 1994] notes that, for any Mersenne prime $M_q = 2^q - 1$ with $q \equiv \pm 1 \pmod 8$, there may be a primitive trinomial of degree M_q, giving rise to a Fibonacci generator with period at least M_q. A known working example is $q = 132049$, giving a long period indeed!

8.27. Though Definition 8.3.1 is rather technical, and though the study of discrepancies D_N, D_N^* remains difficult and incomplete to this day, there do exist some interesting discrepancy bounds of a general character. One such is the Leveque theorem on sequences $P = (x_0, x_1, \ldots)$ of points, each $x_j \in [0,1]$. The elegant statement is [Kuipers and Niederreiter 1974]

$$D_N \le \left(\frac{6}{\pi^2} \sum_{h=1}^{\infty} \frac{1}{h^2} \left| \frac{1}{N} \sum_{n=0}^{N-1} e^{2\pi i h x_n} \right|^2 \right)^{1/3}.$$

One notes that this bound is, remarkably, best possible in one sense: The sequence $P = (0, 0, \ldots, 0)$ actually gives equality. A research problem is to find interesting or useful sequences for which the Leveque bound can actually be computed. For example, what happens in the Leveque formula if the P sequence is generated by a linear-congruential generator (with each x_n normalized, say, via division by the modulus)? It is of interest that knowledge of Fourier sums can be brought to bear in this way on quasi-Monte Carlo studies.

8.28. An interesting and difficult open problem in the qMC field is the following. Whereas low-discrepancy qMC sequences are characterized by the bound

$$D_N^* = O\left(\frac{\ln^D N}{N} \right),$$

the best that is known as a *lower* bound for large general dimension D is [Veach 1997]

$$D_N^* \geq C(D) \left(\frac{\ln^{D/2} N}{N} \right).$$

The hard problem is to endeavor to close the gap between the powers $\ln^{D/2}$ and $\ln^D$. This work is important, since for very high dimensions D the ln error factors can be prohibitive.

8.29. Work out a theory to explain the experiments of [Wolf 1997] by attempting to derive his power law (8.6). (Note that there is no *a priori* guarantee that some deep theory is at work; the claimed law could be an artifact based on the particular numerical regions studied!) Consider, for example, the (large-k) asymptotic behavior of the following integral as a continuous approximation to the discrete transform:

$$I(k) = \int_a^b \frac{e^{ikx}}{\ln(c+x)}\, dx,$$

where a, b, c are fixed positive real constants. Can one explain the experimental $1/k^{1.64}$ power law (which would be for $|I|^2$) in this way?

8.30. Here we indicate some very new directions indicated in recent literature pertaining to the Riemann hypothesis (RH). The research options below could have appeared in Chapter 1, where we outlined some consequences of the RH, but because of a strong interdisciplinary flavor in what follows, the description belongs here just as well.

Consider these RH equivalences as research directions, primarily computational but always potentially theoretical:

(1) There is an older, Riesz condition [Titchmarsh 1986, Section 14.32] that is equivalent to the RH, namely,

$$\sum_{n=1}^{\infty} \frac{(-x)^n}{\zeta(2n)(n-1)!} = O\left(x^{1/4+\varepsilon}\right).$$

Note the interesting feature that only integer arguments of ζ appear. One question is this: Can there be any value whatsoever in numerical evaluations of the sum? If there be any value at all, methods for so-called "recycled" evaluations of ζ come into play. These are techniques for evaluating huge sets of ζ values having the respective arguments in arithmetic progression [Borwein et al. 2000].

(2) The recent work of [Balazard et al. 1999] proves that

$$I = \int \frac{\ln|\zeta(s)|}{|s|^2}\, ds = 2\pi \sum_{\mathrm{Re}(\rho)>1/2} \ln\left|\frac{\rho}{1-\rho}\right|,$$

where the line integral is carried out over the critical line, and ρ denotes any zero in the critical strip, but *to the right* of the critical line as indicated, counting multiplicity. Thus the simple statement "$I = 0$" is equivalent to the RH. One task is to plot the behavior of $I(T)$, which is the integral I restricted to $\mathrm{Im}(s) \in [-T, T]$, and look for evident convergence $I(T) \to 0$, possibly giving a decay estimate. Another question mixes theory and computation: If there is a *single* errant zero $\rho = \sigma + it$ with $\sigma > 1/2$ (and its natural reflections), and if the integral is numerically computed to some height T and with some appropriate precision, what, if anything, can be said about the placement of that single zero? A challenging question is: Even if the RH is true, what is a valid positive α such that

$$I(T) = O(T^{-\alpha}) \ ?$$

It has been conjectured [Borwein et al. 2000] that $\alpha = 2$ is admissible.

(3) Some new equivalences of the RH involve the standard function

$$\xi(s) = \frac{1}{2}s(s-1)\pi^{-s/2}\Gamma(s/2)\zeta(s).$$

The tantalizing result of [Pustyl'nikov 1999] says that a condition applicable at a *single* point $s = 1/2$ as

$$\frac{d^n\xi}{ds^n}\left(\frac{1}{2}\right) > 0,$$

for every $n = 2, 4, 6, \ldots$, is equivalent to the RH. The interesting computational exercise would be to calculate some vast number of such derivatives. A single negative derivative would destroy the RH. Yet another criterion equivalent to the RH is that of [Lagarias 1999]:

$$\mathrm{Re}\left(\frac{\xi'(s)}{\xi(s)}\right) > 0$$

whenever $\mathrm{Re}(s) > 1/2$. Again some graphical or other computational means of analysis is at least interesting. Then there is the work of [Li 1997], [Bombieri and Lagarias 1999] to the effect that the RH is equivalent to the positivity property

$$\lambda_n = \sum_\rho \left(1 - \left(1 - \frac{1}{\rho}\right)^n\right) > 0$$

holding for each $n = 1, 2, 3, \ldots$. The λ_n constants can be cast in terms of derivatives of $\ln \xi(s)$, but this time, all such evaluated at $s = 1$. Again various computational avenues are of interest.

Further details, some computational explorations of these, and yet other new RH equivalences appear in [Borwein et al. 2000].

8.31. Investigate alternative factoring and discrete-logarithm algorithms for quantum Turing machines (QTMs). Here are some (unguaranteed) ideas.

The Pollard–Strassen method of Section 5.5 uses fast algorithms to deterministically uncover factors of N in $O(N^{1/4})$ operations. However, the usual approach to the required polynomial evaluations is FFT-like, and in practice often does involve FFTs. Is there a way to go deeper into the Pollard–Strassen method, using the inherent massive parallelism of QTMs in order to effect an interesting deterministic algorithm?

Likewise, we have seen exercises involving parallelization of Pollard-rho, ECM, QS, NFS factoring, and it is a good rule that whenever parallelism reveals itself, there is some hope of a QTM implementation.

As for DL problems, the rho and lambda methods admit of parallelism; indeed, the DL approach in [Shor 1999] is very much like the collision methods we have toured. But there could be a variant that is easier to implement. For example, it is not unreasonable to presume that the very first working QTM DL/factoring solvers might make use of one of the currently less-popular methods, in favor of simplicity. Observe that rho methods involve very little beyond modular squaring and adding. (As with many factoring algorithm candidates for QTM implementation, the eventual gcd operations could just be classical.) What is more, at the very heart of rho methods lives the phenomenon of periodicity, and as we have seen, QTMs are periodicity detectors *par excellence.*

Chapter 9

FAST ALGORITHMS FOR LARGE-INTEGER ARITHMETIC

In this chapter we explore the galaxy of "fast" algorithms that admit of applications in prime number and factorization computations. In modern times, it is of paramount importance to be able to manipulate multiple-precision integers, meaning integers that in practice, on prevailing machinery, have to be broken up into pieces, with machine operations to involve those pieces, with a view to eventual reassembly of desired results. Although multiple-precision addition and subtraction of integers is quite common in numerical studies, we assume that notions of these very simple fundamental operations are understood, and start with multiplication, which is perhaps the simplest arithmetic algorithm whose classical form admits of genuine enhancements.

9.1 Tour of "grammar-school" methods

9.1.1 Multiplication

One of the most common technical aspects of our culture is the classical, or shall we say "grammar-school," method of long multiplication. Though we shall eventually concentrate on fast, modern methods of remarkable efficiency, the grammar-school multiply remains important, especially when the relevant integers are not too large, and itself allows some speed enhancements. In the typical manifestation of the algorithm, one simply writes out, one below the other, the two integers to be multiplied, then constructs a parallelogram of digitwise products. Actually, the parallelogram is a rhombus, and to complete the multiply we need only add up the columns of the rhombus, with carry. If each of x, y to be multiplied has D digits in some given base B (also called the "radix"), then the total number of operations required to calculate xy is $O(D^2)$, because that is how many entries appear in the rhombus. Here, an "operation" is either a multiply or an add of two numbers each of size B. We shall refer to such a fundamental, digitwise, multiply as a "size-B multiply."

A formal exposition of grammar-school multiply is simple but illuminating, especially in view of later enhancements. We start with two definitions:

Definition 9.1.1. The base-B representation of a nonnegative integer x is the shortest sequence of integer digits (x_i) such that each digit satisfies

$0 \le x_i < B$, and

$$x = \sum_{i=0}^{D-1} x_i B^i.$$

Definition 9.1.2. The balanced base-B representation of a nonnegative integer x is the shortest sequence of integer digits (x_i) such that each digit satisfies $-\lfloor B/2 \rfloor \le x_i \le \lfloor (B-1)/2 \rfloor$, and

$$x = \sum_{i=0}^{D-1} x_i B^i.$$

Say we wish to calculate a product $z = xy$ for x, y both nonnegative. Upon contemplation of the grammar-school rhombus, it becomes evident that given x, y in base-B representation, say, we end up summing columns to construct integers

$$w_n = \sum_{i+j=n} x_i y_j, \tag{9.1}$$

where i, j run through all indices in the respective digit lists for x, y. Now the sequence (w_n) is not generally yet the base-B representation of the product z. What we need to do, of course, is to perform the w_n additions with carry. The carry operation is best understood the way we understood it in grammar school: A column sum w_n affects not only the final digit z_n, but sometimes higher-order digits beyond this. Thus, for example, if w_0 is equal to $B + 5$, then z_0 will be 5, but a 1 must be added into z_1; that is, a carry occurs.

These notions of carry are, of course, elementary, but we have stated them because such considerations figure strongly into modern enhancements to this basic multiply. In actual experience, the carry considerations can be more delicate and, for the programmer, more troublesome than any other part of the algorithm.

9.1.2 Squaring

From the computational perspective, the connection between multiplication and squaring is interesting. We expect the operation xx to involve generally more redundancy than an arbitrary product xy, so that squaring should be easier than general multiplication. Indeed, this intuition turns out to be correct. Say that x has D digits in base B representation, and note that (9.1) can be rewritten for the case of squaring as

$$w_n = \sum_{i=0}^{n} x_i x_{n-i}, \tag{9.2}$$

where $n \in [0, D-1]$. But this sum for w_n generally has reflection symmetry, and we can write

$$w_n = 2 \sum_{i=0}^{\lfloor n/2 \rfloor} x_i x_{n-i} - \delta_n, \tag{9.3}$$

where δ_n is 0 for n odd, else $x_{n/2}^2$ for n even. It is clear that each column component w_n involves about half the size-B multiplies required for the general multiplication algorithm. Of course, final carry operations must be performed on the w_n, to get the final digits z_n of the product $z = x^2$, but in most practical instances, this squaring is indeed roughly twice as fast as a multiple-precision multiply. There exist in the literature some very readable expositions of the squaring algorithm and related algorithms. See, for example, [Menezes et al. 1997].

There is an elegant, if simple, argument showing that general multiplication has no *more* than twice the complexity of squaring. One invokes the identity

$$4xy = (x+y)^2 - (x-y)^2, \tag{9.4}$$

which indicates that a multiplication can be effected by two squarings and a divide by four, this final divide presumed trivial (as, say, a right-shift by two bits). This observation is not just academic, for in certain practical scenarios this algebraic rule may be exploited (see Exercise 9.6).

9.1.3 Div and mod

Div and mod operations are omnipresent in prime-number and factorization studies. These operations often occur in combination with multiplication, in fact, this symbiosis is exploited in some of the algorithms we shall describe. It is quite common that one spends computation effort on operations such as $xy \pmod p$, for primes p, or in factorization studies $xy \pmod N$ where N is to be factored.

It is a primary observation that the mod operation can hinge on the div operation. We shall use, as before, the notation $x \bmod N$ to denote the operation that results in the least nonnegative residue of $x \pmod N$, while the greatest integer in x/N, denoted by $\lfloor x/N \rfloor$, is the div result. (In some computer languages these operations are written "$x\%N$" and "x div N," respectively, while in others the integer divide "x/N" means just div, while in yet others the div is "Floor[x/N]," and so on.) For integers x and positive integers N, a basic relation in our present notation is

$$x \bmod N = x - N\lfloor x/N \rfloor. \tag{9.5}$$

Note that this relation is equivalent to the quotient–remainder decomposition $x = qN + r$, with q, r being respectively the div and mod results under consideration. So the div operation begets the mod, and we can proceed with algorithm descriptions for div.

Analogous to "grammar-school" multiplication is, of course, the elementary method of long division. It is fruitful to contemplate even this simple long division algorithm, with a view to enhancements. In the normal execution of long division in a given base B, the divisor N is first justified to the left, with respect to the dividend x. That is to say, a power B^b of the base is found such that $m = B^b N \le x < B^{b+1}N$. Then one finds $\lfloor x/m \rfloor$, which quotient is guaranteed to be in the interval $[1, B-1]$. The quotient here is, of course, the leading base-B digit of the final div result. One then replaces x with $x - m\lfloor x/m \rfloor$, and divides m by B, that is, shifts m down by one digit, and so on recursively. This sketch shows us right off that for certain bases B, things are relatively simple. In fact, if one adopts binary representations ($B = 2$), then a complete div algorithm can be effected such that there are no multiplies at all. The method can actually be of practical interest, especially on machinery that has addition, subtraction, bit-shifting (left-shift means multiply-by-2, right-shift means divide-by-2), but little else in the way of operations. Explicitly, we proceed as follows:

Algorithm 9.1.3 (Classical binary divide). Given positive integers $x \geq N$, this algorithm performs the div operation, returning $\lfloor x/N \rfloor$. (See Exercise 9.7 for the matter of also returning the value $x \bmod N$.)

1. [Initialize]
   Find the unique integer b such that $2^b N \leq x < 2^{b+1} N$;
   // This can be done by successive left-shifts of the binary representation of N, or better, by comparing the bit lengths of x, N and possibly doing an extra shift.
   $m = 2^b N$; $c = 0$;
2. [Loop over b bits]
   for($0 \leq j \leq b$) {
   $c = 2c$;
   $a = x - m$;
   if($a \geq 0$) {
   $c = c + 1$;
   $x = a$;
   }
   $m = m/2$;
   }
   return c;

A similar binary approach can be used to effect the common "mul-mod" operation $(xy) \bmod N$, where we have adapted the treatment in [Arazi 1994]:

Algorithm 9.1.4 (Binary mul-mod). We are given positive integers x, y with $0 \leq x, y < N$. This algorithm returns the composite operation $(xy) \bmod N$. We assume the base-2 representation of Definition 9.1.1 for x, so that the binary bits of x are $(x_0, \ldots, x_{D-1})$, with $x_{D-1} > 0$ being the high bit.

1. [Initialize]
   $s = 0$;
2. [Loop over D bits]
   for($D - 1 \geq j \geq 0$) {
   $s = 2s$;
   if($s \geq N$) $s = s - N$;
   if($x_j == 1$) $s = s + y$;

```
        if(s ≥ N) s = s − N;
    }
    return s;
```

The binary divide and mul-mod algorithms, though illuminating, suffer from a basic practical shortcoming: One is not taking due advantage of multiple-bit arithmetic as is commonly available on any reasonably powerful computer. One would like to perform multiple-bit operations within machine registers, rather than just operating one bit at a time. For this reason, larger bases than $B = 2$ are usually used, and many modern div implementations invoke "Algorithm D," see [Knuth 1981, p. 257], which is a finely tuned version of the classical long division. That algorithm is a good example of one that has more pseudocode complexity than does our binary Algorithm (9.1.3), yet amounts to a great deal of optimization in actual programs.

9.2 Enhancements to modular arithmetic

The classical div and mod algorithms discussed in Section 9.1.3 all involve some sort of explicit divide operation. For the binary algorithms given, this division is trivial; that is, if $0 \le a < 2b$, then $\lfloor a/b \rfloor$ is of course either 0 or 1. In the case of Knuth's Algorithm D for higher bases than $B = 2$, one is compelled to estimate small div results. But there exist more modern algorithms for which no explicit division of any kind is required. The advantage of these methods to the computationalist is twofold. First, complete number-theoretical programs can be written without relatively complicated long division; and second, the optimization of all the arithmetic can be focused onto just one aspect, namely multiplication.

9.2.1 Montgomery method

An observation of [Montgomery 1985] has turned out to be important in the computational field, especially in situations where modular powers $(x^y) \bmod N$ are to be calculated with optimal speed (and, as we see later, the operands are not too overwhelmingly large). Observe, first of all, that "naive" multiply-mod takes one multiply and one divide (not counting subtractions), and so the spirit of the Montgomery method—as with other methods discussed in this chapter—is to lower or, if we are lucky, remove the difficulty of the divide step.

The Montgomery method, which is a generalization of an old method of Hensel for computing inverses of 2-adic numbers, stems from the following theorem, leading to efficient means for the computation of quantities $(xR^{-1}) \bmod N$, for certain conveniently chosen R:

Theorem 9.2.1 (Montgomery). *Let N, R be coprime positive integers, and define $N' = (-N^{-1}) \bmod R$. Then for any integer x, the number*

$$y = x + N((xN') \bmod R)$$

is divisible by R, with

$$y/R \equiv xR^{-1} \pmod{N}. \tag{9.6}$$

Furthermore, if $0 \le x < RN$, *the difference* $y/R - ((xR^{-1}) \bmod N)$ *is either* 0 *or* N.

As we shall see, Theorem 9.2.1 will be most useful when there are several or many multiplications modulo N to be performed, such as in a powering ladder, in which case the computation of the auxiliary number N' is only a one-time charge for the entire calculation. When N is odd and R is a power of 2, which is often the case in applications, the "mod R" operation is trivial, as is the division by R to get y. In addition, there is an alternative way to compute N' using Newton's method; see Exercise 9.12. It may help in the case N odd and R a power of 2 to cast the basic Montgomery operation in the language of bit operations. Let $R = 2^s$, let & denote the bitwise "and" operation, and let $>> c$ denote "right-shift by c bits." Then the left-hand side of equation (9.6) can be cast as

$$y/R = (x + N * ((x * N')\&(R-1))) \;>>\; s, \tag{9.7}$$

in which the two required multiplies are explicit.

So now, for $0 \le x < RN$, we have a way to calculate $(xR^{-1}) \bmod N$ with a small number (two) of multiplies. This is not quite the mod result $x \bmod N$ of course, but the Montgomery method applies well to the calculation of powers $(x^y) \bmod N$. The reason is that multiplication by R^{-1} or R on the residue system of $\{x : 0 \le x < N\}$ results in a complete residue system (mod N). Thus, powering arithmetic can be performed in a different residue system, with one initial multiply-mod operation and successive calls to a Montgomery multiplication, to yield results (mod N). To make these ideas precise, we adopt the following definition:

Definition 9.2.2. For $\gcd(R, N) = 1$ and $0 \le x < N$, the (R, N)-residue of x is $\overline{x} = (xR) \bmod N$.

Definition 9.2.3. The Montgomery product of two integers a, b is $M(a, b) = (abR^{-1}) \bmod N$.

Then the required facts can be collected in the following theorem:

Theorem 9.2.4 (Montgomery rules). *Let* R, N *be as in Definition 9.2.2, and* $0 \le a, b < N$. *Then* $a \bmod N = M(\overline{a}, 1)$ *and* $M(\overline{a}, \overline{b}) = \overline{ab}$.

This theorem gives rise to the Montgomery powering technique. For example, an example corollary of the theorem is that

$$M(M(M(\overline{x}, \overline{x}), \overline{x}), 1) = x^3 \bmod N. \tag{9.8}$$

To render the notion of general Montgomery powering explicit, we next give the relevant algorithms.

Algorithm 9.2.5 (Montgomery product). This algorithm returns $M(c,d)$ for integers $0 \le c,d < N$, with N odd, and $R = 2^s > N$.

1. [Montgomery mod function M]
   $M(c,d)$ {
   $x = cd$;
   $z = y/R$; // From Theorem 9.2.1.
2. [Adjust result]
   if($z \ge N$) $z = z - N$;
   return z;
   }

The [Adjust result] step in this algorithm always works because $cd < RN$ by hypothesis. The only importance of the choice that R be a power of two is that fast arithmetic may be employed in the evaluation of $z = y/R$.

Algorithm 9.2.6 (Montgomery powering). This algorithm returns $x^y \bmod N$, for $0 \le x < N$, $y > 0$, and R chosen as in Algorithm 9.2.5. We denote by $(y_0, \ldots, y_{D-1})$ the binary bits of y.

1. [Initialize]
   $\overline{x} = (xR) \bmod N$; // Via some divide/mod method.
   $\overline{p} = R \bmod N$; // Via some divide/mod method.
2. [Power ladder]
   for($D - 1 \ge j \ge 0$) {
   $\overline{p} = M(\overline{p}, \overline{p})$; // Via Algorithm 9.2.5.
   if($y_j == 1$) $\overline{p} = M(\overline{p}, \overline{x})$;
   } // Now $\overline{p}$ is $\overline{x}^y$.
3. [Final extraction of power]
   return $M(\overline{p}, 1)$;

Later in this chapter we shall have more to say about general power ladders; the ladder here is exhibited primarily to show how one may call the $M()$ function to advantage.

The speed enhancements of an eventual powering routine all center on the $M()$ function, in particular on the computation of $z = y/R$. We have noted that to get z, two multiplies are required, as in equation (9.7). But the story does not end here; in fact, the complexity of the Montgomery mod operation can be brought (asymptotically, large N) down to that of one size-N multiply. (To state it another way, the composite operation $M(x * y)$ asymptotically requires two size-N multiplies, which can be thought of as one for the "$*$" operation.) The details of the optimizations are intricate, involving various manifestations of the inner multiply loops of the $M()$ function [Koç et al. 1997], [Bosselaers et al. 1994]. But these details stem at least in part from a wasted operation in equation (9.7): The right-shifting effectively destroys some of the bits generated by the two multiplies. We shall see this shifting phenomenon again in the next section. In actual program implementations of Montgomery's scheme, one can assign a word-size base $B = 2^b$, so that a convenient value $R = B^k$ may be used, whence the z value in Algorithm 9.2.5 can be obtained by looping k times and doing arithmetic (mod B) that is particularly convenient for the machine. Explicit word-oriented loops that achieve the optimal asymptotic complexity are laid out nicely in [Menezes et al. 1997].

9.2.2 Newton methods

We have seen in Section 9.1 that the div operation may be effected via additions, subtractions, and bit-shifts, although, as we have also seen, the algorithm can be bested by moving away from the binary paradigm into the domain of general base representations. Then we saw that the technique of Montgomery mod gives us an asymptotically efficient means for powering with respect to a fixed modulus. It is interesting, perhaps at first surprising, that general div and mod may be effected via multiplications alone; that is, even the small div operations attendant to optimized div methods are obviated, as are the special precomputations of the Montgomery method.

One approach to such a general div and mod scheme is to realize that the classical Newton method for solving equations may be applied to the problem of reciprocation. Let us start with reciprocation in the domain of real numbers. If one is to solve $f(x) = 0$, one proceeds with an (adroit) initial guess for x, call this guess x_0, and iterates

$$x_{n+1} = x_n - f(x_n)/f'(x_n), \tag{9.9}$$

for $n = 0, 1, 2 \ldots$, whence—if the initial guess x_0 is good enough—the sequence (x_n) converges to the desired solution. So to reciprocate a *real* number $a > 0$, one is trying to solve $1/x - a = 0$, so that an appropriate iteration would be

$$x_{n+1} = 2x_n - ax_n^2. \tag{9.10}$$

Assuming that this Newton iteration for reciprocals is successful (see Exercise 9.13), we see that the real number $1/a$ can be obtained to arbitrary accuracy with multiplies alone. To calculate a general real division b/a, one simply multiplies b by the reciprocal $1/a$, so that general division in real numbers can be done in this way via multiplies alone.

But can the Newton method be applied to the problem of integer div? Indeed it can, provided that we proceed with care in the definition of a generalized reciprocal for integer division. We first introduce a function $B(N)$, defined for nonnegative integers N as the number of bits in the binary representation of N, except that $B(0) = 0$. Thus, $B(1) = 1, B(2) = B(3) = 2$, and so on. Next we establish a generalized reciprocal; instead of reciprocals $1/a$ for real a, we consider a generalized reciprocal of integer N as the integer part of an appropriate large power of 2 divided by N.

Definition 9.2.7. The generalized reciprocal $R(N)$ is defined for positive integers N as $\lfloor 4^{B(N-1)}/N \rfloor$.

The reason for the particular power in the definition is to allow our eventual general div algorithm to function. Next, we give a method for rapid computation of $R(N)$, based on multiplies, adds, subtracts alone:

Algorithm 9.2.8 (Generalized reciprocation).
This algorithm returns $R(N)$ for positive integer N.

1. [Initialize]
   $b = B(N-1)$; $r = 2^b$; $s = r$;
2. [Perform discrete Newton iteration]
   $r = 2r - \lfloor N \lfloor r^2/2^b \rfloor /2^b \rfloor$;
   if($r \le s$) goto [Adjust result];
   $s = r$;
   goto [Perform discrete Newton iteration];
3. [Adjust result]
   $y = 4^b - Nr$;
   while($y < 0$) {
      $r = r - 1$;
      $y = y + N$;
   }
   return r;

Note that Algorithm 9.2.8 involves a possible "repair" of the final return value, in the form of the while($y < 0$) loop. This is a key to making the algorithm precise, as we see in the proof of the following theorem:

Theorem 9.2.9 (Generalized reciprocal iteration). *The reciprocation Algorithm 9.2.8 works; that is, the returned value is $R(N)$.*

Proof. We have

$$2^{b-1} < N \le 2^b.$$

Let $c = 4^b/N$, so that $R(N) = \lfloor c \rfloor$. Let

$$f(r) = 2r - \left\lfloor \frac{N}{2^b} \left\lfloor \frac{r^2}{2^b} \right\rfloor \right\rfloor,$$

and let $g(r) = 2r - Nr^2/4^b = 2r - r^2/c$. Since deleting the floor functions in the definition of $f(r)$ gives us $g(r)$, and since $N/2^b \le 1$, we have

$$g(r) \le f(r) < g(r) + 2$$

for every r.

Since $g(r) = c - (c-r)^2/c$, we have

$$c - (c-r)^2/c \le f(r) < c - (c-r)^2/c + 2.$$

We conclude that $f(r) < c + 2$ for all r. Further, if $r < c$, then

$$f(r) \ge g(r) = 2r - r^2/c > r.$$

Thus, the sequence of iterates $2^b, f(2^b), f(f(2^b)), \ldots$ that the algorithm produces is strictly increasing until a value s is reached with $c \le s < c+2$. The number r sent to step [Adjust result] is $r = f(s)$. If $c \ge 4$, we also have $c \le r < c+2$. But $c \ge 4$ unless $N = 1$ or 2. In these cases, in fact whenever N is a power of 2, the algorithm terminates immediately with the value $r = N$. Thus, the algorithm always terminates with the number $\lfloor c \rfloor$, as claimed. □

We remark that the number of steps through the Newton iteration in Algorithm 9.2.8 is $O(\ln(b+1)) = O(\ln\ln(N+2))$. In addition, the number of iterations for the while loop in step [Adjust result] is at most 2.

Armed with the iteration for the generalized reciprocal, we can proceed to develop a mod operation that itself involves only multiplies, adds, subtracts, and binary shifts.

Algorithm 9.2.10 (Division-free mod). This algorithm returns $x \bmod N$ and $\lfloor x/N \rfloor$, for any nonnegative integer x. The only precalculation is to have established the generalized reciprocal $R = R(N)$. This precalculation may be done via Algorithm 9.2.8.

1. [Initialize]
   $s = 2(B(R) - 1)$;
2. [Perform reduction loop]
   $d = \lfloor xR/2^s \rfloor$;
   $x = x - Nd$;
   if($x \ge N$) {
      $x = x - N$;
      $d = d + 1$;
   }
   if($x < N$) return (x, d); // x is the mod, d is the div.
   goto [Perform reduction loop];

This algorithm is essentially the Barrett method [Barrett 1987], although it is usually stated for a commonly encountered range on x, namely, $0 \le x < N^2$. But we have lifted this restriction, by recursively using the basic formula

$$x \bmod N \sim x - N\lfloor xR/2^s \rfloor, \tag{9.11}$$

where by "$\sim$" we mean that for appropriate choice of s, the error in this relation is a small multiple of N. There are many enhancements possible to Algorithm 9.2.10, where we have chosen a specific number of bits s by which one is to right-shift. There are other interesting choices for s; indeed, it has been observed [Bosselaers et al. 1994] that there are certain advantages to "splitting up" the right-shifts like so:

$$x \bmod N \sim x - N\lfloor R\lfloor x/2^{b-1} \rfloor/2^{b+1} \rfloor, \tag{9.12}$$

where $b = B(R) - 1$. In particular, such splitting can render the relevant multiplications somewhat simpler. In fact, one sees that

$$\left\lfloor R \left\lfloor x/2^{b-1} \right\rfloor / 2^{b+1} \right\rfloor = \lfloor x/N \rfloor - j \tag{9.13}$$

for $j = 0, 1$, or 2. Thus using the left-hand side for d in Algorithm 9.2.10 involves at most two passes through the while loop. And there is an apparent savings in time, since the length of x can be about $2b$, and the length of R about b. Thus the multiplication xR in Algorithm 9.2.10 is about $2b \times b$ bits, while the multiplication inherent in (9.12) is only about $b \times b$ bits. Because a certain number of the bits of xR are destined to be shifted into oblivion (a shift completely obscures the relevant number of lower-order bits), one can intervene into the usual grammar-school multiply loop, effectively cutting the aforementioned rhombus into a smaller tableau of values. With considerations like this, it can be shown that for $0 \le x < N^2$, the complexity of the $x \bmod N$ operation is asymptotically (large N) the same as a size-N multiply. Alternatively, the complexity of the common operation $(xy) \bmod N$, where $0 \le x, y < N$, is that of two size-N multiplies.

Studies have been carried out for the classical long divide, (Algorithm D [Knuth 1981]), Montgomery and Barrett methods [Bosselaers et al. 1994], [Montgomery 1985], [Arazi 1994], [Koç et al. 1997]. There would seem to be no end to new div-mod algorithms; for example, there is a sign estimation technique of [Koç and Hung 1997], suitable for cryptographic operations (such as exponentiation) when operands are large. While both the Montgomery and (properly refined) Barrett methods are asymptotically of the same complexity, specific implementations of the methods reveal ranges of operands for which a particular approach is superior. In cryptographic applications, the Montgomery method is sometimes reported to be slightly superior to the Barrett method. One reason for this is that reaching the asymptotically best complexity for the Montgomery method is easier than for the Barrett method, the latter requiring intervention into the loop detail. However, there are exceptions; for example, [De Win et al. 1998] ended up adopting the Barrett method for their research purposes, presumably because of its ease of implementation (at the slightly suboptimal level), and its essential competitive equality with the Montgomery method. It is also the case that the inverses required in the Montgomery method can be problematic for very large operands. There is also the fact that if one wants just *one* mod operation (as opposed to a long exponentiation ladder), the Montgomery method is contraindicated. It would appear that a very good choice for general, large-integer arithmetic is the symbiotic combination of our Algorithms 9.2.8 and 9.2.10. In factorization, for example, one usually performs $(xy) \bmod N$ so very often for a stable N, that a single calculation of the generalized reciprocal $R(N)$ is all that is required to set up the division-free mod operations.

We mention briefly some new ideas in the world of divide/mod algorithms. One idea is due to G. Woltman, who found ways to enhance the Barrett divide Algorithm 9.2.10 in the (practically speaking) tough case when x is much greater than a relatively small N. One of his enhancements is to change precision modes in such cases. Another new development is an interesting Karatsuba-like recursive divide, discovered and implemented by [Burnikel and Ziegler 1998]. Their method has the interesting property that the complexities of finding the div or just a mod result are not quite the same.

Newton methods apply beyond the division problem. Just one example is the important computation of $\lfloor\sqrt{N}\rfloor$. One may employ a (real domain) Newton iteration for $\sqrt{a}$ in the form

$$x_{n+1} = \frac{x_n}{2} + \frac{a}{2x_n}, \tag{9.14}$$

to forge an algorithm for integer parts of square roots:

Algorithm 9.2.11 (Integer part of square root). This algorithm returns $\lfloor\sqrt{N}\rfloor$ for positive integer N.

1. [Initialize]
   $x = 2^{\lceil B(N)/2\rceil}$;
2. [Perform Newton iteration]
   $y = \lfloor(x + \lfloor N/x\rfloor)/2\rfloor$;
   if($y \ge x$) return x;
   $x = y$;
   goto [Perform Newton iteration];

We may use Algorithm 9.2.11 to test whether a given positive integer N is a square. After $x = \lfloor\sqrt{N}\rfloor$ is computed, we do one more step and check whether $x^2 = N$. This equation holds if and only if N is a square. Of course, there are other ways to rule out very quickly whether N is a perfect square, for example to test instances of $\left(\frac{N}{p}\right)$ for various small primes p, or the residue of N modulo 8.

It can be argued that Algorithm 9.2.11 requires $O(\ln\ln N)$ iterations to terminate. There are many interesting complexity issues with this and other Newton method applications. Specifically, it is often lucrative to change dynamically the working precision as the Newton iteration progresses, or to modify the very Newton loops (see Exercises 9.14 and 9.15).

9.2.3 Moduli of special form

Considerable efficiency in the mod operation can be achieved when the modulus N is of special form. The Barrett method of the previous section is fast because it exploits mod 2^q arithmetic. In this section we shall see that if the modulus N is close to a power of 2, one can exploit the binary nature of modern computers and carry out the arithmetic very efficiently. In particular, forms

$$N = 2^q + c,$$

where $|c|$ is in some sense "small" (but c is allowed to be negative), admit efficient mod N operations. These enhancements are especially important in the studies of Mersenne primes $p = 2^q - 1$ and Fermat numbers $F_n = 2^{2^n} + 1$, although the techniques we shall describe apply equally well to general moduli $2^q \pm 1$, any q. That is, whether or not the modulus N has additional properties of primality or special structure is of no consequence for the mod algorithm of this section. A relevant result is the following:

Theorem 9.2.12 (Special-form modular arithmetic). *For $N = 2^q + c$, c an integer, q a positive integer, and for any integer x,*

$$x \equiv (x \bmod 2^q) - c\lfloor x/2^q \rfloor \pmod{N}. \tag{9.15}$$

Furthermore, in the Mersenne case $c = -1$, multiplication by 2^k modulo N is equivalent to left-circular shift by k bits (so if $k < 0$, this is right-circular shift). For the Fermat case $c = +1$, multiplication by 2^k, k positive, is equivalent to $(-1)^{\lfloor k/q \rfloor}$ times the left-circular shift by k bits, except that the excess shifted bits are to be negated and carry-adjusted.

As they are easiest to analyze, let us discuss the final statements of the theorem first. Since

$$2^k = 2^{k \bmod q} 2^{q\lfloor k/q \rfloor},$$

and also $2^q \equiv -c \pmod{N}$, the statements are really about $k \in [1, q-1]$ and negatives of such k. As examples, take $N = 2^{17} - 1 = 131071 = 11111111111111111_2$, $x = 8977 = 10001100010001_2$, and consider the product $2^5 x \pmod{N}$. This will be the left-circular shift of x by 5 bits, or $110001000100010_2 = 25122$, which is the correct result. Incidentally, these results on multiplication by powers of 2 are relevant for certain number-theoretical transforms and other algorithms. In particular, discrete Fourier transform arithmetic in the ring $\mathbf{Z}_n$ with $n = 2^m + 1$ can proceed—on the basis of shifting rather than explicit multiplication—when the root in question is a power of 2.

The first result of Theorem 9.2.12 allows us to calculate $x \bmod N$ very rapidly, on the basis of the "smallness" of c. Let us first give an example of the computation of $x = 13000$ modulo the Mersenne prime $N = 2^7 - 1 = 127$. It is illuminating to cast in binary: $13000 = 11001011001000_2$, then proceed via the theorem to split up x easily into two parts whenever it exceeds N (all congruences here are with respect to modulus N):

$$x \equiv 11001011001000 \bmod 10000000 + \lfloor 11001011001000/10000000 \rfloor$$

$$\equiv 1001000 + 1100101 \equiv 10101101 \equiv 101101 + 1 \equiv 101110.$$

As the result $101110_2 = 46 < N$, we have achieved the desired value of $13000 \bmod 127 = 46$. The procedure is thus especially simple for the Mersenne cases $N = 2^q - 1$; namely, one takes the "upper" bits of x (meaning the bits from the 2^q position and up, inclusive) and adds these to the "lower" bits (meaning the lower q bits of x). The general procedure runs as follows, where we adopt for convenience the bitwise "and" operator & and right-shift >>, left-shift << operators:

Algorithm 9.2.13 (Fast mod operation for special-form moduli).
Assume modulus $N = 2^q + c$, with $B(|c|) < q$. This algorithm returns $x \bmod N$ for $x > 0$. The method is generally more efficient for smaller $|c|$.

1. [Perform reduction]

```
    while(B(x) > q) {
        y = x >> q;                    // Right-shift does ⌊x/2^q⌋.
        x = x - (y << q);          // Or x = x&(2^q - 1), or x = x mod 2^q.
        x = x - cy;
    }
    if(x == 0) return x;
2. [Adjust]
    s = sgn(x);                        // Defined as -1, 0, 1 as x <, =, > 0.
    x = |x|;
    if(x ≥ N) x = x - N;
    if(s < 0) x = N - x;
    return x;
```

It is not hard to show that this algorithm terminates and gives the result $x \bmod N$.

Because the method involves nothing but "small" multiplications (by c), applications are widespread. Modern discoveries of new Mersenne primes have used this mod method in the course of extensive Lucas–Lehmer primality testing. There is even a patented encryption scheme based on elliptic curves over fields $\mathbf{F}_{p^k}$, where $p = 2^q + c$, and if extra efficiency is desired, $p \equiv -1 \pmod 4$ (for example, p can be any Mersenne prime, or a prime $2^q + 7$, and so on), with elliptic algebra performed on the basis of essentially negligible mod operations [Crandall 1994b]. Such fields have been called optimal extension fields (OEFs), and further refinements can be achieved by adroit choice of the exponent k and irreducible polynomial for the F_{p^k} arithmetic. It is also true of such elliptic curves that curve order can be assessed more quickly by virtue of the fast mod operation. Yet another application of the special mod reduction is in the factorization of Fermat numbers. The method has been used in the recent discoveries of new factors of the F_n for $n = 13, 15, 16, 18$ [Brent et al. 2000]. For such large Fermat numbers, machine time is so extensive that any algorithmic enhancements, whether for mod or other operations, are always welcome. In recent times the character of even larger F_n has been assessed in this way, where now the Pepin primality test involves a great many (mod F_n) operations. The proofs that F_{22}, F_{24} are composite used the special-form mod of this section [Crandall et al. 1995], [Crandall et al. 1999], together with fast multiplication discussed later in the chapter.

It is interesting that one may generalize the special-form fast arithmetic yet further. Consider numbers of the Proth form:

$$N = k \cdot 2^q + c.$$

We next give a fast modular reduction technique of [Gallot 1999], which is suitable in cases where k and c are low-precision (e.g., single-word) parameters:

Algorithm 9.2.14 (Fast mod operation for Proth moduli). Assume modulus $N = k \cdot 2^q + c$, with bit length $B(|c|) < q$ (and c can be negative or zero).

This algorithm returns $x \bmod N$ for $0 < x < N^2$. The method is generally more efficient for smaller k, $|c|$.

1. [Define a useful shift-add function n]
```
    n(y) {
        return Ny;        // But calculate rapidly, as: Ny = ((ky) << q) + cy.
    }
```
2. [Approximate the quotient]
```
    y = ⌊(x >> q)/k⌋;
    t = n(y);
    if(c < 0) goto [Polarity switch];
    while(t > x) {
        t = n(y);
        y = y - 1;
    }
    return x - t;
```
3. [Polarity switch]
```
    while(t ≤ x) {
        y = y + 1;
        t = n(y);
    }
    y = y - 1;
    t = n(y);
    return x - t;
```

This kind of clever reduction is now deployed in software that has achieved significant success in the discoveries of, as just two examples, new factors of Fermat numbers, and primality proofs for Proth primes.

9.3 Exponentiation

Exponentiation, or powering, is especially important in prime number and factorization studies, for the simple reason that so many known theorems involve the operation x^y, or most commonly $x^y \pmod N$. In what follows, we give various algorithms that efficiently exploit the structure of the exponent y, and sometimes the structure of x. We have glimpsed already in Section 2.1.2, Algorithm 2.1.5, an important fact: While it is certainly true that something like $(x^y) \bmod N$ can be evaluated with $(y-1)$ successive multiplications $\pmod N$, there is generally a much better way to compute powers. This is to use what is now a commonplace computational technique, the powering ladder, which can be thought of as a nonrecursive (or "unrolled") realization of equivalent, recursive algorithms. But one can do more, via such means as preprocessing the bits of the exponent, using alternative base expansions for the exponent, and so on. Let us first summarize the categories of powering ladders:

(1) Recursive powering ladder (Algorithm 2.1.5).

(2) Left-right and right-left "unrolled" binary ladders.

(3) Windowing ladders, to take advantage of certain bit patterns or of alternative base expansions, a simple example of which being what is essentially a ternary method in Algorithm 7.2.7, step [Loop over bits ...], although one can generally do somewhat better [Müller 1997], [De Win et al. 1998], [Crandall 1999b].

(4) Fixed-x ladders, to compute x^y for various y but fixed x.

(5) Addition chains and Lucas ladders, as in Algorithm 3.5.7, interesting references being such as [Montgomery 1992b], [Müller 1997].

(6) Modern methods based on actual compression of exponent bit-streams, as in [Yacobi 1999].

The current section starts with basic binary ladders (and even for these, various options exist); then we turn to the windowing, alternative-base, and fixed-x ladders.

9.3.1 Basic binary ladders

We next give two forms of explicit binary ladders. The first, a "left-right" form (equivalent to Algorithm 2.1.5), is comparable in complexity (except when arguments are constrained in certain ways) to a second, "right-left" form.

Algorithm 9.3.1 (Binary ladder exponentiation (left-right form)). This algorithm computes x^y. We assume the binary expansion $(y_0, \ldots, y_{D-1})$ of $y > 0$, where $y_{D-1} = 1$ is the high bit.

1. [Initialize]
   $z = x$;

2. [Loop over bits of y, starting with next-to-highest]
   for($D - 2 \ge j \ge 0$) {
      $z = z^2$; // For modular arithmetic, do $\bmod N$ here.
      if($y_j == 1$) $z = zx$; // For modular arithmetic, do $\bmod N$ here.
   }
   return z;

This algorithm constructs the power x^y by running through the bits of the exponent y. Indeed, the number of squarings is $(D - 1)$, and the number of operations $z = z * x$ is clearly one less than the number of 1 bits in the exponent y. Note that the operations turn out to be those of Algorithm 2.1.5. A mnemonic for remembering which of the left-right or right-left ladder forms is equivalent to the recursive form is to note that both Algorithms 9.3.1 and 2.1.5 involve multiplications exclusively by the steady multiplier x.

But there is a kind of complementary way to effect the powering. This alternative is exemplified in the relation

$$x^{13} = x * (x^2)^2 * (x^4)^2,$$

where there are again 2 multiplications and 3 squarings (because x^4 was actually obtained as the middle term $(x^2)^2$). In fact, in this example we see more directly the binary expansion of the exponent. The general formula would be

$$x^y = x^{\sum y_j 2^j} = x^{y_0}(x^2)^{y_1}(x^4)^{y_2}\cdots, \tag{9.16}$$

where the y_j are the bits of y. The corresponding algorithm is a "right-left" ladder in which we keep track of successive squarings of x:

Algorithm 9.3.2 (Binary ladder exponentiation (right-left form)). This algorithm computes x^y. We assume the binary expansion $(y_0, \dots, y_{D-1})$ of $y > 0$, where $y_{D-1} = 1$ is the high bit.

1. [Initialize]
   $z = x$; $a = 1$;
2. [Loop over bits of y, starting with lowest]
```
for(0 ≤ j < D − 1) {
    if(y_j == 1) a = za;        // For modular arithmetic, do mod N here.
    z = z^2;                    // For modular arithmetic, do mod N here.
}
return az;
```

This scheme can be seen to involve also $(D-1)$ squarings, and (except for the trivial multiply when $a = z * 1$ is first invoked) has the same number of multiplies as did the previous algorithm.

Even though the operation counts agree on the face of it, there is a certain advantage to the first form given, Algorithm 9.3.1, for the reason that the operation $z = zx$ involves a fixed multiplicand, x. Thus for example, if $x = 2$ or some other small integer, as might be the case in a primality test where we raise a small integer to a high power (mod N), the multiply step can be fast. In fact, for $x = 2$ we can substitute the operation $z = z + z$, avoiding multiplication entirely for that step of the algorithm. Such an advantage is most telling when the exponent y is replete with binary 1's.

These observations lead in turn to the issue of asymptotic complexity for ladders. This is a fascinating—and in many ways open—field of study. Happily, though, most questions about the fundamental binary ladders above can be answered. Let us adopt the heuristic notation that S is the complexity of squaring (in the relevant algebraic domain for exponentiation) and M is the complexity of multiplication. Evidently, the complexity C of one of the above ladders is asymptotically

$$C \sim (\lg y)S + HM,$$

where H denotes the number of 1's in the exponent y. Since we expect about "half 1's" in a random exponent, the average-case complexity is thus

$$C \sim (\lg y)S + (\tfrac{1}{2}\lg y)M.$$

Note that using (9.4) one can often achieve $S \sim M/2$ so reducing the expression for the average-case complexity of the above ladders to $C \sim$

$(\lg y)M$. The estimate $S \sim M/2$ is not a universal truth. For one thing, such an estimate assumes that modular arithmetic is not involved, just straight nonmodular squaring and multiplication. But even in the nonmodular world, there are issues. For example, with FFT multiplication (for very large operands, as described later in this chapter), the S/M ratio can be more like 2/3. With some practical (modular, grammar-school) implementations, the ratio S/M is about 0.8, as reported in [Cohen et al. 1998]. Whatever subroutines one uses, it is of course desirable to have fewer arithmetic operations to perform. As we shall see in the following section, it is possible to achieve further operation reduction.

9.3.2 Enhancements to ladders

In factorization studies and cryptography it is a rule of thumb that power ladders are used much of the time. In factorization, the so-called stage one of many methods involves almost nothing but exponentiation (in the case of ECM, elliptic multiplication is the analogue to exponentiation). In cryptography, the generation of public keys from private ones involves exponentiation, as do digital signatures and so on. It is therefore important to optimize powering ladders as much as possible, as these ladder operations dominate the computational effort in the respective technologies.

One interesting method for ladder enhancement is sometimes referred to as "windowing." Observe that if we expand not in binary but in base 4, and we precompute powers x^2, x^3, then every time we encounter *two* bits of the exponent y, we can multiply by one of $1 = x^0, x^1, x^2, x^3$ and then square twice to shift the current register to the left by two bits. Consider for example the task of calculating x^{79}, knowing that $79 = 1001111_2 = 1033_4$. If we express the exponent $y = 79$ in base 4, we can do the power as

$$x^{79} = \left(x^{4^2} x^3\right)^4 x^3,$$

which takes $6S + 2M$ (recall nomenclature S, M for square and multiply). On the other hand, the left-right ladder Algorithm 9.3.1 does the power this way:

$$x^{79} = \left(\left(\left(x^{2^3} x\right)^2 x\right)^2 x\right)^2 x,$$

for a total effort of $6S + 4M$, more than the effort for the base-4 method. We have not counted the time to precompute x^2, x^3 in the latter method, and so the benefit is not so readily apparent. But a benefit would be seen in most cases if the exponent 79 were larger, as in many cryptographic applications.

There are many detailed considerations not yet discussed, but before we touch upon those let us give a fairly general windowing ladder that contains most of the applicable ideas:

Algorithm 9.3.3 (Windowing ladder). This algorithm computes x^y. We assume a base-($B = 2^b$) expansion (as in Definition 9.1.1), denoted by

$(y_0, \ldots, y_{D-1})$ of $y > 0$, with high digit $y_{D-1} \neq 0$, so each digit satisfies $0 \le y_i < B$. We also assume that the values $\{x^d \,:\, 1 < d < B; d \text{ odd}\}$ have been precomputed.

1. [Initialize]
   $z = 1$;
2. [Loop over digits]
   for($D-1 \ge i \ge 0$) {
     Express $y_i = 2^c d$, where d is odd or zero;
     $z = z(x^d)^{2^c}$; // x^d from storage.
     if($i > 0$) $z = z^{2^b}$;
   }
   return z;

To give an example of why only odd powers of x need to be precomputed, let us take the example of $y = 262 = 406_8$. Looking at this base-8 representation, we see that

$$x^{262} = \left(\left(x^4\right)^8\right)^8 x^6,$$

but if x^3 has been precomputed, we can insert that x^3 at the proper juncture, and Algorithm 9.3.3 tells us to exponentiate like so:

$$x^{262} = \left(\left(\left(x^4\right)^8\right)^4 x^3\right)^2.$$

Thus, the precomputation is relegated to odd powers only. Another way to exemplify the advantage is in base 16 say, for which each of the 4-bit sequences: 1100, 0110, 0011 in any exponent can be handled via the use of x^3 and the proper sequencing of squarings.

Now, as to further detail, it is possible to allow the "window"—essentially the base B—to change as we go along. That is, one can look ahead during processing of the exponent y, trying to find special strings for a little extra efficiency. One "sliding-window" method is presented in [Menezes et al. 1997]. It is also possible to use our balanced-base representation, Definition 9.1.2, to advantage. If we constrain the digits of exponent y to be

$$-\lfloor B/2 \rfloor \le y_i \le \lfloor (B-1)/2 \rfloor,$$

and precompute odd powers x^d where d is restricted within the range of these digit values, then significant advantages accrue, provided that the inverse powers are available. In the case of elliptic multiplication, let us say we desire "exponentiation" $[k]P$, where P is a point, k the exponent. We need to precompute, then, only the multiples

$$\{[d]P \,:\, 1 < d < \lfloor B/2 \rfloor; d \text{ odd}\},$$

because negations $[-d]P$ are immediate, by the rules of elliptic algebra. In this way, one can fashion highly efficient windowing schemes for elliptic multiplication. See Exercise 9.77 for yet more considerations.

Ignoring precomputation, it can be inferred that in Algorithm 9.3.3 with base $B = 2^b$ the asymptotic (large-y) requirement is $Db \sim \lg y$ squarings (i.e., one squaring for each binary bit of y). This is, of course, no gain over the squarings required in the basic binary ladders. But the difference lies in the multiplication count. Whereas in the basic binary ladders the (asymptotic) number of multiplications is the number of 1's, we now only need at most one multiplication per b bits; in fact, we only need $1 - 2^{-b}$ of these on average, because of the chance of a zero digit in random base-B expansions. Thus, the average-case asymptotic complexity for the windowing algorithm is

$$C \sim (\lg y)S + (1 - 2^{-b})\frac{\lg y}{b}M,$$

which when $b = 1$ is equivalent to the previous estimate $C \sim (\lg y)S + (\frac{1}{2}\lg y)M$ for the basic binary ladders. Note though as the window size b increases, the burden of multiplications becomes negligible. It is true that precomputation considerations are paramount, but in practice, a choice of $b = 3$ or $b = 4$ will indeed reduce noticeably the ladder computations.

Along the lines of the previous remarks concerning precomputation, an interesting ladder enhancement obtains in the case that the number x is to be reused. That is, say we wish to exponentiate x^y for many different y values, with x fixed. We can compute and store fixed powers of the fixed x, and use them to advantage.

Algorithm 9.3.4 (Fixed-x ladder for x^y). This algorithm computes x^y. We assume a base-B (not necessarily binary) expansion $(y_0, \ldots, y_{D-1})$ of $y > 0$, with high digit $y_{D-1} > 0$. We also assume that the (total of $(B-1)(D-1)$) values

$$\{x^{iB^j} \ : i \in [1, B-1]; \ j \in [1, D-1]\}$$

have been precomputed.

1. [Initialize]
   $z = 1$;
2. [Loop over digits]
   for$(0 \le j < D)$ $z = z x^{y_j B^j}$;
   return z;

This algorithm clearly requires, beyond precomputation, an operation count

$$C \sim DM \sim \frac{\lg y}{\lg B}M,$$

so the fact of a "stable" value for x really can yield high efficiency, because of the $(\lg B)^{-1}$ factor. Depending on precise practical setting and requirements, there exist yet further enhancements, including the use of less extensive lookup tables (i.e., using only the stored powers such as x^{B^j}), loosening of the restrictions on the ranges of the for() loops depending on the range of values of the y digits in base B (in some situations not every possible digit will occur), and so on. Note that if we do store only the reduced set of powers x^{B^j}, the step [Loop over digits] will have nested for() loops. There also exist fixed-y algorithms using so-called addition chains, so that when the exponent is stable some enhancements are possible. Both fixed-x and fixed-y forms find applications in cryptography. If public keys are generated as fixed x values raised to secret y values, for example, the fixed-x enhancements can be beneficial. Similarly, if a public key (as $x = g^h$) is to be raised often to a key power y, then the fixed-y methods may be invoked for extra efficiency.

9.4 Enhancements for gcd and inverse

In Section 2.1.1 we discussed the great classical algorithms for gcd and inverse. Here we explore more modern methods, especially methods that apply when the relevant integers are very large, or when some operations (such as shifts) are relatively efficient.

9.4.1 Binary gcd algorithms

There is a genuine enhancement of the Euclid algorithm worked out by D. Lehmer in the 1930s. The method exploits the fact that not every implied division in the Euclid loop requires full precision, and statistically speaking there will be many single-precision (i.e., small operand) div operations. We do not lay out the Lehmer method here (for details see [Knuth 1981]), but observe that Lehmer showed how to enhance an old algorithm to advantage in such tasks as factorization.

In the 1960s it was observed by R. Silver and J. Terzian [Knuth 1981], and independently by [Stein 1967], that a gcd algorithm can be effected in a certain binary fashion. The following relations indeed suggest an elegant algorithm:

Theorem 9.4.1 (Silver, Terzian, and Stein). *For integers x, y,*
If x, y are both even, then $\gcd(x, y) = 2\gcd(x/2, y/2)$;
If x is even and y is not, then $\gcd(x, y) = \gcd(x/2, y)$;
(As per Euclid) $\gcd(x, y) = \gcd(x - y, y)$;
If u, v are both odd, then $|u - v|$ is even and less than $\max\{u, v\}$.

These observations give rise to the following algorithm:

Algorithm 9.4.2 (Binary gcd). The following algorithm returns the greatest common divisor of two positive integers x, y. For any positive integer m, let $v_2(m)$ be the number of low-order 0's in the binary representation of m, that is, we have $2^{v_2(m)} \| m$. (Note that $m/2^{v_2(m)}$ is the largest odd divisor of m, and can be computed with a shift into oblivion of the low-order zeros.)

1. [2's power in gcd]
   $\beta = \min\{v_2(x), v_2(y)\}$; // $2^\beta \| \gcd(x, y)$
   $x = x/2^{v_2(x)}$;
   $y = y/2^{v_2(y)}$;

2. [Binary gcd]
   while($x \neq y$) {
     $(x, y) = (\min\{x, y\}, |y - x|/2^{v_2(|y-x|)})$;
   }
   return $2^\beta x$;

In actual practice on most machinery, the binary algorithm is often faster than the Euclid algorithm; and as we have said, Lehmer's enhancements may also be applied to this binary scheme.

But there are other, more modern, enhancements; in fact, gcd enhancements seem to keep appearing in the literature. There is a "k-ary" method due to Sorenson, in which reductions involving $k > 2$ as a modulus are performed. There is also a newer extension of the Sorenson method that is claimed to be, on a typical modern machine that possesses hardware multiply, more than 5 times faster than the *binary* gcd we just displayed [Weber 1995]. The Weber method is rather intricate, involving several special functions for nonstandard modular reduction, yet the method should be considered seriously in any project for which the gcd happens to be a bottleneck.

It is of interest that the Sorenson method has variants for which the complexity of the gcd is $O(n^2/\ln n)$ as opposed to the Euclidean $O(n^2)$ [Sorenson 1994]. In addition, the Sorenson method has an extended form for obtaining not just gcd but inverse as well.

One wonders whether this efficient binary technique can be extended in the way that the classical Euclid algorithm can. Indeed, there is also an extended binary gcd that provides inverses. [Knuth 1981] attributes the method to M. Penk:

Algorithm 9.4.3 (Binary gcd, extended for inverses). For positive integers x, y, this algorithm returns an integer triple (a, b, g) such that $ax + by = g = \gcd(x, y)$. We assume the binary representations of x, y, and use the exponent β as in Algorithm 9.4.2.

1. [Initialize]
   $x = x/2^\beta; y = y/2^\beta$;
   $(a, b, h) = (1, 0, x)$;
   $(v_1, v_2, v_3) = (y, 1 - x, y)$;
   if(x even) $(t_1, t_2, t_3) = (1, 0, x)$;
     else {
       $(t_1, t_2, t_3) = (0, -1, -y)$;
       goto [Check even];
     }
2. [Halve t_3]
   if(t_1, t_2 both even) $(t_1, t_2, t_3) = (t_1, t_2, t_3)/2$;
     else $(t_1, t_2, t_3) = (t_1 + y, t_2 - x, t_3)/2$;
3. [Check even]
   if(t_3 even) goto [Halve t_3];

4. [Reset max]
   if($t_3 > 0$) $(a, b, h) = (t_1, t_2, t_3)$;
      else $(v_1, v_2, v_3) = (y - t_1, -x - t_2, -t_3)$;
5. [Subtract]
   $(t_1, t_2, t_3) = (a, b, h) - (v_1, v_2, v_3)$;
   if($t_1 < 0$) $(t_1, t_2) = (t_1 + y, t_2 - x)$
   if($t_3 \neq 0$) goto [Halve t_3];
   return $(a, b, 2^\beta h)$;

Like the basic binary gcd algorithm, this one tends to be efficient in actual machine implementations. When something is known as to the character of either operand (for example, say y is prime) this and related algorithms can be enhanced (see Exercises).

9.4.2 Special inversion algorithms

Variants on the inverse-finding, extended gcd algorithms have appeared over the years, in some cases depending on the character of the operands x, y. One example is the inversion scheme of [Thomas et al. 1986] for $x^{-1} \bmod p$, for primes p. Actually, the algorithm works for unrestricted moduli (returning either a proper inverse or zero if the inverse does not exist), but the authors were concentrating on moduli p for which a key quantity $\lfloor p/z \rfloor$ within the algorithm can be easily computed.

Algorithm 9.4.4 (Modular inversion). For modulus p (not necessarily prime) and $x \not\equiv 0 \pmod p$, this algorithm returns $x^{-1} \bmod p$.

1. [Initialize]
   $z = x \bmod p$;
   $a = 1$;
2. [Loop]
   while($z \neq 1$) {
      $q = -\lfloor p/z \rfloor$; // Algorithm is best when this is fast.
      $z = p + qz$;
      $a = (qa) \bmod p$;
   }
   return a; // $a = x^{-1} \bmod p$.

This algorithm is conveniently simple to implement, and furthermore (for some ranges of primes), is claimed to be somewhat faster than the extended Algorithm 2.1.4. Incidentally, the authors of this algorithm also give an interesting method for rapid calculation of $\lfloor p/z \rfloor$ when $p = 2^q - 1$ is specifically a Mersenne prime.

Yet other inversion methods focus on the specific case that p is a Mersenne prime. The following is an interesting attempt to exploit the special form of the modulus:

Algorithm 9.4.5 (Inversion modulo a Mersenne prime). For $p = 2^q - 1$ prime and $x \not\equiv 0 \pmod{p}$, this algorithm returns $x^{-1} \bmod p$.

1. [Initialize]
   $(a, b, y, z) = (1, 0, x, p)$;
2. [Relational reduction]
   Find e such that $2^e \| y$;
   $y = y/2^e$; // Shift off trailing zeros.
   $a = (2^{q-e} a) \bmod p$; // Circular shift, by Theorem 9.2.12.
   if($y == 1$) return a;
   $(a, b, y, z) = (a + b, a, y + z, y)$;
   goto [Relational reduction];

9.4.3 Recursive gcd for very large operands

It turns out that the bit-complexity $O(\ln^2 N)$ for evaluating the gcd of two size-N numbers can be, via a recursive reduction scheme [Schönhage 1971], [Aho et al. 1974, pp. 300–310], [Bürgisser et al. 1997, p. 98], brought down to

$$O(M(\ln N) \ln\ln N) = O\left(\ln N (\ln\ln N)^2 \ln\ln\ln N\right),$$

where $M(b)$ denotes the bit-complexity for multiplication of two b-bit integers, the final complexity expression on the right arising from known bounds on M, as discussed later. For the moment, we observe that like various other algorithms we have encountered—such as pre-conditioned CRT—the recursive gcd approach cannot really use grammar-school multiplication to advantage.

Indeed, we have found in practice that this recursive gcd scheme outperforms any other alternative (such as the binary gcd forms with or without Lehmer enhancements) in and above the region $x, y \approx 2^{2^{16}}$ (although this "breakover" threshold depends strongly on machinery and on various options such as the choice of an alternative classical gcd algorithm for small arguments). Because the recursive gcd is so valuable for large-integer work, we shall exhibit an effective algorithm display, briefly touching next upon the effect of recursion.

The basic idea is to recursively compute a 2×2 matrix, call it M, such that $M(x, y)^T = (0, \gcd(x, y))^T$. (The pseudocode to follow does not specifically name M, rather is constructing this matrix via recursive modification of a certain matrix G.) The matrix M is a product of 2×2 matrices corresponding to single steps in the usual Euclid algorithm, and is computed recursively in the algorithm given below. Details can be found in [Aho et al. 1974, pp. 300–310], [Bürgisser et al. 1997], [Cesari 1998]. In fact, the latter reference given describes a novel, parallel approach to such very-large-gcd operations. The first half (approximately) of the matrix chain is computed by the function *hgcd*(b, x, y), where the parameter b tells how many bits to chop off the arguments x, y. Thus the actual gcd (and the matrix M) can be computed by repeatedly calling *hgcd*() to calculate an explicit matrix G, to get $G(x, y)^T = (x', y')^T$, where x', y' are, respectively, half the size of x, y.

Here is some detail in regard to the important "half-gcd" function. Suppose that x, y have n bits each. The value $hgcd(0, x, y)$ is computed by first calling $hgcd(n/2, x, y)$ to find a matrix G_1 such that $(x', y')^T = G_1(x, y)^T$ where each of x', y' has $3n/4$ bits, and then evaluating $hgcd(n/4, x', y')$, so chopping off $n/4$ bits of x', y' to find a matrix G_2 such that $(x'', y'')^T = G_2(x', y')^T$ where each of x'', y'' has $n/2$ bits. Thus $G = G_2G_1$ is the matrix desired in the original $hgcd()$ call atop this paragraph. The individual Euclid algorithm step in $hgcd()$ turns out to be necessary in order for the algorithm to "make progress." A careful analysis of the complexity yields the aforementioned bound.

We have drawn on an implementation by [Buhler 1991] to effect the following algorithm display:

Algorithm 9.4.6 (Recursive gcd). For nonnegative integers x, y this algorithm returns $\gcd(x, y)$. The top-level function $rgcd()$ calls a recursive $hgcd()$, with a classical (such as a Euclid) function $cgcd()$ invoked at recursion bottom. There is a global matrix G, other interior variables being local (in the usual sense for recursive procedures).

1. [Initialize]
   $lim = 2^{256}$; // Breakover threshold for $cgcd()$; adjust for efficiency.
   $prec = 32$; // Breakover bit length for $shgcd()$; adjust for efficiency.
2. [Set up small-gcd function $shgcd$ to return a matrix]
   $shgcd(x, y)$ { // Short gcd, with variables u, v, q, A local.
   $A = \begin{pmatrix} A_1 & A_2 \\ A_3 & A_4 \end{pmatrix} = \begin{pmatrix} 1 & 0 \\ 0 & 1 \end{pmatrix}$;
   $(u, v) = (x, y)$;
   while($v^2 > x$) {
   $q = \lfloor u/v \rfloor$;
   $(u, v) = (v, u \bmod v)$;
   $(A_1, A_3) = (A_3, A_1 - qA_3)$;
   $(A_2, A_4) = (A_4, A_2 - qA_4)$;
   }
   return A;
   }
3. [Set up recursive procedure $hgcd$ to modify global matrix G]
   $hgcd(b, x, y)$ { // Variables u, v, q, m, C are local.
   if($y == 0$) return;
   $u = \lfloor x/2^b \rfloor$;
   $v = \lfloor y/2^b \rfloor$;
   $m = B(u)$; // B is as usual the bit-length function.
   if($m < prec$) {
   $G = shgcd(u, v)$;
   return;
   }
   $m = \lfloor m/2 \rfloor$;

```
        hgcd(m, u, v);                                          // Recurse.
        (u, v)^T = G(u, v)^T;                        // Matrix-vector multiply.
        if(u < 0) (u, G1, G2) = (-u, -G1, -G2);
        if(v < 0) (v, G3, G4) = (-v, -G3, -G4);
        if(u < v) (u, v, G1, G2, G3, G4) = (v, u, G3, G4, G1, G2);
        if(v != 0) {
            (u, v) = (v, u);
            q = floor(v/u);
            G = ((0, 1), (1, -q)) G;                 // Matrix-matrix multiply.
            v = v - qu;
            m = floor(m/2);
            C = G;
            hgcd(m, u, v);                                      // Recurse.
            G = GC;
        }
        return;
    }
4. [Establish the top-level function rcgcd.]
    rgcd(x, y) {               // Top-level function, with variables u, v local.
        (u, v) = (x, y);
5. [Reduce arguments]
        (u, v) = (|u|, |v|);                // Absolute-value each component.
        if(u < v) (u, v) = (v, u);
        if(v < lim) goto [Branch];
        G = ((1, 0), (0, 1));
        hgcd(0, u, v);
        (u, v)^T = G(u, v)^T;
        (u, v) = (|u|, |v|);
        if(u < v) (u, v) = (v, u);
        if(v < lim) goto [Branch];
        (u, v) = (v, u mod v);
        goto [Reduce arguments];
6. [Branch]
        return cgcd(u, v);     // Recursion done, branch to alternative gcd.
    }
```

To clarify the practical application of the algorithm, one chooses the "breakover" parameters *lim* and *prec*, whence the greatest common divisor of x, y is to be calculated by calling the overall function $rgcd(x, y)$. We remark that G. Woltman has managed to implement Algorithm 9.4.6 in a highly memory-efficient way, essentially by reusing certain storage and carrying out other careful bookkeeping. Technically, he reports the ability to effect a random gcd with respect to F_{24} in under an hour on a modern PC, while a classical gcd of such magnitude can consume *days* of machine time. So the algorithm, though intricate, certainly has its rewards, especially in the search of factors of very large numbers, say arguments in the region $x, y \approx F_{20}$ and above.

9.5 Large-integer multiplication

When numbers have, say, hundreds or thousands (even millions) of decimal digits, there are modern methods for multiplication. In practice, one finds that the classical "grammar-school" methods just cannot effect multiplication in certain desired ranges. This is because, of course, the bit complexity of grammar-school multiply of two size-N numbers is $O\left(\ln^2 N\right)$. It turns out that by virtue of modern transform and convolution techniques, this complexity can be brought down to

$$O(\ln N(\ln\ln N)(\ln\ln\ln N)),$$

as we discuss in more detail later in this chapter.

The art of large-integer arithmetic has, especially in modern times, sustained many revisions. Just as with the fast Fourier transform (FFT) engineering literature itself, there seems to be no end to the publication of new approaches, new optimizations, and new applications for computational number theory. The forest is sufficiently thick that we have endeavored in this section to render an overview rather than an encyclopedic account of this rich and exotic field. An interesting account of multiplication methods from a theoretical point of view is [Bernstein 1997a], and modern implementations are discussed, with historical references, in [Crandall 1994b, 1996a].

9.5.1 Karatsuba and Toom–Cook methods

The classical multiplication methods can be applied on parts of integers to speed up large-integer multiplication, as observed by Karatsuba. His recursive scheme assumes that numbers be represented in split form

$$x = x_0 + x_1 W,$$

with $x_0, x_1 \in [0, W-1]$, which is equivalent to base-W representation, except that here the base will be about half the size of x itself. Note that x is therefore a "size-W^2" integer. For two integers x, y of this approximate size, the Karatsuba relation is

$$xy = \frac{t+u}{2} - v + \frac{t-u}{2}W + vW^2, \tag{9.17}$$

where

$$t = (x_0 + x_1)(y_0 + y_1),$$
$$u = (x_0 - x_1)(y_0 - y_1),$$
$$v = x_1 y_1,$$

and we obtain xy, which is originally a size-W^2 multiply, for the price of only three size-W multiplies (and some final carry adjustments, to achieve base-W representation of the final result). This is in principle an advantage, because if grammar-school multiply is invoked throughout, a size-W^2 multiply should be four, not three times as expensive as a size-W one. It can be shown that if one applies the Karatsuba relation to t, u, v themselves, and so on recursively, the asymptotic complexity for a size-N multiply is

$$O\left((\ln N)^{\ln 3/\ln 2}\right)$$

bit operations, a theoretical improvement over grammar-school methods. We say "theoretical improvement" because computer implementations will harbor so-called overhead, and the time to arrange memory and recombine subproducts and so on might rule out the Karatsuba method as a viable alternative. Still, it is often the case in practice that the Karatsuba approach does, in fact, outperform the grammar-school approach over a machine- and implementation-dependent range of operands.

But a related method, the Toom–Cook method, reaches the theoretical boundary of $O\left(\ln^{1+\epsilon} N\right)$ bit operations for the *multiplicative part* of size-N multiplication—that is, ignoring all the additions inherent in the method. However, there are several reasons why the method is not the final word in the art of large-integer multiply. First, for large N the number of additions is considerable. Second, the complexity estimate presupposes that multiplications by constants (such as 1/2, which is a binary shift, and so on) are inexpensive. Certainly multiplications by small constants are so, but the Toom–Cook coefficients grow radically as N increases. Still, the method is of theoretical interest and does have its practical applications, such as fast multiplication on machines whose fundamental word multiply is especially sluggish with respect to addition. The Toom–Cook method hinges on the idea that given two polynomials

$$x(t) = x_0 + x_1 t + \ldots + x_{D-1} t^{D-1}, \tag{9.18}$$

$$y(t) = y_0 + y_1 t + \ldots + y_{D-1} t^{D-1}, \tag{9.19}$$

the polynomial product $z(t) = x(t)y(t)$ is completely determined by its values at $2D-1$ separate t values, for example by the sequence of evaluations $(z(j))$, $j \in [1-D, D-1]$:

Algorithm 9.5.1 (Symbolic Toom–Cook multiplication). Given D, this algorithm generates the (symbolic)Toom–Cook scheme for multiplication of (D-digit)-by-(D-digit) integers.

1. [Initialize]
    Form two symbolic polynomials $x(t), y(t)$ each of degree $(D-1)$, as in equation (9.18);
2. [Evaluation]

Evaluate symbolically $z(j) = x(j)y(j)$ for each $j \in [1-D, D-1]$, so that each $z(j)$ is cast in terms of the original coefficients of the x and y polynomials;

3. [Reconstruction]

Solve symbolically for the coefficients z_j in the following linear system of $(2D-1)$ equations:

$z(t) = \sum_{k=0}^{2D-2} z_k t^k, \quad t \in [1-D, D-1]$;

4. [Report scheme]

Report a list of the $(2D-1)$ relations, each relation casting z_j in terms of the original x, y coefficients;

The output of this algorithm will be a set of formulae that give the coefficients of the polynomial product $z(t) = x(t)y(t)$ in terms of the coefficients of the original polynomials. But this is precisely what is meant by integer multiplication, if each polynomial corresponds to a D-digit representation in a fixed base B.

To underscore the Toom–Cook idea, we note that all of the Toom–Cook multiplies occur in the [Evaluation] step of Algorithm 9.5.1. We give next a specific multiplication algorithm that requires five such multiplies. The previous, symbolic, algorithm was used to generate the actual relations of this next algorithm:

Algorithm 9.5.2 (Explicit $D = 3$ Toom–Cook integer multiplication). For integers x, y given in base B as

$$x = x_0 + x_1 B + x_2 B^2,$$

$$y = y_0 + y_1 B + y_2 B^2,$$

this algorithm returns the base-B digits of the product $z = xy$, using the theoretical minimum of $2D - 1 = 5$ multiplications for acyclic convolution of length-3 sequences.

1. [Initialize]

$r_0 = x_0 - 2x_1 + 4x_2$;
$r_1 = x_0 - x_1 + x_2$;
$r_2 = x_0$;
$r_3 = x_0 + x_1 + x_2$;
$r_4 = x_0 + 2x_1 + 4x_2$;
$s_0 = y_0 - 2y_1 + 4y_2$;
$s_1 = y_0 - y_1 + y_2$;
$s_2 = y_0$;
$s_3 = y_0 + y_1 + y_2$;
$s_4 = y_0 + 2y_1 + 4y_2$;

2. [Toom–Cook multiplies]

for($0 \le j < 5$) $t_j = r_j s_j$;

3. [Reconstruction]

$z_0 = t_2$;
$z_1 = t_0/12 - 2t_1/3 + 2t_3/3 - t_4/12$;
$z_2 = -t_0/24 + 2t_1/3 - 5t_2/4 + 2t_3/3 - t_4/24$;
$z_3 = -t_0/12 + t_1/6 - t_3/6 + t_4/12$;
$z_4 = t_0/24 - t_1/6 + t_2/4 - t_3/6 + t_4/24$;

4. [Adjust carry]
carry = 0;
for($0 \le n < 5$) {
$v = z_n + carry$;
$z_n = v \bmod B$;
$carry = \lfloor v/B \rfloor$;
}
return $(z_0, z_1, z_2, z_3, z_4, carry)$;

Now, as opposed to the Karatsuba method, in which a size-B^2 multiply is brought down to that of three size-B ones for, let us say, a "gain" of 4/3, Algorithm 9.5.2 does a size-B^3 multiply in the form of five size-B ones, for a gain of 9/5. When either algorithm is used in a recursive fashion (for example, the step [Toom–Cook multiplies] is done by calling the same, or another, Toom–Cook algorithm recursively), the complexity of multiplication of two size-N integers comes down to

$$O\left((\ln N)^{\ln(2D-1)/\ln D}\right),$$

small multiplies (meaning of a fixed size independent of N), which complexity can, with sufficiently high Toom–Cook degree $d = D - 1$, be brought down below any given complexity estimate of $O\left(\ln^{1+\epsilon} N\right)$ small multiplies. However, it is to be noted forcefully that this complexity ignores the addition count, as well as the constant-coefficient multiplies (see Exercises 9.38, 9.78 and Section 9.5.8).

The Toom–Cook method can be recognized as a scheme for acyclic convolution, which together with other types of convolutions, we address later in this chapter. For more details on Karatsuba and Toom–Cook methods, the reader may consult [Knuth 1981], [Crandall 1996a], [Bernstein 1997a].

9.5.2 Fourier transform algorithms

Having discussed multiplication methods that enjoy complexities as low as $O\left(\ln^{1+\epsilon} N\right)$ small fixed multiplications (but perhaps unfortunate addition counts), we shall focus our attention on a class of multiplication schemes that enjoy low counts of all operation types. These schemes are based on the notion of the discrete Fourier transform (DFT), a topic that we now cover in enough detail to render the subsequent multiply algorithms accessible.

At this juncture we can think of a "signal" simply as a sequence of elements, in order to forge a connection between transform theory and the field of signal processing. Throughout the remainder of this chapter, signals might be sequences of polynomial coefficients, or sequences in general, and will be denoted by $x = (x_n)$, $n \in [0, D-1]$ for some "signal length" D.

The first essential notion is that multiplication is a kind of convolution. We shall make that connection quite precise later, observing for the moment that the DFT is a natural transform to employ in convolution problems. For the DFT has the unique property of converting convolution to a less expensive dyadic product. We start with a definition:

Definition 9.5.3 (The discrete Fourier transform (DFT)). Let x be a signal of length D consisting of elements belonging to some algebraic domain in which D^{-1} exists, and let g be a primitive D-th root of unity in that domain; that is, $g^k = 1$ if and only if $k \equiv 0 \pmod D$. Then the discrete Fourier transform of x is that signal $X = DFT(x)$ whose elements are

$$X_k = \sum_{j=0}^{D-1} x_j g^{-jk}, \tag{9.20}$$

with the inverse $DFT^{-1}(X) = x$ given by

$$x_j = \frac{1}{D} \sum_{k=0}^{D-1} X_k g^{jk}. \tag{9.21}$$

That the transform DFT^{-1} is well-defined as the correct inverse is left as an exercise. There are several important manifestations of the DFT:

Complex-field DFT: $x, X \in \mathbf{C}^D$, g a primitive D-th root of 1 such as $e^{2\pi i/D}$;

Finite-field DFT: $x, X \in \mathbf{F}_{p^k}^D$, g a primitive D-th root of 1 in the same field;

Integer-ring DFT: $x, X \in \mathbf{Z}_N^D$, g a primitive D-th root of 1 in the ring, D^{-1}, g^{-1} exist.

It should be pointed out that the above are common examples, yet there are many more possible scenarios. As just one extra example, one may define a DFT over quadratic fields (see Exercise 9.51).

In the first instance of complex fields, the practical implementations involve floating-point arithmetic to handle complex numbers (though when the signal has only real elements, significant optimizations apply, as we shall see). In the second, finite-field, cases one uses field arithmetic with all terms reduced (mod p). The third instance, the ring-based DFT, is sometimes applied simultaneously for $N = 2^n - 1$ and $N' = 2^n + 1$, in which cases the assignments $g = 2$ and $D = n, D' = 2n$, respectively, can be made when n is coprime to both N, N'.

It should be said that there exists a veritable menagerie of alternative transforms, many of them depending on basis functions other than the complex exponential basis functions of the traditional DFT; and often, such alternatives admit of fast algorithms, or assume real signals, and so on. Though such transforms lie beyond the scope of the present book, we observe that some of them are also suited for the goal of convolution, so we name a few: The Walsh–Hadamard transform, for which one needs no multiplication, only addition; the discrete cosine transform (DCT), which is a real-signal, real-multiplication analogue to the DFT; various wavelet transforms, which sometimes admit of very fast ($O(N)$ rather than $O(N \ln N)$) algorithms; real-valued FFT, which uses either cos or sin in real-only summands; the real-signal Hartley transform, and so on. Various of these options are discussed in [Crandall 1994b, 1996a].

Just to clear the air, we hereby make explicit the almost trivial difference between the DFT and the celebrated fast Fourier transform (FFT). The FFT is an operation belonging to the general class of divide-and-conquer algorithms, and which calculates the DFT of Definition 9.5.3. The FFT will typically appear in our algorithm layouts in the form $X = FFT(x)$, where it is understood that the DFT is being calculated. Similarly, an operation $FFT^{-1}(x)$ returns the inverse DFT. We make the distinction explicit because "FFT" is in some sense a misnomer: The DFT is a certain sum—an algebraic quantity—yet the FFT is an algorithm. Here is a heuristic analogy to the distinction: In this book, the equivalence class $x \pmod N$ are theoretical entities, whereas the *operation* of reducing x modulo p we have chosen to write a little differently, as $x \bmod p$. By the same token, within an algorithm the notation $X = FFT(x)$ means that we are performing an FFT operation on the signal X; and this operation gives, of course, the result $DFT(x)$. (Yet another reason to make the almost trivial distinction is that we have known students who incorrectly infer that an FFT is some kind of "approximation" to the DFT, when in fact, the FFT is sometimes *more* accurate then a literal DFT summation, in the sense of roundoff error, mainly because of reduced operation count for the FFT.)

The basic FFT algorithm notion has been traced all the way back to some observations of Gauss, yet some authors ascribe the birth of the modern theory to the Danielson–Lanczos identity, applicable when the signal length D is even:

$$DFT(x) = \sum_{j=0}^{D-1} x_j g^{-jk} = \sum_{j=0}^{D/2-1} x_{2j} \left(g^2\right)^{-jk} + g^{-k} \sum_{j=0}^{D/2-1} x_{2j+1} \left(g^2\right)^{-jk}. \tag{9.22}$$

A beautiful identity indeed: A DFT sum for signal length D is split into two sums, each of length $D/2$. In this way the Danielson–Lanczos identity ignites a recursive method for calculating the transform. Note the so-called twiddle factors g^{-k}, which figure naturally into the following recursive form of FFT. In this and subsequent algorithm layouts we denote by $len(x)$ the length of a signal x. In addition, when we perform element concatenations of the form $(a_j)_{j \in J}$ we mean the result to be a natural, left-to-right, element concatenation as the increasing index j runs through a given set J. Similarly, $U \cup V$ is a signal having the elements of V appended to the right of the elements of U.

Algorithm 9.5.4 (FFT, recursive form). Given a length-$(D = 2^d)$ signal x whose DFT (Definition 9.5.3) exists, this algorithm calculates said transform via a single call $FFT(x)$. We employ the signal-length function $len()$, and within the recursion the root g of unity is to have order equal to current signal length.

1. [Recursive FFT function]

```
FFT(x) {
    n = len(x);
    if(n == 1) return x;
    m = n/2;
    X = (x_{2j})_{j=0}^{m-1};                    // The even part of x.
    Y = (x_{2j+1})_{j=0}^{m-1};                  // The odd part of x.
    X = FFT(X);
    Y = FFT(Y);                                  // Two recursive calls of half length.
    U = (X_{k mod m})_{k=0}^{n-1};
    V = (g^{-k} Y_{k mod m})_{k=0}^{n-1};        // Use root g of order n.
    return U + V;                                // Realization of identity (9.22).
}
```

A little thought shows that the number of operations in the algebraic domain of interest is

$$O(D \ln D),$$

and this estimate holds for both multiplies and add/subtracts. The $D \ln D$ complexity is typical of divide-and-conquer algorithms, another example of which would be the several methods for rapid sorting of elements in a list. This recursive form is instructive, and does have its applications, but the overwhelming majority of FFT implementations use a clever loop structure first achieved by [Cooley and Tukey 1965]. The Cooley–Tukey algorithm uses the fact that if the elements of the original length-$(D = 2^d)$ signal x are given a certain "bit-scrambling" permutation, then the FFT can be carried out with convenient nested loops. The scrambling intended is reverse-binary reindexing, meaning that x_j gets replaced by x_k, where k is the reverse-binary representation of j. For example, for signal length $D = 2^5$, the new element x_5 after scrambling is the old x_{20}, because the binary reversal of $5 = 00101_2$ is $10100_2 = 20$. Note that this bit-scrambling of indices could in principle be carried out via sheer manipulation of indices to create a new, scrambled list; but it is often more efficient to do the scrambling in place, by using a certain sequence of two-element transpositions. It is this latter scheme that appears in the next algorithm.

A most important observation is that the Cooley–Tukey scheme actually allows the FFT to be performed *in place*, meaning that an original signal x is replaced, element by element, with the DFT values. This is an extremely memory-efficient way to proceed, accounting for a great deal of the popularity of the Cooley–Tukey and related forms. With bit-scrambling properly done, the overall Cooley–Tukey scheme yields an in-place, in-order (meaning natural DFT order) FFT. Historically, the phrase "decimation in time" is attributed to the Cooley–Tukey form, the phrase meaning that as in the Danielson–Lanczos splitting identity (9.22), we cut up (decimate) the time domain—the index on the original signal. The Gentleman–Sande FFT falls into the "decimation in frequency" class, for which a similar game is played on the k index of the transform elements X_k.

Algorithm 9.5.5 (FFT, in-place, in-order loop forms with bit-scramble). Given a ($D = 2^d$)-element signal x, the functions herein perform an FFT via nested loops. The two essential FFTs are laid out as decimation-in-time (Cooley–Tukey) and decimation-in-frequency (Gentleman–Sande) forms. Note that these forms can be applied symbolically, or in number-theoretical transform mode, by identifying properly the root of unity and the ring or field operations.

1. [Cooley–Tukey, decimation-in-time FFT]

```
FFT(x) {
    scramble(x);
    n = len(x);
    for(m = 1; m < n; m = 2m) {               // m ascends over 2-powers.
        for(0 ≤ j < m) {
            a = g^(-jn/(2m));
            for(i = j; i < n; i = i + 2m)
                (x_i, x_{i+m}) = (x_i + a x_{i+m}, x_i - a x_{i+m});
        }
    }
    return x;
}
```

2. [Gentleman–Sande, decimation-in-frequency FFT]

```
FFT(x) {
    n = len(x);
    for(m = n/2; m ≥ 1; m = m/2) {            // m descends over 2-powers.
        for(0 ≤ j < m) {
            a = g^(-jn/(2m));
            for(i = j; i < n; i = i + 2m)
                (x_i, x_{i+m}) = (x_i + x_{i+m}, a(x_i - x_{i+m}));
        }
    }
    scramble(x);
    return x;
}
```

3. [In-place *scramble* procedure]

```
scramble(x) {                   // In-place, reverse-binary element scrambling.
    n = len(x);
    j = 0;
    for(0 ≤ i < n - 1) {
        if(i < j) (x_i, x_j) = (x_j, x_i);                   // Swap elements.
        k = ⌊n/2⌋;
```

```
            while(k ≤ j) {
                j = j − k;
                k = ⌊k/2⌋;
            }
            j = j + k;
        }
    return;
}
```

It is to be noted that when one performs a convolution in the manner we shall exhibit later, the scrambling procedures are not needed, provided that one performs required FFTs in a specific order. Correct is Gentleman–Sande form (with scrambling procedure omitted) first, Cooley–Tukey form (without initial scrambling) second. This works out because, of course, scrambling is an operation of order two.

Happily, in cases where scrambling is not desired, or when contiguous memory access is important (e.g., on vector computers), there is the Stockham FFT, which *avoids* bit-scrambling and also has an innermost loop that runs essentially consecutively through data memory. The cost of all this is that one must use an extra copy of the data. The typical implementations of the Stockham FFT are elegant [Van Loan 1992], but there is a particular variant that has proved quite useful on modern vector machinery. This special variant is the "ping-pong" FFT, because one goes back and forth between the original data and a separate copy. The following algorithm display is based on a suggested design of [Papadopoulos 1999]:

Algorithm 9.5.6 (FFT, "ping-pong" variant, in-order, no bit-scramble). Given a ($D = 2^d$)-element signal x, a Stockham FFT is performed, but with the original x and external data copy y used in alternating fashion. We interpret X, Y below as pointers to the (complex) signals x, y, respectively, but operating under the usual rules of pointer arithmetic; e.g., $X[0]$ is the first complex datum of x initially, but if 4 is added to pointer X, then $X[0] = x_4$, and so on. If exponent d is even, pointer X has the FFT result, else pointer Y has it.

1. [Initialize]

```
    J = 1;
    X = x; Y = y;                         // Assign memory pointers.
```

2. [Outer loop]

```
    for(d ≥ i > 0) {
        m = 0;
        while(m < D/2) {
            a = e^{−2πim/D};
            for(J ≥ j > 0) {
                Y[0] = X[0] + X[D/2];
                Y[J] = a(X[0] − X[D/2]);
                X = X + 1;
                Y = Y + 1;
```

```
            }
            Y = Y + J;
            m = m + J;
        }
        J = 2 * J;
        X = X - D/2;
        Y = Y - D;
        (X, Y) = (Y, X);                                    // Swap pointers!
    }
3. [Make ping-pong parity decision]
    if(d even) return (complex data at X);
    return (complex data at Y);
```

The useful loop aspect of this algorithm is the fact that the loop variable j runs contiguously (from J down), and so on a vector machine one may process chunks of data all at once, by picking up, then putting back, data as vectors.

Incidentally, to perform an inverse FFT is extremely simple, once the forward FFT is implemented. One approach is simply to look at Definition 9.5.3 and observe that the root g can be replaced by g^{-1}, with a final overall normalization $1/D$ applied to an inverse FFT. But when complex fields are used, so that $g^{-1} = g^*$, the procedure for FFT^{-1} can be, if one desires, just a sequence:

```
x = x*;                                        // Conjugate the signal.
X = FFT(x);                 // The usual FFT, with usual root g.
X = X*/D;                        // Final conjugate and normalize.
```

Though these Cooley–Tukey and Gentleman–Sande FFTs are most often invoked over the complex numbers, so that the root is $g = e^{2\pi i/D}$, say, they are useful also as number-theoretical transforms, with operations carried out in a finite ring or field. In either the complex or finite-field cases, it is common that a signal to be transformed has all real elements, in which case we call the signal "pure-real." This would be so for complex signal $x \in C^D$ but such that $x_j = a_j + 0i$ for each $j \in [0, D-1]$. It is important to observe that the analogous signal class can occur in certain fields, for example $\mathbf{F}_{p^2}$ when $p \equiv 3 \pmod 4$. For in such fields, every element can be represented as $x_j = a_j + b_j i$, and we can say that a signal $x \in \mathbf{F}_{p^2}^D$ is pure-real if and only if every b_j is zero. The point of the pure-real signal class designation is that in general, an FFT for pure-real signals has about 1/2 the usual complexity. This makes sense from an information-theoretical standpoint: Indeed, there is "half as much" data in a pure-real signal. A basic way to cut down thus the FFT complexity for pure-real signals is to embed half of a pure-real signal in the imaginary parts of a signal of half the length, i.e., to form a complex signal

$$y_j = x_j + i x_{j+D/2}$$

for $j \in [0, D/2-1]$. Note that signal y now has length $D/2$. One then performs a full, half-length, complex FFT and then uses some reconstruction formulae to recover the correct DFT of the original signal x. An example of this pure-real signal approach for number-theoretical transforms as applied to cyclic convolution is embodied in Algorithm 9.5.21, with split-radix symbolic pseudocode given in [Crandall 1997b] (see Exercise 9.52 for discussion of the negacyclic scenario).

Incidentally, there are yet lower-complexity FFTs, called split-radix FFTs, which employ an identity more complicated than the Danielson–Lanczos formula. And there is even a split-radix, pure-real-signal FFT due to Sorenson that is quite efficient and in wide use [Crandall 1994b]. The vast "FFT forest" is replete with specially optimized FFTs, and whole texts have been written in regard to the structure of FFT algorithms; see, for example, [Van Loan 1992].

Even at the close of the 20th century there continue to be, every year, a great many published papers on new FFT optimizations. Because our present theme is the implementation of FFTs for large-integer arithmetic, we close this section with one more algorithm: a "parallel," or "piecemeal," FFT algorithm that is quite useful in at least two practical settings. First, when signal data are particularly numerous, the FFT must be performed in limited memory. In practical terms, a signal might reside on disk memory, and exceed a machine's physical random-access memory. The idea is to "spool" pieces of the signal off the larger memory, process them, and combine in just the right way to deposit a final FFT in storage. Because computations occur in large part on separate pieces of the transform, the algorithm can also be used in a parallel setting, with each separate processor handling a respective piece of the FFT. The algorithm following has been studied by various investigators [Agarwal and Cooley 1986], [Swarztrauber 1987], [Ashworth and Lyne 1988], [Bailey 1990], especially with respect to practical memory usage. It is curious that the essential ideas seem to have originated with [Gentleman and Sande 1966]. Perhaps, in view of the extreme density and proliferation of FFT research, one might forgive investigators for overlooking these origins for two decades.

The parallel-FFT algorithm stems from the observation that a length-$(D = WH)$ DFT can be performed by tracking over rows and columns of an $H \times W$ (height times width) matrix. Everything follows from the following algebraic reduction of the DFT X of x:

$$\begin{aligned}
X = DFT(x) &= \left(\sum_{j=0}^{D-1} x_j g^{-jk} \right)_{k=0}^{D-1} \\
&= \left(\sum_{J=0}^{W-1} \sum_{M=0}^{H-1} x_{J+MW} g^{-(J+MW)(K+NH)} \right)_{K+NH=0}^{D-1} \\
&= \left(\sum_{J=0}^{W-1} \left(\sum_{M=0}^{H-1} x_{J+MW} g_H^{-MK} \right) g^{-JK} g_W^{-JN} \right)_{K+NH=0}^{D-1},
\end{aligned}$$

where g, g_H, g_W are roots of unity of order WH, H, W, respectively, and the indices $(K + NH)$ have $K \in [0, H-1]$, $N \in [0, W-1]$. The last double sum here can be seen to involve the FFTs of rows and columns of a certain matrix, as evidenced in the explicit algorithm following:

Algorithm 9.5.7 (Parallel, "four-step" FFT). Let x be a signal of length $D = WH$. For algorithmic efficiency we consider the input signal x to be a matrix T arranged in "columnwise order"; i.e., for $j \in [0, W-1]$ the j-th column of T contains the (originally contiguous) H elements $(x_{jH+M})_{M=0}^{H-1}$. Then, conveniently, each FFT operation of the overall algorithm occurs on some row of some matrix (the k-th row vector of a matrix U will be denoted by $U^{(k)}$). The final DFT resides likewise in columnwise order.

1. [H in-place, length-W FFTs, each performed on a row of T]
   for($0 \le M < H$) $T^{(M)} = DFT\left(T^{(M)}\right)$;
2. [Transpose and twist the matrix]
   $(T_{JK}) = (T_{KJ} g^{-JK})$;
3. [W in-place, length-H FFTs, each performed on a row of the new T]
   for($0 \le J < W$) $T^{(J)} = DFT\left(T^{(J)}\right)$;
4. [Return $DFT(x)$ as elements in columnwise order]
   return T; // T_{MJ} is now $DFT(x)_{JH+M}$.

Note that whatever scheme is used for the transpose (see Exercise 9.54) can also be used to convert lexicographically arranged input data x into the requisite columnwise format, and likewise for converting back at algorithm's end to lexicographic DFT format. In other words, if the input data is assumed to be stored initially lexicographically, then two more transpositions can be placed, one before and one after the algorithm, to render Algorithm 9.5.7 a standard length-WH FFT. A small worked example is useful here. Algorithm 9.5.7 wants, for a length-$N = 4 = 2 \cdot 2$ FFT, and so primitive fourth root of unity $g = e^{2\pi i/4} = i$, the input data in columnwise order like so:

$$T = \begin{pmatrix} x_0 & x_2 \\ x_1 & x_3 \end{pmatrix}.$$

The first algorithm step is to do ($H = 2$) row FFTs each of length ($W = 2$), to get

$$T = \begin{pmatrix} x_0 + x_2 & x_0 - x_2 \\ x_1 + x_3 & x_1 - x_3 \end{pmatrix}.$$

Then we transpose, and twist via dyadic multiply by the phase matrix

$$(g^{-JK}) = \begin{pmatrix} 1 & 1 \\ 1 & -i \end{pmatrix}$$

to yield

$$T = \begin{pmatrix} x_0 + x_2 & x_1 + x_3 \\ x_0 - x_2 & -i(x_1 - x_3) \end{pmatrix},$$

whence a final set of row FFTs gives

$$T = \begin{pmatrix} X_0 & X_2 \\ X_1 & X_3 \end{pmatrix}$$

where $X_k = \sum_j x_j g^{-jk}$ are the usual DFT components, and we note that the final form here for T is again in columnwise order.

Incidentally, if one wonders how this differs from a *two*-dimensional FFT such as an FFT in the field of image processing, the answer to that is simple: This four-step (or six-step, if pre- and post transposes are invoked to start with and end up with standard row-ordering) format involves that internal "twist," or phase factor, in step [Transpose and twist the matrix]. A two-dimensional FFT does not involve the phase-factor twisting step; instead, one simply takes FFTs of all rows in place, then all columns in place.

Of course, with respect to repeated applications of Algorithm 9.5.7 the efficient option is simply this: Always store signals and their transforms in the columnwise format. Furthermore, one can establish a rule that for signal lengths $N = 2^n$, we factor into matrix dimensions as $W = H = \sqrt{N} = 2^{n/2}$ for n even, but $W = 2H = 2^{(n+1)/2}$ for n odd. Then the matrix is square or almost square. Furthermore, for the *inverse* FFT, in which everything proceeds as above but with FFT^{-1} calls and the twisting phase uses g^{+JK}, with a final division by N, one can conveniently assume that the width and height for this inverse case satisfy $W' = H'$ or $H' = 2W'$, so that in such as convolution problems the output matrix of the forward FFT is what is expected for the inverse FFT, even when said matrix is nonsquare. Actually, for convolutions *per se* there are other interesting optimizations due to J. Papadopoulos, such as the use of DIF/DIT frameworks and bit-scrambled powers of g; and a very fast large-FFT implementation of Mayer, in which one *never* transposes, using instead a fast, memory-efficient columnwise FFT stage [Crandall et al. 1999].

One interesting byproduct of this approach is that one is moved to study the basic problem of matrix transposition. The treatment of [Bailey 1989] gives an interesting small example of the [Fraser 1976] algorithm for efficient transposition of a stored matrix, while [Van Loan 1992, p. 138] indicates how active, really, is the ongoing study of fast transpose. Such an algorithm has applications in other aspects of large-integer arithmetic, for example see Section 9.5.7.

9.5.3 Convolution theory

Let x denote a signal $(x_0, x_1, \ldots, x_{D-1})$, where for example, the elements of x could be the digits of Definitions (9.1.1) or (9.1.2) (although we do not *a priori* insist that the elements be digits; the theory to follow is fairly general). We start by defining fundamental convolution operations on signals. In what follows, we assume that signals x, y have been assigned the same length (D) of elements. In all the summations of this next definition, indices i, j each run over the set $\{0, \ldots, D-1\}$:

Definition 9.5.8. The cyclic convolution of two length-D signals x, y is a signal denoted $z = x \times y$ having D elements given by

$$z_n = \sum_{i+j \equiv n \pmod D} x_i y_j,$$

while the negacyclic convolution of x, y is a signal $v = x \times_- y$ having D elements given by

$$v_n = \sum_{i+j=n} x_i y_j - \sum_{i+j=D+n} x_i y_j,$$

and the acyclic convolution of x, y is a signal $u = x \times_A y$ having $2D$ elements given by

$$u_n = \sum_{i+j=n} x_i y_j,$$

for $n \in \{0, \ldots, 2D-2\}$, together with the assignment $u_{2D-1} = 0$. Finally, the half-cyclic convolution of x, y is the length-D signal $x \times_H y$ consisting of the first D elements of the acyclic convolution u.

These fundamental convolutions are closely related, as is seen in the following result. In such statements we interpret the sum of two signals $c = a + b$ in elementwise fashion, that is, $c_n = a_n + b_n$ for relevant indices n. Likewise, a scalar-signal product qa, with q a number and a a signal, is the signal (qa_n). We shall require the notion of the splitting of signals (of even length) into halves, so we denote by $L(a), H(a)$, respectively, the lower-indexed and higher-indexed halves of a. That is, from $c = a \cup b$ the natural, left-right, concatenation of two signals of equal length, we shall have $L(c) = a$ and $H(c) = b$.

Theorem 9.5.9. *Let signals x, y have the same length D. Then the various convolutions are related as follows (it is assumed that in the relevant domain to which signal elements belong, 2^{-1} exists):*

$$x \times_H y = \frac{1}{2}((x \times y) + (x \times_- y)).$$

Furthermore,

$$x \times_A y = (x \times_H y) \cup \frac{1}{2}((x \times y) - (x \times_- y)).$$

Finally, if the length D is even and $x_j, y_j = 0$ for $j \geq D/2$, then

$$L(x) \times_A L(y) = x \times y = x \times_- y.$$

These interrelations allow us to use certain algorithms more universally. For example, a pair of algorithms for cyclic and negacyclic can be used to extract both the half-cyclic or the acyclic, and so on. In the final statement of the theorem, we have introduced the notion of "zero padding," which in practice amounts to appending D zeros to signals already of length D, so that the signals' acyclic convolution is identical to the cyclic (or the negacyclic) convolution of the two padded sequences.

The connection between convolution and the DFT of the previous section is evident in the following celebrated theorem, wherein we refer to the dyadic operator $*$, under which a signal $z = x * y$ has elements $z_n = x_n y_n$:

Theorem 9.5.10 (Convolution theorem). *Let signals x, y have the same length D. Then the cyclic convolution of x, y satisfies*

$$x \times y = DFT^{-1}(DFT(x) * DFT(y)),$$

which is to say

$$(x \times y)_n = \frac{1}{D} \sum_{k=0}^{D-1} X_k Y_k \, g^{kn}.$$

Armed with this mighty theorem we can effect the cyclic convolution of two signals with three transforms (one of them being the inverse transform), or the cyclic convolution of a signal with itself with two transforms. As the known complexity of the DFT is $O(D \ln D)$ operations in the field, the dyadic product implicit in Theorem 9.5.10, being $O(D)$, is asymptotically negligible.

A direct and elegant application of FFTs for large-integer arithmetic is to perform the DFTs of Theorem 9.5.10 in order to effect multiplication via acyclic convolution. This essential idea—pioneered by Strassen in the 1960s and later optimized by [Schönhage and Strassen 1971] (see Section 9.5.6)—runs as follows. If an integer x is represented as a length-D signal consisting of the (base-B) digits, and the same for y, then the integer product xy is an acyclic convolution of length $2D$. Though Theorem 9.5.10 pertains to the cyclic, not the acyclic, we nevertheless have Theorem 9.5.9, which allows us to use zero-padded signals and then perform the cyclic. This idea leads, in the case of complex field transforms, to the following scheme, which is normally applied using floating-point arithmetic, with DFT's done via fast Fourier transform (FFT) techniques:

Algorithm 9.5.11 (Basic FFT multiplication). Given two nonnegative integers x, y, each having at most D digits in some base B (Definition 9.1.1), this algorithm returns the base-B representation of the product xy. (The FFTs normally employed herein would be of the floating-point variety, so one must beware of roundoff error.)

1. [Initialize]
   Zero-pad both of x, y until each has length $2D$, so that the cyclic convolution of the padded sequences will contain the acyclic convolution of the unpadded ones;
2. [Apply transforms]
   $X = DFT(x)$; // Perform transforms via efficient FFT algorithm.
   $Y = DFT(y)$;

3. [Dyadic product]
   $Z = X * Y$;
4. [Inverse transform]
   $z = DFT^{-1}(Z)$;
5. [Round digits]
   $z = round(z)$; // Elementwise rounding to nearest integer.
6. [Adjust carry in base B]
   $carry = 0$;
   for($0 \le n < 2D$) {
      $v = z_n + carry$;
      $z_n = v \bmod B$;
      $carry = \lfloor v/B \rfloor$;
   }
7. [Final digit adjustment]
   Delete leading zeros, with possible $carry > 0$ as a high digit of z;
   return z;

This algorithm description is intended to be general, conveying only the principles of FFT multiplication, which are transforming, rounding, carry adjustment. There are a great many details left unsaid, not to mention a great number of enhancements, some of which we address later. But beyond these minutiae there is one very strong caveat: The accuracy of the floating-point arithmetic must always be held suspect. A key step in the general algorithm is the elementwise rounding of the z signal. If floating-point errors in the FFTs are too large, an element of the convolution z could get rounded to an incorrect value.

One immediate practical enhancement to Algorithm 9.5.11 is to employ the balanced representation of Definition 9.1.2. It turns out that floating-point errors are significantly reduced in this representation [Crandall and Fagin 1994], [Crandall et al. 1999]. This phenomenon of error reduction is not completely understood, but certainly has to do with the fact of generally smaller magnitude for the digits, plus, perhaps, some cancellation in the DFT components because the signal of digits has a statistical mean (in engineering terms, "DC component") that is very small, due to the balancing.

Before we move on to algorithmic issues such as further enhancements to the FFT multiply and the problem of pure-integer convolution, we should mention that convolutions can appear in number-theoretical work quite outside the large-integer arithmetic paradigm. We give two examples to end this subsection; namely, convolutions applied to sieving and to regularity results on primes.

Consider the following theorem, which is reminiscent of (although obviously much less profound than) the celebrated Goldbach conjecture:

Theorem 9.5.12. *Let $N = 2\cdot 3\cdot 5\cdots p_m$ be a product of consecutive primes. Then every sufficiently large even n is a sum $n = a+b$ with each of a, b coprime to N.*

It is intriguing that this theorem can be proved, without too much trouble, via convolution theory. (We should point out that there are also proofs using CRT ideas, so we are merely using this theorem to exemplify applications of discrete convolution methods (see Exercise 9.41).) The basic idea is to consider a special signal y defined by $y_j = 1$ if $\gcd(j, N) = 1$, else $y_j = 0$, with the signal given some cutoff length D. Now the acyclic convolution $y \times_A y$ will tell us precisely which $n < D$ of the theorem have the $a + b$ representations, and furthermore, the n-th element of the acyclic is precisely the number of such representations of n.

As a brief digression, we should note here that the original Goldbach conjecture is true if a different signal of infinite length, namely

$$G = (1, 1, 1, 0, 1, 1, 0, 1, 1, 0, \ldots),$$

where the 1's occur at indices $(p-3)/2$ for the odd primes $p = 3, 5, 7, 11, 13, \ldots$, has the property that the acyclic $G \times_A G$ has no zero elements. In this case the n-th element of the acyclic is precisely the number of Goldbach representations of $2n + 6$.

Back to Theorem 9.5.12: It is advantageous to study the length-N DFT Y of the aforementioned signal y. This DFT turns out to be a famous sum:

$$Y_k(N) = c_N(k) = \sum_{\gcd(j,N)=1} e^{\pm 2\pi i jk/N}, \tag{9.23}$$

where j is understood to run over those elements in the interval $[0, N-1]$ that are coprime to N, so the sign choice in the exponent doesn't matter, while $c_N(k)$ is the standard notation for the Ramanujan sum, which sum is already known to enjoy intriguing multiplicative properties [Hardy and Wright 1979]. In fact, the appearance of the Ramanujan sum in Section 1.4.4 suggests that it makes sense for c_N also to have some application in discrete convolution studies. We leave the proof of Theorem 9.5.12 to the reader (see Exercise 9.41), but wish to make several salient points. First, the sum in relation (9.23) can itself be thought of as a result of "sieving" out finite sums corresponding to the divisors of N. This gives rise to interesting series algebra. Second, it is remarkable that the *cyclic* length-N convolution of y with itself can be given a closed form. The result is

$$(y \times y)_n = \varphi_2(N, n) = \prod_{p|N} (p - \theta(n, p)), \tag{9.24}$$

where $\theta(n, p)$ is 1 if $p|n$, else 2. Thus, for $0 \le n < N$, this product expression is the exact number of representations of *either* n or $n+N$ as $a+b$ with both a, b coprime to N. As discussed in the exercises, to complete this line of reasoning one must invoke negacyclic convolution ideas (or some other means such as sieving) to show that the representations of $n + N$ are, for an appropriate range n, less than those of n itself. These observations will, after some final arguments, prove Theorem 9.5.12.

Now to yet another application of convolution. In 1847 E. Kummer discovered that if $p > 2$ is a regular prime, then Fermat's last theorem, that

$$x^p + y^p = z^p$$

has no Diophantine solution with $xyz \neq 0$, holds. (We note in passing that FLT is now a genuine theorem of A. Wiles, but the techniques here predated that work and still have application to such remaining open problems as the Vandiver conjecture.) Furthermore, p is regular if it does not divide any of the numerators of the even-index Bernoulli numbers

$$B_2, B_4, \ldots, B_{p-3}.$$

There is an elegant relation due to Shokrollahi in 1994; see [Buhler et al. 2000], that gives a congruence for precisely these Bernoulli numbers:

Theorem 9.5.13. *Let g be a primitive root of the odd prime p, and set:*

$$c_j = \left\lfloor \frac{(g^{-1} \bmod p)(g^j \bmod p)}{p} \right\rfloor g^{-j}$$

for $j \in [0, p-2]$. Then for $k \in [1, (p-3)/2]$ we have

$$\sum_{j=0}^{p-2} c_j g^{2kj} \equiv \left(1 - g^{2k}\right) \frac{B_{2k}}{2kg} \pmod{p}. \tag{9.25}$$

We see that Shokrollahi's relation involves a length-$(p-1)$ DFT, with the operant field being $\mathbf{F}_p$. One could proceed with an FFT algorithm, except that there are two problems with that approach. First, the best lengths for standard FFTs are powers of two; and second, one cannot use floating-point arithmetic, especially when the prime p is large, unless the precision is extreme (and somehow guaranteed). But we have the option of performing a DFT itself via convolution (see Algorithm 9.6.6), so the Shokrollahi procedure for determining regular primes; indeed, for finding precise irregularity indices of any prime, can be effected via power-of-two length convolutions. As we shall see later, there are "symbolic FFT" means to do this, notably in Nussbaumer convolution, which avoids floating-point arithmetic and so is suitable for pure-integer convolution. These approaches—Shokrollahi identity and Nussbaumer convolution—have been used together to determine all regular primes $p < 12000000$ [Buhler et al. 2000].

9.5.4 Discrete weighted transform (DWT) methods

One variant of DFT-based convolution that has proved important for modern primality and factorization studies (and when the relevant integers are large, say in the region of $2^{1000000}$ and beyond) is the discrete weighted transform (DWT). This transform is defined as follows:

Definition 9.5.14 (Discrete weighted transform (DWT)). Let x be a signal of length D, and let a be a signal (called the weight signal) of the same length, with the property that every a_j is invertible. Then the discrete weighted transform $X = DWT(x, a)$ is the signal of elements

$$X_k = \sum_{j=0}^{D-1} (a * x)_j g^{-jk}, \tag{9.26}$$

with the inverse $DWT^{-1}(X, a) = x$ given by

$$x_j = \frac{1}{Da_j} \sum_{k=0}^{D-1} X_k g^{jk}. \tag{9.27}$$

Furthermore, the weighted cyclic convolution of two signals is the signal $z = x \times_a y$ having

$$z_n = \frac{1}{a_n} \sum_{j+k \equiv n \pmod D} (a * x)_j (a * y)_k. \tag{9.28}$$

It is clear that the DWT is simply the DFT of the dyadic product signal $a * x$ consisting of elements $a_j x_j$. The considerable advantage of the DWT is that particular weight signals give rise to useful alternative convolutions. In some cases, the DWT eliminates the need for the zero padding of the standard FFT multiplication Algorithm 9.5.11. We first state an important result:

Theorem 9.5.15 (Weighted convolution theorem). *Let signals x, y and weight signal a have the same length D. Then the weighted cyclic convolution of x, y satisfies*

$$x \times_a y = DWT^{-1}(DWT(x, a) * DWT(y, a), a),$$

that is to say,

$$(x \times_a y)_n = \frac{1}{Da_n} \sum_{k=0}^{D-1} (X * Y)_k g^{kn}.$$

Thus FFT algorithms may be applied now to weighted convolution. In particular, one may compute not just the cyclic, but also the negacyclic, convolution in this manner, because the specific choice of weight signal

$$a = \left(A^j\right), \quad j \in [0, D-1]$$

yields, when A is a primitive $2D$-th root of unity in the field, the identity:

$$x \times_- y = x \times_a y, \tag{9.29}$$

which means that the weighted cyclic in this case is the negacyclic. Note that when the D-th root g has a square root in the field, as is the case with the complex field arithmetic, we can simply assign $A^2 = g$ to effect the negacyclic. Another interesting instance of generator A, namely when A is a primitive $4D$-th root of unity, gives the so-called right-angle convolution [Crandall 1996a].

These observations lead in turn to an important algorithm that has been used to advantage in modern factorization studies. By using the DWT, the method obviates zero padding entirely. Consider the problem of multiplication of two numbers, modulo a Fermat number $F_n = 2^{2^n} + 1$. This operation can happen, of course, a great number of times in attempts to factor an F_n. There are at least three ways to attempt $(xy) \bmod F_n$ via convolution of length-D signals where D and a power-of-two base B are chosen such that $F_n = B^D + 1$:

(1) Zero-pad each of x, y up to length $2D$, perform cyclic convolution, do carry adjust as necessary, take the result $\pmod{F_n}$.

(2) Perform length-D weighted convolution, with weight generator A a primitive $(2D)$-th root of unity, do carry adjust as necessary.

(3) Create length-$(D/2)$ "fold-over" signals, as $x' = L(x) + iH(x)$ and similarly for a y', employ a weighted convolution with generator A a primitive $(4D)$-th root of unity, do carry adjust.

Method (1) could, of course, involve Algorithm 9.5.11, with perhaps a fast Fermat-mod of Section 9.2.3; but one could instead use a pure-integer Nussbaumer convolution discussed later. Method (2) is the negacyclic approach, in which the weighted convolution can be seen to be multiplication $\pmod{F_n}$; that is, the mod operation is "free" (see Exercises). Method (3) is the right-angle convolution approach, which also gives the mod operation for free (see Exercises). Note that neither method (2) nor method (3) involves zero-padding, and that method (3) actually halves the signal lengths (at the expense of complex arithmetic). We focus on method (3), to state the following algorithm, which, as with Algorithm 9.5.11, is often implemented in a floating-point paradigm:

Algorithm 9.5.16 (DWT multiplication modulo Fermat numbers). For a given Fermat number $F_n = 2^{2^n} + 1$, and positive integers $x, y \not\equiv -1 \pmod{F_n}$, this algorithm returns $(xy) \bmod F_n$. We choose B, D such that $F_n = B^D + 1$, with the inputs x, y interpreted as length-D signals of base-B digits. We assume that there exists a primitive $4D$-th root of unity, A, in the field.

1. [Initialize]
   $E = D/2$; // Halve the signal length and "fold-over" the signals.
   $x = L(x) + iH(x)$; // Length-E signals.
   $y = L(y) + iH(y)$;
   $a = (1, A, A^2, \ldots, A^{E-1})$; // Weight signal.
2. [Apply transforms]
   $X = DWT(x, a)$; // Via an efficient length-E FFT algorithm.
   $Y = DWT(y, a)$;
3. [Dyadic product]
   $Z = X * Y$;

```
4. [Inverse transform]
      z = DWT^{-1}(Z, a);
5. [Unfold signal]
      z = Re(z) ∪ Im(z);                  // Now z will have length D.
6. [Round digits]
      z = round(z);            // Elementwise rounding to nearest integer.
7. [Adjust carry in base B]
      carry = 0;
      for(0 ≤ n < D) {
         v = z_n + carry;
         z_n = v mod B;
         carry = ⌊v/B⌋;
      }
8. [Final modular adjustment]
      Include possible carry > 0 as a high digit of z;
      z = z mod F_n;
                  // Via another 'carry' loop or via special-form mod methods.
      return z;
```

Note that in the steps [Adjust carry in base B] and [Final modular adjustment] the logic depends on the digits of the reconstructed integer z being positive. We say this because there are efficient variants using balanced-digit representation, in which variants care must be taken to interpret negative digits (and negative carry) correctly.

This algorithm was used in the discoveries of new factors of F_{13}, F_{15}, F_{16}, and F_{18} [Brent et al. 2000] (see the Fermat factor tabulation in Section 1.3.2), and also to establish the composite character of F_{22}, F_{24}, and of various cofactors for other F_n [Crandall et al. 1995], [Crandall et al. 1999]. In more recent times, [Woltman 2000] has implemented the algorithm to forge highly efficient factoring software for Fermat numbers (see remarks following Algorithm 7.4.4).

Another DWT variant has been used in the discovery of the recent Mersenne primes $2^{1398269}-1, 2^{2976221}-1, 2^{3021377}-1, 2^{6972593}-1, 2^{13466917}-1$ (see Table 1.2), the last of which being the largest known explicit prime as of this writing. For these discoveries, a network of volunteer users ran extensive Lucas–Lehmer tests that involve vast numbers of squarings modulo $p = 2^q - 1$. The algorithm variant in question has been called the irrational-base discrete weighted transform (IBDWT) [Crandall and Fagin 1994], [Crandall 1996a] for the reason that a special digit representation reminiscent of irrational-base expansion is used, which representation amounts to a discrete rendition of an attempt to expand in an irrational base. Let $p = 2^q - 1$ and observe first that if an integer x be represented in base $B = 2$ as

$$x = \sum_{j=0}^{q-1} x_j 2^j,$$

equivalently, x is the length-q signal (x_j); and similarly for an integer y, then the cyclic convolution $x \times y$ has, without carry, the digits of $(xy) \bmod p$. Thus, in principle, the standard FFT multiply could be effected in this way, modulo Mersenne primes, without zero-padding. However, there are two problems with this approach. First, the arithmetic is merely bitwise, not exploiting typical machine advantages of word arithmetic. Second, one would have to invoke a length-q FFT. This can certainly be done (see Exercises), but power-of-two lengths are usually more efficient, definitely more prevalent.

It turns out that both of the obstacles to a not-zero-padded Mersenne multiply-mod can be overcome, if only we could somehow represent integers x in the *irrational* base $B = 2^{q/D}$, with $1 < D < q$ being some power of two. This is because the representation

$$x = \sum_{j=0}^{D-1} x_j 2^{qj/D},$$

and similarly for y (where the digits in base B are generally irrational also), leads to the equivalence, without carry, of $(xy) \bmod p$ and $x \times y$. But now the signal lengths are powers of two, and the digits, although not integers, are some convenient word size. It turns out to be possible to mimic this irrational base expansion, by using a certain variable-base representation according to the following theorem:

Theorem 9.5.17 (Crandall). *For $p = 2^q - 1$ (p not necessarily prime) and integers $0 \le x, y < p$, choose signal length $1 < D < q$. Interpret x as the signal $(x_0, \dots, x_{D-1})$ from the variable-base representation*

$$x = \sum_{j=0}^{D-1} x_j 2^{\lceil qj/D \rceil} = \sum_{j=0}^{D-1} x_j 2^{\sum_{i=1}^{j} d_i},$$

where

$$d_i = \lceil qi/D \rceil - \lceil q(i-1)/D \rceil,$$

and each digit x_j is in the interval $[0, 2^{d_{j+1}} - 1]$, and all of this similarly for y. Define a length-D weight signal a by

$$a_j = 2^{\lceil qj/D \rceil - qj/D}.$$

Then the weighted cyclic convolution $x \times_a y$ is a signal of integers, equivalent without carry to the variable base representation of $(xy) \bmod p$.

This theorem is proved and discussed in [Crandall and Fagin 1994], [Crandall 1996a], the only nontrivial part being the proof that the elements of the weighted convolution $x \times_a y$ are actually integers. The theorem leads immediately to

Algorithm 9.5.18 (IBDWT multiplication modulo Mersenne numbers). For a given Mersenne number $p = 2^q - 1$ (need not be prime), and positive integers x, y, this algorithm returns—via floating-point FFT—the variable-base representation of $(xy) \bmod p$. Herein we adopt the nomenclature of Theorem 9.5.17, and assume a signal length $D = 2^k$ such that $\lfloor 2^{q/D} \rfloor$ is an acceptable word size (small enough that we avoid unacceptable numerical error).

1. [Initialize base representations]
   Create the signal x as the collection of variable-base digits (x_j), as in Theorem 9.5.17, and do the same for y;
   Create the weight signal a, also as in Theorem 9.5.17;
2. [Apply transforms]
   $X = DWT(x, a)$; // Perform via floating-point length-D FFT algorithm.
   $Y = DWT(y, a)$;
3. [Dyadic product]
   $Z = X * Y$;
4. [Inverse transform]
   $z = DWT^{-1}(Z, a)$;
5. [Round digits]
   $z = round(z)$; // Elementwise rounding to nearest integer.
6. [Adjust carry in variable base]
   $carry = 0$;
   for($0 \le n < len(z)$) {
      $B = 2^{d_{n+1}}$; // Size of place-n digits.
      $v = z_n + carry$;
      $z_n = v \bmod B$;
      $carry = \lfloor v/B \rfloor$;
   }
7. [Final modular adjustment]
   Include possible $carry > 0$ as a high digit of z;
   $z = z \bmod p$; // Via carry loop or special-form mod.
   return z;

As this scheme is somewhat intricate, an example is appropriate. Consider multiplication modulo the Mersenne number $p = 2^{521} - 1$. We take $q = 521$ and choose signal length $D = 16$. Then the signal d of Theorem 9.5.17 can be seen to be

$$d = (33, 33, 32, 33, 32, 33, 32, 33, 33, 32, 33, 32, 33, 32, 33, 32),$$

and the weight signal will be

$$a = \left(1, 2^{7/16}, 2^{7/8}, 2^{5/16}, 2^{3/4}, 2^{3/16}, 2^{5/8}, 2^{1/16}, 2^{1/2}, 2^{15/16}, 2^{3/8}, 2^{13/16}, 2^{1/4}, 2^{11/16}, 2^{1/8}, 2^{9/16}\right).$$

In a typical floating-point FFT implementation, this a signal is, of course, given inexact elements. But in Theorem 9.5.17 the weighted convolution (as calculated approximately, just prior to the [Round digits] step of Algorithm 9.5.18) consists of exact integers. Thus, the game to be played is to choose signal length D to be as small as possible (the smaller, the faster the FFTs that do the DWT), while not allowing the rounding errors to give incorrect elements of z. Rigorous theorems on rounding error are hard to come by, although there are some observations—some rigorous and some not so—in [Crandall and Fagin 1994] and references therein. More modern treatments include the very useful book [Higham 1996] and the work of [Percival 2000] on generalized IBDWT; see Exercise 9.49.

9.5.5 Number-theoretical transform methods

The DFT of Definition 9.5.3 can be defined over rings and fields other than the traditional complex field. Here we give some examples of transforms over finite rings and fields. The primary observation is that over a ring or field, the DFT defining relations (9.20) and (9.21) need no modification whatever, as long as we understand the requisite operations to occur (legally) in the algebraic domain at hand. In particular, a number-theoretical DFT of length D supports cyclic convolution of length D, via the celebrated convolution Theorem 9.5.10, whenever both D^{-1} and g, a primitive D-th root of unity, exist in the algebraic domain. With these constraints in mind, number-theoretical transforms have attained a solid niche, in regard to fast algorithms in the field of digital signal processing. Not just raw convolution, but other interesting applications of such transforms can be found in the literature. A typical example is the use of number-theoretical transforms for classical algebraic operations [Yagle 1995], while yet more applications are summarized in [Madisetti and Williams 1997].

Our first example will be the case that the relevant domain is $\mathbf{F}_p$. For a prime p and some divisor $d|p-1$ let the field be $\mathbf{F}_p$ and consider the relevant transform to be

$$X_k = \sum_{j=0}^{(p-1)/d-1} x_j h^{-jk} \bmod p, \tag{9.30}$$

where h is an element of multiplicative order $(p-1)/d$ in $\mathbf{F}_p$. Note that the mod operation can in principle be taken either after individual summands, or for the whole sum, or in some combination of these, so that for convenience we simply append the symbols "mod p" to indicate that a transform element X_k is to be reduced to lie in the interval $[0, p-1]$. Now the inverse transform is

$$x_j = -d \sum_{k=0}^{(p-1)/d-1} X_k h^{jk} \bmod p, \tag{9.31}$$

whose prefactor is just $((p-1)/d)^{-1} \bmod p \equiv -d$. These transforms can be used to provide increased precision for convolutions. The idea is to establish each convolution element (mod p_r) for some convenient set of primes $\{p_r\}$, whence the exact convolution can be reconstructed using the Chinese remainder theorem.

Algorithm 9.5.19 (Integer convolution on a CRT prime set). Given two signals x, y each of length $N = 2^m$ having integer elements bounded by $0 \le x_j, y_j < M$, this algorithm returns the cyclic convolution $x \times y$ via the CRT with distinct prime moduli $p_1, p_2, \ldots, p_q$.

1. [Initialize]
   Find a set of primes of the form $p_r = a_r N + 1$ for $r = 1, \ldots, q$ such that $\prod p_r > NM^2$;
   for($1 \le r \le q$) {
      Find a primitive root g_r of p_r;
      $h_r = g_r^{a_r} \bmod p_r$; // h_r is an N-th root of 1.
   }
2. [Loop over primes]
   for($1 \le r \le q$) {
      $h = h_r$; $p = p_r$; $d = a_r$; // Preparing for DFTs.
      $X^{(r)} = DFT(x)$; // Via relation (9.30).
      $Y^{(r)} = DFT(y)$;
3. [Dyadic product]
      $Z^{(r)} = X^{(r)} * Y^{(r)}$;
4. [Inverse transforms]
      $z^{(r)} = DWT^{-1}(Z^{(r)})$; // Via relation (9.31).
   }
5. [Reconstruct elements]
   From the now known relations $z_j \equiv z_j^{(r)} \pmod{p_r}$ find each (unambiguous) element z_j in $[0, NM^2)$ via CRT reconstruction, using such as Algorithm 2.1.7 or 9.5.25;
   return z;

What this algorithm does is allow us to invoke length-2^m FFTs for the DFT and its inverse, except that only integer arithmetic is to be used in the usual FFT butterflies (and of course the butterflies are continually reduced $(\bmod\ p_r)$ during the FFT calculations). This scheme has been used to good effect by [Montgomery 1992a] in various factorization implementations. Note that if the forward DFT (9.30) is performed with a decimation-in-frequency (DIF) algorithm, and the reverse DFT (9.31) with a DIT algorithm, there is no need to invoke the *scramble* function of Algorithm 9.5.5 in either of the FFT functions shown there.

A second example of useful number-theoretical transforms has been called the discrete Galois transform (DGT) [Crandall 1996a], with relevant field $\mathbf{F}_{p^2}$ for $p = 2^q - 1$ a Mersenne prime. The delightful fact about such fields is that the multiplicative group order is

$$|\mathbf{F}_{p^2}^*| = p^2 - 1 = 2^{q+1}(2^{q-1} - 1),$$

so that in practice, one can find primitive roots of unity of orders $N = 2^k$ as long as $k \le q+1$. We can thus define discrete transforms of such lengths, as

$$X_k = \sum_{j=0}^{N-1} x_j h^{-jk} \bmod p, \tag{9.32}$$

where now all arithmetic is presumed, due to the known structure of $\mathbf{F}_{p^2}$ for primes $p \equiv 3 \pmod 4$, to involve complex (Gaussian) integers (mod p) with

$$\begin{aligned} N &= 2^k, \\ x_j &= \mathrm{Re}(x_j) + i\,\mathrm{Im}(x_j), \\ h &= \mathrm{Re}(h) + i\,\mathrm{Im}(h), \end{aligned}$$

the latter being an element of multiplicative order N in $\mathbf{F}_{p^2}$, with the transform element X_k itself being a Gaussian integer (mod p). Happily, there is a way to find immediately an element of suitable order, thanks to the following result of [Creutzburg and Tasche 1989]:

Theorem 9.5.20 (Creutzburg and Tasche). *Let $p = 2^q - 1$ be a Mersenne prime with q odd. Then*

$$g = 2^{2^{q-2}} + i(-3)^{2^{q-2}}$$

is an element of order 2^{q+1} in $\mathbf{F}^*_{p^2}$.

These observations lead to the following integer convolution algorithm, in which we indicate the enhancements that can be invoked to reduce the complex arithmetic. In particular, we exploit the fact that integer signals are real, so the imaginary components of their elements vanish in the field, and thus the transform lengths are halved:

Algorithm 9.5.21 (Convolution via DGT (Crandall)). Given two signals x, y each of length $N = 2^k \ge 2$ and whose elements are integers in the interval $[0, M]$, this algorithm returns the integer convolution $x \times y$. The method used is convolution via "discrete Galois transform" (DGT).

1. [Initialize]
   Choose a Mersenne prime $p = 2^q - 1$ such that $p > NM^2$ and $q > k$;
   Use Theorem 9.5.20 to find an element g of order 2^{q+1};
   $h = g^{2^{q+2-k}}$; // h is now an element of order $N/2$.
2. [Fold signals to halve their lengths]
   $x = \left(x_{2j} + i x_{2j+1}\right)$, $j = 0, \ldots, N/2 - 1$;
   $y = \left(y_{2j} + i y_{2j+1}\right)$, $j = 0, \ldots, N/2 - 1$;
3. [Length-$N/2$ transforms]
   $X = DFT(x)$; // Via, say, split-radix FFT $\pmod p$, root h.
   $Y = DFT(y)$;
4. [Special dyadic product]

for($0 \le k < N/2$) {
$Z_k = (X_k + X_{-k}^*)(Y_k + Y_{-k}^*) + 2(X_k Y_k - X_{-k}^* Y_{-k}^*) - h^{-k}(X_k - X_{-k}^*)(Y_k - Y_{-k}^*)$;
}

5. [Inverse length-$N/2$ transform]
$z = \frac{1}{4} DFT^{-1}(Z)$; // Via split-radix FFT (mod p) with root h.
6. [Unfold signal to double its length]
$z = \big((\mathrm{Re}(z_j), \mathrm{Im}(z_j))\big)$, $j = 0, \ldots, N/2 - 1$;
return z;

To implement this algorithm, one needs only a complex (integer only!) FFT (mod p), complex multiplication (mod p), and a binary ladder for powering in the field. The split-radix FFT indicated in the algorithm, though it is normally used in reference to standard floating-point FFT's, can nevertheless be used because "i" is defined [Crandall 1997b].

There is one more important aspect of the DGT convolution: All mod operations are with respect to Mersenne primes, and so an implementation can enjoy the considerable speed advantage we have previously encountered for such special cases of the modulus.

9.5.6 Schönhage method

The pioneering work of [Schönhage and Strassen 1971], [Schönhage 1982], based on Strassen's ideas for using FFTs in large-integer multiplication, focuses on the fact that a certain number-theoretical transform is possible—using exclusively integer arithmetic—in the ring $\mathbf{Z}_{2^m+1}$. This is sometimes called a Fermat number transform (FNT) (see Exercise 9.53) and can be used within a certain negacyclic convolution approach as follows (we are grateful to P. Zimmermann for providing a clear exposition of the method, from which description we adapted our rendition here):

Algorithm 9.5.22. [Fast multiplication $\pmod{2^n+1}$ (Schönhage)] Given two integers $0 \le x, y < 2^n + 1$, this algorithm returns the product $xy \bmod (2^n + 1)$.

1. [Initialize]
Choose FFT size $D = 2^k$ dividing n;
Writing $n = DM$, set a recursion length $n' \ge 2M + k$ such that D divides n', i.e., $n' = DM'$;
2. [Decomposition]
Split x and y into D parts of M bits each, and store these parts, considered as residues modulo $(2^{n'}+1)$, in two respective arrays $A_0, \ldots, A_{D-1}$ and $B_0, \ldots, B_{D-1}$, taking care that an array element could in principle have $n' + 1$ bits later on;
3. [Prepare DWT by weighting the A, B signals]
for($0 \le j < D$) {

```
        A_j = (2^{jM'} A_j) mod (2^{n'} + 1);
        B_j = (2^{jM'} B_j) mod (2^{n'} + 1);
    }
4. [In-place, symbolic FFTs]
    A = DFT(A);                  // Use 2^{2M'} as D-th root mod(2^{n'} + 1).
    B = DFT(B);
5. [Dyadic stage]
    for(0 ≤ j < D) A_j = A_j B_j mod (2^{n'} + 1);
6. [Inverse FFT]
    A = DFT(A);                  // Inverse via index reversal, next loop.
7. [Normalization]
    for(0 ≤ j < D) {                                  // A_D defined as A_0.
        C_j = A_{D-j}/2^{k+jM'} mod (2^{n'} + 1);     // Reverse and twist.
8. [Adjust signs]
        if(C_j > (j + 1)2^{2M}) C_j = C_j - (2^{n'} + 1);
    }                                                 // C_j now possibly negative.
9. [Composition]
    Perform carry operations as in steps [Adjust carry in base B] for B = 2^M
        (the original decomposition base) and [Final modular adjustment] of
        Algorithm 9.5.16 to return the desired sum:
        xy mod (2^n + 1) = \sum_{j=0}^{D-1} C_j 2^{jM} mod (2^n + 1);
```

Note that in the [Decomposition] step, A_{D-1} or B_{D-1} may equal 2^M and have $M + 1$ bits in the case where x or y equal 2^n. In step [Prepare DWT ...], each multiply can be done using shifts and subtractions only, as $2^{n'} \equiv -1 (\bmod 2^{n'} + 1)$. In step [Dyadic stage], one can use any multiplication algorithm, for example a grammar-school stage, Karatsuba algorithm, or this very Schönhage algorithm recursively. In step [Normalization], the divisions by a power of two again can be done using shifts and subtractions only. Thus the only multiplication *per se* is in step [Dyadic stage], and this is why the method can attain, in principle, such low complexity. Note also that the two FFTs required for the negacyclic result signal C can be performed in the order DIF, DIT, for example by using parts of Algorithm 9.5.5 in proper order, thus obviating the need for any bit-scrambling procedure.

As it stands, Algorithm 9.5.22 will multiply two integers modulo any Fermat number, and such application is an important one, as explained in other sections of this book. For *general* multiplication of two integers x and y, one may call the Schönhage algorithm with $n \ge \lceil \lg x \rceil + \lceil \lg y \rceil$, and zero-padding x, y accordingly, whence the product $xy \bmod 2^n + 1$ equals the integer product. (In a word, the negacyclic convolution of appropriately zero-padded sequences is the acyclic convolution—the product in essence.) In practice, Schönhage suggests using what he calls "suitable numbers," i.e., $n = \nu 2^k$ with $k - 1 \le \nu \le 2k - 1$. For example, $688128 = 21 \cdot 2^{15}$ is a suitable number. Such numbers enjoy the property that if $k = \lceil n/2 \rceil + 1$, then $n' = \lceil \frac{\nu+1}{2} \rceil 2^k$

is also a suitable number; here we get indeed $n' = 11 \cdot 2^8 = 2816$. Of course, one loses a factor of two initially with respect to modular multiplication, but in the recursive calls all computations are performed modulo some $2^M + 1$, so the asymptotic complexity is still that reported in Section 9.5.8.

9.5.7 Nussbaumer method

It is an important observation that a cyclic convolution of some even length D can be cast in terms of a pair of convolutions, a cyclic and a negacyclic, each of length D. The relevant identity is

$$2(x \times y) = [(u_+ \times v_+) + (u_- \times_- v_-)] \cup [(u_+ \times v_+) - (u_- \times_- v_-)], \quad (9.33)$$

where u, v signals depend in turn on half-signals:

$$u_\pm = L(x) \pm H(x),$$
$$v_\pm = L(y) \pm H(y).$$

This recursion formula for cyclic convolution can be proved via polynomial algebra (see Exercise 9.43). The recursion relation together with some astute algebraic observations led [Nussbaumer 1981] to an efficient convolution scheme devoid of floating-point transforms. The algorithm is thus devoid of rounding-error problems, and often, therefore, is the method of choice for rigorous machine proofs involving large-integer arithmetic.

Looking longingly at the previous recursion, it is clear that if only we had a fast negacyclic algorithm, then a cyclic convolution could be done directly, much like that which an FFT performs via decimation of signal lengths. To this end, let R denote a ring in which 2 is cancelable; i.e., $x = y$ whenever $2x = 2y$. (It is intriguing that this is all that is required to "ignite" Nussbaumer convolution.) Assume a length $D = 2^k$ for negacyclic convolution, and that D factors as $D = mr$, with $m|r$. Now, negacyclic convolution is equivalent to polynomial multiplication (mod $t^D + 1$) (see Exercises), and as an operation can in a certain sense be "factored" as specified in the following:

Theorem 9.5.23 (Nussbaumer). *Let $D = 2^k = mr$, $m|r$. Then negacyclic convolution of length-D signals whose elements belong to a ring R is equivalent, in the sense that polynomial coefficients correspond to signal elements, to multiplication in the polynomial ring*

$$S = R[t]/\left(t^D + 1\right).$$

Furthermore, this ring is isomorphic to

$$T[t]/\left(z - t^m\right),$$

where T is the polynomial ring

$$T = R[z]/\left(z^r + 1\right).$$

Finally, $z^{r/m}$ is an m-th root of -1 in T.

Nussbaumer's idea is to use the root of -1 in a manner reminiscent of our DWT, to perform a negacyclic convolution.

Let us exhibit explicit polynomial manipulations to clarify the import of Theorem 9.5.23. Let

$$x(t) = x_0 + x_1 t + \cdots + x_{D-1} t^{D-1},$$

and similarly for signal y, with the x_j, y_j in R. Note that $x \times_- y$ is equivalent to multiplication $x(t)y(t)$ in the ring S. Now decompose

$$x(t) = \sum_{j=0}^{m-1} X_j(t^m) t^j,$$

and similarly for $y(t)$, and interpret each of the *polynomials* X_j, Y_j as an element of ring T; thus

$$X_j(z) = x_j + x_{j+m} z + \cdots + x_{j+m(r-1)} z^{r-1},$$

and similarly for the Y_j. It is evident that the (total of) $2m$ X, Y polynomials can be stored in two arrays that are (r, m)-transpositions of x, y arrays respectively. Next we multiply $x(t)y(t)$ by performing the cyclic convolution

$$Z = (X_0, X_1, \ldots, X_{m-1}, 0, \ldots, 0) \times (Y_0, Y_1, \ldots, Y_{m-1}, 0, \ldots, 0),$$

where each operand signal here has been zero-padded to total length $2m$. The key point here is that Z can be evaluated by a *symbolic* DFT, using what we know to be a primitive $2m$-th root of unity, namely $z^{r/m}$. What this means is that the usual FFT butterfly operations now involve mere shuttling around of polynomials, because multiplications by powers of the primitive root just translate coefficient polynomials. In other words the polynomial arithmetic now proceeds along the lines of Theorem 9.2.12, in that multiplication by a power of the relevant root is equivalent to a kind of shift operation.

At a key juncture of the usual DFT-based convolution method, namely the dyadic (elementwise) multiply step, the dyadic operations can be seen to be themselves length-r negacyclic convolutions. This is evident on the observation that each of the polynomials X_j, Y_j has degree $(r-1)$ in the variable $z = t^m$, and so $z^r = t^D = -1$. To complete the Z convolution, a final, inverse DFT, with root $z^{-r/m}$, is to be used. The result of this zero-padded convolution is seen to be a product in the ring S:

$$x(t)y(t) = \sum_{j=0}^{2m-2} Z_j(t^m) t^j, \tag{9.34}$$

from which we extract the negacyclic elements of $x \times_- y$ as the coefficients of the powers of t.

Algorithm 9.5.24 (Nussbaumer convolution, cyclic and negacyclic). Assume length-($D = 2^k$) signals x, y whose elements belong to a ring R, which ring also admits of cancellation-by-2. This algorithm returns either the cyclic ($x \times y$) or negacyclic ($x \times_- y$) convolution. Inside the negacyclic function neg is a "small" negacyclic routine $smallneg$, for example a grammar-school or Karatsuba version, which is called below a certain length threshold.

1. [Initialize]
   $r = 2^{\lceil k/2 \rceil}$;
   $m = D/r$; // Now m divides r.
   $blen = 16$; // Tune this small-negacyclic breakover length to taste.
2. [Cyclic convolution function cyc, recursive]
   $cyc(x, y)$ {
   By calling half-length cyclic and negacyclic convolutions, return the desired cyclic, via identity (9.33);
   }
3. [Negacyclic convolution function neg, recursive]
   $neg(x, y)$ {
   if($len(x) \le blen$) return $smallneg(x, y)$;
4. [Transposition step]
   Create a total of $2m$ arrays X_j, Y_j each of length r;
   Zero-pad the X, Y collections so each collection has $2m$ polynomials;
   Using root $g = z^{r/m}$, perform (symbolically) two length-$2m$ DFTs to get the transforms $\hat{X}, \hat{Y}$;
5. [Recursive dyadic operation]
   for($0 \le h < 2m$) $\hat{Z}_h = neg(\hat{X}_h, \hat{Y}_h)$;
6. [Inverse transform]
   Using root $g = z^{-r/m}$, perform (symbolically) a length-($2m$) inverse DWT on $\hat{Z}$ to get Z;
7. [Untranspose and adjust]
   Working in the ring S (i.e., reduce polynomials according to $t^D = -1$) find the coefficients z_n of t^n, $n \in [0, D-1]$, from equation (9.34);
   return (z_n); // Return the negacyclic of x, y.
   }

Detailed implementation of Nussbaumer's remarkable algorithm can be found in [Crandall 1996a], where enhancements are discussed. One such enhancement is to obviate the zero-padding of the X, Y collections (see Exercise 9.66). Another is to recognize that the very formation of the X_j, Y_j amounts to a transposition of a two-dimensional array, and memory can be reduced significantly by effective such transposition "in place." [Knuth 1981] has algorithms for in-place transposition. Also of interest is the algorithm [Fraser 1976] mentioned in connection with Algorithm 9.5.7.

9.5.8 Complexity of multiplication algorithms

In order to summarize the complexities of the aforementioned fast multiplication methods, let us clarify the nomenclature. In general, we speak of operands (say x, y) to be multiplied, of size $N = 2^n$, or n binary bits, or D digits, all equivalently in what follows. Thus for example, if the digits are in base $B = 2^b$, we have

$$Db \approx n$$

signifying that the n bits of an operand are split into D signal elements. This symbolism is useful because we need to distinguish between bit- and operation-complexity bounds.

Recall that the complexity of grammar-school, Karatsuba, and Toom–Cook multiplication schemes all have the form $O(n^\alpha) = O(\ln^\alpha N)$ *bit* operations for all the involved multiplications. (We state things this way because in the Toom–Cook case one must take care to count bit operations due to the possibly significant addition count.) So for example, $\alpha = 2$ for grammar-school methods, Karatsuba and Toom–Cook methods lower this α somewhat, and so on.

Then we have the basic Schönhage–Strassen FFT multiplication Algorithm 9.5.11. Suddenly, the natural description has a different flavor, for we know that the complexity must be

$$O(D \ln D)$$

operations, and as we have said, these are usually, in practice, floating-point operations (both adds and multiplies are bounded in this fashion). Now the *bit* complexity is not $O((n/b)\ln(n/b))$—that is, we cannot just substitute $D = n/b$ in the operation-complexity estimate—because floating-point arithmetic on larger digits must, of course, be more expensive. When these notions are properly analyzed we obtain the Strassen bound of

$$O(n(C\ln n)(C\ln\ln n)(C\ln\ln\ln n)\cdots)$$

bit operations for the basic FFT multiply, where C is a constant and the $\ln\ln\cdots$ chain is understood to terminate when it falls below 1. Before we move ahead with other estimates, we must point out that even though this bit complexity is not asymptotically optimal, some of the greatest achievements in the general domain of large-integer arithmetic have been achieved with this basic Schönhage–Strassen FFT, and yes, using floating-point operations.

Now, the Schönhage Algorithm 9.5.22 gets neatly around the problem that for a fixed number of signal digits D, the digit operations (small multiplications) must get more complex for larger operands. Analysis of the recursion within the algorithm starts with the observation that at top recursion level, there are two DFTs (but very simple ones—only shifting and adding occur) and the dyadic multiply. Detailed analysis yields the best-known complexity bound of

$$O(n(\ln n)(\ln\ln n))$$

Algorithm	optimal B	complexity
Basic FFT, fixed-base	...	$O_{\rm op}(D \ln D)$
Basic FFT, variable-base	$O(\ln n)$	$O(n(C \ln n)(C \ln \ln n) \ldots)$
Schönhage	$O(n^{1/2})$	$O(n \ln n \ln \ln n)$
Nussbaumer	$O(n/\ln n)$	$O(n \ln n \ln \ln n)$

Table 9.1 Complexities for fast multiplication algorithms. Operands to be multiplied have n bits each, which during top recursion level are split into $D = n/b$ digits of b bits each, so the digit size (the base) is $B = 2^b$. All bounds are for bit complexity, except that $O_{\rm op}$ means operation complexity.

bit operations, although the Nussbaumer method's complexity, which we discuss next, is asymptotically equivalent.

Next, one can see that (as seen in Exercise 9.67) the complexity of Nussbaumer convolution is

$$O(D \ln D)$$

operations in the R ring. This is equivalent to the complexity of floating-point FFT methods, if ring operations are thought of as equivalent to floating-point operations. However, with the Nussbaumer method there is a difference: One may choose the digit base B with impunity. Consider a base $B \sim n$, so that $b \sim \ln n$, in which case one is effectively using $D = n/\ln n$ digits. It turns out that the Nussbaumer method for integer multiplication then takes $O(n \ln \ln n)$ additions and $O(n)$ multiplications of numbers each having $O(\ln n)$ bits. It follows that the complexity of the Nussbaumer method is asymptotically that of the Schönhage method, i.e., $O(n \ln n \ln \ln n)$ bit operations. Such complexity issues for both Nussbaumer and the original Schönhage–Strassen algorithm are discussed in [Bernstein 1997a].

9.5.9 Application to the Chinese remainder theorem

We described the Chinese remainder theorem in Section 2.1.3, and there gave a method, Algorithm 2.1.7, for reassembling CRT data given some precomputation. We now describe a method that not only takes advantage of preconditioning, but also fast multiplication methods.

Algorithm 9.5.25 (Fast CRT reconstruction with preconditioning). Using the nomenclature of Theorem 2.1.6, we assume fixed moduli $m_0, \ldots, m_{r-1}$ whose product is M, but with $r = 2^k$ for computational convenience. The goal of the algorithm is to reconstruct n from its given residues (n_i). Along the way, tableaux (q_{ij}) of partial products and (n_{ij}) of partial residues are calculated. The algorithm may be reentered with a new n if the m_i remain fixed.

1. [Precomputation]
```
    for(0 ≤ i < r) {                    // Generate the M_i and inverses.
        M_i = M/m_i;
        v_i = M_i^{-1} mod m_i;
```

```
    }
    for(0 ≤ j ≤ k) {                              // Generate partial products.
        for(0 ≤ i ≤ r − 2^j) q_ij = ∏_{a=i}^{i+2^j−1} m_a;
    }
2. [Reentry point for given input residues (n_i)]
    for(0 ≤ i < r) n_i0 = v_i n_i;
3. [Reconstruction loop]
    for(1 ≤ j ≤ k) {
        for(i = 0; i < r; i = i+2^j) n_ij = n_{i,j−1} q_{i+2^{j−1},j−1} + n_{i+2^{j−1},j−1} q_{i,j−1};
    }
4. [Return the unique n in [0, M − 1]]
    return n_0k mod q_0k;
```

Note that the first, precomputation, phase of the algorithm can be done just once, with a particular input of residues (n_i) used for the first time at the initialization phase. Note also that the precomputation of the (q_{ij}) can itself be performed with a fast divide-and-conquer algorithm of the type discussed in Chapter 9 (for example, Exercise 9.74). As an example of the operation of Algorithm 9.5.25, let us take $r = 8 = 2^3$ and eight moduli: $(m_1, \ldots, m_8) = (3, 5, 7, 11, 13, 17, 19, 23)$. Then we use these moduli along with the product $M = \prod_{i=1}^{8} m_i = 111546435$, to obtain at the [Precomputation] phase $M_1, \ldots, M_8$, which are, respectively,

$$37182145, 22309287, 15935205, 10140585, 8580495, 6561555, 5870865, 4849845,$$

$$(v_1, \ldots, v_8) = (1, 3, 6, 3, 1, 11, 9, 17),$$

and the tableau

$$\begin{aligned}(q_{00}, \ldots, q_{70}) &= (3, 5, 7, 11, 13, 17, 19, 23),\\ (q_{01}, \ldots, q_{61}) &= (15, 35, 77, 143, 221, 323, 437),\\ (q_{02}, \ldots, q_{42}) &= (1155, 5005, 17017, 46189, 96577),\\ q_{03} &= 111546435,\end{aligned}$$

where we recall that for fixed j there exist q_{ij} for $i \in [0, r - 2^j]$. It is important to note that all of the computation up through the establishment of the q tableau can be done just once—as long as the CRT moduli m_i are not going to change in future runs of the algorithm. Now, when specific residues n_i of some mystery n are to be processed, let us say

$$(n_1, \ldots, n_8) = (1, 1, 1, 1, 3, 3, 3, 3),$$

we have after the [Reconstruction loop] step, the value

$$n_{0k} = 878271241,$$

which when reduced mod q_{03} is the correct result $n = 97446196$. Indeed, a quick check shows that

$$97446196 \bmod (3, 5, 7, 11, 13, 17, 19, 23) = (1, 1, 1, 1, 3, 3, 3, 3).$$

The computational complexity of Algorithm 9.5.25 is known in the following form [Aho et al. 1974, pp. 294–298], assuming that fast multiplication is used. If each of the r moduli m_i has b bits, then the complexity is

$$O(br \ln r \ln(br) \ln \ln(br))$$

bit operations, on the assumption that all of the precomputation for the algorithm is in hand.

9.6 Polynomial arithmetic

It is an important observation that polynomial multiplication/division is not quite the same as large-integer multiplication/division. However, ideas discussed in the previous sections can be applied, in a somewhat different manner, in the domain of arithmetic of univariate polynomials.

9.6.1 Polynomial multiplication

We have seen that polynomial multiplication is equivalent to acyclic convolution. Therefore, the product of two polynomials can be effected via a cyclic and a negacyclic. One simply constructs respective signals having the polynomial coefficients, and invokes Theorem 9.5.9. An alternative is simply to zero-pad the signals to twice their lengths and perform a single cyclic (or single negacyclic).

But there exist interesting—and often quite efficient—means of multiplying polynomials, if one has a general *integer* multiply algorithm. The method amounts to placing polynomial coefficients strategically within certain large integers, and doing all the arithmetic with one high-precision integer multiply. We give the algorithm for the case that all polynomial coefficients are nonnegative, although this constraint is irrelevant for multiplication in polynomial rings (mod p):

Algorithm 9.6.1 (Fast polynomial multiplication: Binary segmentation). Given two polynomials $x(t) = \sum_{j=0}^{D-1} x_j t^j$ and $y(t) = \sum_{k=0}^{E-1} y_k t^k$ with all coefficients integral and nonnegative, this algorithm returns the polynomial product $z(t) = x(t)y(t)$ in the form of a signal having the coefficients of z.

1. [Initialize]
   Choose b such that $2^b > \max\{D, E\} \ \max\{x_j\} \ \max\{y_k\}$;
2. [Create binary segmentation integers]
   $X = x\left(2^b\right)$;
   $Y = y\left(2^b\right)$;

```
        // These X, Y can be constructed by arranging a binary array of
           sufficiently many 0's, then writing in the bits of each coefficient,
           justified appropriately.
3. [Multiply]
     u = XY;                                          // Integer multiplication.
4. [Reassemble coefficients into signal]
     for(0 ≤ l < D + E − 1) {
         z_l = ⌊u/2^{bl}⌋ mod 2^b;                     // Extract next b bits.
     }
     return z = Σ_{l=0}^{D−E−2} z_l t^l;   // Base-b digits of u are desired coefficients.
```

The method is a good one in the sense that if a large-integer multiply is at hand, there is not very much extra work required to establish a polynomial multiply. It is not hard to show that the bit-complexity of multiplying two degree-D polynomials in $\mathbf{F}_p[X]$, that is, all coefficients are reduced modulo p, is

$$O\left(M\left(D\ln\left(Dp^2\right)\right)\right),$$

where $M(n)$ is as elsewhere the bit-complexity for multiplying two integers of n bits each. Incidentally, if polynomial multiplication in such fields be done via the Nussbaumer Algorithm 9.5.24, one obtains a different bound: $O(M(p)D\ln D)$ bit operations. It is interesting to compare these various estimates for polynomial multiplication (see Exercise 9.70).

9.6.2 Fast polynomial inversion and remaindering

Let $x(t) = \sum_{j=0}^{D-1} x_j t^j$ be a polynomial. If $x_0 \neq 0$, there is a formal inversion

$$1/x(t) = 1/x_0 - (x_1/x_0^2)t + (x_1^2/x_0^3 - x_2/x_0^2)t^2 + \cdots$$

that admits of rapid evaluation, by way of a scheme we have already invoked for reciprocation, the celebrated Newton method. We describe the scheme in the case that $x_0 = 1$, from which case generalizations are easily inferred. In what follows, the notation

$$z(t) \bmod t^k$$

is a polynomial remainder (which we cover later), but in this setting it is simple truncation: The result of the mod operation is a polynomial consisting of the terms of polynomial $z(t)$ through order t^{k-1} inclusive. Let us define, then, a truncated reciprocal,

$$R[x, N] = x(t)^{-1} \bmod t^{N+1}$$

as the series of $1/x(t)$ through degree t^N, inclusive.

Algorithm 9.6.2 (Fast polynomial inversion). Let $x(t)$ be a polynomial with first coefficient $x_0 = 1$. This algorithm returns the truncated reciprocal $R[x, N]$ through a desired degree N.

```
1. [Initialize]
     g(t) = 1;                                     // Degree-zero polynomial.
     n = 1;                                        // Working degree precision.
2. [Newton loop]
     while(n < N + 1) {
          n = 2n;                          // Double the working degree precision.
          if(n > N + 1) n = N + 1;
          h(t) = x(t) mod t^n;                              // Simple truncation.
          h(t) = h(t)g(t) mod t^n;
          g(t) = g(t)(2 - h(t)) mod t^n;                    // Newton iteration.
     }
     return g(t);
```

One point that should be stressed right off is that in principle, an operation $f(t)g(t) \bmod t^n$ is simple truncation of a product (the operands usually themselves being approximately of degree n). This means that within multiplication loops, one need not handle terms of degree higher than indicated. In convolution-theory language, we are therefore doing "half-cyclic" convolutions, so when transform methods are used, there is also gain to be realized because of the truncation.

As is typical of Newton methods, the dynamical precision degree n essentially doubles on each pass of the Newton loop. Let us give an example of the workings of the algorithm. Take

$$x(t) = 1 + t + t^2 + 4t^3$$

and call the algorithm to output $R[x, 8]$. Then the values of $g(t)$ at the end of each pass of the Newton loop come out as

$$\begin{aligned}
&1 - t, \\
&1 - t - 3t^3, \\
&1 - t - 3t^3 + 7t^4 - 4t^5 + 9t^6 - 33t^7, \\
&1 - t - 3t^3 + 7t^4 - 4t^5 + 9t^6 - 33t^7 + 40t^8,
\end{aligned}$$

and indeed, this last output of $g(t)$ multiplied by the original $x(t)$ is $1 + 43t^9 - 92t^{10} + 160t^{11}$, showing that the last output $g(t)$ is correct through $O(t^8)$.

Polynomial remaindering (polynomial mod operation) can be performed in much the same way as some of our mod algorithms for integers used a "reciprocal." However, it is not always possible to divide one polynomial by another and get a unique and legitimate remainder: This can depend on the ring of coefficients for the polynomials. However, if the divisor polynomial has its high coefficient invertible in the ring, then there is no problem with divide and remainder; see the discussion in Section 2.2.1. For simplicity, we shall restrict to the case that the divisor polynomial is monic, that is, the high coefficient is 1, since generalizing is straightforward. Assume $x(t), y(t)$

are polynomials and that $y(t)$ is monic. Then there are unique polynomials $q(t), r(t)$ such that

$$x(t) = q(t)y(t) + r(t),$$

and $r = 0$ or $\deg(r) < \deg(x)$. We shall write

$$r(t) = x(t) \bmod y(t),$$

and view $q(t)$ as the quotient and $r(t)$ as the remainder. Incidentally, for some polynomial operations one demands that coefficients lie in a field, for example in the evaluation of polynomial gcd's, but many polynomial operations do not require field coefficients. Before exhibiting a fast polynomial remaindering algorithm, we establish some nomenclature:

Definition 9.6.3 (Polynomial operations). Let $x(t) = \sum_{j=0}^{D-1} x_j t^j$ be a polynomial. We define the reversal of x by degree d as the polynomial

$$rev(x, d) = \sum_{j=0}^{d} x_{d-j} t^j,$$

where it is understood that $x_j = 0$ for all $j > D - 1$. We also define a polynomial index function as

$$ind(x, d) = \min\{j \,:\, j \ge d;\ x_j \ne 0\},$$

or $ind(x, d) = 0$ if the stated set of j is empty.

For example,

$$rev(1 + 3t^2 + 6t^3 + 9t^5 + t^6, 3) = 6 + 3t + t^3,$$
$$ind(1 + 3t^2 + 6t^3, 1) = 2.$$

A remaindering algorithm can now be given:

Algorithm 9.6.4 (Fast polynomial mod). Let $x(t), y(t)$ be given polynomials with $y(t)$ monic (high coefficient is 1). This algorithm returns the polynomial remainder $x(t) \bmod y(t)$.

1. [Initialize]
   if($\deg(y) == 0$) return 0;
   $d = \deg(x) - \deg(y)$;
   if($d < 0$) return x;
2. [Reversals]
   $X = rev(x, \deg(x))$;
   $Y = rev(y, \deg(y))$;
3. [Reciprocation]
   $q = R[Y, d]$; // Via Algorithm 9.6.2.
4. [Multiplication and reduction]

```
    q = (qX) mod t^{d+1};            // Multiply and truncate after degree d.
    r = X - qY;
    i = ind(r, d+1);
    r = r/t^i;
    return rev(r, deg(x) - i);
```

The proof that this algorithm works is somewhat intricate, but it is clear that the basic idea of the Barrett integer mod is at work here; the calculation $r = X - qY$ is similar to the manipulations done with generalized integer reciprocals in the Barrett method.

The challenge of fast polynomial gcd operations is an interesting one. There is a direct analogue to the Euclid integer gcd algorithm, namely, Algorithm 2.2.2. Furthermore, the complicated recursive Algorithm 9.4.6 is, perhaps surprisingly, actually simpler for polynomials than for integers [Aho et al. 1974, pp. 300–310]. We should point out also that some authors attribute the recursive idea, originally for polynomial gcd's, to [Moenck 1973]. Whatever method used for polynomial gcd, the fast polynomial remaindering scheme of this section can be applied as desired for the internal polynomial mod operations.

9.6.3 Polynomial evaluation

We close this chapter with a brief discussion of polynomial evaluation techniques. The essential problem is to evaluate a polynomial $x(t) = \sum_{j=0}^{D-1} x_j t^j$ at, say, each of n field values $t_0, \ldots, t_{n-1}$. It turns out that the entire sequence $(x(t_0), x(t_1), \ldots, x(t_{n-1}))$ can be evaluated in

$$O\left(n \ln^2 \min\{n, D\}\right)$$

field operations. We shall split the problem into three basic cases:

(1) The arguments $t_0, \ldots, t_{n-1}$ lie in arithmetic progression.

(2) The arguments $t_0, \ldots, t_{n-1}$ lie in geometric progression.

(3) The arguments $t_0, \ldots, t_{n-1}$ are arbitrary.

Of course, case (3) covers the other two, but in (1), (2) it can happen that special enhancements apply.

Algorithm 9.6.5 (Evaluation of polynomial on arithmetic progression). Let $x(t) = \sum_{j=0}^{D-1} x_j t^j$. This algorithm returns the n evaluations $x(a), x(a+d), x(a+2d), \ldots, x(a+(n-1)d)$. (The method attains its best efficiency when n is much greater than D.)

1. [Evaluate at first D points]

```
    for(0 ≤ j < D) e_j = x(a + jd);
```

2. [Create difference tableau]

```
    for(1 ≤ q < D) {
        for(D - 1 ≥ k ≥ q) e_k = e_k - e_{k-1};
```

```
    }
3. [Operate over tableau]
    E_0 = e_0;
    for(1 ≤ q < n) {
        E_q = E_{q-1} + e_1;
        for(1 ≤ k < D - 1) e_k = e_k + e_{k+1};
    }
    return (E_q), q ∈ [0, n - 1];
```

A variant of this algorithm has been used in searches for Wilson primes (see Exercise 9.73, where computational complexity issues are also discussed).

Next, assume that evaluation points lie in geometric progression, say $t_k = T^k$ for some constant T, so we need to evaluate every sum $\sum x_j T^{kj}$ for $k \in [0, D-1]$. There is a so-called Bluestein trick, by which one transforms such sums according to

$$\sum_j x_j T^{kj} = T^{-k^2/2} \sum_j \left(x_j T^{-j^2/2} \right) T^{(-k-j)^2/2},$$

and thus calculates the left-hand sum via the convolution implicit in the right-hand sum. However, in certain settings it is somewhat more convenient to avoid halving the squares in the exponents, relying instead on properties of the triangular numbers $\Delta_n = n(n+1)/2$. Two relevant algebraic properties of these numbers are

$$\Delta_{\alpha+\beta} = \Delta_\alpha + \Delta_\beta + \alpha\beta,$$
$$\Delta_\alpha = \Delta_{-\alpha-1}.$$

A variant of the Bluestein trick can accordingly be derived as

$$\sum_j x_j T^{kj} = T^{\Delta_{-k}} \sum_j \left(x_j T^{\Delta_j} \right) T^{-\Delta_{-(k-j)}}.$$

Now the implicit convolution can be performed using only integral powers of the T constant. Moreover, we can employ an efficient, cyclic convolution by carefully embedding the x signal in a longer, zero-padded signal and reindexing, as in the following algorithm.

Algorithm 9.6.6 (Evaluation of polynomial on geometric progression). Let $x(t) = \sum_{j=0}^{D-1} x_j t^j$, and let T have an inverse in the arithmetic domain. This algorithm returns the sequence of values $\left(x(T^k)\right)$, $k \in [0, D-1]$.

```
1. [Initialize]
    Choose N = 2^n such that N ≥ 2D;
    for(0 ≤ j < D) x_j = x_j T^{Δ_j};                // Weight the signal x.
    Zero-pad x = (x_j) to have length N;
    y = (T^{-Δ_{N/2-j-1}}), j ∈ [0, N - 1];    // Create symmetrical signal y.
2. [Length-N cyclic convolution]
    z = x × y;
```

3. [Final assembly of evaluation results]
   return $\left(x(T^k)\right) = \left(T^{\Delta_{k-1}} z_{N/2+k-1}\right)$, $k \in [0, D-1]$;

We see that a single convolution serves to evaluate all of the values $x(T^k)$ at once. It is clear that the complexity of the entire evaluation is $O(D \ln D)$ field operations. One important observation is that an actual DFT is just such an evaluation over a geometric progression; namely, the DFT of (x_j) is the sequence $\left(x(g^{-k})\right)$, where g is the appropriate root of unity for the transform. So Algorithm 9.6.6 is telling us that evaluations over geometric progressions are, except perhaps for the minor penalty of zero-padding and so on, essentially of FFT complexity given only that g is invertible. It is likewise clear that any FFT can be embedded in a convolution of power-of-two length, and so require at most three FFTs of that padded length (note that in some scenarios the y signal's symmetry allows further optimization).

The third, and most general, case of polynomial evaluation starts from the observation that polynomial remaindering can be used to decimate the evaluation procedure. Say that $x(t)$ has degree $D-1$ and is to be evaluated at the points $t_0, t_1, \ldots, t_{D-1}$. Let us simplify by assuming that d is a power of two. If we define two polynomials, each of essentially half the degree of x, by

$$y_0(t) = (t - t_0)(t - t_1) \ldots (t - t_{D/2-1}),$$

$$y_1(t) = (t - t_{D/2})(t - t_{D/2+1}) \ldots (t - t_{D-1}),$$

then we can write the original polynomial in quotient–remainder form as

$$x(t) = q_0(t)y_0(t) + r_0(t) = q_1(t)y_1(t) + r_1(t).$$

But this means that a desired evaluation $x(t_j)$ is either $r_0(t_j)$ (for $j < D/2$) or $r_1(t_j)$ (for $j \geq D/2$). So the problem of evaluating the degree-$(D-1)$ polynomial x comes down to two copies of the simpler problem: Evaluate a degree-(about $D/2$) polynomial at about $D/2$ points. The recursive algorithm runs as follows:

Algorithm 9.6.7 (Evaluation of a polynomial at arbitrary points). Let $x(t) = \sum_{j=0}^{D-1} x_j t^j$. This algorithm, via a recursive function *eval*, returns all the values of $x(t_j)$ for arbitrary points $t_0, \ldots, t_{D-1}$. Let T denote the sequence $(t_0, \ldots, t_{D-1})$. For convenience, we assume $D = 2^k$, yet simple options will generalize to other D (see Exercise 9.76).

1. [Set breakover]
   $\delta = 4$; // Or whatever classical evaluation threshold is best.

2. [Recursive *eval* function]
   $eval(x, T)$ {
     $d = len(x)$;

3. [Check breakover threshold for recursion exit]
   // Next, use literal evaluation at the t_i in small cases.
   if($len(T) \leq \delta$) return $(x(t_0), x(t_1), \ldots, x(t_{d-1}))$;

4. [Split the signal into halves]
   $u = L(T)$; // Low half of signal.
   $v = H(T)$; // High half.
5. [Assemble half-polynomials]
   $w(t) = \prod_{m=0}^{d/2-1}(t - u_m)$;
   $z(t) = \prod_{m=0}^{d/2-1}(t - v_m)$;
6. [Modular reduction]
   $a(t) = x(t) \bmod w(t)$;
   $b(t) = x(t) \bmod z(t)$;
   return $eval(a, u) \cup eval(b, v)$;

}

Note that in the calculations of $w(t), z(t)$ the intent is that the product must be expanded, to render w, z as signals of coefficients. The operations to expand these products must be taken into account in any proper complexity estimate for this evaluation algorithm (see Exercise 9.75). To lend support to the reader who desires to try this general evaluation Algorithm 9.6.7, let us give an example of its workings. Consider the task of calculating the number 64! not by the usual, sequential multiplication of successive integers but by evaluating the polynomial

$$\begin{aligned} x(t) &= t(1+t)(2+t)(3+t)(4+t)(5+t)(6+t)(7+t) \\ &= 5040t + 13068t^2 + 13132t^3 + 6769t^4 + 1960t^5 322t^6 + 28t^7 + t^8 \end{aligned}$$

at the 8 points

$$T = (1, 9, 17, 25, 33, 41, 49, 57)$$

and then taking the product of the eight evaluations to get the factorial. As the algorithm is fully recursive, tracing is nontrivial. However, if we assign $b = 2$, say, in the step [Set breakover] and print out the half-polynomials w, z and polynomial-mod results a, b right after these entities are established, then our output should look as follows. On the first pass of *eval* we obtain

$$\begin{aligned} w(t) &= 3825 - 4628t + 854t^2 - 52t^3, \\ z(t) &= 3778929 - 350100t + 11990t^2 - 180t^3 + t^4, \\ a(t) &= x(t) \bmod w(t) \\ &= -14821569000 + 17447650500t - 2735641440t^2 + 109600260t^3, \\ b(t) &= x(t) \bmod z(t) \\ &= -791762564494440 + 63916714435140t - 1735304951520t^2 \\ &\quad + 16010208900t^3, \end{aligned}$$

and for each of a, b there will be further recursive passes of *eval*. If we keep tracing in this way the subsequent passes reveal

$$w(t) = 9 - 10t + t^2,$$

$$z(t) = 425 - 42t + t^2,$$
$$a(t) = -64819440 + 64859760t,$$
$$b(t) = -808538598000 + 49305458160t,$$

and, continuing in recursive order,

$$w(t) = 1353 - 74t + t^2,$$
$$z(t) = 2793 - 106t + t^2,$$
$$a(t) = -46869100573680 + 1514239317360t,$$
$$b(t) = -685006261415280 + 15148583316720t.$$

There are no more recursive levels (for our example choice $b = 2$) because the *eval* function will break over to some classical method such as an easy instance of Horner's rule and evaluate these last $a(t), b(t)$ values directly, each one at four $t = t_i$ values. The final returned entity from *eval* turns out to be the sequence

$$(x(t_0), \ldots, x(t_7)) = (40320, 518918400,\ 29654190720,\ 424097856000,\ 3100796899200,\ 15214711438080,\ 57274321104000,\ 178462987637760).$$

Indeed, the product of these eight values is exactly 64!, as expected. One should note that in such a "product" operation—where evaluations are eventually all multiplied together—the last phase of the *eval* function need not return a union of two signals, but may instead return the *product* $eval(a, u) * eval(b, v)$. If that is the designer's choice, then the step [Check breakover threshold ...] must also return the product of the indicated $x(t_i)$.

Incidentally, polynomial coefficients do not necessarily grow large as the above example seems to suggest. For one thing, when working on such as a factoring problem, one will typically be reducing all coefficients modulo some N, at every level. And there is a clean way to handle the problem of evaluating $x(t)$ of degree D at some smaller number of points, say at $t_0, \ldots, t_{n-1}$ with $n < D$. One can simply calculate a new polynomial s as the remainder

$$s(t) = x(t) \bmod \left(\prod_{j=0}^{n-1} (t - t_j) \right),$$

whence evaluation of s (whose degree is now about n) at the n given points t_i will suffice.

9.7 Exercises

9.1. Show that both the base-B and balanced base-B representations are unique. That is, for any nonnegative integer x, there is one and only one collection of digits corresponding to each definition.

9.2. Although this chapter has started with multiplication, it is worthwhile to look at least once at simple addition and subtraction, especially in view of signed arithmetic.

(1) Assuming a base-B representation for each of two nonnegative integers x, y, give an explicit algorithm for calculating the sum $x + y$, digit by digit, so that this sum ends up also in base-B representation.

(2) Invoke the notion of signed-integer arithmetic, by arguing that to get general sums and differences of integers of any signs, all one needs is the summation algorithm of (1), and one other algorithm, namely, to calculate the difference $x - y$ when $x \geq y \geq 0$. (That is, every add/subtract problem can be put into one of two forms, with an overall sign decision on the result.)

(3) Write out complete algorithms for addition and subtraction of integers in base B, with signs arbitrary.

9.3. Assume that each of two nonnegative integers x, y is given in balanced base-B representation. Give an explicit algorithm for calculating the sum $x + y$, digit by digit, but always staying entirely within the balanced base-B representation for the sum. Then write out a such a self-consistent multiply algorithm for balanced representations.

9.4. It is known to children that multiplication can be effected via addition alone, as in $3 \cdot 5 = 5 + 5 + 5$. This simple notion can actually have practical import in some scenarios (actually, for some machines, especially older machines where word multiply is especially costly), as seen in the following tasks, where we study how to use storage tricks to reduce the amount of calculation during a large-integer multiply. Consider the multiplication of D-digit, base-$(B = 2^b)$ integers of size 2^n, so that $n \approx bD$. For the tasks below, define a "word" operation (word multiply or word add) as one involving two size-B operands (each having b bits).

(1) Argue first that standard grammar-school multiply, whereby one constructs via word multiplies a parallelogram and then adds up the columns via word adds, requires $O(D^2)$ word multiplies and $O(D^2)$ word adds.

(2) Noting that there can be at most B possible rows of the parallelogram, argue that all possible rows can be precomputed in such a way that the full multiply requires $O(BD)$ word multiplies and $O(D^2)$ word adds.

(3) Now argue that the precomputation of all possible rows of the parallelogram can be done with successive additions and no multiplies of any kind, so that the overall multiply can be done in $O(D^2 + BD)$ word adds.

(4) Argue that the grammar-school paradigm of task (1) above can be done with $O(n)$ bits of temporary memory. What, then, are the respective memory requirements for tasks (2), (3)?

If one desires to create an example program, here is a possible task: Express large integers in base $B = 256 = 2^8$ and implement via machine task (2) above, using a 256-integer precomputed lookup table of possible rows to create the usual parallelogram. Such a scheme may well be slower than other large-integer methods, but as we have intimated, a machine with especially slow word multiply can benefit from these ideas.

9.5. Write out an explicit algorithm (or an actual program) that uses the w_n relation (9.3) to effect multiple-precision squaring in about half a multiple-precision multiply time. Note that you do not need to subtract out the term δ_n explicitly, if you elect instead to modify slightly the i sum. The basic point is that the grammar-school rhombus is effectively cut (about) in half. This exercise is not as trivial as it may sound—there are precision considerations attendant on the possibility of huge column sums.

9.6. Use the identity (9.4) to write a program that calculates any product xy for each of x, y having at most 15 binary bits, using only table lookups, add/subtracts, shifts, and involving no more than 2^{21} bits of table storage. (Hint: The identity of the text can be used after one computes a certain lookup table.)

9.7. Modify the binary divide algorithm (9.1.3) so that the value $x \bmod N$ is also returned. Note that one could just use equation (9.5), but there is a way to use the local variables of the algorithm itself, and avoid the multiply by N.

9.8. Prove that Arazi's prescription (Algorithm 9.1.4) for simple modular multiplication indeed returns the value $(xy) \bmod N$.

9.9. Work out an algorithm similar to Algorithm 9.1.3 for bases $B = 2^k$, for $k > 1$. Can this be done without explicit multiplies?

9.10. Prove Theorem 9.2.1. Then prove an extension: that the difference $y/R - (xR^{-1}) \bmod N$ is one of $\{0, N, 2N, \ldots, (1 + \lfloor x/(RN) \rfloor)N\}$.

9.11. Prove Theorem 9.2.4. Then develop and prove a corollary for powering, of which equation (9.8) would be the special case of cubing.

9.12. In using the Montgomery rules, one has to precompute the residue $N' = (-N^{-1}) \bmod R$. In the case that $R = 2^s$ and N is odd, show that the Newton iteration (9.10) with a set at $-N$, with initial value $-N \bmod 8$, and the iteration thought of as a congruence modulo R, quickly converges to N'. In particular, show how the earlier iterates can be performed modulo smaller powers of 2, so that the total work involved, assuming naive multiplication and squaring, can be effected with about 4/3 of an s-bit multiply and about 1/3 of an s-bit square operation. Since part of each product involved is obliterated by the mod reduction, show how the work involved can be reduced further. Contrast this method with a traditional inverse calculation.

9.13. We have indicated that Newton iterations, while efficient, involve adroit choices of initial values. For the reciprocation of real numbers, equation

(9.10), describe rigorously the range of initial guesses for a given positive real a, such that the Newton iteration indeed causes x to converge to $1/a$.

9.14. We have observed that with Newton iteration one may "divide using multiplication alone." It turns out that one may also take square roots in the same spirit. Consider the coupled Newton iteration

```
x = y = 1;
do {
    x = x/2 + (1 + a)y/2;
    y = 2y − xy^2;
    y = 2y − xy^2;
}
```

where "do" simply means one repeats what is in the braces for some appropriate total iteration count. Note that the duplication of the y iteration is intentional! Show that this scheme *formally* generates the binomial series of $\sqrt{1+a}$ via the variable x. How many correct terms obtain after k iterations of the do loop?

Next, calculate some real-valued square roots in this way, noting the important restriction that $|a|$ cannot be too large, lest divergence occur (the formal correctness of the resulting series in powers of a does not, of course, automatically guarantee convergence).

Then, consider this question: Can one use these ideas to create an algorithm for extracting integer square roots? This could be a replacement for Algorithm 9.2.11; the latter, we note, does involve explicit division. On this question it may be helpful to consider, for given n to be square-rooted, such as $\sqrt{n/4^q} = 2^{-q}\sqrt{n}$ or some similar construct, to keep convergence under control.

Incidentally, it is of interest that the standard, real-domain, Newton iteration for the *inverse* square root automatically has division-free form, yet we appear to be compelled to invoke such as the above coupled-variable expedient for a positive fractional power.

9.15. Generalize Algorithm 9.2.11 to the problem of finding higher roots; i.e., values of $\lfloor N^{1/d} \rfloor$ for integers $d > 2$. Invoke the extended algorithm to find the smallest prime p that is greater than two million and also enjoys the property that $\lfloor p^{1/2} \rfloor + \lfloor p^{1/3} \rfloor + \lfloor p^{1/5} \rfloor$ is prime.

9.16. The Cullen numbers are $C_n = n2^n + 1$. Write a Montgomery powering program specifically tailored to find composite Cullen numbers, via relations such as $2^{C_n - 1} \not\equiv 1 \pmod{C_n}$. For example, within the powering algorithm for modulus $N = C_{245}$ you would be taking say $R = 2^{253}$ so that $R > N$. You could observe, for example, that C_{141} is a base-2 pseudoprime in this way (it is actually a prime). A much larger example of a Cullen prime is Wilfrid Keller's C_{18496}. For more on Cullen numbers see Exercise 1.80.

9.17. Say that we wish to evaluate 1/3 using the Newton reciprocation of the text (among real numbers, so that the result will be $0.3333\ldots$). For initial guess $x_0 = 1/2$, prove that for positive n the n-th iterate x_n is in fact

$$x_n = \frac{2^{2^n} - 1}{3 \cdot 2^{2^n}},$$

in this way revealing the quadratic-convergence property of a successful Newton loop. The fact that a closed-form expression can even be given for the Newton iterates is interesting in itself. Such closed forms are rare—can you find any others?

9.18. Work out the asymptotic complexity of Algorithm 9.2.8, in terms of a size-N multiply, and assuming all the shifting enhancements discussed in the text. Then give the asymptotic complexity of the composite operation $(xy) \bmod N$, for $0 \le x, y < N$, in the case that the generalized reciprocal is not yet known. What is the complexity for $(xy) \bmod N$ if the reciprocal is known? (This should be asymptotically the same as the composite Montgomery operation $(xy) \bmod N$ if one ignores the precomputations attendant to the latter.) Incidentally, in actual programs that invoke the Newton–Barrett ideas, one can place within the general mod routine a check to see whether the reciprocal is known, and if it is not, then the generalized reciprocal algorithm is invoked, and so on.

9.19. Work out the asymptotic complexity of Algorithm 9.2.13 for given x, N in terms of a count of multiplications by integers c of various sizes. For example, assuming some grammar-school variant for multiplication, the bit-complexity of an operation yc would be $O(\ln y \ln c)$. Answer the interesting question: At what size of $|c|$ (compared to $N = 2^q + c$) is the special form reduction under discussion about as wasteful as some other prevailing schemes (such as long division, or the Newton–Barrett variants) for the mod operation? Incidentally, the most useful domain of applicability of the method is the case that c is one machine word in size.

9.20. Simplify algorithm 9.4.2 in the case that one does *not* need an extended solution $ax + by = g$, rather needs only the inverse itself. (That is, not all the machinations of the algorithm are really required.)

9.21. Implement the recursive gcd Algorithm 9.4.6. Optimize the breakover parameters *lim* and *prec* for maximum speed in the calculation of $rgcd(x, y)$ for each of x, y of various (approximately equal) sizes. You should be able to see $rgcd()$ outperforming $cgcd()$ in the region of, very roughly speaking, thousands of bits. (Note: Our display of Algorithm 9.4.6 is done in such a way that if the usual rules of global variables, such as matrix G, and variables local to procedures, such as the variables x, y in $hgcd()$ and so on, are followed in the computer language, then transcription from our notation to a working program should not be too tedious.)

9.22. Prove that Algorithm 9.2.10 works. Furthermore, work out a version that uses the shift-splitting idea embodied in the relation (9.12) and comments following. A good source for loop constructs in this regard is [Menezes et al. 1997].

9.23. Prove that Algorithm 9.2.11 works. It helps to observe that x is definitely decreasing during the iteration loop. Then prove the $O(\ln\ln N)$ estimate for the number of steps to terminate. Then invoke the idea of changing precision at every step, to show that the bit-complexity of a properly tuned algorithm can be brought down to $O\left(\ln^2 N\right)$. Many of these ideas date back to the treatment of [Alt 1979].

9.24. How general can be the initialization of x in Algorithm 9.2.11?

9.25. Write out a (very) simple algorithm that uses Algorithm 9.2.11 to determine whether a given integer N is a square. Note that there are much more efficient ways of approaching the problem, for example first ruling out the square property modulo some small primes [Cohen 2000].

9.26. Implement Algorithm 9.2.13 within a Lucas–Lehmer test, to prove or disprove primality of various Mersenne numbers $2^q - 1$. Note that with the special form mod reduction, one does not even need general multiplication for Lucas–Lehmer tests; just squaring will do.

9.27. Prove that Algorithm 9.2.13 works, that is, terminates with the correct returned result.

9.28. Work out an algorithm for fast mod operation with respect to moduli of the form

$$p = 2^a + 2^b + \cdots + 1,$$

where the existing exponents (binary-bit positions) $a, b, \ldots$ are sparse; i.e., a small fraction of the bits of p are 1's. Work out also a generalization in which minus signs are allowed, e.g., $p = 2^a \pm 2^b \pm \cdots \pm 1$, with the existing exponents still being sparse. You may find the relevant papers of [Solinas 1999] and [Johnson and Menezes 1999] of interest in this regard.

9.29. Some computations, such as the Fermat number transform (FNT) and other number-theoretical transforms, require multiplication by powers of two. On the basis of Theorem 9.2.12, work out an algorithm that for modulus $N = 2^m + 1$, quickly evaluates $(x2^r) \bmod N$ for $x \in [0, N-1]$ and any (positive or negative) integer r. What is desired is an algorithm that quickly performs the carry adjustments to which the theorem refers, rendering all bits of the desired residue in standard, nonnegative form (unless, of course, one prefers to stay with a balanced representation or some other paradigm that allows negative digits).

9.30. Work out the symbolic powering relation of the type (9.16), but for the scheme of Algorithm 9.3.1.

9.31. Prove that Algorithm 7.2.4 works. It helps to track through small examples, such as $n = 0011_2$, for which $m = 1001_2$ (and so we have intentionally padded n to have four bits). Compare the complexity with that of a trivial modification, suitable for elliptic curve arithmetic, to the "left-right" ladder, Algorithm 9.3.1, to determine whether there is any real advantage in the "add-subtract" paradigm.

9.32. For the binary gcd and extended binary algorithms, show how to enhance performance by removing some of the operations when, say, y is prime and we wish to calculate $x^{-1} \bmod y$. The key is to note that after the [Initialize] step of each algorithm, knowledge that y is odd allows the removal of some of the internal variables. In this way, end up with an inversion algorithm that inputs x, y and requires only four internal variables to calculate the inverse of x.

9.33. Can Algorithm 9.4.4 be generalized to composite p?

9.34. Prove that Algorithms 9.4.4 and 9.4.5 work. For the latter algorithm, it may help to observe how one inverts a pure power of two modulo a Mersenne prime.

9.35. In the spirit of the special-case mod Algorithm 9.2.13, which relied heavily on bit shifting, recast Algorithm 9.4.5 to indicate the actual shifts required in the various steps. In particular, not only the mod operation but multiplication by a power of two is especially simple for Mersenne prime moduli, so use these simplifications to rewrite the algorithm.

9.36. Can one perform a gcd on two numbers each of size N in polynomial time (i.e., time proportional to some power $\lg^\alpha N$), using a polynomial number of parallelized processors (i.e., $\lg^\beta N$ of them)? An interesting reference is [Cesari 1998], where it is explained that it is currently unknown whether such a scheme is possible.

9.37. Write out a clear algorithm for a full integer multiplication using the Karatsuba method. Make sure to show the recursive nature of the method, and also to handle properly the problem of carry, which must be addressed when any final digits overflow the base size.

9.38. Show that a Karatsuba-like recursion on the $(D = 3)$ Toom–Cook method (i.e., recursion on Algorithm 9.5.2) yields integer multiplication of two size-N numbers in what is claimed in the text, namely, $O((\ln N)^{\ln 5/\ln 3})$ word multiplies. (All of this assumes that we count neither additions nor the constant multiplies as they would arise in every recursive [Reconstruction] step of Algorithm 9.5.2.)

9.39. Recast the [Initialize] step of Algorithm 9.5.2 so that the r_i, s_i can be most efficiently calculated.

9.40. We have seen that an acyclic convolution of length N can be effected in $2N-1$ multiplies (aside from multiplications by constants; e.g., a term such as $4x$ can be done with left-shift alone, no explicit multiply). It turns out that a cyclic convolution can be effected in $2N - d(N)$ multiplies, where d is the standard divisor function (the number of divisors of n), while a negacyclic can be effected in $2N - 1$ multiplies. (These wonderful results are due chiefly to S. Winograd; see the older but superb collection [McClellan and Rader 1979].) Here are some explicit related problems:

(1) Show that two complex numbers $a + bi, c + di$ may be multiplied via only three real multiplies.

(2) Work out an algorithm that performs a length-4 negacyclic in nine multiplies, but with all constant mul or div operations being by powers of two (and thus, mere shifts). The theoretical minimum is, of course, seven multiplies, but such a nine-mul version has its advantages.

(3) Use Toom–Cook ideas to develop an explicit length-4 negacyclic scheme that does require only seven multiplies.

(4) Can one use a length-$(D > 2)$ negacyclic to develop a Karatsuba-like multiply that is asymptotically better than $O\left((\ln D)^{\ln 3/\ln 2}\right)$?

(5) Show how to use a Walsh–Hadamard transform to effect a length-16 cyclic convolution in 43 multiplies [Crandall 1996a]. Though the theoretical minimum multiply count for this length is 27, the Walsh–Hadamard scheme has no troublesome constant coefficients. The scheme also appears to be a kind of bridge between Winograd complexities (linear in N) and transform-based complexities ($N \ln N$). Indeed, 43 is not even as large as $16 \lg 16$. Incidentally, the true complexity of the Walsh–Hadamard scheme is still unknown.

9.41. Prove Theorem 9.5.12 by way of convolution ideas, along the following lines. Let $N = 2 \cdot 3 \cdot 5 \cdots p_m$ be a consecutive prime product, and define

$$r_N(n) = \#\{(a, b) \;:\; a + b = n;\ \gcd(a, N) = \gcd(b, N) = 1;\ a, b \in [1, N-1]\},$$

that is, $r_N(n)$ is the number of representations we wish to bound below. Now define a length-N signal y by $y_n = 1$ if $\gcd(n, N) = 1$, else $y_n = 0$. Define the cyclic convolution

$$R_N(n) = (y \times y)_n,$$

and argue that for $n \in [0, N-1]$,

$$R_N(n) = r_N(n) + r_N(N + n).$$

In other words, the cyclic convolution gives us the combined representations of n and $N + n$. Next, observe that the Ramanujan sum Y (9.23) is the DFT of y, so that

$$R_N(n) = \frac{1}{N} \sum_{k=0}^{N-1} Y_k^2 e^{2\pi i k n/N}.$$

Now prove that R is multiplicative, in the sense that if $N = N_1 N_2$ with N_1, N_2 coprime, then $R_N(n) = R_{N_1}(n) R_{N_2}(n)$. Conclude that

$$R_N(n) = \varphi_2(N, n),$$

where φ_2 is defined in the text after Theorem 9.5.12. So now we have a closed form for $r_N(n) + r_N(N+n)$. Note that φ_2 is positive if n is even. Next, argue that if $a + b = n$ (i.e., n is representable) then $2N - n$ is also representable. Conclude that if $r_N(n) > 0$ for all even $n \in [N/2+1, N-1]$, then all sufficiently large even integers are representable. This means that all we have to show is that for n even in $[N/2+1, N-1]$, $r_N(n+N)$ is suitably small compared to $\varphi_2(N,n)$. To this end, observe that $a+b = N+n$ implies $b > n$, and consider the count

$$\#\{b \in [n, N] : b \not\equiv 0, n \pmod{N}\}.$$

By estimating that count, conclude that for a suitable absolute constant C and even $n \in [N/2+1, N-1]$

$$r_N(n) \geq C\frac{n}{(\ln\ln N)^2} - 2^{m+1}.$$

This settles Theorem 9.5.12 for large enough products N, and the smaller cases one may require such as $N = 2, 6, 30$ can be handled by inspecting the finite number of cases $n < 2N$.

We note that the theorem can be demonstrated via direct sieving techniques. Another alternative is to use the Chinese remainder theorem with some combinatorics, to get R_N as the φ_2 function. An interesting question is: Can the argument above (for bounding $r_N(N+n)$), which is admittedly a sieve argument of sorts, be completely avoided, by doing instead algebraic manipulations on the negacyclic convolution $y \times_- y$? As we intimated in the text, this would involve the analysis of some interesting exponential sums. We are unaware of any convenient closed form for the negacyclic, but if one could be obtained, then the precise number of representations $n = a+b$ would likewise be cast in closed form.

9.42. Interesting exact results involving sums of squares can be achieved elegantly through careful application of convolution principles. The essential idea is to consider a signal whose elements x_{n^2} are 1's, with all other elements 0's. Let p be an odd prime, and start with the definition

$$\hat{x}_k = \sum_{j=0}^{(p-1)/2} \left(1 - \frac{\delta_{0j}}{2}\right) e^{-2\pi i j^2 k/p},$$

where $\delta_{ij} = 1$ if $i = j$ and is otherwise 0. Show that $\hat{x}_0 = p/2$, while for $k \in [1, p-1]$ we have

$$\hat{x}_k = \frac{\omega_k}{2}\sqrt{p},$$

where $\omega_k = \left(\frac{k}{p}\right), -i\left(\frac{k}{p}\right)$, respectively, as $p \equiv 1, 3 \pmod 4$. The idea is to show all of this as a corollary to Theorem 2.3.7. (Note that the theory of more general Gauss character sums connects with primality testing, as in our Lemma 4.4.1 and developments thereafter.) Now for $n \in [0, p-1]$ define $R_m(n)$ to be the count of m-squares representations

$$a_1^2 + a_2^2 + \cdots + a_m^2 \equiv n \pmod p$$

in integers $a_j \in [0, (p-1)/2]$, except that a representation is given a weight factor of $1/2$ for every zero component a_j. For example, a representation $0^2 + 3^2 + 0^2$ is given a weight factor of $1/4$. By considering an appropriate m-fold convolution of a certain signal with itself, show that

$$R_2(n) = \frac{1}{4}\left(p + \left(\frac{-1}{p}\right)(p\delta_{0n} - 1)\right),$$

$$R_3(n) = \frac{1}{8}\left(p^2 + \left(\frac{-n}{p}\right)p\right),$$

$$R_4(n) = \frac{1}{16}\left(p^3 + p^2\delta_{0n} - p\right).$$

(A typical test case that can be verified by hand is for $p = 23$: $R_4(0) = 12673/16$, and for any $n \not\equiv 0 \pmod p$, $R_4(n) = 759$.)

Now, from these exact relations, conclude:

(1) Any prime $p \equiv 1 \pmod 4$ is a sum of two squares, while $p \equiv 3 \pmod 4$ cannot be (cf. Exercise 5.15).

(2) There exists $0 < m < p$ such that $mp = a^2 + b^2 + c^2 + d^2$.

The result (2) leads quickly to the classical Lagrange theorem that every nonnegative integer is a sum of four squares. One would use, say, the Lagrange descent argument to argue that the smallest m that satisfies the statement (2) is $m = 1$, so that every *prime* is a sum of four squares. A final step then is to prove that if any two integers a, b are representable via four squares, then ab is. These finishing details can be found in [Hardy and Wright 1979].

What can be said about sums of three squares? An interesting challenge would be to use convolution to establish the relatively difficult celebrated theorem of Gauss that "$num = \Delta + \Delta + \Delta$," meaning every nonnegative integer is a sum of three triangular numbers, i.e., numbers of the form $k(k+1)/2, k \geq 0$. This is equivalent to the statement that every integer congruent to $3 \pmod 8$ is a sum of three squares. (In fact the only numbers not admitting of a three-square representation are those of the form $4^a(8b+7)$.) It is unclear how to proceed with such a challenge; for one thing, from the relation above for R_2, any $p \equiv 7 \pmod 8$ enjoys, strangely enough, some representation $mp = a^2 + b^2 + c^2$ with $m < p$. (For example, 7 is not a sum of three squares, but 14 is.)

9.43. Show that cyclic convolution of two length-D signals is equivalent to multiplication of two polynomials; that is,

$$x \times y \equiv x(t)y(t) \pmod{t^D - 1},$$

where "$\equiv$" here means that the elements of the signal on the left correspond to the coefficients of the polynomial on the right. Then show that negacyclic convolution $x \times_- y$ is equivalent to multiplication $\pmod{t^D + 1}$. Using the Chinese remainder theorem for polynomials, use these facts to establish the identity (9.33) that is exploited in Nussbaumer convolution.

9.44. In the spirit of Exercise 9.43, give a polynomial description of the more general weighted convolution $x \times_a y$ where $a = (A^j)$ for some generator A.

9.45. Implement Algorithm 9.5.18, with a view to proving that $p = 2^{521} - 1$ is prime via the Lucas–Lehmer test. The idea is to maintain the peculiar, variable-base representation for everything, all through the primality test. (In other words, the output of Algorithm 9.5.18 is ready-made as input for a subsequent call to the algorithm.) For larger primes, such as the gargantuan new Mersenne prime discoveries, investigators have used run lengths such that q/D, the typical bit size of a variable-base digit, is roughly 16 bits or less. Again, this is to suppress as much as possible the floating-point errors.

9.46. Implement Algorithm 9.5.16 to establish the character of various Fermat numbers, using the Pepin test, that F_n is prime if and only if $3^{(F_n-1)/2} \equiv -1 \pmod{F_n}$. Alternatively, the same algorithm can be used in factorization studies [Brent et al. 2000]. (Note: The balanced representation error reduction scheme mentioned in Exercise 9.55 also applies to this algorithm for arithmetic with Fermat numbers.) This method has been employed for the resolution of F_{22} in 1993 [Crandall et al. 1995] and F_{24} [Crandall et al. 1999].

9.47. Implement Algorithm 9.5.19 to perform large-integer multiplication via cyclic convolution of zero-padded signals. Can the DWT methods be applied to do negacyclic integer convolution via an appropriate CRT prime set?

9.48. Show that if the arithmetic field is equipped with a cube root of unity, then for $D = 3 \cdot 2^k$ one can perform a length-D cyclic convolution by recombining three separate length-2^k convolutions. (See Exercise 9.44 and consider the symbolic factorization of $t^D - 1$ for such D.) This technique has actually been used by G. Woltman in the discovery of new Mersenne primes (he has employed IBDWTs of length $3 \cdot 2^k$).

9.49. Implement the ideas of [Percival 2000], who has generalized Algorithm 9.5.18 for arithmetic modulo Proth numbers $k \cdot 2^n \pm 1$. The essential idea is that working modulo a number $a \pm b$ can be done with good error control, as long as the prime product $\prod_{p|ab} p$ is sufficiently small. In the Percival approach, one generalizes the variable-base representation of Theorem 9.5.17 to involve products over prime powers in the form

$$x = \sum_{j=0}^{D-1} x_j \prod_{p^k \| a} p^{\lceil kj/D \rceil} \prod_{q^m \| b} q^{\lceil -mj/D \rceil + mj/D},$$

for fast arithmetic modulo $a - b$.

Note that the marriage of such ideas with the fast mod operation of Algorithm 9.2.14 would result in an efficient union for computations that need to move away from the restricted theme of Mersenne/Fermat numbers.

Indeed, as evidenced in the generalized Fermat number searches described in [Dubner and Gallot 2000], wedding bells have already sounded.

9.50. In the FFT literature there exists an especially efficient real-signal transform called the Sorenson FFT. This is a split-radix transform that uses $\sqrt{2}$ and a special decimation scheme to achieve essentially the lowest-complexity FFT known for real signals; although in modern times the issues of memory, machine cache, and processor features are so overwhelming that sheer complexity counts have fallen to a lesser status. Now, for the ring $\mathbf{Z}_n$ with $n = 2^m + 1$ and m a multiple of 4, show that a square root of 2 is given by

$$\sqrt{2} = 2^{3m/4} - 2^{m/4}.$$

Then, determine whether a Sorenson transform modulo n can be done simply by using what is now the standard Sorenson routine but with $\sqrt{2}$ interpreted as above. (Detailed coding for a Sorenson real-signal FFT is found in [Crandall 1994b].)

9.51. Study the transform that has the usual DFT form

$$X_k = \sum_{j=0}^{N-1} x_j h^{-jk},$$

except that the signal elements x_j and the root h of order N exist in the field $\mathbf{Q}\left(\sqrt{5}\right)$. This has been called a number-theoretical transform (NTT) over the "golden section quadratic field," because the golden mean $\phi = \left(\sqrt{5} - 1\right)/2$ is in the field. Assume that we restrict further to the ring $\mathbf{Z}[\phi]$ so that the signal elements and the root are of the form $a + b\phi$ with a, b integers. Argue first that a multiplication in the domain takes three integer multiplies. Then consider the field $\mathbf{F}_p\left(\sqrt{5}\right)$ and work out a theory for the possible length of such transforms over that field, when the root is taken to be a power of the golden mean ϕ. Then, consider the transform (N is even)

$$\mathbf{X}_k = \sum_{j=0}^{N/2-1} \mathbf{H}^{-jk} \mathbf{x}_j$$

where the new signal vector is $\mathbf{x}_j = (a_j, b_j)$ and where the original signal component was $x_j = a_j + b_j\phi$ in the field. Here, the matrix H is

$$\mathbf{H} = \begin{pmatrix} 1 & 1 \\ 1 & 0 \end{pmatrix}.$$

Describe in what sense this matrix transform is equivalent to the DFT definition preceding, that the powers of $\mathbf{H}$ are given conveniently in terms of Fibonacci numbers

$$\mathbf{H}^n = \begin{pmatrix} F_{n+1} & F_n \\ F_n & F_{n-1} \end{pmatrix},$$

and that this n-th power can be computed in divide-and-conquer fashion in $O(\ln n)$ matrix multiplications. In conclusion, derive the complexity of this matrix-based number-theoretical transform.

This example of exotic transforms, being reminiscent of the discrete Galois transform (DGT) of the text, appears in [Dimitrov et al. 1995], [Dimitrov et al. 1998], and has actually been proposed as an idea for obtaining meaningful spectra—in a discrete field, no less—of real-valued, real-world data.

9.52. Pursuant to Algorithm 9.5.21 for cyclic convolution, work out a similar algorithm for negacyclic integer convolution via a combined DGT/DWT method, with *halved* run length, meaning you want to convolve two real integer sequences each of length D, via a complex DGT of length $D/2$. You would need to establish, for a relevant weighted convolution of length $D/2$, a $(D/2)$-th root of i in a field $\mathbf{F}_{p^2}$ with p a Mersenne prime. Details that may help in such an implementation can be found in [Crandall 1997b].

9.53. Study the so-called Fermat number transform (FNT) defined by

$$X_k = \sum_{j=0}^{D-1} x_j g^{-jk} \pmod{f_n},$$

where $f_n = 2^n + 1$ and g has multiplicative order D in $\mathbf{Z}_n$. A useful choice is g a power of two, in which case, what are the allowed signal lengths D? The FNT has the advantage that the internal butterflies of a fast implementation involve multiply-free arithmetic, but the distinct disadvantage of restricted signal lengths. A particular question is: Are there useful applications of the FNT in computational number theory, other than the appearance in the Schönhage Algorithm 9.5.22?

9.54. In such as Algorithm 9.5.7 one may wish to invoke an efficient transpose. This is not hard to do if the matrix is *square*, but otherwise, the problem is nontrivial. Note that the problem is again trivial, for any matrix, if one is allowed to copy the original matrix, then write it back in transpose order. However this can involve long memory jumps, which are not necessary, as well as all the memory for the copy.

So, work out an algorithm for a general *in-place* transpose, that is, no matrix copy allowed, trying to keep everything as "local" as possible, meaning you want in some sense minimal memory jumps. Some references are [Van Loan 1992], [Bailey 1990].

9.55. Rewrite Algorithm 9.5.11 to employ balanced-digit representation (Definition 9.1.2). Note that the important changes center on the carry adjustment step. Study the phenomenon raised in the text after the algorithm, namely, that of reduced error in the balanced option. There exist some numerical studies of this, together with some theoretical conjectures (see [Crandall and Fagin 1994], [Crandall et al. 1999] and references therein), but very little is known in the way of error bounds that are both rigorous and pragmatic.

9.56. Show that if $p = 2^q - 1$ with q odd and $x \in \{0, \ldots, p-1\}$, then $x^2 \bmod p$ can be calculated using two size-$(q/2)$ multiplies. Hint: Represent $x = a + b2^{(q+1)/2}$ and relate the result of squaring x to the numbers

$$(a+b)(a+2b) \text{ and } (a-b)(a-2b).$$

This interesting procedure gives nothing really new—because we already know that squaring (in the grammar-school range) is about half as complex as multiplication—but the method here is a different way to get the speed doubling, and furthermore does not involve microscopic intervention into the squaring loops as discussed for equation (9.3).

9.57. Do there always exist primes $p_1, \ldots, p_r$ required in Algorithm 9.5.19, and how does one find them?

9.58. Prove, as suggested by the statement of Algorithm 9.5.19, that any convolution element of $x \times y$ in that algorithm is indeed bounded by NM^2. For application to large-integer multiplication, can one invoke balanced representation ideas, that is, considering any integer (mod p) as lying in $[-(p+1)/2, (p-1)/2]$, to lower the bounding requirements, hence possibly reducing the set of CRT primes?

9.59. For the discrete, prime-based transform (9.30) in cases where g has a square root, $h^2 = g$, answer precisely: What is a closed form for the transform element X_k if the input signal is defined $x = \left(h^{j^2}\right)$, $j = 0, \ldots, p-1$? Noting the peculiar simplicity of the X_k, find an analogous signal x having N elements in the complex domain, for which the usual, complex-valued FFT has a convenient property for the magnitudes $|X_k|$. (Such a signal is called a "chirp" signal and has high value in testing FFT routines, which must, of course, exhibit a numerical manifestation of the special magnitude property.)

9.60. For the Mersenne prime $p = 2^{127} - 1$, exhibit an explicit primitive 64th root of unity $a + bi$ in $\mathbf{F}_{p^2}^*$.

9.61. Show that if $a + bi$ is a primitive root of maximum order $p^2 - 1$ in $\mathbf{F}_{p^2}^*$ (with $p \equiv 3 \pmod 4$, so that "i" exists), then $a^2 + b^2$ must be a primitive root of maximum order $p - 1$ in $\mathbf{F}_p^*$. Is the converse true?

Give some Mersenne primes $p = 2^q - 1$ for which $6 + i$ is a primitive root in $\mathbf{F}_{p^2}^*$.

9.62. Prove that the DGT integer convolution Algorithm 9.5.21 works.

9.63. If the Mersenne prime $p = 2^{89} - 1$ is used in the DGT integer convolution Algorithm 9.5.21 for zero-padded, large-integer multiply, and the elements of signals x, y are interpreted as digits in base $B = 2^{16}$, how large can x, y be? What if balanced digit representation (with each digit in $[-2^{15}, 2^{15} - 1]$) is used?

9.64. Describe how to use Algorithm 9.5.21 with a set of Mersenne primes to effect integer convolution via CRT reconstruction, including the precise manner of reconstruction. (Incidentally, CRT reconstruction for a Mersenne prime set is especially straightforward.)

9.65. Analyze the complexity of Algorithm 9.5.21, with a view to the type of recursion seen in the Schönhage Algorithm 9.5.22, and explain how this compares to the entries of Table 9.1.

9.66. Describe how DWT ideas can be used to obviate the need for zero-padding in Algorithm 9.5.24. Specifically, show how to use not a length-$(2m)$ cyclic, rather a length-m cyclic and a length-m negacyclic. This is possible because we have a primitive m-th root of -1, so a DWT can be used for the negacyclic. Note that this does not significantly change the complexity, but in practice it reduces memory requirements.

9.67. Prove the complexity claim following the Nussbaumer Algorithm 9.5.24 for the $O(D \ln D)$ operation bound. Then analyze the somewhat intricate problem of bit-complexity for the algorithm. One way to start on such bit-complexity analysis is to decide upon the optimal base B, as intimated in the complexity table of Section 9.5.8.

9.68. For odd primes p, the Nussbaumer Algorithm 9.5.24 will serve to evaluate cyclic or negacyclic convolutions $(\bmod\ p)$, that is, for ring R identified with $\mathbf{F}_p$. All that is required is to perform all R-element operations $(\bmod\ p)$, so the structure of the algorithm as given does not change. Use such a Nussbaumer implementation to establish Fermat's last theorem for some large exponents p, by invoking a convolution to effect the Shokrollahi DFT. There are various means for converting DFTs into convolutions. One method is to invoke the Bluestein reindexing trick, another is to consider the DFT to be a polynomial evaluation problem, and yet another is Rader's trick (in the case that signal length is a prime power). Furthermore, convolutions of not-power-of-two length can be embedded in larger, more convenient convolutions (see [Crandall 1996a] for a discussion of such interplay between transforms and convolutions). You would use Theorem 9.5.13, noting first that the DFT length can be brought down to $(p-1)/2$. Then evaluate the DFT via a cyclic convolution of power-of-two length by invoking the Nussbaumer method $(\bmod\ p)$. Aside from the recent and spectacular theoretical success of A. Wiles in proving the "last theorem," numerical studies have settled all exponents $p < 12000000$ [Buhler et al. 2000]. Incidentally, the largest prime to have been shown regular via the Shokrollahi criterion is $p = 671008859$ [Crandall 1996a].

9.69. Implement Algorithm 9.6.1 for multiplication of polynomials with coefficients $(\bmod\ p)$. Such an implementation is useful in, say, the Schoof algorithm for counting points on elliptic curves, for in that method, one has not only to multiply large polynomials, but create powering ladders that rely on the large-degree-polynomial multiplies.

9.70. Prove both complexity claims in the text following Algorithm 9.6.1. Describe under what conditions, e.g., what D, p ranges, or what memory constraints, and so on, which of the methods indicated—Nussbaumer convolution or binary-segmentation method—would be the more practical.

For further analysis, you might consider the Shoup method for polynomial multiplication [Shoup 1995], which is a CRT-convolution-based method, which will have its own complexity formula. To which of the two above methods does the Shoup method compare most closely, in complexity terms?

9.71. Say that polynomials $x(t), y(t)$ have coefficients (mod p) and degrees $\approx N$. For Algorithm 9.6.4, which calls Algorithm 9.6.2, what is the asymptotic bit complexity of the polynomial mod operation $x \bmod y$, in terms of p and N? (You need to make an assumption about the complexity of the integer multiplication for products of coefficients.) What if one is, as in many *integer* mod scenarios, doing many polynomial mods with the same modulus polynomial $y(t)$, so that one has only to evaluate the truncated inverse $R[y, \ ]$ once?

9.72. Here we explore another relation for Bernoulli numbers (mod p). Prove the theorem that if $p \geq 5$ is prime, a is coprime to p, and we define $d = -p^{-1} \bmod a$, then for even m in $[2, p-3]$,

$$\frac{B_m}{m}(a^m - 1) \equiv \sum_{j=0}^{p-1} j^{m-1}(dj \bmod a) \pmod{p}.$$

Then establish the corollary that

$$\frac{B_m}{m}(2^{-m} - 1) \equiv \frac{1}{2} \sum_{j=1}^{(p-1)/2} j^{m-1} \pmod{p}.$$

Now achieve the interesting conclusion that if $p \equiv 3 \pmod{4}$, then $B_{(p+1)/2}$ cannot vanish (mod p).

Such summation formulae have some practical value, but more computationally efficient forms exist, in which summation indices need cover only a fraction of the integers in the interval $[0, p-1]$, see [Wagstaff 1978], [Tanner and Wagstaff 1987].

9.73. Prove that Algorithm 9.6.5 works. Then modify the algorithm for a somewhat different problem, which is to evaluate a polynomial given in product form

$$x(t) = t(t+d)(t+2d) \cdots (t+(n-1)d),$$

at a single given point t_0. The idea is to choose some optimal $G < n$, and start with a loop

$$\mathsf{for}(0 \leq j < G) \quad a_j = \prod_{q=0}^{G-1}(t_0 + (q + Gj)d);$$

Arrive in this way at an algorithm that requires $O(G^2 + n/G)$ multiplies and $O(n + G^2)$ adds to find $x(t_0)$. Show that by recursion on the partial product in the for() loop above (which partial product is again of the type handled by the overall algorithm), one can find $x(t_0)$ in $O(n^{\phi+\epsilon})$ multiplies, where $\phi = \left(\sqrt{5}-1\right)/2$ is the golden mean. In this scenario, what is the total count of adds?

Finally, use this sort of algorithm to evaluate large factorials, for example to verify primality of some large p by testing whether $(p-1)! \equiv -1 \pmod{p}$. The basic idea is that the evaluations of

$$(t+1)(t+2)\cdots(t+m)$$

at points $\{0, m, 2m, \ldots, (m-1)m\}$ do yield, when multiplied all together, $(m^2)!$. Searches for Wilson primes have used this technique with all arithmetic performed $\pmod{p^2}$ [Crandall et al. 1997].

9.74. Say that a polynomial $x(t)$ is known in product form, that is,

$$x(t) = \prod_{k=0}^{D-1} (t - t_k),$$

with the field elements t_k given. By considering the accumulation of pairwise products, show that x can be expressed in coefficient form $x(t) = x_0 + x_1 t + \cdots + x_{D-1}t^{D-1}$ in $O\left(D \ln^2 D\right)$ field operations.

9.75. Prove that Algorithm 9.6.7 works, and prove the complexity estimate of $O\left(D \ln^2 D\right)$ field operations. (See Exercise 9.74.)

9.76. Investigate ways to relax the restriction that D be a power of two in Algorithm 9.6.7. One way, of course, is just to assume that the original polynomial has a flock of zero coefficients (and perforce, that the evaluation point set T has power-of-two length), and pretend the degree of x is thus one less than a power of two. But another is to change the step [Check breakover threshold ...] to test just whether $len(T)$ is *odd*. These kinds of approaches will ensure that halving of signals can proceed during recursion.

9.8 Research problems

9.77. As we have intimated, the enhancements to power ladders can be intricate, in many respects unresolved. In this exercise we tour some of the interesting problems attendant on such enhancements.

When an inverse is in hand (alternatively, when point negations are available in elliptic algebra), the add/subtract ladder options make the situation more interesting. The add/subtract ladder Algorithm 7.2.4, for example, has an interesting "stochastic" interpretation, as follows. Let x denote a real number in $(0, 1)$ and let y be the fractional part of $3x$; i.e., $y = 3x - \lfloor 3x \rfloor$. Then denote the exclusive-or of x, y by

$$z = x \wedge y,$$

meaning z is obtained by an exclusive-or of the bit streams of x and y together. Now investigate this conjecture: If x, y are chosen at random, then with probability 1, one-third of the binary bits of z are ones. If true, this conjecture means that if you have a squaring operation that takes time S, and a multiply operation that takes time M, then Algorithm 7.2.4 takes about time $(S + M/3)b$, when the relevant operands have b binary bits. How does this compare with the standard binary ladders of Algorithms 9.3.1, 9.3.2? How does it compare with a base-($B = 3$) case of the general windowing ladder Algorithm 9.3.3? (In answering this you should be able to determine whether the add/subtract ladder is equivalent or not to some windowing ladder.)

Next, work out a theory of *precise* squaring and addition counts for practical ladders. For example, a more precise complexity estimate for he left-right binary ladder is

$$C \sim (b(y) - 1)S + (o(y) - 1)M,$$

where the exponent y has $b(y)$ total bits, of which $o(y)$ are 1's. Such a theory should be extended to the windowing ladders, with precomputation overhead not ignored. In this way, describe quantitatively what sort of ladder would be best for a typical cryptography application; namely, x, y have say 192 bits each and x^y is to be computed modulo some 192-bit prime.

Next, implement an elliptic multiplication ladder in base $B = 16$, which means as in Algorithm 9.3.3 that four bits at a time of the exponent are processed. Note that, as explained in the text following the windowing ladder algorithm, you would need only the following point multiples: $P, 3P, 5P, 7P$. Of course, one should be precomputing these small multiples also in an efficient manner.

Next, study yet other ladder options (and this kind of extension to the exercise reveals just how convoluted is this field of study) as described in [Müller 1997], [De Win et al. 1998], [Crandall 1999b] and references therein. As just one example of attempted refinements, some investigators have considered exponent expansions in which there is some guaranteed number of 0's interposed between other digits. Then, too, there is the special advantage inherent in highly compressible exponents [Yacobi 1999], such study being further confounded by the possibility of base-dependent compressibility. It is an interesting research matter to ascertain the precise relation between the compressibility of an exponent and the optimal efficiency of powering to said exponent.

9.78. In view of complexity results such as in Exercise 9.38, it would seem that a large-D version of Toom–Cook could, with recursion, be brought down to what is essentially an ideal bit complexity $O\left(\ln^{1+\epsilon} N\right)$. However, as we have intimated, the additions grow rapidly. Work out a theory of Toom–Cook addition counts, and discuss the tradeoffs between very low multiplication complexity and overwhelming complexity of additions. Note also the existence of addition optimizations, as intimated in Exercise 9.39.

This is a difficult study, but of obvious practical value. For example, there is nothing *a priori* preventing us from employing different, alternating Toom–Cook schemes within a single, large recursive multiply. Clearly, to optimize such a mixed scheme one should know something about the interplay of the multiply and add counts, as well as other aspects of overhead. Yet another such aspect is the shifting and data shuttling one must do to break up an integer into its Toom–Cook coefficients.

9.79. How far should one be able to test numerically the Goldbach conjecture by considering the acyclic convolution of the signal

$$G = (1, 1, 1, 0, 1, 1, 0, 1, 1, 0, \ldots)$$

with itself? (Here, as in the text, the signal element G_n equals 1 if and only if $2n + 3$ is prime.) What is the computational complexity for this convolution-based approach for the settling of Goldbach's conjecture for all even numbers not exceeding x? Note that the conjecture has been settled for all even numbers up to $x = 4 \cdot 10^{14}$ [Richstein 2001]. We note that explicit FFT-based computations up to 10^8 or so have indeed been performed [Saouter 1999]. Here is an interesting question: Can one resolve Goldbach representations via pure-integer convolution on arrays of b-bit integers (say $b = 16$ or 32), with prime locations signified by 1 bits, knowing in advance that two prime bits lying in one integer is a relatively rare occurrence?

9.80. One can employ convolution ideas to analyze certain higher-order additive problems in rings $\mathbf{Z}_N$, and perhaps in more complicated settings leading into interesting research areas. Note that Exercise 9.42 deals with sums of squares. But when higher powers are involved, the convolution and spectral manipulations are problematic.

To embark on the research path intended herein, start by considering a k-th powers exponential sum (the square and cubic versions appear in Exercise 1.63), namely

$$U_k(a) = \sum_{x=0}^{N-1} e^{2\pi i a x^k/N}.$$

Denote by $r_s(n)$ the number of representations of n as a sum of s k-th powers in $\mathbf{Z}_N$. Prove that whereas

$$\sum_{n=0}^{N-1} r_s(n) = N^s,$$

it also happens that

$$\sum_{n=0}^{N-1} r_s(n)^2 = \frac{1}{N} \sum_{a=0}^{N-1} |U_k(a)|^{2s}.$$

It is this last relation that allows some interesting bounds and conclusions. In fact, the spectral sum of powers $|U|^{2s}$, if bounded above, will allow *lower* bounds to be placed on the number of representable elements of $\mathbf{Z}_N$. In other words, upper bounds on the spectral amplitude $|U|$ effectively "control" the representation counts across the ring, to analytic advantage.

Next, as an initial foray into the many research options, use the ideas and results of Exercises 1.42, 1.63 to show that a positive constant c exists such that for p prime, more than a fraction c of the elements of $\mathbf{Z}_p$ are sums of two cubes. Admittedly, we have seen that the theory of elliptic curves completely settles the two-cube question—even for rings $\mathbf{Z}_N$ with N composite—in the manner of Exercise 7.20, but the idea of the present exercise is to use the convolution and spectral notions alone. How high can you force c for, say, sufficiently large primes p? One way to proceed is first to show from the "$p^{3/4}$" bound of Exercise 1.63 that every element of $\mathbf{Z}_p$ is a sum of 5 cubes, then to obtain sharper results by employing the best-possible "$p^{1/2}$" bound. And what about this spectral approach for composite N? In this case one may employ, for appropriate Fourier indices a, an "$N^{2/3}$" bound (see for example [Vaughan 1997, Theorem 4.2]).

Now try to find a simple proof of the theorem: If N is prime, then for every k there exist positive constants c_k, ϵ_k such that for $a \not\equiv 0 \pmod N$ we have

$$|U_k(a)| < c_k N^{1-\epsilon_k}.$$

Then, show from this that for any k there is a fixed s (independent of everything except k) such that every element of $\mathbf{Z}_N$, prime N, is a sum of s k-th powers. Such bounds as the above on $|U|$ are not too hard to establish, using recursion on the Weyl expedient as used for the cubic case in Exercise 1.63. (Some of the references below explain how to do more work, to achieve $\epsilon_k \approx 1/k$, in fact.)

Can you show the existence of the fixed s for composite N? Can you establish explicit values for s for various k (recall the "4,5" dichotomy for the cubic case)? In such research, you would have to find upper bounds on general U sums, and indeed these can be obtained; see [Vinogradov 1985], [Ellison and Ellison 1985], [Nathanson 1996], [Vaughan 1997]. However, the hard part is to establish *explicit* s, which means explicit bounding constants need to be tracked; and many references, for theoretical and historical reasons, do not bother with such detailed tracking.

One of the most fascinating aspects of this research area is the fusion of theory and computation. That is, if you have bounding parameters c_k, ϵ_k for k-th power problems as above, then you will likely find yourself in a situation where theory is handling the "sufficiently large" N, yet you need computation to handle all the cases of N from the ground up to that theory threshold. Computation looms especially important, in fact, when the constant c_k is large or, to a lesser extent, when ϵ_k is small. In this light, the great efforts of 20th-century analysts to establish general bounds on exponential sums can now be viewed from a computational perspective.

These studies are, of course, reminiscent of the literature on the celebrated Waring conjecture, which conjecture claims representability by a fixed number s of k-th powers, but among the nonnegative integers (e.g., the Lagrange four-square theorem of Exercise 9.42 amounts to proof of the $k = 2$, $s = 4$ subcase of the general Waring conjecture). The issues in this full Waring scenario are different, because for one thing the exponential sums are to be taken not over all ring elements but only up to index $x \approx \lfloor N^{1/k} \rfloor$ or thereabouts, and the bounding procedures are accordingly more intricate. In spite of such obstacles, a good research extension would be to establish the classical Waring estimates on s for given k—which estimates historically involve continuous integrals—using discrete convolution methods *alone*. (In 1909 D. Hilbert proved the Waring conjecture via an ingenious combinatorial approach, while the incisive and powerful continuum methods appear in many references, e.g., [Hardy 1966], [Nathanson 1996], [Vaughan 1997].) Incidentally, many Waring-type questions for finite fields have been completely resolved; see for example [Winterhof 1998].

9.81. Is there a way to handle large convolutions without DFT, by using the kind of matrix idea that underlies Algorithm 9.5.7? That is, you would be calculating a convolution in small pieces, with the usual idea in force: The signals to be convolved can be stored on massive (say disk) media, while the computations proceed in relatively small memory (i.e., about the size of some matrix row/column).

Along these lines, design a standard three-FFT convolution for arbitrary signals, except do it in matrix form reminiscent of Algorithm 9.5.7, yet do *not* do unnecessary transposes. Hint: Arrange for the first FFT to leave the data in such a state that after the usual dyadic (spectral) product, the inverse FFT can start right off with row FFTs.

Incidentally, E. Mayer has worked out FFT schemes that do no transposes of any kind; rather, his ideas involve columnwise FFTs that avoid common memory problems. See [Crandall et al. 1999] for Mayer's discussion.

9.82. A certain prime suggested by [Craig-Wood 1998], namely

$$p = 2^{64} - 2^{32} + 1,$$

has advantageous properties in regard to CRT-based convolution. Investigate some of these advantages, for example by stating the possible signal lengths for number-theoretical transforms modulo p, exhibiting a small-magnitude element of order 64 (such elements might figure well into certain FFT structures), and so on.

9.83. Here is a surprising result: Length-8 cyclic convolution modulo a Mersenne prime can be done via only *eleven* multiplies. It is surprising because the Winograd bound would be $2 \cdot 8 - 4 = 12$ multiplies, as in Exercise 9.40. Of course, the resolution of this paradox is that the Mersenne mod changes the problem slightly.

To reveal the phenomenon, first establish the existence of an 8th root of unity in $\mathbf{F}_{p^2}$, with p being a Mersenne prime and the root being symbolically simple enough that DGTs can be performed without explicit integer multiplications. Then consider the length-8 DGT, used to cyclically convolve two integer signals x, y. Next, argue that the transforms X, Y have sufficient symmetry that the dyadic product $X * Y$ requires two real multiplies and three complex multiplies. This is the requisite count of 11 muls.

An open question is: Are there similar "violations" of the Winograd bound for lengths greater than eight?

9.84. Study the interesting observations of [Yagle 1995], who notes that matrix multiplication involving $n \times n$ matrices can be effected via a convolution of length n^3. This is not especially surprising, since we cannot do an arbitrary length-n convolution faster than $O(n \ln n)$ operations. However, Yagle saw that the indicated convolution is *sparse*, and this leads to interesting developments, touching, even, on number-theoretical transforms.

Appendix

BOOK PSEUDOCODE

All algorithms in this book are written in a particular pseudocode form describable, perhaps, as a "fusion of English and C languages." The motivations for our particular pseudocode design have been summarized in the Preface, where we have indicated our hope that this "mix" will enable all readers to understand, and programmers to code, the algorithms. Also in the Preface we indicated a network source for *Mathematica* implementations of the book algorithms.

That having been said, the purpose of this Appendix is to provide not a rigorous compendium of instruction definitions, for that would require something like an entire treatise on syntactical rules as would be expected to appear in an off-the-shelf C reference. Instead, we give below some explicit examples of how certain pseudocode statements are to be interpreted.

English, and comments

For the more complicated mathematical manipulations within pseudocode, we elect for English description. Our basic technical motivation for allowing "English" pseudocode at certain junctures is evident in the following example. A statement in the C language

```
if((n== floor(n)) && (j == floor(sqrt(j))*floor(sqrt(j)))) ...,
```

which really means "if n is an integer and j is a square," we might have cast in this book as

$$\text{if}(n, \sqrt{j} \in \mathbf{Z}) \ldots$$

That is, we have endeavored to put "chalkboard mathematics" within conditionals. We have also adopted a particular indentation paradigm. If we had allowed (which we have not) such English as:

> For all pseudoprimes in S, apply equation (X); Apply equation (Y);

then, to the aspiring programmer, it might be ambiguous whether equation (Y) were to be applied for all pseudoprimes, or just once, after the loop on equation (X). So the way we wish such English to appear, assuming the case that equation (Y) is indeed applied only once after looping, is like so:

> For all pseudoprimes in S, apply equation (X);
> Apply equation (Y);

Because of this relegation of English statements to their own lines, the interpretation that equation (Y) is to be invoked once, after the pseudoprime loop, is immediate. Accordingly, when an English statement is sufficiently long that it wraps around, we have adopted reverse indentation, like so:

```
Find a random t ∈ [0, p − 1] such that t^2 − a is a quadratic nonresidue
    (mod p), via Algorithm 2.3.5;
x = (t + √(t^2 − a))^((p+1)/2);
. . .;
```

In this last example, one continually chooses random integers t in the stated range until one is found with the required condition, and then one goes to the next step, which calls for a single calculation and the assignment of letter x to the result of the calculation.

Also in English will be comments throughout the book pseudocode. These take the following form (and are right-justified, unlike pseudocode itself):

```
x = (t + √(t^2 − a))^((p+1)/2);                    // Use F_{p^2} arithmetic.
```

The point is, a comment prefaced with "//" is *not* to be executed as pseudocode. For example, the above comment is given as a helpful hint, indicating perhaps that to execute the instruction one would first want to have a subroutine to do $\mathbf{F}_{p^2}$ arithmetic. Other comments clarify the pseudocode's nomenclature, or provide further information on how actually to carry out the executable statement.

Assignment of variables, and conditionals

We have elected *not* to use the somewhat popular assignment syntax $x := y$, rather, we set x equal to y via the simple expedient $x = y$. (Note that in this notation for assignment used in our pseudocode, the symbol "=" does not signify a symmetric relation: The assignment $x = y$ is not the same instruction as the assignment $y = x$.) Because assignment appears on the face of it like equality, the *conditional* equality $x == y$ means we are not assigning, merely testing whether x and y are equal. (In this case of testing conditional equality, the symbol "==" is indeed symmetric.) Here are some examples of our typical assignments:

```
x = 2;                                      // Variable x gets the value 2.
x = y = 2;                             // Both x and y get the value 2.
F = { };                                    // F becomes the empty set.
(a, b, c) = (3, 4, 5);  // Variable a becomes 3, b becomes 4, c becomes 5.
```

Note the important rule that simultaneous (vector) assignment assumes first the full evaluation of the vector on the right side of the equation, followed by the forcing of values on the left-hand side. For example, the assignment

```
(x, y) = (y^2, 2x);
```

means that the right-hand vector is evaluated for all components, then the left-hand vector is forced in all components. That is, the example is equivalent to the chain (technically, we assume neither of x, y invokes hidden functions)

```
t = x;                                    //Variable t is temporary here.
x = y^2;
y = 2t;
```

and it is quite evident by comparison how visually efficient is the single-line vector assignment. Note, too, that the composite assignments

```
x = y^2;
y = 2x;
```

and

```
y = 2x;
x = y^2;
```

are both different than the vector assignment, and different from each other.

Because our text adheres to the rule that ordered sequences are symbolized by parentheses (as in (x_n)) while sets use braces (as in $\{X, a, \alpha\}$), we assign sequences, vectors, and so on with a style consistent with the text; e.g., $\vec{v} = (0, 1, 0)$ is an ordered assignment, whereas a set of three polynomials might be assigned as $S = \{x^2 + 1, x, x^3 - x\}$ and the order is unimportant. Moreover, $S = \{x^2 + 1, x, x, x^3 - x\}$ is exactly the same assignment, since set notation does not record multiplicity. Note that the distinction between sequence and set assignment is important, in view of the liberal use of braces in modern languages. In the *Mathematica* language, braces denote "lists" and these in turn can be manipulated as either vectors (sequences) or sets, with vector-algebraic (such as matrix-vector) and set-theoretical (such as union, intersection) operators available. Likewise, the C language allows assignment of data records via braces, as in "float $x[3] = \{1.1, 2.2, 3.3\}$;" which would fill a vector x in ordered fashion. In this latter case, our pseudocode would say instead $x = (1.1, 2.2, 3.3)$.

The internal conditionals in if() statements often use classical mathematical notation, but not always. Let us exemplify conditional syntax like so:

```
if(x == y) task();   // Testing equality of x, y, without changing either.
if(x ≥ y) task();        // Testing whether x is greater than or equal to y.
if(x|y) task();                                // Testing whether x divides y.
if(x ≡ y (mod p)) task();  // Testing whether x, y congruent modulo p.
```

Note that a congruence conditional does *not* take the form $x \equiv\equiv y \pmod p$, because there need not be any confusion with assignment in such cases. However, it may be possible to have the construction $x == y \bmod p$, since as is explained in the text, the notation $y \bmod p$ refers to the integer $y - p\lfloor y/p \rfloor$. Thus, it may be that x is equal to this integer, or it may be that we wish to assign x this value (in which case we would write $x = y \bmod p$).

Another conditional form is the while() statement, exemplified by

```
while(x ≠ 0) {
        task1();
        task2();
```

```
        ...;
    }
```

which means that x is checked for zero value upon entry of the whole loop, then all the interior tasks are to be performed until, after some complete pass of the interior tasks, x is found to be zero and that ends the while() looping.

Operations that change the value of a single variable include these:

$x = x + c;$ // x is increased by c.
$x = cx;$ // x is multiplied by c.
$x = x << 3;$ // Shift (integer only) x left by 3 bits, same as $x = 8x$.
$x = x >> 3;$ // Shift right, same as $x = \lfloor x/8 \rfloor$.
$x = x \wedge 37;$ // Exclusive-or bits of x with $0\ldots0100101$ binary.
$x = x \ \& \ 37;$ // And bits of x with $0\ldots0100101$ binary.

For() loops

The for() loop is ubiquitous in this book, in being our primary automaton for executing tasks repeatedly. Again we defer to a set of examples, not trying rigorously to exhaust all possible forms of English-C loops possible, rather, covering some of the common styles appearing in the book.

for($a \le x < b$) task(); // For all integers $x \in [a, b)$, *ascending* order.
for($a \ge x \ge b$) task(); // For all integers $x \in [b, a]$, *descending* order.
for($x \in [a, b)$) task(); // For all integers $x \in [a, b)$, *ascending* order.

Note that the relative magnitudes of a, b in the above are assumed correct to imply the ascending or descending order; e.g., if a loop starts for($a \ge \ldots$), then b should not exceed a (or if it does, the loop is considered empty). Note also that the first and third for() examples above are equivalent; we are just saying that the third form is allowed under our design rules. Note further that neither a nor b is necessarily an integer. This is why we cannot have put in a comment in the first for() example above like: "For $x = a, a+1, a+2, \ldots, b-1$," although such a comment does apply if both a, b are integers with $a < b$. Along such lines, an example of how for() conditionals get into more traditional mathematical notation is

for($1 \le a$ and $a^2 \le m$) task(); // Perform task for $a = 1, 2, \ldots \lfloor \sqrt{m} \rfloor$.

of which Algorithm 7.5.8 is an example. Other examples of mixed English-C are:

for(prime $p|F$) task(); // Perform for all primes p dividing F.
//Note: p values are in ascending order, unless otherwise specified.
for($p \in \mathcal{P}$, $p \in \mathcal{S}$) task(); // Perform for all primes $p \in \mathcal{S}$; observe order.
for(odd $j \in [1, C]$) task(); // Perform for $j = 1, 3, 5, \ldots$ not exceeding C.

Algorithms 3.2.1, 4.1.7, 7.4.4 involve such for() constructs as those above. For more general looping constraints, we have elected to adopt the standard C syntax, especially when the running variable is supposed to jump by nontrivial amounts. We exemplify this general case by for($j = q$; $j < B$; $j = j + p$) task(); // C-style loop form.

Assuming q is an integer, the above loop means that j takes on the values $q,\ q+p, q+2p, \ldots, q+kp$, where k is the largest integer strictly less than $(B-q)/p$. Algorithm 3.2.1 is an example of the use of this more general C loop. Incidentally, for nonprogrammers there is a good rule of thumb for dispelling confusion on the question: Exactly when do the innards of this general loop execute? Looking at the for() loop above, we can phrase the rule as: The task() is *never* allowed to execute when the middle conditional is false, i.e. if $j \ge B$ the loop innards will not execute for such a j value and the loop terminates. Another rule is: The incrementing $j = j + p$ occurs *after* a pass of the loop innards (throughout our pseudocode we assume the innards do not further modify the running variable). So one can see that after any pass of the loop's innards, j is increased by p, and then the middle conditional is checked.

Program control

Our pseudocode is to be executed starting at the top, although sometimes we merely place callable functions/procedures there; in such cases the order of placement is irrelevant, and we actually begin execution at the first obvious label that occurs after functions/procedures are defined. In any case we intend the pseudocode statements to follow labels that appear in brackets [ ], like so:

```
3. [Test p for primality]
        Indented statement;
        Indented statement;
        ...;
```

with the statements executed in serial, downward fashion (unless of course there is a goto [Another label]; see below on "goto"). It is important to note that in such a label as [Test p ...] above, we do not intend execution to happen right at the label itself. The label is never an executable statement. (This is much the same as with comments set off by "//" in which tasks are described rather than performed.) In the above example we expect primality testing to occur somewhere below the label, via actual indented statements.

Thus we have given labels "in English," intending them to be thematic of the pseudocode to follow, up to the next label. The serial, downward order is absolute; for example, the above label or any label for that matter can be interpreted as [Next, test p ...]; in the case of function/procedure definitions a label means [Next, define a function/procedure].

In some instances the pseudocode has been simplified by the use of "goto" statements, as in "goto [Test p ...]," which directs us to the indicated label where we start executing in downward order from that new label.

All of our pseudocode loops use braces { and } to denote begin/end of the loop innards. This use of braces is independent of their use to denote sets. Also the use of braces to indicate the operational block for a function or procedure (see next section) is independent of the set notation.

Functions and return/report values

Typically a customized function in our pseudocode is cast in the form

```
func(x) {
    ...;
    ...;
    return y;
}
```

and the idea is the same as in most any modern language: One calls $func(x)$ in the same way one would call a trigonometric function or a square root, with the attained value y. Similarly, a procedure (as opposed to a function) has the same syntax, with no returned value, although certain variables are usually set within a procedure. Also, a return statement is an exit statement, e.g., a sequence

```
if(x ≠ y) return x^3;
return x^4;
```

does not need an "else" structure for the x^4 case, because we always assume the current function/procedure *exits immediately* on any demand by the if() statement here. Likewise, a return statement, when executed, immediately causes exit from within any while() or for() loop.

Finally, we use report statements in the following way. Instead of returning a value from a function/procedure, a report statement simply relays the value—as in printing it, or reporting it to another program—on the fly, as it were. Thus the following function exemplifies the use of report/return (the function assumes a subroutine that evaluates the number-of-divisors function $d(n)$):

```
mycustomπ(x) {          //Report (and count!) all primes not exceeding x.
    c = 0;                          //This c will be the primes count.
    for(2 ≤ n ≤ x) {
        if(d(n) == 2) {
            c = c + 1;
            report n;              //As in "print" n, but keep looping.
        }
    }
    return c;
}
```

Primes will be reported in ascending order, with the return value of function $mycustom\pi(x)$ being the classical $\pi(x)$.

References

[Adleman 1994] L. Adleman. The function field sieve. In L. Adleman and M.-D. Huang, editors, *Algorithmic Number Theory: Proc. ANTS-I, Ithaca, NY*, volume 877 of *Lecture Notes in Computer Science*, pages 108–121. Springer-Verlag, 1994.

[Adleman and Huang 1992] L. Adleman and M.-D. Huang. *Primality testing and abelian varieties over finite fields*, volume 1512 of *Lecture Notes in Mathematics*. Springer-Verlag, 1992.

[Adleman et al. 1983] L. Adleman, C. Pomerance, and R. Rumely. On distinguishing prime numbers from composite numbers. *Ann. of Math.*, 117:173–206, 1983.

[Agarwal and Cooley 1986] R. Agarwal and J. Cooley. Fourier transform and convolution subroutines for the IBM 3090 vector facility. *IBM Journal of Research and Development*, 30:145–162, 1986.

[Aho et al. 1974] A. Aho, J. Hopcroft, and J. Ullman. *The Design and Analysis of Computer Algorithms*. Addison-Wesley, 1974.

[Alford et al. 1994a] W. Alford, A. Granville, and C. Pomerance. There are infinitely many Carmichael numbers. *Ann. of Math.*, 139:703–722, 1994.

[Alford et al. 1994b] W. Alford, A. Granville, and C. Pomerance. On the difficulty of finding reliable witnesses. In L. Adleman and M.-D. Huang, editors, *Algorithmic Number Theory: Proc. ANTS-I, Ithaca, NY*, volume 877 of *Lecture Notes in Computer Science*, pages 1–16. Springer-Verlag, 1994.

[Alford and Pomerance 1995] W. Alford and C. Pomerance. Implementing the self-initializing quadratic sieve on a distributed network. In *Number-theoretic and algebraic methods in computer science (Moscow, 1993)*, pages 163–174. World Scientific, 1995.

[Alt 1979] H. Alt. Square rooting is as difficult as multiplication. *Computing*, 21:221–232, 1979.

[Apostol 1986] T. Apostol. *Introduction to Analytic Number Theory,* 3rd printing. Springer-Verlag, 1986.

[Arazi 1994] B. Arazi. On primality testing using purely divisionless operations. *The Computer Journal*, 37:219–222, 1994.

[Archibald 1949] R. Archibald. Outline of the history of mathematics. *Amer. Math. Monthly*, 56, 1949. The second Herbert Ellsworth Slaught Memorial Paper: supplement to no. 1 issue, 114 pp.

[Artjuhov 1966/67] M. Artjuhov. Certain criteria for the primality of numbers connected with the little Fermat theorem (Russian). *Acta Arith.*, 12:355–364, 1966/67.

[Ashworth and Lyne 1988] M. Ashworth and A. Lyne. A segmented FFT algorithm for vector computers. *Parallel Computing*, 6:217–224, 1988.

[Atkin 1986] A. Atkin. Schoof's algorithm. Unpublished manuscript, 1986.

[Atkin 1988] A. Atkin. The number of points on an elliptic curve modulo a prime (i). Unpublished manuscript, 1988.

[Atkin 1992] A. Atkin. The number of points on an elliptic curve modulo a prime (ii). Unpublished manuscript, 1992.

[Atkin and Bernstein 1999] A. Atkin and D. Bernstein. Prime sieves using binary quadratic forms, 1999. `http://pobox.com/~djb/papers/primesieves.dvi`.

[Atkin and Morain 1993a] A. Atkin and F. Morain. Finding suitable curves for the elliptic curve method of factorization. *Math. Comp.*, 60:399–405, 1993.

[Atkin and Morain 1993b] A. Atkin and F. Morain. Elliptic curves and primality proving. *Math. Comp.*, 61:29–68, 1993.

[Bach 1985] E. Bach. *Analytic Methods in the Analysis and Design of Number-Theoretic Algorithms*. A 1984 ACM Distinguished Dissertation. The MIT Press, 1985.

[Bach 1990] E. Bach. Explicit bounds for primality testing and related problems. *Math. Comp.*, pages 355–380, 1990.

[Bach 1991] E. Bach. Toward a theory of Pollard's rho method. *Inform. and Comput.*, 90:139–155, 1991.

[Bach 1997a] E. Bach. The complexity of number-theoretic constants. *Inform. Process. Lett.*, 62:145–152, 1997.

[Bach 1997b] E. Bach. Comments on search procedures for primitive roots. *Math. Comp.*, 66(220):1719–1727, 1997.

[Bach and Shallit 1996] E. Bach and J. Shallit. *Algorithmic Number Theory*, volume I. MIT Press, 1996.

[Baillie and Wagstaff 1980] R. Baillie and S. Wagstaff Jr. Lucas pseudoprimes. *Math. Comp.*, 35:1391–1417, 1980.

[Bailey 1990] D. Bailey. FFTs in external or hierarchical memory. *J. Supercomp.*, 4:23–35, 1990.

[Bailey and Crandall 2001a] D. Bailey and R. Crandall. On the random character of fundamental constant expansions, *Experiment. Math.*, 10:175–190, 2001.

[Bailey and Crandall 2001b] D. Bailey and R. Crandall. Random generators and normal numbers, 2001. `http://hpcf.nersc.gov/ dhbailey/dhbpapers`.

[Balasubramanian and Nagaraj 1997] R. Balasubramanian and S. Nagaraj. Density of Carmichael numbers with three prime factors. *Math. Comp.*, 66:1705–1708, 1997.

[Balazard et al. 1999] M. Balazard, E. Saias, and M. Yor. Notes sur la fonction ζ de Riemann. II. *Adv. Math.*, 143:284–287, 1999.

[Balog 1989] A. Balog. On a variant of the Piatetski-Shapiro prime number theorem. In *Groupe de travail en théorie analytique et élementaire des nombres, 1987–1988*, volume 89-01 of *Publ. Math. Orsay*, pages 3–11. Univ. Paris XI, Orsay, 1989.

[Barrett 1987] P. Barrett. Implementing the Rivest Shamir and Adleman public key encryption algorithm on a standard digital signal processor. In A. Odlyzko, editor, *Advances in Cryptology, Proc. Crypto '86*, volume 263 of *Lecture Notes in Computer Science*, pages 311–323. Springer-Verlag, 1987.

[Bateman et al. 1989] P. Bateman, J. Selfridge, and S. Wagstaff Jr. The new Mersenne conjecture. *Amer. Math. Monthly*, 96:125–128, 1980.

[Bays and Hudson 2000a] C. Bays and R. Hudson. Zeroes of Dirichlet L-functions and irregularities in the distibution of primes. *Math. Comp.*, 69:861–866, 2000.

[Bays and Hudson 2000b] C. Bays and R. Hudson. A new bound for the smallest x with $\pi(x) > \mathrm{li}\,(x)$. *Math. Comp.*, 69:1285–1296, 2000.

[Bernstein 1997a] D. Bernstein. Multidigit multiplication for mathematicians, 1997.

[Bernstein 1997b] D. Bernstein. Multidigit modular multiplication with the explicit chinese remainder theorem, 1997. `ftp://koobera.math.uic.edu/pub/papers/m3.dvi`.

[Bernstein 1998] D. Bernstein. Bounding smooth integers (extended abstract). In [Buhler 1998], pages 128–130.

[Berry 1997] M. Berry. Quantum chaology. *Proc. Roy. Soc. London Ser. A*, 413:183–198, 1987.

[Blackburn and Teske 1999] S. Blackburn and E. Teske. Baby-step giant-step algorithms for non-uniform distributions. Unpublished manuscript, 1999.

[Bleichenbacher 1996] D. Bleichenbacher. *Efficiency and security of cryptosystems based on number theory.* PhD thesis, Swiss Federal Institute of Technology Zürich, 1996.

[Blum et al. 1986] L. Blum, M. Blum, and M. Shub. A simple unpredictable pseudorandom number generator. *SIAM J. Comput.*, 15:364–383, 1986.

[Bombieri and Iwaniec 1986] E. Bombieri and H. Iwaniec. On the order of $\zeta(1/2 + it)$. *Ann. Scuola Norm. Sup. Pisa Cl. Sci. (4)*, 13:449–472, 1986.

[Bombieri and Lagarias 1999] E. Bombieri and J. Lagarias. Complements to Li's criterion for the Riemann hypothesis. *J. Number Theory*, 77:274–287, 1999.

[Boneh 1999] D. Boneh. Twenty years of attacks on the RSA cryptosystem. *Notices Amer. Math. Soc.*, 46:203–213, 1999.

[Boneh and Venkatesan 1998] D. Boneh and R. Venkatesan. Breaking RSA may not be equivalent to factoring. In *Advances in Cryptology, Proc. Eurocrypt '98*, volume 1514 of *Lecture Notes in Computer Science*, pages 25–34, 1998.

[Borwein 1991] P. Borwein. On the irrationality of $\sum(1/(q^n + r))$. *J. Number Theory*, 37:253–259, 1991.

[Borwein and Borwein 1987] J. Borwein and P. Borwein. *Pi and the AGM: A Study in Analytic Number Theory and Computational Complexity.* John Wiley and Sons, 1987.

[Borwein et al. 2000] J. Borwein, D. Bradley, and R. Crandall. Computational strategies for the Riemann zeta function. *J. Comp. App. Math.*, 121:247–296, 2000.

[Bosma and van der Hulst 1990] W. Bosma and M.-P. van der Hulst. *Primality proving with cyclotomy.* PhD thesis, University of Amsterdam, 1990.

[Bosselaers et al. 1994] A. Bossalaers, R. Govaerts, and J. Vandewalle. Comparison of three modular reduction functions. In D. Stinson, editor, *Advances in Cryptology, Proc. Crypto '93*, number 773 in Lecture Notes in Computer Science, pages 175–186. Springer-Verlag, 1994.

[Boyle et al. 1995] P. Boyle, M. Broadie, and P. Glasserman. Monte Carlo methods for security pricing. Unpublished manuscript, June 1995.

[Bratley and Fox 1988] P. Bratley and B. Fox. ALGORITHM 659: Implementing Sobol's quasirandom sequence generator. *ACM Trans. Math. Soft.*, 14:88–100, 1988.

[Bredihin 1963] B. Bredihin. Applications of the dispersion method in binary additive problems. *Dokl. Akad. Nauk. SSSR*, 149:9–11, 1963.

[Brent 1979] R. Brent. On the zeros of the Riemann zeta function in the critical strip. *Math. Comp.*, 33:1361–1372, 1979.

[Brent 1994] R. Brent. On the period of generalized Fibonacci recurrences. *Math. Comp.*, 63:389–401, 1994.

[Brent 1999] R. Brent. Factorization of the tenth Fermat number. *Math. Comp.*, 68:429–451, 1999.

[Brent et al. 1993] R. Brent, G. Cohen, and H. te Riele. Improved techniques for lower bounds for odd perfect numbers. *Math. Comp.*, 61:857–868, 1993.

[Brent et al. 2000] R. Brent, R. Crandall, K. Dilcher, and C. van Halewyn. Three new factors of Fermat numbers. *Math. Comp.*, 69: 1297–1304, 2000.

[Brent and Pollard 1981] R. Brent and J. Pollard. Factorization of the eighth Fermat number. *Math. Comp.*, 36:627–630, 1981.

[Bressoud and Wagon 2000] D. Bressoud and S. Wagon. *A Course in Computational Number Theory.* Key College Publishing, 2000.

[Brillhart et al. 1981] J. Brillhart, M. Filaseta, and A. Odlyzko. On an irreducibility theorem of A. Cohn. *Canad. J. Math.*, 33:1055–1059, 1981.

[Brillhart et al. 1988] J. Brillhart, D. Lehmer, J. Selfridge, B. Tuckerman, and S. Wagstaff Jr. *Factorizations of* $b^n \pm 1$, $b = 2, 3, 5, 6, 7, 10, 11, 12$ *up to high powers*. Second edition, volume 22 of *Contemporary Mathematics*. Amer. Math. Soc., 1988.

[Buchmann et al. 1997] J. Buchmann, M. Jacobson Jr, and E. Teske. On some computational problems in finite groups. *Math. Comp.*, 66:1663–1687, 1997.

[Buell and Young 1988] D. Buell and J. Young. The twentieth Fermat number is composite. *Math. Comp.*, 50:261–263, 1988.

[Buhler 1991] J. Buhler, 1991. Private communication.

[Buhler 1998] J. Buhler, editor. *Algorithmic Number Theory: Proc. ANTS-III, Portland, OR*, volume 1423 of *Lecture Notes in Computer Science*. Springer-Verlag, 1998.

[Buhler 2000] J. Buhler, R. Crandall, R. Ernvall, T. Metsänkylä, and M. Shokrollahi. Irregular primes and cyclotomic invariants to 12 million. *J. Symbolic Comput.*, 11:1–8, 2000.

[Buhler et al. 1993] J. Buhler, H. Lenstra Jr, and C. Pomerance. Factoring integers with the number field sieve. In A. Lenstra and H. Lenstra Jr, editors, *The development of the number field sieve*, volume 1554 of *Lecture Notes in Mathematics*, pages 50–94. Springer-Verlag, 1993.

[Bürgisser et al. 1997] P. Bürgisser, M. Clausen, and M. Shokrollahi. *Algebraic Complexity Theory*. Springer-Verlag, 1997.

[Burnikel and Ziegler 1998] C. Burnikel and J. Ziegler. Fast recursive division, 1998. `http://www.mpi-sb.mpg.de/~ziegler/TechRep.ps.gz`.

[Burthe 1996] R. Burthe. Further investigations with the strong probable prime test. *Math. Comp.*, 65:373–381, 1996.

[Burthe 1997] R. Burthe. Upper bounds for least witnesses and generating sets. *Acta Arith.*, 80:311–326, 1997.

[Caldwell 1999] C. Caldwell. Website for prime numbers, 1999. `http://www.utm.edu/research/primes`.

[Canfield et al. 1983] E. Canfield, P. Erdős, and C. Pomerance. On a problem of Oppenheim concerning "factorisatio numerorum". *J. Number Theory*, 17:1–28, 1983.

[Cassels 1966] J. Cassels. Diophantine equations with special reference to elliptic curves. *J. London Math. Soc.*, 41:193–291, 1966.

[Cesari 1998] G. Cesari. Parallel implementation of Schönhage's integer GCD algorithm. In [Buhler 1998], pages 64–76.

[Chen 1966] J. Chen. On the representation of a large even integer as the sum of a prime and the product of at most two primes. *Kexue Tongbao*, 17:385–386, 1966.

[Cheng 1999] Y. Cheng. An explicit upper bound for the Riemann zeta-function near the line $\sigma = 1$. *Rocky Mountain J. Math.*, 29:115–140, 1999.

[Cochrane 1987] T. Cochrane. On a trigonometric inequality of Vinogradov. *J. Number Theory*, 27:9–16, 1987.

[Cohen 2000] H. Cohen. *A Course in Computational Algebraic Number Theory*, volume 138 of *Graduate Texts in Mathematics*. Springer-Verlag, 2000.

[Cohen et al. 1998] H. Cohen, A. Miyaji, and T. Ono. Efficient elliptic curve exponentiation using mixed coordinates. In *Advances in Cryptology, Proc. Asiacrypt '98*, volume 1514 of *Lecture Notes in Computer Science*, pages 51–65. Springer-Verlag, 1998.

[Contini 1997] S. Contini. *Factoring integers with the self-initializing quadratic sieve*. Masters thesis, U. Georgia, 1997.

[Cooley and Tukey 1965] J. Cooley and J. Tukey. An algorithm for the machine calculation of complex Fourier series. *Math. Comp.*, 19:297–301, 1965.

[Copeland and Erdős 1946] A. Copeland and P. Erdős. Note on normal numbers. *Bull. Amer. Math. Soc.*, 52:857–860, 1946.

[Coppersmith 1993] D. Coppersmith. Modifications to the number field sieve. *J. Cryptology*, 6:169–180, 1993.

[Coppersmith 1997] D. Coppersmith. Small solutions to polynomial equations, and low exponent RSA vulnerabilities. *J. Cryptology*, 10:233–260, 1997.

[Coppersmith et al. 2000] D. Coppersmith, N. Howgrave-Graham, and S. Nagaraj. Divisors in residue classes, constructively. Unpublished manuscript, May 2000.

[Couveignes 1993] J.-M. Couveignes. Computing a square root for the number field sieve. In A. Lenstra and H. Lenstra Jr, editors, *The development of the number field sieve*, volume 1554 of *Lecture Notes in Mathematics*, pages 95–102. Springer-Verlag, 1993.

[Couveignes and Morain 1994] J.-M. Couveignes and F. Morain. Schoof's algorithm and isogeny cycles. In L. Adleman and M.-D. Huang, editors, *Algorithmic Number Theory: Proc. ANTS-I, Ithaca, NY*, volume 877 of *Lecture Notes in Computer Science*, pages 43–58. Springer-Verlag, 1994.

[Couveignes et al. 1996] J.-M. Couveignes, L. Dewaghe, and F. Morain. Isogeny cycles and the Schoof–Atkin–Elkies algorithm. Unpublished manuscript, 1996.

[Cox 1989] D. Cox. *Primes of the Form $x^2 + ny^2$*. John Wiley and Sons, 1989.

[Craig-Wood 1998] N. Craig-Wood, 1998. Private communication.

[Crandall 1994a] R. Crandall. Method and apparatus for public key exchange in a cryptographic system., 1994. U.S. Patents #5159632 (1992), #5271061 (1993), #5463690 (1994).

[Crandall 1994b] R. Crandall. *Projects in Scientific Computation*. TELOS/Springer-Verlag, 1994.

[Crandall 1996a] R. Crandall. *Topics in Advanced Scientific Computation.* TELOS/Springer-Verlag, 1996.

[Crandall 1996b] R. Crandall. Method and apparatus for Digital Signature Authentication, 1996. U. S. Patent #5581616.

[Crandall 1997a] R. Crandall. The challenge of large numbers. *Scientific American*, pages 58–62, February 1997.

[Crandall 1997b] R. Crandall. Integer convolution via split-radix fast Galois transform, 1997. `http://www.perfsci.com`.

[Crandall 1998] R. Crandall. Recycled (simultaneous) evaluations of the Riemann zeta function. Unpublished manuscript, 1998.

[Crandall 1999a] R. Crandall. Applications of space-filling curves. Unpublished manuscript, 1999.

[Crandall 1999b] R. Crandall. Fast algorithms for elliptic curve cryptography. Unpublished manuscript, 1999.

[Crandall 1999c] R Crandall. Alternatives to the Riemann–Siegel formula. Unpublished manuscript, 1999.

[Crandall 1999d] R. Crandall. Parallelization of Pollard-rho factorization, 1999. `http://www.perfsci.com`.

[Crandall et al. 1997] R. Crandall, K. Dilcher, and C. Pomerance. A search for Wieferich and Wilson primes. *Math. Comp.*, 66:433–449, 1997.

[Crandall et al. 1995] R. Crandall, J. Doenias, C. Norrie, and J. Young. The twenty-second Fermat number is composite. *Math. Comp.*, 64 210:863–868, 1995.

[Crandall and Fagin 1994] R. Crandall and B. Fagin. Discrete weighted transforms and large integer arithmetic. *Math. Comp.*, 62:305–324, 1994.

[Crandall et al. 1999] R. Crandall, E. Mayer, and J. Papadopoulos. The twenty-fourth Fermat number is composite, 1999. `http://www.perfsci.com`.

[Crandall and Garst 2001] R. Crandall, Method and apparatus for fast elliptic encryption with direct embedding, U. S. Patent #6307935, 2001.

[Creutzburg and Tasche 1989] R. Creutzburg and M. Tasche. Parameter determination for complex number-theoretic transforms using cyclotomic polynomials. *Math. Comp.*, 52:189–200, 1989.

[Damgård et al. 1993] I. Damgård, P. Landrock, and C. Pomerance. Average case error estimates for the strong probable prime test. *Math. Comp.*, 61:177–194, 1993.

[Darmon and Granville 1995] H. Darmon and A. Granville. On the equations $z^m = F(x, y)$ and $Ax^p + By^q = Cz^r$. *Bull. London Math. Soc.*, 27:513–543, 1995.

[Davenport 1980] H. Davenport. *Multiplicative Number Theory* (second edition). Springer-Verlag, 1980.

[Davis 1973] M. Davis. Hilbert's tenth problem is unsolvable. *Amer. Math. Monthly*, 80:233–269, 1973.

[De Win et al. 1998] E. De Win, S. Mister, B. Preneel, and M. Wiener. On the performance of signature schemes based on elliptic curves. In [Buhler 1998], pages 252–266.

[Deléglise and Rivat 1996] M. Deléglise and J. Rivat. Computing $\pi(x)$: the Meissel, Lehmer, Lagarias, Miller, Odlyzko method. *Math. Comp.*, 65:235–245, 1996.

[Deléglise and Rivat 1998] M. Deléglise and J. Rivat. Computing $\psi(x)$. *Math. Comp.*, 67:1691–1696, 1998.

[Deshouillers et al. 1998] J.-M. Deshouillers, H. te Riele, and Y. Saouter. New experimental results concerning the Goldbach conjecture. In [Buhler 1998], pages 204–215.

[Deuring 1941] M. Deuring. Die Typen der Multiplikatorenringe elliptischer Funktionenkörper. *Abh. Math. Sem. Hansischen Univ.*, 14:197–272, 1941.

[Deutsch 1982] D. Deutsch. Is there a fundamental bound on the rate at which information can be processed? *Phys. Rev. Lett.*, 42:286–288, 1982.

[Deutsch 1985] D. Deutsch. Quantum theory, the Church–Turing principle, and the universal quantum computer. *Proc. Roy. Soc. London Ser. A*, 400:97–117, 1985.

[Dickson 1904] L. Dickson. A new extension of Dirichlet's theorem on prime numbers. *Messenger of Math.*, 33:155–161, 1904.

[Diffie and Hellman 1976] W. Diffie and M. Hellman. New directions in cryptography. *IEEE Trans. Inform. Theory*, 22:644–654, 1976.

[Dilcher 1999] K. Dilcher. Nested squares and evaluation of integer products, 1999. `http://www.mscs.dal.ca/~dilcher/Preprints/nested.ps`.

[Dimitrov et al. 1995] V. Dimitrov, T. Cooklev, and B. Donevsky. Number theoretic transforms over the golden section quadratic field. *IEEE Trans. Sig. Proc.*, 43:1790–1797, 1995.

[Dimitrov et al. 1998] V. Dimitrov, G. Jullien, and W. Miller. A residue number system implementation of real orthogonal transforms. *IEEE Trans. Sig. Proc.*, 46:563–570, 1998.

[Ding et al. 1996] C. Ding, D. Pei, and A. Salomaa. *Chinese Remainder Theorem: Applications in Computing, Coding, Cryptography*. World Scientific, 1996.

[Dixon 1981] J. Dixon. Asymptotically fast factorization of integers. *Math. Comp.*, 36:255–260, 1981.

[Dress and Olivier 1999] F. Dress and M. Olivier. Polynômes prenant des valeurs premières. *Experiment. Math.*, 8:319–338, 1999.

[Dubner et al. 1998] H. Dubner, T. Forbes, N. Lygeros, M. Mizony, and P. Zimmermann. Ten consecutive primes in arithmetic progression, 1998. http://listserv.nodak.edu/archives/nmbrthry.html.

[Dubner and Gallot 2000] H. Dubner and Y. Gallot. Distribution of generalized Fermat numbers. Unpublished manuscript, 2000.

[Dudon 1987] J. Dudon. The golden scale. *Pitch*, I/2:1–7, 1987.

[Edwards 1974] H. Edwards. *Riemann's Zeta Function.* Academic Press, 1974.

[Ekert and Jozsa 1996] A. Ekert and R Jozsa. Quantum computation and Shor's factoring algorithm. *Rev. Mod. Phys.*, 68:733–753, 1996.

[Elkenbracht-Huizing 1997] M. Elkenbracht-Huizing. *Factoring integers with the Number Field Sieve.* PhD thesis, University of Leiden, 1997.

[Elkies 1991] N. Elkies. Explicit isogenies. Unpublished manuscript, 1991.

[Elkies 1997] N. Elkies. Elliptic and modular curves over finite fields and related computational issues. In J. Teitelbaum, editor, *Computational Perspectives on Number Theory (Chicago, IL, 1995)*, volume 7 of *AMS/IP Stud. Adv. Math.*, pages 21–76. Atkin Conference, Amer. Math. Soc., 1998.

[Ellison and Ellison 1985] W. Ellison and F. Ellison. *Prime Numbers.* John Wiley and Sons, 1985.

[Erdős 1948] P. Erdős. On arithmetical properties of Lambert series. *J. Indian Math. Soc. (N.S.)*, 12:63–66, 1948.

[Erdős 1950] P. Erdős. On almost primes. *Amer. Math. Monthly*, 57:404–407, 1950.

[Erdős and Pomerance 1986] P. Erdős and C. Pomerance. On the number of false witnesses for a composite number. *Math. Comp.*, 46:259–279, 1986.

[Erdős et al. 1988] P. Erdős, P. Kiss, and A. Sárközy. A lower bound for the counting function of Lucas pseudoprimes. *Math. Comp.*, 51:315–323, 1988.

[Escott et al. 1998] A. Escott, J. Sager, A. Selkirk, and D. Tsapakidis. Attacking elliptic curve cryptosystems using the parallel Pollard rho method. *RSA Cryptobytes*, 4(2):15–19, 1998.

[Estermann 1952] T. Estermann. *Introduction to Modern Prime Number Theory.* Cambridge University Press, 1952.

[Faber and Faber 2000] Faber and Faber Publishing Company. $1,000,000 challenge to prove Goldbach's conjecture, 2000. `http://www.faber.co.uk/faber/million_dollar.asp`.

[Faure 1981] H. Faure. Discrépances de suites associées à un système de numération (en dimension un). *Bull. Soc, Math. France*, 109:143–182, 1981.

[Faure 1982] H. Faure. Discrépances de suites associées à un système de numération (en dimension s). *Acta Arith.*, 41:337–351, 1982.

[Feynman 1982] R. Feynman. Simulating physics with computers. *Intl. J. Theor. Phys.*, 21(6/7):467–488, 1982.

[Feynman 1985] R. Feynman. Quantum mechanical computers. *Optics News*, II:11–20, 1985.

[Forbes 1999] T. Forbes. Prime k-tuplets, 1999. `http://www.ltkz.demon.co.uk/ktuplets.htm`.

[Fraser 1976] D. Fraser. Array permutation by index-digit permutation. *J. ACM*, 23:298–309, 1976.

[Friedlander and Iwaniec 1998] J. Friedlander and H. Iwaniec. The polynomial $X^2 + Y^4$ captures its primes. *Ann. of Math.*, 148:945–1040, 1998.

[Friedlander et al. 2001] J. Friedlander, C. Pomerance, and I. Shparlinski. Period of the power generator and small values of Carmichael's function. *Math. Comp.* 70:1591–1605, 2001.

[Furry 1942] W. Furry. Number of primes and probability considerations. *Nature*, 150:120–121, 1942.

[Gabcke 1979] W. Gabcke. *Neue Herleitung und explizite Restabschätzung der Riemann–Siegel Formel.* PhD thesis, Georg-August-Universität zu Göttingen, 1979.

[Gallot 1999] Y. Gallot, 1999. Private communication.

[Galway 1998] W. Galway, 1998. Private communication.

[Galway 2000] W. Galway. *Analytic computation of the prime-counting function.* PhD thesis, U. Illinois at Urbana-Champaign, 2000.

[Gardner 1977] M. Gardner. Mathemtical games: a new kind of cipher that would take millions of years to break. *Scientific American*, August 1977.

[Gentleman and Sande 1966] W. Gentleman and G. Sande. Fast Fourier transforms—for fun and profit. In *Proc. AFIPS*, volume 29, pages 563–578, 1966.

[Goldwasser and Kilian 1986] S. Goldwasser and J. Kilian. Almost all primes can be quickly certified. In *Proc. 18th Annual ACM Symposium on the Theory of Computing*, pages 316–329, 1986.

[Goldwasser and Micali 1982] S. Goldwasser and S. Micali. Probabilistic encryption and how to play mental poker keeping secret all mental information. In *Proc. 14th Annual ACM Symposium on the Theory of Computing*, pages 365–377, 1982.

[Golomb 1956] S. Golomb. Combinatorial proof of Fermat's 'little theorem'. *Amer. Math. Monthly*, 63, 1956.

[Golomb 1982] S. Golomb. *Shift Register Sequences,* (revised version). Aegean Park Press, 1982.

[Gong et al. 1999] G. Gong, T. Berson, and D. Stinson. Elliptic curve pseudorandom sequence generators. In *Proc. Sixth Annual Workshop on Selected Areas in Cryptography*, Kingston, Canada, August 1999.

[Gordon 1993] D. Gordon. Discrete logarithms in $GF(p)$ via the number field sieve. *SIAM J. Discrete Math.*, 16:124–138, 1993.

[Gordon and Pomerance 1991] D. Gordon and C. Pomerance. The distribution of Lucas and elliptic pseudoprimes. *Math. Comp.*, 57:825–838, 1991. Corrigendum *ibid.* 60:877, 1993.

[Gordon and Rodemich 1998] D. Gordon and G. Rodemich. Dense admissible sets. In [Buhler 1998], pages 216–225.

[Graham and Kolesnik 1991] S. Graham and G. Kolesnik. *Van der Corput's method of exponential sums*, volume 126 of *Lecture Note Series.* Cambridge University Press, 1991.

[Grantham 1998] J. Grantham. A probable prime test with high confidence. *J. Number Theory*, 72:32–47, 1998.

[Grantham 2001] J. Grantham. Frobenius pseudoprimes. *Math. Comp.* 70:873–891, 2001.

[Guy 1976] R. Guy. How to factor a number. In *Proceedings of the Fifth Manitoba Conference on Numerical Mathematics (Univ. Manitoba, Winnipeg, Man., 1975)*, volume 16 of *Congressus Numerantium*, pages 49–89, 1976.

[Guy 1994] R. Guy. *Unsolved Problems in Number Theory.* Second edition, volume I of *Problem Books in Mathematics. Unsolved Problems in Intuitive Mathematics.* Springer-Verlag, 1994.

[Hafner and McCurley 1989] J. Hafner and K. McCurley. A rigorous subexponential algorithm for computation of class groups. *J. Amer. Math. Soc.*, 2:837–850, 1989.

[Halberstam and Richert 1974] H. Halberstam and H.-E. Richert. *Sieve Methods*, volume 4 of *London Mathematical Society Monographs.* Academic Press, 1974.

[Hardy 1966] G. Hardy. *Collected Works of G. H. Hardy*, Vol. I. Clarendon Press, Oxford, 1966.

[Hardy and Wright 1979] G. Hardy and E. Wright. *An Introduction to the Theory of Numbers.* Fifth edition. Clarendon Press, Oxford, 1979.

[Håstad et al. 1999] J. Håstad, R. Impagliazzo, L. Levin, and M. Luby. A pseudorandom generator from any one-way function. *SIAM J. Computing*, 28:1364–1396, 1999. `http://www.siam.org/journals/sicomp/28-4/24470.html`.

[Hensley and Richards 1973] D. Hensley and I. Richards. Primes in intervals. *Acta Arith.*, 25:375–391, 1973/74.

[Hey 1999] T. Hey. Quantum computing. *Computing and Control Engineering*, 10(3):105–112, 1999.

[Higham 1996] N. Higham. *Accuracy and Stability of Numerical Algorithms.* SIAM, 1996.

[Hildebrand 1988a] A. Hildebrand. On the constant in the Pólya–Vinogradov inequality. *Canad. Math. Bull.*, 31:347–352, 1988.

[Hildebrand 1988b] A. Hildebrand. Large values of character sums. *J. Number Theory*, 29:271–296, 1988.

[Honaker 1998] G. Honaker, 1998. Private communication.

[Hooley 1976] C. Hooley. *Applications of Sieve Methods to the Theory of Numbers*, volume 70 of *Cambridge Tracts in Mathematics*. Cambridge University Press, 1976.

[Ivić 1985] A. Ivić. *The Riemann Zeta-Function*. John Wiley and Sons, 1985.

[Izu et al. 1998] T. Izu, J. Kogure, M. Noro, and K. Yokoyama. Efficient implementation of Schoof's algorithm. In *Advances in Cryptology, Proc. Asiacrypt '98*, volume 1514 of *Lecture Notes in Computer Science*, pages 66–79. Springer-Verlag, 1998.

[Jaeschke 1993] G. Jaeschke. On strong pseudoprimes to several bases. *Math. Comp.*, 61:915–926, 1993.

[Joe 1999] S. Joe. An average L_2 discrepancy for number-theoretic rules. *SIAM J. Numer. Anal.*, 36:1949–1961, 1999.

[Johnson and Menezes 1999] D. Johnson and A. Menezes. The elliptic curve digital signature algorithm (ECDSA), 1999. Technical Report CORR99-31, Dept. of Combinatorics and Optimization, University of Waterloo, `http://www.cacr.math.uwaterloo.ca`.

[Jurišić and Menezes 1997] A. Jurišić and A. Menezes. Elliptic curves and cryptography. *Dr. Dobb's Journal*, pages 26–36, April 1997.

[Kaczorowski 1984] J. Kaczorowski. On sign changes in the remainder-term of the prime-number formula. I. *Acta Arith.*, 44:365–377, 1984.

[Kaliski 1988] B. Kaliski Jr. *Elliptic Curves and Cryptography: a Pseudorandom Bit Generator and other Tools*. PhD thesis, Massachusetts Institute of Technology, 1988.

[Kaliski 1991] B. Kaliski Jr. One-way permutations on elliptic curves. *J. Cryptology*, 3:187–199, 1991.

[Keller 1999] W. Keller. Prime factors $k.2^n + 1$ of Fermat numbers F_m and complete factoring status, 1999. `http://www.prothsearch.net/fermat.html`.

[Knuth 1981] D. Knuth. *Seminumerical Algorithms* (Second edition), volume 2 of *The Art of Computer Programming*. Addison-Wesley, 1981.

[Knuth and Trabb Pardo 1976] D. Knuth and L. Trabb Pardo. Analysis of a simple factorization algorithm. *Theoret. Comput. Sci.*, 3:321–348, 1976-77.

[Koblitz 1987] N. Koblitz. Elliptic curve cryptosystems. *Math. Comp.*, 48:203–209, 1987.

[Koblitz 1994] N. Koblitz. *A Course in Number Theory and Cryptography*. Springer-Verlag, 1994.

[Koç et al. 1997] Ç. Koç, T. Acar, and B. Kaliski Jr. Analysis and comparison of several algorithms for Montgomery multiplication. Unpublished manuscript, 1997.

[Koç and Hung 1997] Ç. Koç and C. Hung. Fast algorithm for modular reduction. *IEEE Proc.: Computers and Digital Techniques*, 145(4), 1998.

[Kocis and White 1997] L. Kocis and W. Whiten. Computational investigations of low-discrepancy sequences. *ACM Trans. Math. Soft.*, 23:266–294, 1997.

[Konyagin and Pomerance 1997] S. Konyagin and C. Pomerance. On primes recognizable in deterministic polynomial time. In *The Mathematics of Paul Erdős, I*, volume 13 of *Algorithms and Combinatorics*, pages 176–198. Springer-Verlag, 1997.

[Korobov 1992] N. Korobov, *Exponential Sums and their Applications*, Kluwer Academic Publishers, 1992.

[Kuipers and Niederreiter 1974] L. Kuipers and H. Niederreiter. *Uniform Distribution of Sequences*. John Wiley and Sons, 1974.

[Lagarias 1990] J. Lagarias. Pseudorandom number generators in cryptography and number theory. In C. Pomerance, editor, *Cryptology and computational number theory*, volume 42 of *Proc. Sympos. Appl. Math.*, pages 115–143. Amer. Math. Soc., 1990.

[Lagarias 1999] J. Lagarias. On a positivity property of the Riemann ξ-function. *Acta Arith.*, 89:217–234, 1999.

[Lagarias et al. 1985] J. Lagarias, V. Miller, and A. Odlyzko. Computing $\pi(x)$: the Meissel-Lehmer method. *Math. Comp.*, 44:537–560, 1985.

[Lagarias and Odlyzko 1987] J. Lagarias and A. Odlyzko. Computing $\pi(x)$: an analytic method. *J. Algorithms*, 8:173–191, 1987.

[Languasco 2000] A. Languasco. Some refinements of error terms estimates for certain additive problems with primes. *J. Number Theory*, 81:149–161, 2000.

[L'Ecuyer and Simard 1999] P. L'Ecuyer and R. Simard. Beware of linear congruential generators with multipliers of the form $a = \pm 2^q \pm 2^r$. *ACM Trans. Math. Soft.*, 25:367–374, 1999.

[Lehman 1974] R. Lehman. Factoring large integers. *Math. Comp.*, 28:637–646, 1974.

[Lehmer 1964] E. Lehmer. On the infinitude of Fibonacci pseudo-primes. *Fibonacci Quart.*, 2:229–230, 1964.

[Lenstra 1983] A. Lenstra. Factoring polynomials over algebraic number fields. In *Computer algebra (London, 1983)*, volume 162 of *Lecture Notes in Computer Science*, pages 245–254, 1983.

[Lenstra and Lenstra 1993] A. Lenstra and H. Lenstra Jr, editors. *The development of the number field sieve*, volume 1554 of *Lecture Notes in Mathematics*. Springer-Verlag, 1993.

[Lenstra et al. 1982] A. Lenstra, H. Lenstra Jr, and L. Lovasz. Factoring polynomials with rational coefficients. *Math. Ann.*, 261:515–534, 1982.

[Lenstra et al. 1993a] A. Lenstra, H. Lenstra Jr, M. Manasse, and J. Pollard. The factorization of the ninth Fermat number. *Math. Comp.*, 61:319–349, 1993.

[Lenstra and Manasse 1994] A. Lenstra and M. Manasse. Factoring with two large primes. *Math. Comp.*, 63:785–798, 1994.

[Lenstra 1981] H. Lenstra Jr. Primality testing algorithms (after Adleman, Rumely and Williams). In *Seminar Bourbaki 33 (1980/81)*, volume 901 of *Lecture Notes in Mathematics*, 1981. exp. 576.

[Lenstra 1984] H. Lenstra Jr. Divisors in residue classes. *Math. Comp.*, 42:331–340, 1984.

[Lenstra 1985] H. Lenstra Jr. Galois theory and primality testing. In *Orders and their applications (Oberwolfach, 1984)*, volume 1142 of *Lecture Notes in Mathematics*, pages 169–189. Springer-Verlag, 1985.

[Lenstra 1987] H. Lenstra Jr. Factoring integers with elliptic curves. *Ann. of Math.*, 2:649–673, 1987.

[Lenstra 1991] H. Lenstra Jr, 1991. Private communication.

[Lenstra et al. 1993b] H. Lenstra Jr, J. Pila, and C. Pomerance. A hyperelliptic smoothness test. I. *Philos. Trans. Roy. Soc. London Ser. A*, 345:397–408, 1993. Special issue compiled and edited by R. Vaughan: Theory and applications of numbers without large prime factors.

[Lenstra and Pomerance 1992] H. Lenstra Jr and C. Pomerance. A rigorous time bound for factoring integers. *J. Amer. Math. Soc.*, 5:483–516, 1992.

[Li 1997] X. Li. The positivity of a sequence of numbers and the Riemann hypothesis. *J. Number Theory*, 65:325–333, 1997.

[Lindqvist and Peetre 1997] P. Lindqvist and J. Peetre, On the remainder in a series of Mertens, *Exposition. Math.*, 15:467–478, 1997.

[Lim and Lee 1997] C. Lim and P. Lee. A key recovery attack on discrete log-based schemes using a prime order subgroup. In *Advances in Cryptology, Proc. Crypto '97*, volume 1294 of *Lecture Notes in Computer Science*, pages 249–265. Springer-Verlag, 1997.

[Long 1981] D. Long. Random equivalence of factorization and computation of orders, 1981. Princeton U. Dept. Elec. Eng. and Comp. Sci. Technical Report 284.

[Lovorn 1992] R. Lovorn. *Rigorous, subexponential algorithms for discrete logarithms over finite fields.* PhD thesis, U. Georgia, 1992.

[Lovorn Bender and Pomerance 1998] R. Lovorn Bender and C. Pomerance. Rigorous discrete logarithm computations in finite fields via smooth polynomials. In J. Teitelbaum, editor, *Computational Perspectives on Number Theory (Chicago, IL, 1995)*, volume 7, pages 221–232. Atkin Conference, Amer. Math. Soc., 1998.

[Madisetti and Williams 1997] V. Madisetti and D. Williams, editors. *The Digital Signal Processing Handbook.* CRC Press, 1997.

[Marcus 1977] D. Marcus. *Number Fields.* Springer-Verlag, 1977.

[Marsaglia 1991] G. Marsaglia. The mathematics of random number generators. In S. Burr, editor, *The Unreasonable Effectiveness of Number Theory,* volume 46 of *Proc. Sympos. Appl. Math.*, pages 73–90. American Math. Soc., 1991.

[Matijasevič 1971] Y. Matijasevič. Diophantine representations of the set of prime numbers. *Dokl. Akad. Nauk SSSR*, 12:354–358, 1971.

[Mauldin 1999] R. Mauldin. The Beal conjecture and prize, 1999. `http://www.math.unt.edu/~mauldin/beal.html`.

[McClellan and Rader 1979] J. McClellan and C. Rader. *Number Theory in Digital Signal Processing.* Prentice-Hall, 1979.

[McKee 1996] J. McKee. Turning Euler's factoring method into a factoring algorithm. *Bull. London Math. Soc.*, 28:351–355, 1996.

[McKee 1999] J. McKee. Speeding Fermat's factoring method. *Math. Comp.*, 68:1729–1737, 1999.

[Menezes et al. 1993] A. Menezes, T. Okamoto, and S. Vanstone. Reducing elliptic curve logarithms to a finite field. *IEEE Trans. Inform. Theory*, 39:1639–1646, 1993.

[Menezes et al. 1997] A. Menezes, P. van Oorschot, and S. Vanstone. *Handbook of Applied Cryptography.* CRC Press, 1997.

[Mignotte 2001] M. Mignotte. Catalan's equation just before 2000. In M. Jutila and T. Metsänkylä, editors, *Number Theory (Turku, 1999).* de Gruyter, 247–254.

[Miller 1976] G. Miller. Riemann's hypothesis and tests for primality. *J. Comput. System Sci.*, 13:300–317, 1976.

[Miller 1987] V. Miller. Use of elliptic curves in cryptography. In H. Williams, editor, *Advances in Cryptology, Proc. Crypto '85*, volume 218 of *Lecture Notes in Computer Science*, pages 417–426. Springer-Verlag, 1987.

[Mills 1947] W. Mills. A prime-representing function. *Bull. Amer. Math. Soc.*, 53:604, 1947.

[Moenck 1973] R. Moenck. Fast computation of GCDs. In *Proc. 5th Annual ACM Symposium on the Theory of Computing*, pages 142–151, 1973.

[Monier 1980] L. Monier. Evaluation and comparison of two efficient probabilistic primality testing algorithms. *Theoret. Comput. Sci.*, 12:97–108, 1980.

[Montgomery 1985] P. Montgomery. Modular multiplication without trial division. *Math. Comp.*, 44:519–521, 1985.

[Montgomery 1987] P. Montgomery. Speeding the Pollard and elliptic curve methods of factorization. *Math. Comp.*, 48:243–264, 1987.

[Montgomery 1992a] P. Montgomery. *An FFT Extension of the Elliptic Curve Method of Factorization.* PhD thesis, University of California, Los Angeles, 1992.

[Montgomery 1992b] P. Montgomery. Evaluating recurrences of form $X_{m+n} = f(X_m, X_n, X_{m-n})$ via Lucas chains. Unpublished manuscript, 1992.

[Montgomery 1994] P. Montgomery. Square roots of products of algebraic numbers. In W. Gautschi, editor, *Mathematics of Computation 1943–1993*, volume 48 of *Proc. Sympos. Appl. Math.*, pages 567–571. Amer. Math. Soc., 1994.

[Montgomery 1995] P. Montgomery. A block Lanczos algorithm for finding dependencies over $GF(2)$. In *Advances in Cryptology, Eurocrypt '95*, volume 921 of *Lecture Notes in Computer Science*, pages 106–120, 1995.

[Montgomery and Silverman 1990] P. Montgomery and R. Silverman. An FFT extension to the $P - 1$ factoring algorithm. *Math. Comp.*, 54:839–854, 1990.

[Morain 1990] F. Morain. *Courbes elliptiques et tests de primalité.* PhD thesis, Université Claude Bernard-Lyon I, 1990.

[Morain 1992] F. Morain. Building cyclic elliptic curves modulo large primes. Unpublished manuscript, 1992.

[Morain 1995] F. Morain. Calcul du nombre de points sur une courbe elliptique dans un corps fini: aspects algorithmiques. *J. Théor. Nombres Bordeaux*, 7:255–282, 1995. Les Dix-huitèmes Journées Arithmétiques (Bordeaux, 1993).

[Morain 1998] F. Morain. Primality proving using elliptic curves: an update. In [Buhler 1998], pages 111–127.

[Morain 2000] F. Morain. Quelques records actuels en théorie des nombres, 2000. `http://www.lix.polytechnique.fr/Labo/Francois.Morain`.

[Morrison and Brillhart 1975] M. Morrison and J. Brillhart. A method of factoring and the factorization of F_7. *Math. Comp.*, 29:183–205, 1975. Collection of articles dedicated to Derrick Henry Lehmer on the occasion of his seventieth birthday.

[Müller 1997] V. Müller. Efficient algorithms for multiplication on elliptic curves. Unpublished manuscript, October 1997.

[Müller and Paulus 1998] V. Müller and S. Paulus. On the generation of cryptographically strong elliptic curves. Preprint, 1998.

[Murphy 1998] B. Murphy. Modelling the yield of number field sieve polynomials. In [Buhler 1998], pages 137–150.

[Murphy 1999] B. Murphy. *Polynomial selection for the number field sieve integer factorisation algorithm.* PhD thesis, Australian National University, 1999.

[Namba 1984] M. Namba. *Geometry of Projective Algebraic Curves*, volume 88 of *Monographs and Textbooks in Pure and Applied Mathematics.* Marcel Dekker, 1984.

[Narkiewicz 1986] W. Narkiewicz. *Classical Problems in Number Theory.* PWN-Polish Scientific Publishers, 1986.

[Nathanson 96] M. Nathanson. *Additive Number Theory: The Classical Bases*, volume 164 of *Graduate Texts in Mathematics.* Springer-Verlag, 1996.

[Nguyen 1998] P. Nguyen. A Montgomery-like square root for the number field sieve. In [Buhler 1998], pages 151–168.

[Nicely 1999] T. Nicely. Enumeration to 1.6×10^{15} of the twin primes and Brun's constant, 1999. `http://lasi.lynchburg.edu/Nicely_T/Public/twins/twins2.htm`.

[Niederreiter 1992] H. Niederreiter. *Random Number Generation and Quasi-Monte-Carlo Methods*, volume 63 of *CBMS-NSF Regional Conference Series in Applied Mathematics*. SIAM, 1992.

[Niven et al. 1991] I. Niven, H. Zuckerman, and H. Montgomery. *An Introduction to the Theory of Numbers.* Fifth edition. John Wiley and Sons, 1991.

[Nussbaumer 1981] H. Nussbaumer. *Fast Fourier Transform and Convolution Algorithms.* Springer-Verlag, 1981.

[Odlyzko 1985] A. Odlyzko. Discrete logarithms in finite fields and their cryptographic significance. In *Advances in Cryptology, Proc. Eurocrypt '84*, volume 209 of *Lecture Notes in Computer Science*, pages 224–313, 1985.

[Odlyzko 1987] A. Odlyzko. On the distribution of spacings between zeros of the zeta function. *Math. Comp.*, 48:273–308, 1987.

[Odlyzko 1992] A. Odlyzko. The 10^{20}-th zero of the Riemann zeta function and 175 million of its neighbors, 1992. `http://www.research.att.com/~amo`.

[Odlyzko 1994] A. Odlyzko. Analytic computations in number theory. In W. Gautschi, editor, *Mathematics of Computation 1943–1993*, volume 48 of *Proc. Sympos. Appl. Math.*, pages 441–463. Amer. Math. Soc., 1994.

[Odlyzko 2000a] A. Odlyzko. Discrete logarithms: The past and the future. *Designs, Codes, and Cryptography*, 19:129–145, 2000.

[Odlyzko 2000b] A. Odlyzko. The zeros of the Riemann zeta function: the 10^{22}-nd zero and 10 billion of its neighbors. Unpublished manuscript, 2000.

[Odlyzko and te Riele 1985] A. Odlyzko and H. te Riele. Disproof of the Mertens conjecture. *J. Reine Angew. Math.*, 357:138–160, 1985.

[Odlyzko and Schönhage 1988] A. Odlyzko and A. Schönhage. Fast algorithms for multiple evaluations of the Riemann zeta-function. *Trans. Amer. Math. Soc.*, 309:797–809, 1988.

[Oesterlé 1985] J. Oesterlé. Nombres de classes des corps quadratiques imaginaires. In *Seminar Bourbaki (1983/84)*, Astérisque No. 121-122, pages 309–323, 1985.

[Owen 1995] A. Owen. Randomly permuted (t, m, s)-nets and (t, m, s)-sequences. In *Monte Carlo and Quasi-Monte Carlo Methods in Scientific Computing*, volume 106 of *Lecture Notes in Statistics*, pages 299–317. Springer-Verlag, 1995.

[Owen 1997a] A. Owen. Monte Carlo variance of scrambled net quadrature. *SIAM J. Numer. Anal.*, 34:1884–1910, 1997.

[Owen 1997b] A. Owen. Scrambled net variance for integrals of smooth functions. *Ann. Statist.*, 25:1541–1562, 1997.

[Padma and Venkataraman 1996] R. Padma and S. Venkataraman. Elliptic curves with complex multiplication and a character sum. *J. Number Theory*, 61:274–282, 1996.

[Papadopoulos 1999] J. Papadopoulos, 1999. Private communication.

[Papageorgiu 1997] A. Papageorgiu and J. Traub. Faster evaluation of multidimensional integrals. Unpublished manuscript, 1997.

[Parberry 1970] E. Parberry. On primes and pseudo-primes related to the Fibonacci sequence. *Fibonacci Quart.*, 8:49–60, 1970.

[Park and Miller 1988] S. Park and K. Miller. Random number generators: good ones are hard to find. *Comm. ACM*, 31:1192–1201, 1988.

[Paskov and Traub 1995] S. Paskov and J. Traub. Faster valuation of financial derivatives. *J. Portfolio Management*, 22:113–120, 1995.

[Patel and Sundaram 1998] S. Patel and G. Sundaram. An efficient discrete log pseudo random generator. In H. Krawczyk, editor, *Advances in Cryptology, Proc. Crypto '98*, volume 1462 of *Lecture Notes in Computer Science*, pages 304–317. Springer-Verlag, 1998.

[Paulos 1995] J. Paulos. High 5 jive. *Forbes*, 156:102, October 1995.

[Paun et al. 2000] G. Paun, G. Rozenberg, and A. Salomaa. *DNA Computing: New Computing Paradigms*. Springer-Verlag, 1998.

[Peralta 1993] R. Peralta. A quadratic sieve on the n-dimensional hypercube. In *Advances in Cryptology, Proc. Crypto '92*, volume 740 of *Lecture Notes in Computer Science*. Springer-Verlag, 1993.

[Peralta and Okamoto 1996] R. Peralta and E. Okamoto. Faster factoring of integers of a special form. *IEICE Transactions on Fundamentals of Electronics, Communications and Computer Sciences*, E79-A:489–493, 1996.

[Percival 2000] C. Percival. Rapid multiplication modulo the sum and difference of highly composite numbers. Unpublished manuscript, 2000.

[Peterson 2000] I. Peterson. Great computations. *Science News*, 157(10):152–153, March 4, 2000.

[Pinch 1993] R. Pinch. The Carmichael numbers up to 10^{15}. *Math. Comp.*, 61:381–391, 1993.

[Pollard 1974] J. Pollard. Theorems on factorization and primality testing. *Proc. Cambridge Philos. Soc.*, 76:521–528, 1974.

[Pollard 1975] J. Pollard. A Monte Carlo method for factorization. *Nordisk Tidskr. Informationsbehandling (BIT)*, 15:331–334, 1975.

[Pollard 1978] J. Pollard. Monte Carlo methods for index computation (mod p). *Math. Comp.*, 32:918–924, 1978.

ustyl'nkov 1999] L. Pustyl'nikov. On a property of the classical zeta-function associated with the Riemann conjecture on zeros. *Russian Math. Surveys*, 54:162–163, 1999.

Rabin 1976] M. Rabin. Probabilistic algorithms. In *Algorithms and Complexity (Proc. Sympos., Carnegie-Mellon Univ., Pittsburgh, PA, 1976)*, pages 21–39. Academic Press, 1976.

[Rabin 1980] M. Rabin. Probabilistic algorithm for testing primality. *J. Number Theory*, 12:128–138, 1980.

[Ramaré 1995] O. Ramaré. On Šnirel'man's constant. *Ann. Scuola Norm. Sup. Pisa Cl. Sci. (4)*, 22:645–706, 1995.

[Ramaré and Rumely 1996] O. Ramaré and R. Rumely. Primes in arithmetic progressions. *Math. Comp.*, 65:397–425, 1996.

[Ribenboim 1994] P. Ribenboim. *Catalan's Conjecture: Are 8 and 9 the Only Consecutive Powers?* Academic Press, 1994.

[Ribenboim 1996] P. Ribenboim. *The New Book of Prime Number Records.* Springer-Verlag, 1996.

[Richstein 2001] J. Richstein. Verifying the Goldbach conjecture up to $4 \cdot 10^{14}$. *Math. Comp.*, 70:1745–1749, 2001.

[Rishi et al. 1984] D. Rishi, J. Parnami, and A. Rajwade. Evaluation of a cubic character sum using the $\sqrt{-19}$ division points of the curve $Y^2 = X^3 - 2^3 \cdot 19X + 2 \cdot 19^2$. *J. Number Theory*, 19:184–194, 1984.

[Rivest et al. 1978] R. Rivest, A. Shamir, and L. Adleman. A method for obtaining digital signatures and public-key cryptosystems. *Comm. ACM*, 21:120–126, 1978.

[Rose 1988] H. Rose. *A Course in Number Theory.* Clarendon Press, Oxford, 1988.

[Rosser 1939] J. Rosser. The n-th prime is greater than $n \log n$. *Proc. London Math. Soc.*, 45:21–44, 1939.

[Rosser and Schoenfeld 1962] J. Rosser and L. Schoenfeld. Approximate formulas for some functions of prime numbers. *Illinois J. Math.*, 6:64–94, 1962.

[Rotkiewicz 1973] A. Rotkiewicz. On the pseudoprimes with respect to the Lucas sequences. *Bull. Acad. Polon. Sci. Sér. Sci. Math. Astronom. Phys.*, 21:793–797, 1973.

[Rumely 1993] R. Rumely. Numerical computations concerning the ERH. *Math. Comp.*, 61:415–440, S17–S23, 1993.

[Ruzsa 1999] I. Ruzsa. Erdős and the integers. *J. Number Theory*, 79:115–163, 1999.

[Saouter 1998] Y. Saouter. Checking the odd Goldbach conjecture up to 10^{20}. *Math. Comp.*, 67:863–866, 1998.

[Saouter 1999] Y. Saouter. Computations of Goldbach's partitions up to $128 \cdot 10^6$ with FFT. Unpublished manuscript, April 1999.

[Pomerance 1981] C. Pomerance. On the distribution of pseudopri *Comp.*, 37:587–593, 1981.

[Pomerance 1982] C. Pomerance. Analysis and comparison of some factoring algorithms. In H. Lenstra Jr and R. Tijdeman, ed. *Computational methods in number theory, Part I*, volume 15 *Centre Tracts*, pages 89–139. Math. Centrum, 1982.

[Pomerance 1985] C. Pomerance. The quadratic sieve factoring algorit *Advances in cryptology, Proc. Eurocrypt '84*, volume 209 of *Le* *in Computer Science*, pages 169–182. Springer-Verlag, 1985.

[Pomerance 1986] C. Pomerance. On primitive divisors of Mersenne num *Acta Arith.*, 46:355–367, 1986.

[Pomerance 1987a] C. Pomerance. Very short primality proofs. *Math. Com* 48:315–322, 1987.

[Pomerance 1987b] C. Pomerance. Fast, rigorous factorization and discrete logarithm algorithms. In *Discrete Algorithms and Complexity*, pages 119–143. Academic Press, 1987.

[Pomerance 1996a] C. Pomerance. Multiplicative independence for random integers. In *Analytic Number Theory, Vol. 2 (Allerton Park, IL, 1995)*, volume 139 of *Progr. Math.*, pages 703–711. Birkhäuser, 1996.

[Pomerance 1996b] C. Pomerance. A tale of two sieves. *Notices Amer. Math. Soc* 43:1473–1485, 1996.

[Pomerance and Smith 1992] C. Pomerance and J. Smith. Reduction of huge, sparse matrices over finite fields via created catastrophes. *Experiment. Math.*, 1:89–94, 1992.

[Pomerance et al. 1988] C. Pomerance, J. Smith, and R. Tuler. A pipeline architecture for factoring large integers with the quadratic sieve algorithm. *SIAM J. Comput.*, 17:387–403, 1988. Special issue on cryptography.

[Prachar 1978] K. Prachar. *Primzahlverteilung*, volume 91 of *Grundlehren der Mathematischen Wissenschaften*. Springer-Verlag, 1978. Reprint of the 1957 edition.

[Pratt 1975] V. Pratt. Every prime has a succinct certificate. *SIAM J. Comput.*, 4:214–220, 1975.

[Preskill 1999] J. Preskill. Course notes, Phys 229, Calif. Inst. of Tech., 1999. `www.theory.caltech.edu/people/preskill/ph229/`.

[Press et al. 1996] W. Press, S. Teukolsky, W. Vettering, and B. Flannery. *Numerical Recipes in C.* Cambridge University Press, 1996.

[Pritchard 1981] P. Pritchard. A sublinear additive sieve for finding prime numbers. *Comm. ACM*, 24:18–23, 1981.

[Pritchard et al. 1995] P. Pritchard, A. Moran, and A. Thyssen. Twenty-two primes in arithmetic progression. *Math. Comp.*, 64:1337–1339, 1995.

[Satoh and Araki 1998] T. Satoh and K. Araki. Fermat quotients and the polynomial time discrete log algorithm for anomalous elliptic curves. *Comment. Math. Univ. St. Paul.*, 47:81–92, 1998. Errata, *ibid.* 48:1999, 211-213.

[Schinzel and Sierpiński 1958] A. Schinzel and W. Sierpiński. Sur certaines hypothèses concernant les nombres premiers. *Acta Arith.*, 4:185–208, 1958. Erratum, *ibid.* 5:259, 1958.

[Schirokauer et al. 1996] O. Schirokauer, D. Weber, and T. Denny. Discrete logarithms: the effectiveness of the index calculus method. In *Algorithmic Number Theory: Proc. ANTS II, Talence, France*, volume 1122 of *Lecture Notes in Computer Science*, pages 337–361. Springer-Verlag, 1996.

[Schmidt 1972] W. Schmidt. Irregularities of distribution. VII. *Acta Arith.*, 21:45–50, 1972.

[Schneier 1996] B. Schneier. *Applied Cryptography.* John Wiley and Sons, 1996.

[Schoenfeld 1976] L. Schoenfeld. Sharper bounds for the Chebyshev functions $\theta(x)$ and $\psi(x)$. II. *Math. Comp.*, 30:337–360, 1976. Corrigendum, *ibid.* 30:900, 1976.

[Schönhage 1971] A. Schönhage. Schnelle Berechnung von Kettenbruchentwicklungen. *Acta Informatica*, 1:139–144, 1971.

[Schönhage 1982] A. Schönhage. Asymptotically fast algorithms for the numerical multiplication and division of polynomials with complex coefficients. In *Computer Algebra, EUROCAM '82, Marseille*, volume 144 of *Lecture Notes in Computer Science*, pages 3–15. Springer-Verlag, 1982.

[Schönhage and Strassen 1971] A. Schönhage and V. Strassen. Schnelle Multiplikation grosser Zahlen. *Computing (Arch. Elektron. Rechnen)*, 7:281–292, 1971.

[Schoof 1982] R. Schoof. Quadratic fields and factorization. In H. Lenstra Jr and R. Tijdeman, editors, *Computational methods in number theory, Part I*, volume 154 of *Math. Centre Tracts*, pages 235–286. Math. Centrum, 1982.

[Schoof 1985] R. Schoof. Elliptic curves over finite fields and the computation of square roots mod *p*. *Math. Comp.*, 44:483–494, 1985.

[Schoof 1995] R. Schoof. Counting points on elliptic curves over finite fields. *J. Théor. Nombres Bordeaux*, 7:219–254, 1995. Les Dix-huitèmes Journées Arithmétiques (Bordeaux, 1993).

[Schroeder 1999] M. Schroeder. *Number Theory in Science and Communication*, volume 7 of *Springer Series in Information Sciences*. Springer-Verlag, 1999. Corrected printing of the third (1997) edition.

[Scott 1999] M. Scott, 1999. Private communication.

[Selfridge and Hurwitz 1964] J. Selfridge and A. Hurwitz. Fermat numbers and Mersenne numbers. *Math. Comp.*, 18:146–148, 1964.

[Semaev 1998] I. Semaev. Evaluation of discrete logarithms in a group of *p*-torsion points of an elliptic curve in characteristic *p*. *Math. Comp.*, 67:353–356, 1998.

[Seroussi et al. 1999] G. Seroussi, N. Smart, and I. Blake. *Elliptic Curves in Cryptography*, volume 265 of *London Math. Soc. Lecture Note Series*. Cambridge University Press, 1999.

[Shamir 1999] A. Shamir. Factoring large numbers with the TWINKLE device (extended abstract). In Ç. Koç and C. Paar, editors, *Cryptographic Hardware and Embedded Systems, First International Workshop, CHES '99, Worcester, MA*, volume 1717 of *Lecture Notes in Computer Science*, pages 2–12. Springer-Verlag, 1999.

[Shanks 1971] D. Shanks. Class number, a theory of factorization, and genera. In *1969 Number Theory Institute, Stony Brook, N.Y.*, volume 20 of *Proc. Sympos. Pure Math.*, pages 415–440. Amer. Math. Soc., 1971.

[Shanks and Schmid 1966] D. Shanks and L. Schmid. Variations on a theorem of Landau. Part I. *Math. Comp.*, 20:551–569, 1966.

[Shlesinger 1986] M. Shlesinger. On the Riemann hypothesis: a fractal random walk approach. *Physica*, 138A:310–319, 1986.

[Shor 1994] P. Shor. Algorithms for quantum computation: discrete logarithms and factoring. In *Proc. 35th Annual Symp. Found. Comp. Sci.*, pages 124–134, 1994.

[Shor 1999] P. Shor. Polynomial-time algorithms for prime factorization and discrete logarithms on a quantum computer. *SIAM Review*, 41:303–332, 1999.

[Shoup 1992] V. Shoup. Searching for primitive roots in finite fields. *Math. Comp.*, 58:369–380, 1992.

[Shoup 1995] V. Shoup. A new polynomial factorization algorithm and its implementation. *J. Symbolic Comput.*, 20:363–397, 1995.

[Silverman 1986] J. Silverman. *The Arithmetic of Elliptic Curves*, volume 106 of *Graduate Texts in Mathematics*. Springer-Verlag, 1986.

[Silverman and Wagstaff 1993] R. Silverman and S. Wagstaff Jr. A practical analysis of the elliptic curve factoring algorithm. *Math. Comp.*, 61:445–462, 1993.

[Sloan and Wozniakowski 1998] I. Sloan and H. Wozniakowski. When are quasi-Monte Carlo algorithms efficient for high dimensional integrals? *Complexity*, 14:1–33, 1998.

[Smart 1999] N. Smart. The discrete logarithm problem on elliptic curves of trace one. *J. Cryptology*, 12:193–196, 1999.

[Solinas 1998] J. Solinas. Standard specifications for public key cryptography. Annex A: Number-theoretic background. *IEEE P1363 Draft,*, 1998. `http://grouper.ieee.org/groups/1363/`.

[Solinas 1999] J. Solinas. Generalized Mersenne numbers, 1999. `http://www.cacr.math.uwaterloo.ca/techreports/1999/tech_reports99.html`.

[Sorenson 1994] J. Sorenson. Two fast GCD algorithms. *J. Algorithms*, 16:110–144, 1994.

[Srinivasan 1995] A. Srinivasan. *Compuations of Class Numbers of Quadratic Fields.* PhD thesis, U. Georgia, 1995.

[Stein 1967] J. Stein. Computational problems associated with Racah algebra. *J. Comp. Phys.*, 1:397–405, 1967.

[Strassen 1977] V. Strassen. Einige Resultate über Berechnungskomplexität. *Jber. Deutsch. Math.-Verein.*, 78:1–8, 1976/77.

[Stuart 1996] I. Stuart. The magic of seven: signifies creation, the sum of the spiritual three and the material four. *British Medical Journal*, 313(7072), December 21 1996.

[Sun and Sun 1992] Z.-H. Sun and Z.-W. Sun. Fibonacci numbers and Fermat's last theorem. *Acta Arith.*, 60:371–388, 1992.

[Swarztrauber 1987] P. Swarztrauber. Multiprocessor FFTs. *Parallel Computing*, 5:197–210, 1987.

[Tanner and Wagstaff 1987] J. Tanner and S. Wagstaff Jr. New congruences for the Bernoulli numbers. *Math. Comp.*, 48:341–350, 1987.

[Tatuzawa 1952] T. Tatuzawa. On a theorem of Siegel. *Jap. J. Math.*, 21:163–178, 1951-1952.

[Teitelbaum 1998] J. Teitelbaum. Euclid's algorithm and the Lanczos method over finite fields. *Math. Comp.*, 67:1665–1678, 1998.

[Terr 2000] D. Terr. A modification of Shanks' baby-step giant-step algorithm. *Math. Comp.*, 69:767–773, 2000.

[Teske 1998] E. Teske. Speeding up Pollard's rho method for computing discrete logarithms. In [Buhler 1998], pages 541–554.

[Tezuka 1995] S. Tezuka. *Uniform Random Numbers: Theory and Practice.* Kluwer Academic Publishers, 1995.

[Thomas et al. 1986] J. Thomas, J. Keller, and G. Larsen. The calculation of multiplicative inverses over $GF(P)$ efficiently where P is a Mersenne prime. *IEEE Trans. Comp.*, C-35:478–482, 1986.

[Titchmarsh 1986] E. Titchmarsh and D. Heath-Brown. *The Theory of the Riemann Zeta-function.* Oxford University Press, 1986.

[Trevisan and Carvalho 1993] V. Trevisan and J. Carvalho. The composite character of the twenty-second Fermat number. *J. Supercomputing*, 9:179–182, 1995.

[van de Lune et al. 1986] J. van de Lune, H. te Riele, and D. Winter. On the zeros of the Riemann zeta function in the critical strip. IV. *Math. Comp.*, 46:667–681, 1986.

[van der Corput 1922] J. van der Corput. Verscharfung der Abschätzungen beim Teilerproblem. *Math. Ann.*, 87:39–65, 1922.

[van del Pol 1947] B. van der Pol. An electro-mechanical investigation of the Riemann zeta function in the critical strip. *Bull. Amer. Math. Soc.*, 53, 1947.

[Van Loan 1992] C. Van Loan. *Computational Frameworks for the Fast Fourier Transform*, volume 10 of *Frontiers in Applied Mathematics*. SIAM, 1992.

[van Oorschot and Wiener 1999] P. van Oorschot and M. Wiener. Parallel collision search with cryptanalytic applications. *J. Cryptology*, 12:1–28, 1999.

[Vaughan 1977] R. Vaughan. Sommes trigonométriques sur les nombres premiers. *C. R. Acad. Sci. Paris Sér. A-B*, 285:A981–A983, 1977.

[Vaughan 1989] R. Vaughan, A new iterative method in Waring's problem, *Acta Arith.*, 162:1–71, 1989.

[Vaughan 1997] R. Vaughan. *The Hardy–Littlewood Method.* Second edition, volume 125 of *Cambridge Tracts in Mathematics*. Cambridge University Press, 1997.

[Veach 1997] E. Veach. *Robust Monte Carlo methods for light transport simulation.* PhD thesis, Stanford University, 1997.

[Vehka 1979] T. Vehka. Explicit construction of an admissible set for the conjecture that sometimes $\pi(x+y) > \pi(x) + \pi(y)$. *Notices Amer. Math. Soc.*, 26, A-453, 1979.

[Vinogradov 1985] I. Vinogradov. *Ivan Matveevič Vinogradov: Selected Works.* Springer-Verlag, 1985. L. Faddeev, R. Gamkrelidze, A. Karacuba, K. Mardzhanishvili, and E. Miščenko, editors.

[Vladimirov et al. 1994] V. Vladimirov, I. Volovich, and E. Zelenov. *p-adic Analysis and Mathematical Physics*, volume 1 of *Series on Soviet and East European Mathematics*. World Scientific, 1994.

[Wagstaff 1978] S. Wagstaff Jr. The irregular primes to 125000. *Math. Comp.*, 32:583–591, 1978.

[Wagstaff 1993] S. Wagstaff Jr. Computing Euclid's primes. *Bull. Inst. Combin. Appl.*, 8:23–32, 1993.

[Warren 1995] B. Warren. An interesting group of combination-product sets produces some very nice dissonances. *The Journal of the Just Intonation Network*, 9(1):1, 4–9, 1995.

[Watkins 2000] M. Watkins. *Class Numbers of Imaginary Quadratic Fields.* PhD thesis, U. Georgia, 2000.

[Watt 1989] N. Watt. Exponential sums and the Riemann zeta-function. II. *J. London Math. Soc.*, 39, 1989.

[Weber 1995] K. Weber. The accelerated GCD algorithm. *ACM Trans. Math. Soft.*, 21:111–122, 1995.

[Weiss 1963] E. Weiss. *Algebraic Number Theory.* McGraw-Hill, 1963.

[Weisstein 2000] E. Weisstein. World of mathematics, 2000. `http://www.mathworld.com`.

[Wellin 1998] P. Wellin, 1998. Private communication.

[Weyl 1916] H. Weyl. Über die Gleichverteilung von Zahlen mod. Eins. *Math. Ann.*, 77, 1916.

[Wiedemann 1986] D. Wiedemann. Solving sparse linear equations over finite fields. *IEEE Trans. Inform. Theory*, 32:54–62, 1986.

[Wieferich 1909] A. Wieferich. Zum letzten Fermat'schen Theorem. *J. Reine Angew. Math.*, 136:293–302, 1909.

[Wiener 1990] M. Wiener. Cryptanalysis of short RSA secret exponents. *IEEE Trans. Inform. Theory*, 36:553–558, 1990.

[Williams 1998] H. Williams. *Edouard Lucas and Primality Testing*, volume 22 of *Canadian Mathematics Society Series of Monographs and Advanced Texts*. John Wiley and Sons, 1998.

[Williams and Clearwater 1998] C. Williams and S. Clearwater. *Explorations in Quantum Computing*. TELOS/Springer-Verlag, 1998.

[Williams and Shallit 1993] H. Williams and J. Shallit. Factoring integers before computers. In W. Gautschi, editor, *Mathematics of Computation 1943–1993*, volume 48 of *Proc. Sympos. Appl. Math.*, pages 481–531. Amer. Math. Soc., 1994.

[Winterhof 1998] A. Winterhof, On Waring's problem in finite fields, *Acta Arith.*, 87:171–177, 1998.

[Wolf 1997] M. Wolf. $1/f$ noise in the distribution of prime numbers. *Physica A*, 241:493–499, 1997.

[Woltman 2000] G. Woltman. Great Internet Mersenne prime search (GIMPS), 2000. `http://www.mersenne.org`.

[Wozniakowski 1991] H. Wozniakowski. Average case complexity of multivariate integration. *Bull. Amer. Math. Soc. (N.S.)*, 24:185–194, 1991.

[Wu 1997] P. Wu. Multiplicative, congruential random-number generators. *ACM Trans. Math. Soft.*, 23:255–265, 1997.

[Yacobi 1999] Y. Yacobi. Fast exponentiation using data compression. *SIAM J. Comput.*, 28:700–703, 1999.

[Yagle 1995] A. Yagle. Fast algorithms for matrix multiplication using pseudo-number-theoretic transforms. *IEEE Trans. Sig. Proc*, 43:71–76, 1995.

[Yan et al. 1991] J. Yan, A. Yan, and B. Yan. Prime numbers and the amino acid code: analogy in coding properties. *J. Theor. Biol.*, 151(3):333–341, 1991.

[Yoshimura 1997] J. Yoshimura. The evolutionary origins of periodical cicadas during ice ages. *American Naturalist*, 149(1):112–124, 1997.

[Yu 1996] G. Yu. The differences between consecutive primes. *Bull. London Math. Soc.*, 28:242–248, 1996.

[Zhang 1998] M. Zhang. Factorization of the numbers of the form $m^3 + c_2m^2 + c_1m + c_0$. In [Buhler 1998], pages 131–136.

[Zimmermann 2000] P. Zimmermann. The ECMNET project, 2000. `http://www.loria.fr/~zimmerma/records/ecmnet.html`.

[Zinoviev 1997] D. Zinoviev. On Vinogradov's constant in Goldbach's ternary problem. *J. Number Theory*, 65:334–358, 1997.

Index